Nanobiotechnology in Diagnosis, Drug Delivery, and Treatment

Nanobiotechnology in Diagnosis, Drug Delivery, and Treatment

Edited by

Mahendra Rai

Nanobiotechnology Laboratory
Department of Biotechnology
Sant Gadge Baba Amravati University
Amravati, Maharashtra, India

Mehdi Razzaghi-Abyaneh

Department of Mycology
Pasteur Institute of Iran
Tehran, Iran

Avinash P. Ingle

Department of Biotechnology
Engineering School of Lorena
University of Sao Paulo
Lorena, Sao Paulo, Brazil

This edition first published 2021

Registered Offices
John Wiley & Sons, Inc., 111 River Street, Hoboken, NJ 07030, USA
John Wiley & Sons Ltd, The Atrium, Southern Gate, Chichester, West Sussex, PO19 8SQ, UK

Editorial Office
The Atrium, Southern Gate, Chichester, West Sussex, PO19 8SQ, UK

For details of our global editorial offices, customer services, and more information about Wiley products visit us at www.wiley.com.

Wiley also publishes its books in a variety of electronic formats and by print-on-demand. Some content that appears in standard print versions of this book may not be available in other formats.

Library of Congress Cataloging-in-Publication Data applied for
ISBN Paperback: 9781119671770

Cover Design: Wiley
Cover Image: © LAGUNA DESIGN/Getty Images

Set in 9.5/12.5pt STIXTwoText by SPi Global, Pondicherry, India

SKY43DC9F39-8867-42BF-BF69-C8618AFC8F17_102920

Contents

List of Contributors

Avinash P. Ingle
Department of Biotechnology
Engineering School of Lorena
University of Sao Paulo
Lorena, Sao Paulo
Brazil

Patrycja Golińska
Department of Microbiology
Nicolaus Copernicus University
Lwowska, Torun
Poland

Alka Yadav
Nanobiotechnology Laboratory, Department of Biotechnology
Sant Gadge Baba
Amravati University
Amravati, Maharashtra
India

Mehdi Razzaghi-Abyaneh
Department of Mycology
Pasteur Institute of Iran
Tehran
Iran

Mrunali Patel
Department of Pharmaceutics, Ramanbhai Patel College of Pharmacy
Charotar University of Science and Technology (CHARUSAT)
Changa, Gujarat
India

Rashmin Patel
Department of Pharmaceutics, Ramanbhai Patel College of Pharmacy
Charotar University of Science and Technology (CHARUSAT)
Changa, Gujarat
India

Yulia Plekhanova
Laboratory of Biosensors
G.K. Skryabin Institute of Biochemistry and Physiology of Microorganisms
Russian Academy of Sciences
Moscow Region
Russia

Anatoly Reshetilov
Laboratory of Biosensors
G.K. Skryabin Institute of Biochemistry and Physiology of Microorganisms
Russian Academy of Sciences
Moscow Region
Russia

Mahendra Rai
Nanobiotechnology Laboratory
Department of Biotechnology
Sant Gadge Baba Amravati University
Amravati, Maharashtra
India

Irina A. Shurygina
Irkutsk Scientific Center of Surgery and Traumatology
Bortsov Revolutsii st.
Irkutsk
Russia

Michael G. Shurygin
Irkutsk Scientific Center of Surgery and Traumatology
Bortsov Revolutsii st., Irkutsk
Russia

Magdalena Wypij
Department of Microbiology
Nicolaus Copernicus University
Torun
Poland

Sougata Ghosh
Department of Microbiology
School of Science, RK University
Rajkot, Gujarat
India

Rohini Kitture
Springer Nature Scientific Technology and Publishing Solutions
Magarpatta City
Pune, Maharashtra
India

Fahimeh Charbgoo
Pharmaceutical Research Center
Pharmaceutical Technology Institute, Mashhad University of Medical Sciences
Mashhad
Iran

and

DWI – Leibniz Institute for Interactive Materials
Aachen
Germany

Seyed Mohammad Taghdisi
Targeted Drug Delivery Research Center
Pharmaceutical Technology Institute Mashhad University of Medical Sciences
Mashhad
Iran

and

Department of Pharmaceutical Biotechnology
School of Pharmacy, Mashhad University of Medical Sciences
Mashhad
Iran

Rezvan Yazdian-Robati
Molecular and Cell Biology Research Center
Faculty of Medicine, Mazandaran University of Medical Sciences
Sari
Iran

Khalil Abnous
Pharmaceutical Research Center
Pharmaceutical Technology Institute Mashhad University of
Medical Sciences
Mashhad
Iran

and

Department of Medicinal Chemistry
School of Pharmacy, Mashhad University of Medical Sciences
Mashhad
Iran

Mohammad Ramezani
Pharmaceutical Research Center
Pharmaceutical Technology Institute Mashhad University of Medical Sciences
Mashhad
Iran

Mona Alibolandi
Pharmaceutical Research Center
Pharmaceutical Technology Institute Mashhad University of Medical Sciences
Mashhad
Iran

Dilesh Jagdish Singhavi
Department of Pharmaceutics, Institute of Pharmaceutical Education and Research
Borgaon (Meghe)
Wardha, Maharashtra
India

Shagufta Khan
Department of Pharmaceutics, Institute of Pharmaceutical Education and Research
Borgaon (Meghe)
Wardha
Maharashtra
India

Dipak Maity
Department of Chemical Engineering
Institute of Chemical Technology Mumbai, Indian Oil Campus, IIT KGP Extension Centre,
Bhubaneswar, Odisha
India

Atul Sudame
Department of Mechanical Engineering
Shiv Nadar University
Greater Noida, Uttar Pradesh
India

Ganeshlenin Kandasamy
Department of Biomedical Engineering
Vel Tech Rangarajan Dr. Sagunthala R&D Institute of Science and Technology
Chennai, Tamil Nadu
India

Yasser Shahzad
Department of Pharmacy
COMSATS University Islamabad,
Lahore
Pakistan

Abid Mehmood Yousaf
Department of Pharmacy
COMSATS University Islamabad,
Lahore
Pakistan

Talib Hussain
Department of Pharmacy
COMSATS University Islamabad, Lahore
Lahore
Pakistan

Syed A.A. Rizvi
School of Pharmacy
Hampton University
Virginia
USA

Parinaz Nezhad-Mokhtari
Department of Medical Nanotechnology
Faculty of Advanced Medical Sciences University of Medical Sciences
Tabriz
Iran

Fatemeh Salahpour-Anarjan
Department of Medical Nanotechnology
Faculty of Advanced Medical Sciences University of Medical Sciences
Tabriz
Iran

Armin Rezanezhad
Faculty of Mechanical Engineering, Department of Materials Science and Engineering
University of Tabriz
Tabriz
Iran

Abolfazl Akbarzadeh
Department of Medical Nanotechnology
Faculty of Advanced Medical Sciences and Stem Cell Research Center
University of Tabriz
Tabriz
Iran

and

Stem Cell Research
University of Tabriz

Divya Koilparambil
School of Biosciences
Mahatma Gandhi University
Kottayam, Kerala
India

Sherin Varghese
School of Biosciences
Mahatma Gandhi University
Kottayam, Kerala
India

Jisha Manakulam Shaikmoideen
School of Biosciences
Mahatma Gandhi University
Kottayam, Kerala
India

Priti Paralikar
Nanobiotechnology Laboratory, Department of Biotechnology
SGB Amravati University
Amravati, Maharashtra
India

Erasmo Gámez-Espinosa
Center for Research and Development in Painting Technology
National University of La Plata
La Plata
Buenos Aires, Argentina

Leyanet Barberia-Roque
Center for Research and Development in Painting Technology
National University of La Plata
La Plata
Buenos Aires, Argentina

Natalia Bellotti
Center for Research and Development in Painting Technology
National University of La Plata
La Plata
Buenos Aires, Argentina

and

Faculty of Natural Sciences and Museum
National University of La Plata
La Plata
Argentina

Jacqueline Teixeira da Silva
Laboratory of Nanobiotechnology,
Institute of Tropical Pathology and Public Health
Federal University of Goiás
Goiânia
Brazil

Andre Correa Amaral
Laboratory of Nanobiotechnology, Institute of Tropical Pathology and Public Health
Federal University of Goiás
Goiânia
Brazil

Smitha Vijayan
School of Bioscience
Mahatma Gandhi University
Kottayam, Kerala
India

Mohammad Imani
Novel Drug Delivery Systems Department
Iran Polymer and Petrochemical Institute
Tehran
Iran

Azam Dehghan
Novel Drug Delivery Systems Department
Iran Polymer and Petrochemical Institute
Tehran
Iran

Thais Francine Ribeiro Alves
Laboratory of Biomaterials and Nanotechnology, LaBNUS
University of Sorocaba
Sorocaba
Brazil

Fernando Batain
Laboratory of Biomaterials and Nanotechnology, LaBNUS
University of Sorocaba
Sorocaba
Brazil

Cecília Torqueti de Barros
Laboratory of Biomaterials and Nanotechnology, LaBNUS
University of Sorocaba
Sorocaba
Brazil

Kessie Marie Moura Crescencio
Laboratory of Biomaterials and Nanotechnology, LaBNUS
University of Sorocaba
Sorocaba
Brazil

Venâncio Alves do Amaral
Laboratory of Biomaterials and Nanotechnology, LaBNUS
University of Sorocaba
Sorocaba
Brazil

Mariana Silveira de Alcântara Chaud
Department of Cardiology, Municipal Clinics Hospital José Alencar
São Bernardo do Campo
Brazil

Décio Luís Portella
Laboratory of Biomaterials and Nanotechnology, LaBNUS
University of Sorocaba
Sorocaba
Brazil

and

Department of Surgery and Plastic Surgery
Pontifical Catholic University of São Paulo
Sorocaba
Brazil

Marco Vinícius Chaud
Laboratory of Biomaterials and Nanotechnology, LaBNUS
University of Sorocaba
Sorocaba
Brazil

Indarchand Gupta
Department of Biotechnology
Government Institute of Science
Nipat-Niranjan Nagar
Aurangabad, Maharashtra
India

and

Nanobiotechnology Laboratory, Department of Biotechnology
SGB Amravati University
Amravati, Maharashtra
India

Preface

In the last two decades, nanotechnology has revolutionized different areas of research as well as life. It has demonstrated its promising power to resolve the majority of medical problems like infectious diseases, cancer, genetic disorders, neurodegenerative diseases, etc. Nanomedicine, which refers to highly specific medical intervention at the scales of 100 nm or less for diagnosis, prevention, and treatment of diseases, is one of the most promising areas of nanotechnology.

The major face of nanomedicine is "drug delivery," which is considered as one of the most promising functions of nanotechnology and can maneuver molecules and supramolecular structures to create devices with programmed functions. The current drug delivery systems are distributed into the nano- and microscale systems which mainly involve the use of nanoparticles, liposomes, polymeric micelles (nanovehicles), dendrimers, nanocrystals, microchips, microtherapeutic systems, aptamer-incorporated nanoparticles, and novel microparticles of size 100 nm. These technologies will be expanded in the future to establish efficient nano/microdrug delivery systems in a manner which will provide future necessities of medicine in terms of diagnosis and treatment of infectious diseases, with special focus on those microbes which are emerging as drug-resistant.

In this book, there are 17 chapters that are broadly focused on the recent advances in nano-based drug delivery systems, diagnosis, and the role of various nanomaterials in the management of infectious diseases and non-infectious disorders such as cancers and other malignancies and their role in future medicine. Chapter 1 is an introductory chapter that presents the role of nanotechnology as revolutionary science as far as drug delivery, diagnosis, and treatment of diseases are concerned. Chapter 2 focuses on the role of selenium nanocomposites in diagnosis, drug delivery, and treatment of diseases like cancer, Alzheimer's disease, diabetes, and many others. Chapter 3 includes the application of a variety of nanomaterials in the diagnosis and management of gastrointestinal tract disorders. Chapter 4 explains the concept of nanotheranostics in detail and its role in effective monitoring of drug response, targeted drug delivery, enhanced drug accumulation in the target tissues, sustained as well as the triggered release of drugs, reduction in adverse effects, etc. Chapter 5 discusses the application of aptamer-incorporated nanoparticle systems for drug delivery, which is considered as one of the most compelling medicinal platforms of nanotechnology. The aptamer-incorporated nanoparticle systems are reported to have a promising impact on the delivery of different kinds of therapeutics, and also have great promise for improving the therapeutic index and pharmacokinetics of several drugs. Chapter 6 incorporates the application of nanotechnology in transdermal drug delivery. Moreover, a detailed explanation given on the interaction of the skin and nanoparticles will be helpful to enhance the reader's understanding of new concepts and the use of drug delivery carriers in transdermal delivery. Chapter 7 examines the application of superparamagnetic iron oxide nanoparticle-based drug delivery in cancer therapeutics. Nowadays, superparamagnetic iron oxide nanoparticles have attracted a great deal of attention from researchers all over the world due to their strong magnetic properties, which provide an added advantage when they are used in biomedical applications. Chapter 8 emphasizes one of the most novel concepts, i.e. application of virus-like nanoparticles in the delivery of the cancer therapeutics. It is well known that viruses have a unique ability to coordinate with host cellular components and processes for their survival and multiplication. The ability of self-replication and transduction property makes the viruses potential vectors for the delivery of small molecules and protein therapeutics. Chapter 9 is also about the revolutionary applications of magnetic nanoparticles as future cancer theranostics. Chapter 10 is dedicated to the utilization of chitosan nanoparticles as novel antimicrobial agents. Actually, due to certain toxicological effects of some nanomaterials like metallic nanoparticles, polymeric nanoparticles like chitosan nanoparticles have gained more attention due to their biodegradable nature. Chapter 11 covers various aspects related to sulfur nanoparticles, such as their biosynthesis, antibacterial applications, and possible mechanisms involved in their action. Chapter 12 emphasizes the role of nanotechnology in the management of indoor fungi. The problem of indoor fungi is one of the most important public health concerns because they are responsible for a wide range of mild to severe diseases like allergies, asthma, etc. Chapter 13 deliberates the role of nanotechnology in antifungal therapy. Considering the alarming increase in resistance in a variety of fungi and infection caused due to such fungi, this chapter is very interesting. Chapter 14 discusses the application of conjugated nanoparticles of chitosan and biogenic silver in antimicrobial and anticancer perspectives. The development of such novel conjugated nanoparticles is required to reduce or eradicate the problem of nanomaterials toxicity. Chapter 15 is focused on one of the important diseases, leishmaniasis. Leishmaniasis is a very dreadful disease, and available therapeutic strategies are not very effective in the management of this disease. In this context, the application of different nanomaterials as a part of its treatment strategies would be a novel alternative. Chapter 16 is about theranostics and vaccines; in this chapter, authors focused on their current status and future expectations. Finally, Chapter 17 is focused on the most important and debatable concept, i.e. toxicity of nanomaterials. There is no doubt that nanomaterials bring a revolution in biomedical science, and hence they are widely used in various biomedical applications and products. But, it is also true that increased use of nanomaterials also possesses an elevated risk of toxicity. Therefore, in this context, this chapter is very important. This chapter covers several aspects like factors affecting the toxicity of nanomaterials, why there is a necessity to evaluate the toxicity of nanomaterials, recent advances in in vitro and in vivo toxicity, and how the toxicity of nanomaterials can be managed.

Overall, this book comprises very informative chapters written by one or more specialists, experts in their particular topic. In this way, we would like to offer a rich guide for doctors, researchers in this field, undergraduate or graduate students of various disciplines like microbiology, biotechnology, nanotechnology, pharmaceutical biotechnology, pharmacology, pharmaceutics, nanomedicine, tissue engineering, biomaterials, etc., and allied subjects. In addition, this book is useful for people working in various industries, regulatory bodies, and nanotechnological organizations.

We would like to thank all the contributors for their outstanding efforts to provide state-of-the-art information on the subject matter of their respective chapters. Their efforts will certainly enhance and update the knowledge of the readers about the role of nanotechnology in biomedicine and public health. We also thank everyone in the Wiley team for their constant help and constructive suggestions particularly to Julia, senior editor, for her patience and cooperation. Finally, we would like to thank our colleagues, Professors Chistiane M. Feitosa and Rafael M. Bandeira for their cooperation during the editing of the book.

We hope that the book will be useful for all readers to find the required information on the latest research and advances in the field of biomedical nanotechnology.

Mahendra Rai, India
Mehdi Razzaghi-Abyaneh, Iran
Avinash P. Ingle, Brazil

1

Nanotechnology: A New Era in the Revolution of Drug Delivery, Diagnosis, and Treatment of Diseases

Avinash P. Ingle[1], Patrycja Golińska[2], Alka Yadav[3], Mehdi Razzaghi-Abyaneh[4], Mrunali Patel[5], Rashmin Patel[5], Yulia Plekhanova[6], Anatoly Reshetilov[6], and Mahendra Rai[3]

[1] *Department of Biotechnology, Engineering School of Lorena, University of Sao Paulo, Lorena, SP, Brazil*
[2] *Department of Microbiology, Nicolaus Copernicus University, Lwowska, Torun, Poland*
[3] *Nanobiotechnology Laboratory, Department of Biotechnology, Sant Gadge Baba Amravati University, Amravati, Maharashtra, India*
[4] *Department of Mycology, Pasteur Institute of Iran, Tehran, Iran*
[5] *Department of Pharmaceutics, Ramanbhai Patel College of Pharmacy, Charotar University of Science and Technology (CHARUSAT), Changa, Gujarat, India*
[6] *Laboratory of Biosensors, G.K. Skryabin Institute of Biochemistry and Physiology of Microorganisms, Russian Academy of Sciences, Moscow Region, Russia*

1.1 Introduction

Various factors, including the fast pace of today's world, knowingly or unknowingly have created many concerns in the area of healthcare. Different infectious diseases, depression, hypertension, diabetes, neurodegenerative disorders, cardiovascular diseases, cancers, etc. are a small part of the list of common outcomes associated with a high-speed, stress-filled lifestyle, among other reasons (Petrie et al. 2018). Therefore, early diagnosis and effective treatment are required to manage all of these health conditions. However, it has been a major challenge in recent times. Moreover, the recent noteworthy scientific advancements in the field of nanotechnology have potentially improved medical diagnosis and treatment strategies (Bonnard et al. 2019). In this context, early diagnosis of diseases, even before the presentation of symptoms, and improved imaging systems for internal body structure, etc., in addition to various treatment approaches, have been developed with the help of nanotechnology.

Nanotechnology or nanoscience is referred to as the science which involves the study of materials at the atomic or molecular level. In other words, nanotechnology is defined as the investigation, design, manufacture, synthesis, manipulation, and application of materials, strategies, and structures at a scale of 1–100 nanometers (nm) (Gholami-Shabani et al. 2014). It means to design, manufacture, characterize, and apply structures, devices, and methods through controlled manipulation of shape and size at the nanometric scale which have at least one novel/superior characteristic or property. The word "nano" is derived from the Greek word which means "dwarf." One nanometer is the 1 billionth or 10^{-9} part of a meter. The nano-size range holds so much interest because in this range materials can have diverse and enhanced properties compared with the same material at a larger (bulk) size (Gholami-Shabani et al. 2015; Dudefoi et al. 2018). Materials in the nano-scale differ

Nanobiotechnology in Diagnosis, Drug Delivery, and Treatment, First Edition. Edited by Mahendra Rai, Mehdi Razzaghi-Abyaneh, and Avinash P. Ingle.

significantly from other materials due to the following two major principal reasons: physical effects such as expanded surface area and phenomena are based on "quantum effects" (Gholami-Shabani et al. 2016). These properties can enhance the reactivity, durability, and electrical features and *in vivo* behavior of nanomaterials.

Due to these unique properties of nanomaterials, modern nanotechnology is emerging as potential branch of science that can revolutionize various fields, including biomedicine. Looking at the recent advances in the field of nanotechnology it can be observed that nanotechnology influences almost every facet of everyday life from security to medicine. Nanotechnology and its medical applications are usually seen as having a wide potential to cause benefits to various areas of investigation and applications. Currently, nanotechnology is providing completely novel concepts and approaches in various fields of biomedicine such as diagnosis, drug delivery, and treatment of a wide range of diseases including various serious and life-threatening diseases like cancer, neurodegenerative disorders, cardiovascular diseases, etc.

To date, a variety of nanomaterials have been investigated which play a crucial role in the diagnosis and management of different diseases as mentioned. The nanomaterials which are used in medicine are termed as "nanomedicine." The concept of nanomedicine was first put forward in 1993 by Robert A. Freitas, Jr. Nanomedicine is considered the science of preventing, diagnosing, and treating disease using nanosized particles (Abiodun-Solanke et al. 2014). Various nanomaterials such as organic, inorganic, polymeric, and metallic nanostructures like dendrimers, micelles, solid lipid nanoparticles (SLNs), carbon nanotubes (CNTs), liposomes, niosomes, etc. have been successfully exploited in nanomedicine. Therefore, the use of these nanomaterials in the development of various nanodiagnostic tools (such as microchips, biosensors, nano-robots, nano identification of single-celled structures, and microelectromechanical systems) and therapeutic treatment approaches via target-specific drug delivery has attracted a great deal of attention from the scientific community around the world (Liang et al. 2014; Núñez et al. 2018; Mitragotri and Stayton 2019).

Currently, various diagnostic and therapeutic strategies are in practice which are very complex, time-consuming, and also very costly. However, the recent advances in nanotechnology allow us to provide accurate, sensitive, rapid, and inexpensive diagnostic techniques, as well as treatments for the patients with the least number of possible interventions and without any adverse effects (Leary 2010; Gholami-Shabani et al. 2018).

Usually, drugs function through the whole body before they reach the specific disease-affected zone. In this context, nanotechnology has opened up novel opportunities to deliver specific drugs using various nanomaterials as delivery vehicles. Such nanotechnology-based drug delivery has the ability to achieve effective, precise, and target-specific drug delivery in order to reduce the chances of possible side effects (Gholami-Shabani et al. 2017). Suitable drug-delivery techniques have two fundamentals: the capability to target and to control the drug release. Targeting will ensure high performance of the drug and decrease the side effects, particularly when acting with drugs that are recognized to kill cancer cells but can also kill healthy cells when delivered (Cho et al. 2008). The decrease or prevention of side effects can be effectively achieved by the controlled release of a drug. In this context, nanotechnology-based drug delivery systems provide a healthier diffusion of the drugs inside the body as their size allows delivery through intra venous injection or other methods. The nano-size of these particulate structures also reduces the exciter reactions at the injection spot. Initial attempts to direct cure in a specific set of cells involved conjugation of radioactive materials to antibodies specific to markers shown on the surface of cancer cells (Patra et al. 2018).

Although nanotechnology has shown a number of revolutionary promises in various fields, it is still at the juvenile stage and there are numerous fields to be explored. Therefore, it is believed that in the near future, nanotechnology will help us to understand various aspects of human physiology in a significant way and it will be boon for the betterment of the biomedical field. Considering the huge potential of nanotechnology in biomedicine, in the present chapter we have focused on the general concept of nanotechnology and nanomaterials commonly used as nanomedicine. Moreover, the role of various nanomaterials in diagnosis, drug delivery, and treatment of various diseases has also been presented. In addition, different challenges in the use of nanomaterials as nanomedicine are further discussed.

1.2 Nanomaterials Used in Diagnosis, Drug Delivery, and Treatment of Diseases

From the many studies performed over the last couple of decades, it has been proven that nanotechnology has a huge impact on the development of therapeutics. To date, a variety of organic and inorganic nanomaterials have been developed to encapsulate and deliver therapeutic and imaging agents (Mitragotri and Stayton 2019). These nanomaterials have allowed encapsulation and targeted release of drugs. Some of the nanomaterials-based drugs are already being used in patients, however many others are making excellent progress toward clinical translation. Some of the important organic and inorganic nanomaterials commonly used in diagnosis, drug delivery, and treatment of a wide range of diseases are briefly discussed here.

1.2.1 Inorganic Nanomaterials

Different inorganic nanomaterials discussed in the chapter have been successfully exploited directly or indirectly in the diagnosis and management of various diseases.

1.2.1.1 Colloidal Metal Nanoparticles

Different colloidal metal nanoparticles (Figure 1.1a) have been reported as having potential applications in diagnosis, drug delivery, and treatment of many diseases. However, among the colloidal metal nanoparticles, colloidal gold nanoparticles (GNPs) are recognized as suitable nanocarriers

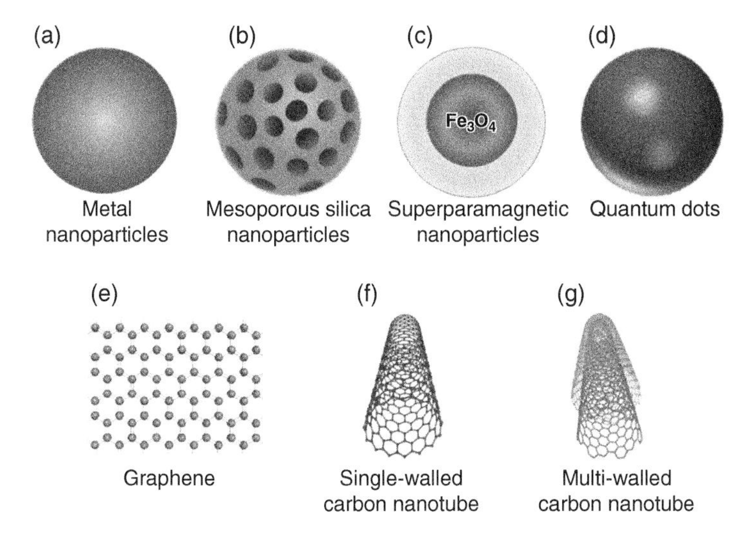

Figure 1.1 Schematic illustration of various inorganic nanomaterials.

for biomedicine, i.e. for the intracellular and *in vivo* delivery of genes, drugs, and as contrast agents because of their easy synthesis, large surface area, and flexible surface chemistry (Lewinski et al. 2008; Jeong et al. 2019). Moreover, these nanoparticles can be easily modified by conjugating smart polymers in order to develop novel drug delivery systems that have the ability to release their payload in response to outside stimuli (Yavuz et al. 2009; An et al. 2010). Furthermore, due to the above-mentioned novel properties and their high molar absorption coefficient, GNPs can be directly or indirectly used for the diagnosis and management of various diseases including photothermal agents in cancer photothermal therapy (Liang et al. 2014). In addition, there are other metal nanoparticles like silver, copper, etc., that were potentially used as novel antimicrobial agents due to their promising antimicrobial properties.

1.2.1.2 Mesoporous Silica Nanoparticles

Mesoporous silica nanoparticles (MSNPs) are another important group of inorganic nanomaterials (Figure 1.1b). These nanoparticles are also considered ideal candidates for biomedical applications due to their controllable morphologies, mesostructures with biocompatibility, and easy functionalization ability (Dykman and Khlebtsov 2012; Liu and Xu 2019). The presence of numerous silanol groups on the surface of MSNPs make them hydrophilic; moreover, their functionalization using a variety of groups helps to achieve controlled holding/release of cargo molecules. In addition, the large internal surface area and pore capacity of mesoporous materials allow a high loading of cargo molecules and also prevent them from escaping into water easily by dissolving in an aqueous environment (Jafari et al. 2019). These advantages enhance the effectiveness of the MSNP-based delivery system and allow a specific amount of drugs to reach their therapeutic target (Li et al. 2012).

1.2.1.3 Superparamagnetic Nanoparticles

Among the inorganic nanoparticles, superparamagnetic nanoparticles (Figure 1.1c) are considered the most unique nanoparticles due to their strong magnetic properties. These nanoparticles were for the first time used in the late 1980s for biomedical applications (Stark et al. 1988). Usually, the core of these nanoparticles consists of metal molecules of nickel, cobalt, or iron oxide (Fe_3O_4 magnetite, which is the most commonly used metal). As mentioned above, superparamagnetic nanoparticles are considered most unique because the surface of these nanoparticles can be easily modified by coating the core with various organic polymers like dextran, starch, alginate, inorganic metals, oxides (silica, alumina), etc. (Núñez et al. 2018).

Superparamagnetic nanoparticles can be promisingly used for the diagnosis of various diseases including cancer (tumors) by conjugating with various bioactive ligands (Anderson et al. 2019). To date, a number of approaches have been developed for the fabrication of superparamagnetic nanoparticles which have the potential ability to distinguish cancerous tissue from healthy tissue. In addition, these nanoparticles can be used for magnetic resonance imaging (MRI) of tumor tissue, cell labeling, and drug delivery in different diseases (Núñez et al. 2018; Anderson et al. 2019).

1.2.1.4 Quantum Dots

Quantum dots (Figure 1.1d) are the class of semiconductor nanoparticles with unique photo-physical properties. Usually, quantum dots have a core/shell structure composed of molecules of various metals like technetium, cadmium selenide, zinc, indium, tantalum, etc. (Medintz et al. 2005; Wang et al. 2019). The most commonly used, commercially available quantum dots contain a cadmium selenide core covered with a zinc-sulfide shell. The core-shell complex is generally encapsulated in a coordinating ligand and an amphiphilic polymer (Gao et al. 2004). Due to unique optical properties, quantum dots have been used as dominant classes of fluorescent imaging probes for various biomedical applications (Núñez et al. 2018).

1.2.1.5 Graphene

Graphene is an atom-thick monolayer of carbon atoms arranged in a two-dimensional honeycomb structure (Figure 1.1e) (Novoselov et al. 2004). Generally, graphene has been extensively used for a wide array of applications in many fields such as quantum physics, nanoelectronic devices, transparent conductors, energy research, catalysis, etc. (Wang et al. 2011; Huang et al. 2012). However, according to recent technological advancements graphene, graphene oxide, and reduced graphene oxide have shown promising applications in the biomedical field and hence have attracted significant interest worldwide (Gonzalez-Rodriguez et al. 2019; Yang et al. 2019). Due to its excellent physicochemical and mechanical properties, single-layered graphene has been widely utilized as a novel nanocarrier for drug and gene delivery in different diseases (Núñez et al. 2018).

1.2.1.6 Carbon Nanotubes (CNTs)

CNTs are cylindrical nanomaterials that consist of rolled-up sheets of graphene (single-layer carbon atoms). Depending on their structure CNTs can be divided into two types, namely single-walled carbon nanotubes (SWCNTs) (Figure 1.1f), which usually have diameter of less than 1 nm, and multi-walled carbon nanotubes (MWCNTs) (Figure 1.1g), which are composed of many concentrically interlinked nanotubes, usually having a diameter size of more than 100 nm (Hsu and Luo 2019). CNTs are considered as one of the stiffest and strongest fibers having novel exceptional characteristics and a unique physicochemical framework, which makes them suitable candidates for efficient delivery of different therapeutic drugs/molecules for various biomedical applications (Vengurlekar and Chaturvedi 2019). Figure 1.1 represents the schematic illustration of various inorganic nanomaterials.

1.2.2 Organic Nanomaterials

1.2.2.1 Polymeric Nanoparticles

Polymeric nanoparticles (Figure 1.2a,b) are colloidal solid particles having a size in the range of 10 nm–1 μm. Based on the preparation method, polymeric nanoparticles can be classified into two types of structures: nanocapsule (Figure 1.2a) and nanosphere (Figure 1.2b). Among these, nanospheres consist of a matrix system which facilitates uniform dispersion of the drug. However, in the case of nanocapsules, the drug is only embedded in a cavity and the cavity is surrounded by a polymeric membrane (Sharma 2019). Among the various organic nanomaterials, polymeric nanoparticles have attracted huge attention over the last few years due to their unique properties and behaviors resulting from their small size. As reported in many studies, these nanoparticles demonstrated potential applications in biomedicine particularly in diagnostics and drug delivery. Polymeric nanoparticles are preferably used as a nanocarrier for the conjugation of various drugs, natural polymers (e.g. natural polymers like chitosan, gelatin, alginate, and albumen), and synthetic polymers (Zhang et al. 2013a,b). Further, they showed significant benefits in treatment because of controlled release of the drug, their ability to combine both therapy and imaging (theranostics), protection of drug molecules due to conjugation, and their target-specific drug delivery (Crucho and Barros 2017).

1.2.2.2 Polymeric Micelles

Polymeric micelle (Figure 1.2c) is the class of organic nanomaterials usually formed by the spontaneous arrangement of amphiphilic block copolymers in aqueous solutions (Kulthe et al. 2012).

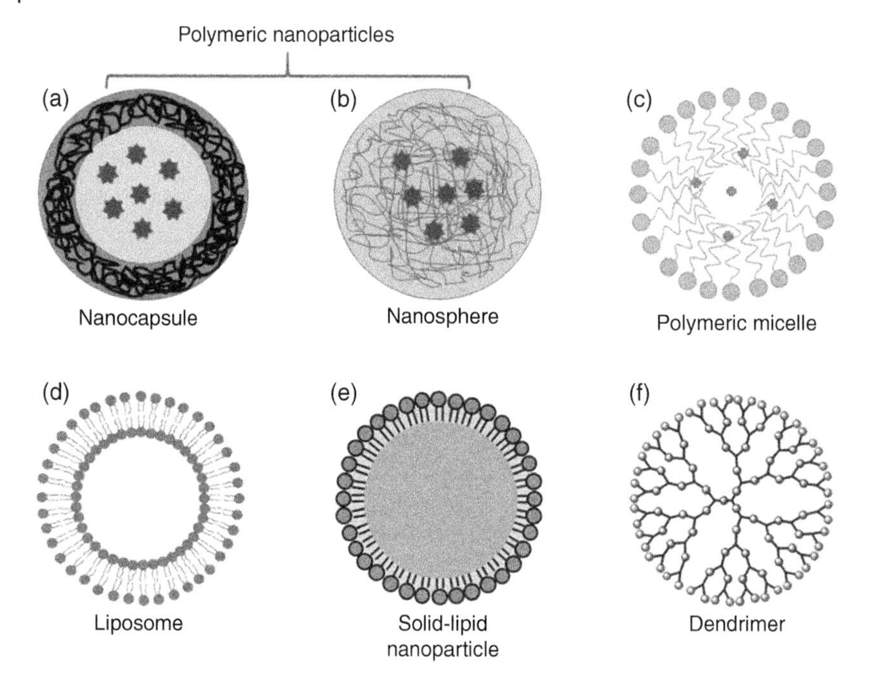

Figure 1.2 Schematic representation of various organic nanomaterials.

Structurally these nanomaterials are composed of the hydrophobic core and hydrophilic shell, which facilitates the loading of various hydrophobic drugs like camptothecin, docetaxel, paclitaxel, etc. into the core to be used as nanocarriers (Singh et al. 2019). The encapsulation of drugs in polymeric micelles enhances their solubility. Various novel properties of polymeric micelles, including their small size, make them promising nanocarriers in drug delivery. Polymeric micelle-mediated drug delivery showed many advantages like easy penetration, target-specific drug delivery, narrow distribution to avoid fast renal excretion, etc. Moreover, polymeric micelles can also be conjugated with targeting ligands which facilitates their uptake by specific cells, thus reducing off-target side effects. Polymeric micelles can be synthesized using two different approaches: (i) convenient solvent-based direct dissolution of polymer followed by dialysis process; or (ii) precipitation of one block by adding a solvent (Patra et al. 2018).

1.2.2.3 Liposomes

The word liposome (Figure 1.2d) has been derived from the two Greek words: lipo ("fat") and soma ("body"), and was described for the first time by British hematologist Alec Bangham in 1961, at the Babraham Institute, in Cambridge (Bangham and Horne 1964). Liposomes are among the group of organic nanomaterials which are comparatively small, spherical vesicles that are amphipathic (contains both hydrophilic and hydrophobic structures). Liposomes are generally composed of phospholipids and spontaneously formed when certain lipids are hydrated in an aqueous media (Singh et al. 2019). The unique structure of liposome (i.e. lipid bilayer[s] structure) facilitates the incorporation of both hydrophilic and hydrophobic drugs, which helps to prevent the rapid decomposition of the incorporated drug and also to release the drug molecules at a specific targeted site (Lujan et al. 2019). Liposomes reported having several promising properties such as small size, lipid bilayer structure, surface charge, biocompatibility, biodegradability, low toxicity, site-specific delivery, etc. Due to all these properties, liposomes have attracted a great deal of attention from researchers all over the world for their use as potential drug delivery systems in various diseases including cancer (El-Hammadi and Arias 2019). Moreover, liposomes have been modified to develop some other phospholipid vesicles which selectively have their applications in the delivery of specific drugs or biomolecules (mostly for transdermal delivery). These vesicles mainly include transferosomes, niosomes, and ethosomes.

1.2.2.4 Transferosomes

Transferosomes are the modified form of liposomes which are considered to be highly elastic and deformable. These modified forms of liposome were developed by Gregor Ceve for the first time in 1990 (Blume and Ceve 1990). Transferosomes are almost similar to liposomes in their basic structural arrangement; the only difference is that the outer layer of transferosomes is more complex in nature compared to liposomes. These nanomaterials reported having enhanced flexibility due to edge activator presence in lipid bilayer (Abdallah 2013). Usually, transferosomes are formed by self-controlled assembly and they are efficient to cross the various transport barriers, and hence are selectively used as carriers for the delivery of drugs and other macromolecules instead of liposomes (Sharma 2019).

1.2.2.5 Niosomes

Niosomes are another kind of liposomes that are supposed to be osmotically active, highly flexible, and comparatively more stable than liposomes (Bartelds et al. 2018; Sharma 2019). These nanomaterials are mainly composed of nonionic surfactants like alkyl ethers, alkyl glyceryl ethers, sorbitan fatty acid esters, and polyoxyethylene fatty acid esters stabilized by cholesterol (CH) (Muzzalupo and Mazzotta 2019). Like liposomes, they also form a lamellar structure in which the hydrophilic heads have oriented outward and the hydrophobic tails point inward or facing the opposite direction to form a bilayer (Sharma 2019). Niosomes are economically viable nanomaterials compared to liposomes and other related nanomaterials; moreover, they possess various novel properties such as their biodegradable, biocompatible, and non-immunogenic nature (Singh et al. 2019).

1.2.2.6 Ethosomes

Ethosomes are also a type of phospholipid vesicles, considered a modified form of liposome mainly composed of ethanol, phospholipids, and water. In addition, some other components can also be included in ethosomes for specific characteristics e.g. polyglycol as a permeation enhancer, cholesterol to increase the stability, and dyes useful for characterization studies (Sharma 2019). These vesicles for the first time were developed by Prof. Elka Touitou around 1997. The simple synthesis process, high efficacy, and nontoxic nature of ethosomes allowed their use in widespread applications related to transdermal delivery. Ethosomes are soft, malleable vesicles tailored for enhanced delivery of active agents (Verma and Pathak 2010). Ethosomes are noninvasive delivery nanocarriers that facilitate penetration of drugs deep in the skin layers and the systemic circulation, and are reported to have higher transdermal flux than liposomes (Godin and Touitou 2003). The presence of ethanol in higher concentrations makes the ethosomes novel and unique, as ethanol is known for its disturbance of skin lipid bilayer organization.

1.2.2.7 Solid Lipid Nanoparticles (SLN)

SLNs (Figure 1.2e) are among the class of lipid nanoparticles having a size in the range of 1–1000 nm, usually have a crystalline lipid core which is stabilized by interfacial surfactants (Sun et al. 2019). These nanomaterials were introduced in 1991 for the first time. SLN are reported as having various novel properties such as easy synthesis, low cost, ability to store various molecules/drugs with high loading capacity, controlled drug release, improved stability, improved biopharmaceutical performance, etc. Therefore, SLN have been preferably used over the other various drug delivery systems like emulsions, liposomes, and polymeric nanoparticles (Pink et al. 2019).

1.2.2.8 Dendrimers

Dendrimers (Figure 1.2f) are highly branched three-dimensional nanomaterials consisting of polymeric branching units attached to a central core through covalent bonding, which are organized in concentric layers and that terminate with several external surface functional groups (Lombardo et al. 2019). Dendrimers are synthetic nanomaterials fabricated by a specific synthesis approach involving a series of different reactions that allow precise control on various parameters like size, shape, and surface chemistry which result in highly monodisperse nanostructures. Like various other nanomaterials described above, it is possible to conjugate suitable drugs or macromolecules like proteins or nucleic acid into the surface of dendrimers in order to use them as potential nanocarrier (Virlan et al. 2016). Dendrimers reportedly enhance the solubility and bioavailability of hydrophobic drugs that are entrapped in their intramolecular cavity or conjugated on their surface. However, various factors such as surface modification, ionic strength, pH, temperature, etc., influence the structural properties of dendrimers (Choudhary et al. 2017). Figure 1.2 represents the schematic illustration of various organic nanomaterials.

1.3 Role of Nanomaterials in Diagnosis, Drug Delivery, and Treatment

All the above-mentioned inorganic and organic nanomaterials are reported as having direct applications as antimicrobial agents, or indirect applications as a nanocarrier for the conjugation of a variety of drugs and other biomolecules in order to develop efficient drug delivery systems for various life-threatening diseases, including cancers.

1.3.1 In Diagnosis

Nanotechnology has provided many useful tools that can be applied to the detection of biomolecules and analyte relevant for diagnostic purposes (Baptista 2014). This new branch of laboratory medicine, termed nanodiagnostics, includes early disease detection even before symptoms' presentation, improved imaging of internal body structure, and ease of diagnostic procedures; determines disease state and any predisposition to such pathology; and identifies the causative organisms by using recently developed methods and techniques of nanotechnology such as microchips, biosensors, nanorobots, nano identification of single-celled structures, and microelectromechanical systems (Figure 1.3) (Jain 2003; Baptista 2014; Jackson et al. 2017). As an evolving field of molecular diagnostics, nanodiagnostics have been positively changing laboratory procedures by providing new ways for patient's sample assessment and early detection of disease biomarkers with increased sensitivity and specificity while nanomaterials used for detection of pathogens or disease biomarkers have been developed and optimized in such way that becomes less nuisance for patients (Jackson et al. 2017; Bejarano et al. 2018). Although nanotechnologies have been applied to diagnostics of several diseases with promising results, the medical imaging and oncology are still the most active areas of development (Bejarano et al. 2018). In recent years, many studies have been directed to the design of new contrast agents allowing easy, reliable, and noninvasive identification of various diseases (Ahmed and Douek 2013).

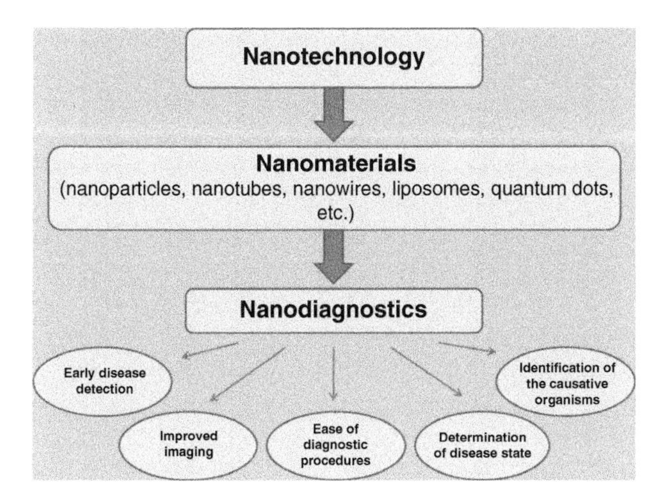

Figure 1.3 Role of nanotechnology and nanomaterials in diagnostics and its advantages.

Superparamagnetic iron oxide nanoparticles (SPIONs) are well known as MRI contrast agents for the study of the pathologically changed tissues, e.g. tumors or atherosclerotic plaque. They can be functionalized with various biomolecules (e.g. hormones, antibodies, cyclic tripeptides) which improve their bioavailability and interaction with specific tissues. Conjugation of SPIONs with biomolecules affecting their binding to the receptors of cancer cells or other types of internalization by cells and strong accumulation of these conjugates in the pathologically changed tissues, e.g. tumors. Therefore, it allows to detect tumors and enhance the negative contrast in the MRI (Chen et al. 2009; Meng et al. 2009; Kievit et al. 2012; Peiris et al. 2012; Bejarano et al. 2018). Similarly, iodinated polymer nanoparticles (Hyafil et al. 2007) or GNPs coated with polyethylene glycol (PEG) (Kim et al. 2007) have been developed as contrast agents for computed tomography (CT) imaging. Another imaging technique that benefits from nanoparticles as contrast agents is photoacoustic imaging, which detects the distribution of optical absorption within the organs (Li and Chen 2015).

As mentioned above, diagnostic imaging techniques have certain limitations, therefore multimodal nanosystems have been developed to overcome these limitations. Multimodal nanosystems combine the properties of different nanoparticles with various imaging techniques for improved detection. These multimodal nanosystems use PET-CT and PET-MRI techniques that combine the sensitivity of positron emission tomography (PET) for metabolism imaging and tracking of labeled cells or cell receptors with the outstanding structural and functional characterization of tissues by MRI and the anatomical precision of CT. The lipid nanoparticles have been labeled with contrast agents and successfully employed in multimodal molecular imaging. These liposomes may be incorporated with gold, iron oxide, or quantum dot nanocrystals for CT, MRI, and fluorescence imaging, respectively (Rajasundari and Hamurugu 2011; Bejarano et al. 2018). Recently it was demonstrated that nanomaterials such as PdCu@Au nanoparticles radiolabeled with ^{64}Cu and functionalized toward target receptors provided a tool for highly accurate PET imaging and photothermal treatment (Pang et al. 2016). Similarly, a ^{89}Zr-labeled liposome encapsulating a near-infrared fluorophore was developed for both PET and optical imaging of cancer (Pérez-Medina et al. 2015).

Many different nanomaterials, namely nanoparticles (e.g. gold nanoparticles), liposomes, nanotubes, nanowires, quantum dots, and nanobots have been developed for nanodiagnostics (Jackson et al. 2017). However, nanomaterials in combination with biomolecules that are used as biosensors have the greatest application as exemplified by a sensor made from densely packed CNTs coated with GNPs or CNTs and silicon nanowires used for detection of oral cancer or various volatile organic compounds present in breath samples of lung and gastric cancer patients, respectively (Beishon 2013; Shehada et al. 2015). Especially, nanowires have been used as a platform for other biomolecules such as antibodies, which are attached to their surface. Such a platform acts as a detector when antibodies interact with biomolecules of a target and as a consequence change their conformation which is picked up as an electrical signal on the nanowire. Therefore, nanowires associated with different antibodies may be used as a device for the detection of variable biomarkers that are produced or released from cells during the disease process. Such nanobiosensors can be used also for monitoring cancer disease, its earlier prediction before full manifestation, or the risk of biochemical relapse (Reimhult and Höök 2015). Therefore, the nanowires may be applied for measurement of RNA expression level of cancer antigens or as platform functionalized with ssDNA to detect mutations related to different types of cancers (Lyberopoulou et al. 2015; Takahashi et al. 2015).

Moreover, nanotechnology plays a crucial role in the devolvement of nanobiosensors which has varied applications in the detection of pathogens and other contaminants present in the products. The standard methods of assaying various substances require, as a rule, trained personnel, special preparation of samples, and expensive reagents; besides, they are time-consuming. The emergence of biosensors that make use of nanotechnologies (nanobiosensors) enabled high-speed diagnosis without worsening the quality, directly on the sampling site, without attracting qualified personnel. The biosensor represents an analytical device containing a biological recognition element (cell, tissue, enzyme, nucleic acid, antibody, etc.) coupled with a signal transducer. Interaction of the biological recognition element with an analyte leads to a change of its physical, chemical, optical or electrical characteristics, which is picked up by a signal transducer (a schematic operating principle of the biosensor is shown in Figure 1.4). The use of nanomaterials leads to decrease the size of the biosensor; to increase its sensitivity, selectivity, reproducibility of the assay, and enable its incorporation into multiplexed, transportable, and portable devices for assessment of food quality (Otles and Yalcin 2012; Ríos-Corripio et al. 2020).

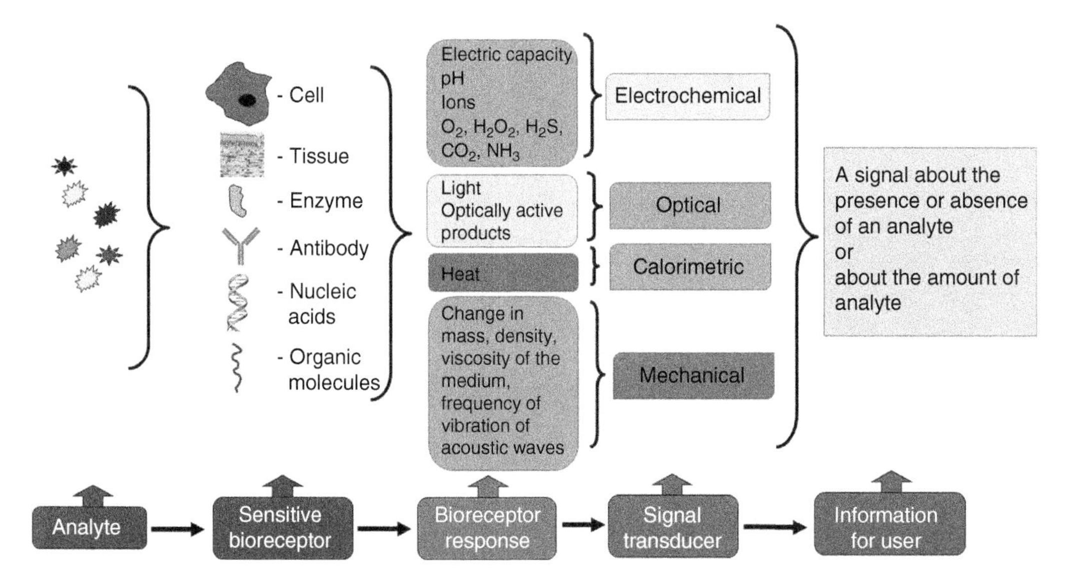

Figure 1.4 Schematic representation of the operating principle of the biosensor.

Nanosensors use nanoparticles of different chemical nature: carbon nanomaterials (graphene, CNTs, carbon fibers, fullerenes, etc.) (Kurbanoglu and Ozkan 2018), nanoparticles of metals (gold, silver, copper, silicon; metal oxides; quantum dots) (Li et al. 2019), and branched polymers (dendrimers) (Abbasi et al. 2014). GNPs are used most often, due to their resistance to oxidation, low toxicity, and ability to amplify the biosensor signal. The application of such particles leads to increased sensitivity and detection limits up to one molecule (Vigneshvar et al. 2016). An important positive point of using nanosensors is also shorter assay time, especially when pathogenic microorganisms in food are detected. There are many reports available which involved the use of biosensors based on nanoparticles for screening for pathogens, toxins, and allergen products in food matrices (Warriner et al. 2014; Inbaraj and Chen 2016; Prakitchaiwattana and Detudom 2017).

Many modern applications of nanosensors are based on the observed changes of color, which occur with solutions of metal nanoparticles in the presence of an analyte of interest. For instance, aggregation of GNPs caused by mercury (II) ions (Hg^{2+}) is the main sensor mechanism resulting in a significant redshift in the absorption band, with color changing to blue (Chansuvarn et al. 2015). Ma et al. (2018) reported that with the presence of tobramycin in the sample, the DNA aptamer will bind exactly to it, separating from GNPs, and their aggregation will lead to a color change from red to purple-blue. According to Ramezani et al. (2015) in the presence of an analyte (tetracycline) GNPs remain stable and do not aggregate under the action of the salt; as a result, the color changes from blue (in the absence of the analyte) to red (in the presence of the analyte). Immunosensors also based on GNPs non-aggregation have been developed for the assay of β-lactams in milk (Chen et al. 2015) and for the simultaneous detection of benzimidazoles and metabolite residues in milk samples (Guo et al. 2018). A general scheme of similar colorimetric nanosensors is given in Figure 1.5.

Kaur et al. (2019) presents not only advantages of using nanoparticles but also restrictions associated with the cost of their production, preparation of samples, sensitivity to other substances occurring in the sample, lack of self-standardization, and validation with real samples, as well as

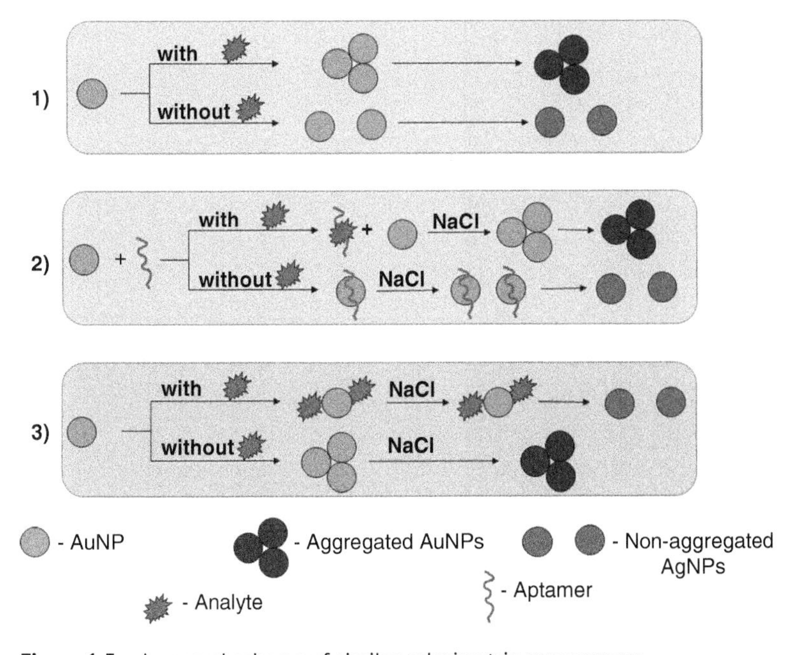

Figure 1.5 A general scheme of similar colorimetric nanosensors.

with the toxicity of used nanomaterials and ways of utilizing spent sensors. Despite the negative aspects of using nanoparticles in sensor systems, this direction is actively developed. Investigators working with nanomaterials have yet to solve many problems associated both with leveling off their negative characteristics and developing methods of fabricating and using nanoparticles in various regions of the economy.

1.3.2 In Drug Delivery and Treatment

Nanotechnologies have enabled novel solutions for the treatment of various diseases. Nano-drug delivery systems present some advantages over conventional (non-targeted) drug delivery systems such as high cellular uptake and reduced side effects (Singh et al. 2019). Figure 1.6 represents the comparison between untargeted and targeted drug delivery systems. The development of drug delivery systems by using nanotechnology for various diseases, particularly for cancer treatment, is making revolutionary changes in treatment methods and handling side effects of chemotherapy (Zhao et al. 2018). Furthermore, nanotechnology allows for selective targeting of disease- and infection-containing cells and malfunctioning cells. Nanotechnology has also opened a new era in

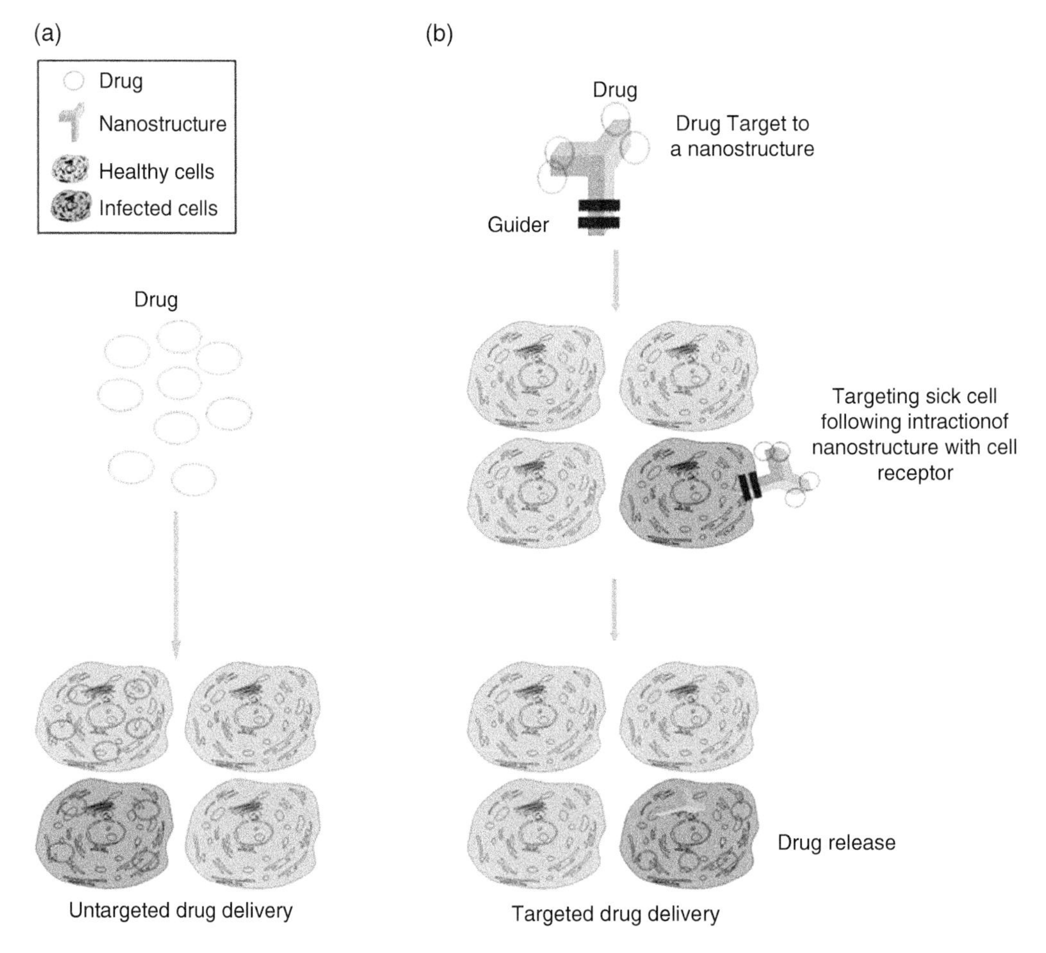

Figure 1.6 Schematic representation of comparison between untargeted and targeted drug delivery systems.

implantable delivery systems such as those used in bone cement, nano-needle patches, etc., which are preferable than using other modes of direction like injections and oral delivery. Therefore, nano-based drug delivery systems have attracted a great of attention due to novel and promising properties like enhanced bioavailability and stability of the drug.

Moreover, antimicrobials are one of the most important therapeutic discoveries in the history of medicine. Some projections suggest that by 2050 the annual deaths caused by multidrug-resistant bacterial infections will reach up to 10 million per year (de Kraker et al. 2016). Nanotechnology provides an innovative platform to address this challenge, because of their small size and various other physical, chemical, and optical properties. As mentioned above, numerous nanomaterials are reported to have significant antimicrobial efficacies and hence such nanomaterials can be used as next-generation antimicrobials against various multidrug-resistant organisms and also in the treatment of different infectious diseases (Rai et al. 2012; Beyth et al. 2015; Rai et al. 2016).

The bioavailability of a drug within the body depends on some factors like the size of the drug molecules and solubility factors (Kesharwani et al. 2018). The conventional dosage system consequently faces some challenges in reaching the target site at an appropriate dose. For example, highly water-soluble drugs cause fluctuations in drug concentration in the body due to high disintegration properties, and also result in quicker clearance of the drug from the bloodstream. However, some medicines are fat-soluble and when such drugs are taken in the form of conventional dosage, they face bioavailability difficulties. Similarly, patients suffering from chronic diseases like diabetes need to take painful insulin injections regularly. Likewise, cancer patients regularly have to undergo powerful chemotherapy, which involves quite severe side effects as the anticancer drugs target cancer cells and normal cells equally. Therefore, proper platforms to deliver the drugs at targeted sites without losing their efficacies, and while limiting the associated side effects, are highly required (Mura et al. 2013).

Many novel technologies for developing effective drug delivery systems came into existence in this context, among which nanotechnology platforms for achieving targeted drug delivery are gaining prominence. Research in this field includes the development of drug nanoparticles, polymeric and inorganic biodegradable nanocarriers for drug delivery, and surface engineering of carrier molecules (Senapati et al. 2018). These nanocarriers help in solubilizing the lipophilic drugs, protecting fragile drugs from enzymatic degradation, pH conditions, etc., and targeting specific sites with triggered release of drug contents.

To date, a wide range of nanomaterials enlisted above has been developed and used for applications in nano-drug delivery. There are many reports on the usage of metallic nanoparticles in drug delivery and diagnostic applications. It mainly includes the applications of silver, gold, and iron-based superparamagnetic nanoparticles as nanocarriers for controlled and targeted delivery of potential drugs and genes for enhanced clinical efficacy. In addition, nanosuspensions or nanodispersions, which are theoretically considered the simplest form of nanomedicine, contain two specific components, the active pharmaceutical ingredients nanoparticle and the adsorbed surface stabilizer(s), which have been also effectively used in the treatment of various diseases (Adeyemi and Sulaiman 2015).

Similarly, polymeric nanoparticles or nanopolymer manufactured through the chemical conjugation of active pharmaceutical ingredients and a water-soluble polymer have been used to develop a polymer-drug or polymer-protein conjugate or pro-drug compounds which are further used in the treatment of a wide range of diseases. The chemical degradation releases the active pharmaceutical ingredients into the bloodstream or the site of disease (Tong et al. 2009). Likewise, various other nanocarrier systems like liposome, SLN, dendrimers, quantum dots, etc., are promisingly used in the development of efficient drug delivery systems for potential management of diseases

(Liang et al. 2014; Núñez et al. 2018; Mitragotri and Stayton 2019). The role of such nano-based drug delivery systems in the treatment of various diseases has been briefly discussed here.

1.3.2.1 Diabetes

Silica nanoparticles have been used in drug delivery applications for years, as they can be adjusted for continuous or triggered drug release (Bharti et al. 2015). Delivery of insulin across intestinal Caco-2 cells using silica nanoparticles was reported. Silica nanoparticles, due to high surface area and selective absorption, can be extensively used in drug delivery systems (Bharti et al. 2015). Polymeric nanoparticles have been harnessed for targeted drug delivery and protection of nucleic acids. Interleukin (IL)-10 and IL-4 entrapped in polymeric nanoparticles were delivered to white blood cells to reduce the T cell response against native islet cells in prediabetic animal models. It was observed that the polymeric nanoparticles were useful in diabetic treatment as it inhibited the development of diabetes in 75% of the animal models (Singh et al. 2019).

1.3.2.2 Cancer

Porous silica nanoparticles have emerged as an efficient delivery system for cancer therapy. Targeted drug therapy requires zero release before reaching the target site (Bharti et al. 2015). Targeted delivery of Doxorubicin (DOX) using silica nanoparticles coated with PEG have been reported. For the study, $120\,mg\,kg^{-1}$ of nanoparticles were injected on a weekly basis for a period of three weeks to a KB-31 xenograft model. The results of the study showed 85% inhibition of tumor by DOX-loaded silica nanoparticles in comparison to DOX drug. A study reported curcumin-containing liposomes conjugated with synthetic RNA aptamer (Apt-CUR-NPs) to target epithelial cell adhesion molecule (EpCAM) protein which is observed in colorectal adenocarcinoma. The Apt-CUR-NPs depicted enhanced bioactivity of curcumin (CUR) after 24 hours compared to free CUR. Apt-CUR-NPs also showed increased binding to HT29 colon cancer cells and cellular uptake. Comparative study of *in vitro* induced cytotoxicity of free CUR and Apt-CUR-NPs in HT29 cell line demonstrated more cytotoxicity of Apt-CUR-NPs in comparison to totally free CUR (cellular viabilities about 58% and 72%, respectively) (Rabiee et al. 2018).

1.3.2.3 Psoriasis

Psoriasis is a distinctive chronic inflammatory disease with a strong genetic makeup. Srisuk et al. (2012) evaluated permeability of Methotrexate (MTX)-entrapped deformable liposomes and devised lipid vesicle from phosphatidylcholine (PC) and oleic acid (OA), and compared with MTX-entrapped conventional liposomes synthesized using PC and cholesterol (CH) by thin-film hydration method. MTX entrapped in PC: CH was observed to be more stable in size and loading. Although, MTX-entrapped PC: OA liposomes increased the skin permeability characterized with higher absorption and flux of MTX diffused across or accumulated in the epidermis and dermis layers of porcine skin (Srisuk et al. 2012; Chandra et al. 2018).

Pinto et al. (2014) also developed and assessed the potential of nanostructured lipid carriers (NLCs) loaded with MTX by hot homogenization technique as a novel technique for topical therapy of psoriasis. The assessment of the *in vitro* skin permeation of MTX in their study showed its capability to go through the skin barrier when loaded within NLCs formulation which confirms the high potential of NLC as carriers for MTX and feasibility for topical delivery.

1.3.2.4 HIV

To deal with HIV progression and prevent development of resistance, a combination of multiple drugs is given which is known as highly active antiretroviral therapy (HAART). Antiretroviral drugs should be able to cross the mucosal epithelial barrier when used orally. It has been reported that nanoparticles conjugated with antiretroviral drugs have the potential to target monocytes and macrophages *in vitro*. Poly(lactic-co-glycolic acid) (PLGA) nanoparticles were entrapped with three antiretroviral drugs, ritonavir, lopinavir, and efavirenz, and analyzed. It was observed that PLGA nanoparticles depicted sustained drug release for over 4 weeks (28 days), whereas the free drugs within 48 hours (2 days) (Rizvi and Saleh 2018). Thus, nanoparticles could be utilized for drug delivery of anti-HIV drugs.

1.3.2.5 Neurodegenerative Diseases

Nerve growth factor (NGF) is very crucial for the survival of neurons and can be used a potential therapeutic agent for neurodegenerative disorders. NGF could not cross the blood-brain barrier (BBB) but can be preferably used as a drug delivery vehicle to transport drugs across BBB. Various studies have reported NGF encapsulated in transferring and cereport-functionalized liposomes improve the permeability across the BBB. Also, resveratrol in combination with lipid core nanocapsules were found to be highly efficient against Aβ-induced neurotoxicity. The drug delivery system was observed to improve short-term and long-term memory (Singh et al. 2019).

1.3.2.6 Blood Pressure (BP) and Hypertension

Polymer nanoparticles like PLGA, poly(lactic acid) (PLA), and chitosan have been used in conjugation with antihypertensive drugs for targeted and controlled drug delivery. A significant and sustained reduction in blood pressure (BP) has been recorded employing nifedipine conjugated with PLGA, PCl nanoparticles. The most important benefit of sustained release of hypertensive drugs is that it regulates BP fluctuations; also, lower doses are required compared to conventional drugs (Singh et al. 2019). Liposomal drug formulations have also been investigated using animal models for hypertension. It was observed that a single intravenous dose of liposomal formulation encapsulated with vasoactive intestinal peptide normalized BP for longer duration compared to non-encapsulated peptide. Thus, successful conjugation of anti-hypertensive drugs with nanoparticles increases drug circulation and also prolongs systemic availability of drugs at required concentrations (Singh et al. 2019).

1.3.2.7 Pulmonary Tuberculosis

Mycobacterium tuberculosis (MTB) is the causative agent for pulmonary tuberculosis causing worldwide deaths. It affects all the parts of the body, but lungs are mostly infected due to inhalation of MTB. Drug dosage for the treatment is generally given orally and repeated doses in high concentration are required for the treatment process. However, drug administration through inhalation is more advantageous and requires lower doses. Encapsulated nanoparticles for drug delivery efficiently penetrate the biological membrane and reach the target site. MSNPs were reported to act as a platform for the delivery of anti-TB drugs. Functionalized MSNPs could be internalized competently and used as controlled drug delivery vehicles (Singh et al. 2019).

1.4 Advantages and Challenges Associated with Nanomaterials Used in Drug Delivery, Diagnosis, and Treatment of Diseases

Nanotechnology is a double-edged sword. On one edge, it has revolutionized medicine as well as the healthcare system by introducing innovative ways of developing new drug delivery systems or reformulation of existing drugs, as well as new devices for diagnosis, monitoring, and treatment of diseases owing to their enhanced permeability, retention, and therapeutic effects; the other edge showed potential health risks (Patel et al. 2015). As discussed earlier, nanomaterials which are considered as building blocks of nanotechnology possess novel and unique properties such as small size, high surface to volume ratio, high performance, etc. All these properties make them suitable candidates for the various biomedical applications discussed previously. On the other hand, sometimes the same properties become responsible for causing harmful effects in human beings. However, while focusing on the significant advantages of nanomaterials, their toxicological aspects are overlooked.

1.4.1 Advantages of Nanomaterials

There are many studies have been performed and some of them are already discussed above which proposed the potential role of various nanomaterials such as liposomes, SLN, polymeric nanoparticles, etc. in specialized drug delivery, in the development of biocompatible nanomaterial prosthetic implants, the metal-containing engineered nanoparticles, etc. for both the imaging and treatment of various diseases including cancers (Wright et al. 2016). Moreover, such nano-scale size materials usually encapsulate therapeutic and/or imaging compounds, popularly known as nanomedicine, in nano-size systems typically with sizes smaller than eukaryotic or prokaryotic cells. They offer immense opportunity in patient-specific, targeted, and regenerative medicine technology with applications such as: regeneration of tissue cell therapy; regeneration of tissue with help of nano-scale biomaterials; active or passive drug release; diagnostic tests; *in vitro* tests with sensors for determination of molecules that react with particular disease (biomarkers); *in vivo* measurements of biomarkers by imaging techniques using nanoparticles as contrast media; and more (Sharma et al. 2018).

Also, nanomaterials on chips, nanorobotics, and magnetic nanoparticles attached to specific antibodies, nano-size empty virus capsids, and magnetic immunoassay are new dimensions of their use in drug delivery. The benefit of nano-scale drug delivery systems, like nanotubes, nanocrystals, fullerenes, nanosphere, nanoparticles, nanoliposomes, dendrimers, nanopores, nanoshells, quantum dots, nanocapsule, nano vaccines, etc., is that they increase the efficacy and efficiency of the loaded drug by delivering a notable array of medications to almost any organ or specific site in the body (Mukherjee et al. 2014). As well, they minimize accumulation in healthy body sites to reduce toxic effects of the drug, as they can reach the specific site through active or passive means providing targeted, controlled, and sustained therapeutic effects. These unique characteristics lead them to generally inaccessible areas such as cancer cells, inflamed tissues, etc., and also provide an opportunity for the peroral route of administration of genes and proteins on account of weakening lymphatic drainage.

Formulation scientists can modify the structure of materials to extremely small scales leading to an increase in surface area relative to volume, and large surface area allows for increased functionalities of these multifunctional nanosized molecules, which consecutively promote selective targeting to the desired sub-cellular targets, avoid destruction by macrophages, effect permeation through barriers, and deliver its components in a controlled way once it gets to the target cells and tissues. They also facilitate passive targeting of actives to the macrophages of the liver and spleen through direct delivery to reticuloendothelial cells and thus permitting a natural system for treating intracellular infections. Their suitability for enhancing the efficacy of drugs with short half-lives is attributable to the long-time spent in circulation and can be used to examine drugs as sustained-release formulations as well as for delivering DNA (Mukherjee et al. 2014; Sharma et al. 2018).

There is no denying the fact that articulating drugs at the nano-scale provides potential advantages with the possibility to modify properties like solubility, drug release profiles, diffusivity, bioavailability, and immunogenicity. The dissolution rate of the drug can be enhanced with an increase in the onset of therapeutic action as well as the reduction in dose and dose-dependent side effects (Patra et al. 2018). Furthermore, an amalgamation of drug therapy and diagnosis, termed "theranostics," at the nano-scale has exceptional applications and can help diagnose the disease, affirm the location, identify the stage of the disease, and provide information about the treatment response.

1.4.2 Challenges Associated with the Use of Nanomaterials

The distinctive behavior of nano-scale materials, as compared to conventional chemicals or biological agents, in biological systems is mainly expected due to their minute size. The minuscule size allows them to enter not only organs, tissues, and cells, but also cell organelles, e.g. mitochondria and nuclei, by crossing various barriers (Auría-Soro et al. 2019; Tang et al. 2019). However, this may drastically modulate the structures of macromolecules, thereby impeding critical biological functions (Patel et al. 2015). They can also initiate blood coagulation pathways and stimulate platelet aggregation resulting in thrombosis. Various mechanisms as proposed by research scientists after evaluating *in vivo* toxicity of the nanomaterials is mainly through the generation of oxidative responses via the formation of free radicals and reactive oxygen species which may cause oxidative stress, inflammation, and damage to DNA, proteins, and membranes, ultimately leading to toxicity. The clearance of these materials by the reticuloendothelial system protects other tissues but engenders oxidative stress in organs such as liver and spleen (Badar et al. 2019).

Several toxicological studies have demonstrated that the toxic effect of nanomaterials is also regulated by their route of entry into the body, such as oral, skin, respiratory route, site of injection, and digestive canal, and further translocation and distribution according to the size determine additional toxic manifestations. The reduction in size leads to an increase in a number of surface atoms and as the surface area increases, it confronts dose-dependent increments in oxidation and DNA-damaging abilities. This also brings about an increase in surface energy further initiating binding of proteins that send a signal to macrophages and in turn engulfs the nanosized particles (Werner et al. 2018). Apart from this, certain unpredictable reactions can also take place inside the body due to unanticipated interactions and behavior of these particles. The research fraternities believe that the toxicity of nanomaterials strongly depends on their physical and chemical properties, such as the shape, size, electric charge, solubility, presence of functional groups, and chemical compositions of the core and shell. Reckoning with these facts of the toxicity and safety of nanosized materials presents a challenge in its clinical translation for drug delivery, diagnosis, and treatment of diseases (Gatoo et al. 2014). In recent years, when nanomaterials are becoming a part of daily life, toxicity concerns should not be ignored and accurate methods must be established to evaluate both the short-term and long-term toxicity analysis of nanosized drug delivery systems. Other significant hurdles faced are various biological challenges, fate and behavior in the environment, biocompatibility, safety, large-scale manufacturing, intellectual property, government regulations, and cost-effectiveness as compared to traditional therapy (Hua et al. 2018).

Although nanomaterials hold great promise in medicine, their production, use, and disposal lead to discharges into air, soil, and aquatic systems. It thus becomes imperative to assess the impact of such material with unknown toxicological properties on environmental health in addition to various pre-processing procedures investigated before disposal. The vagueness associated with the risks of using nanomedicine in living beings cannot be ignored and they cannot be regarded as an ideal form of therapy. The science of using nanosized materials in the field of medicine has dynamically developed in recent times, but the ambiguity surrounding their characterization about safety and toxicity, and the lack of effective regulation with necessary closer cooperation between regulatory agencies, put forward the utmost challenge to evaluate their real impact in the healthcare system.

1.5 Conclusion

Nanotechnology is a novel and emerging technology having enormous applications in the whole biomedical sector, particularly in diagnosis, drug delivery, and treatment of a wide range of diseases such as infectious diseases, neurological disorders, cardiovascular diseases, cancers, etc. Therefore, considering the huge potential of nanotechnology in these fields it is believed that nanotechnology will play a crucial role in revolutionizing the current scenario of biomedicine (diagnostic and therapeutic strategies) and start a new era of nanomedicine. Nanomaterials have attracted huge attention in their use for diagnosis and management of varied diseases due to their novel and unique properties such as small size, high drug loading capacity, and biodegradable and nontoxic nature. In addition, nanomaterials also help to enhance the solubility of non-soluble drug molecules, increase their bioavailability, and reduce the side effects caused by synthetic drugs.

Although nanotechnology in general, and nanomaterials in particular, have great potential in various avenues related to biomedicine, in some cases the cost factor becomes a hindrance in its use. Moreover, certain nanoscale materials lack long-term safety data, therefore such issues need to be addressed. Apart from these challenges, nanomaterials and their approval methods have not been well defined until now, which brings another limitation and takes even more time in developing clinically useful nanotechnology-based drug delivery systems and therapies.

References

Abbasi, E., Aval, S.F., Akbarzadeh, A. et al. (2014). Dendrimers: synthesis, applications, and properties. *Nanoscale Research Letters* 9 (1): 247. https://doi.org/10.1186/1556-276x-9-247.

Abdallah, M.H. (2013). Transfersomes as a transdermal drug delivery system for enhancement the antifungal activity of nystatin. *International Journal of Pharmacy and Pharmaceutical Sciences* 5 (4): 560–567.

Abiodun-Solanke, I.M.F., Ajayi, D.M., and Arigbede, A.O. (2014). Nanotechnology and its application in dentistry. *Annals of Medical and Health Science Research* 4 (3): S171–S177.

Adeyemi, O.S. and Sulaiman, F.A. (2015). Evaluation of metal nanoparticles for drug delivery systems. *Journal of Biomedical Research* 29 (2): 145–149.

Ahmed, M. and Douek, M. (2013). The role of magnetic nanoparticles in the localization and treatment of breast cancer. *BioMed Research International* 2013 (281230): 1–13.

An, X., Zhang, F., Zhu, Y., and Shen, W. (2010). Photo-induced drug release from thermosensitive AuNPs-liposome using a AuNPs-switch. *Chemical Communications* 46: 7202–7204.

Anderson, S.D., Gwenin, V.V., and Gwenin, C.D. (2019). Magnetic functionalized nanoparticles for biomedical, drug delivery and imaging applications. *Nanoscale Research Letters* 14: 188. https://doi.org/10.1186/s11671-019-3019-6.

Auría-Soro, C., Nesma, T., Juanes-Velasco, P. et al. (2019). Interactions of nanoparticles and biosystems: microenvironment of nanoparticles and biomolecules in nanomedicine. *Nanomaterials* 9 (10): 1365. https://doi.org/10.3390/nano9101365.

Badar, A., Pachera, S., Ansari, A.S., and Lohiya, N.K. (2019). Nano based drug delivery systems: present and future prospects. *Nanomedicine and Nanotechnology Journal* 2 (1): 121.

Bangham, A.D. and Horne, R.W. (1964). Negative staining of phospholipids and their structural modification by surface-active agents as observed in the electron microscope. *Journal of Molecular Biology* 8 (5): 660–668.

Baptista, P.V. (2014). Nanodiagnostics: leaving the research lab to enter the clinics? *Diagnosis (Berl)* 1: 305–309.

Bartelds, R., Nematollahi, M.H., Pols, T. et al. (2018). Niosomes, an alternative for liposomal delivery. *PLoS One* 13 (4): e0194179. https://doi.org/10.1371/journal.pone.0194179.

Beishon, M. (2013). Exploiting a nano-sized breach in cancer's defenses. *Cancer World*: 14–21.

Bejarano, J., Navarro-Marquez, M., Morales-Zavala, F. et al. (2018). Nanoparticles for diagnosis and therapy of atherosclerosis and myocardial infarction: evolution toward prospective theranostic approaches. *Theranostics* 8 (17): 4710–4732.

Beyth, N., Houri-Haddad, Y., Domb, A. et al. (2015). Alternative antimicrobial approach: nano-antimicrobial materials. *Evidence-based Complementary and Alternative Medicine* 2015 (246012): 1–16.

Bharti, C., Nagaich, U., Pal, A.K., and Gulati, N. (2015). Mesosporus silica nanoparticles in target drug delivery system: a review. *International Journal of Pharmaceutical Investigation* 5 (3): 124–134.

Blume, G. and Ceve, G. (1990). Liposomes for the sustained drug release *in vivo*. *Biochimica et Biophysica Acta* 1029 (1): 91–97.

Bonnard, T., Gauberti, M., de Lizarrondo, S.M. et al. (2019). Recent advances in nanomedicine for ischemic and hemorrhagic stroke. *Stroke* 50: 1318–1324.

Chandra, A., Joshi, K., and Aggarwal, G. (2018). Topical nano drug delivery for treatment of psoriasis: progressive and novel delivery. *Asian Journal of Pharmaceutics* 12 (3): S835–S848.

Chansuvarn, W., Tuntulani, T., and Imyim, A. (2015). Colorimetric detection of mercury(II) based on gold nanoparticles, fluorescent gold nanoclusters and other gold-based nanomaterials. *TrAC Trends in Analytical Chemistry* 65: 83–96.

Chen, T.J., Cheng, T.H., Chen, C.Y. et al. (2009). Targeted Herceptin dextraniron oxide nanoparticles for noninvasive imaging of HER2/neu receptors using MRI. *Journal of Biological Inorganic Chemistry* 14 (2): 253–260.

Chen, Y., Wang, Y., Liu, L. et al. (2015). A gold immunochromatographic assay for the rapid and simultaneous detection of fifteen β-lactams. *Nanoscale* 7: 16381–16388.

Cho, K., Wang, X.U., Nie, S., and Shin, D.M. (2008). Therapeutic nanoparticles for drug delivery in cancer. *Clinical Cancer Research* 14 (5): 1310–1316.

Choudhary, S., Gupta, L., Rani, S. et al. (2017). Impact of dendrimers on solubility of hydrophobic drug molecules. *Frontiers in Pharmacology* 8: 261. https://doi.org/10.3389/fphar.2017.00261.

Crucho, C.I.C. and Barros, M.T. (2017). Polymeric nanoparticles: a study on the preparation variables and characterization methods. *Materials Science and Engineering C* 80: 771–784.

Dudefoi, W., Villares, A., Peyron, S. et al. (2018). Nanoscience and nanotechnologies for biobased materials, packaging and food applications: new opportunities and concerns. *Innovative Food Science & Emerging Technologies* 46: 107–121.

Dykman, L. and Khlebtsov, N. (2012). Gold nanoparticles in biomedical applications: recent advances and perspectives. *Chemical Society Reviews* 41: 2256–2282.

El-Hammadi, M.M. and Arias, J.L. (2019). An update on liposomes in drug delivery: a patent review (2014-2018). *Expert Opinion on Therapeutic Patents* 29 (11): 891–907.

Gao, X., Cui, Y., Levenson, R.M. et al. (2004). *In vivo* cancer targeting and imaging with semiconductor quantum dots. *Nature Biotechnology* 22 (8): 969–976.

Gatoo, M.A., Naseem, S., Arfat, M.Y. et al. (2014). Physicochemical properties of nanomaterials: implication in associated toxic manifestations. *BioMed Research International* 2014 (498420) https://doi.org/10.1155/2014/498420.

Gholami-Shabani, M., Akbarzadeh, A., Norouzian, D. et al. (2014). Antimicrobial activity and physical characterization of silver nanoparticles green synthesized using nitrate reductase from *Fusarium oxysporum*. *Applied Biochemistry and Biotechnology* 172 (8): 4084–4408.

Gholami-Shabani, M., Shams-Ghahfarokhi, M., Gholami-Shabani, Z. et al. (2015). Enzymatic synthesis of gold nanoparticles using sulfite reductase purified from *Escherichia coli*: a green eco-friendly approach. *Process Biochemistry* 50 (7): 1076–1085.

Gholami-Shabani, M., Imani, A., Shams-Ghahfarokhi, M. et al. (2016). Bioinspired synthesis, characterization and antifungal activity of enzyme-mediated gold nanoparticles using a fungal oxidoreductase. *Journal of the Iranian Chemical Society* 13 (11): 2059–2068.

Gholami-Shabani, M., Gholami-Shabani, Z., Shams-Ghahfarokhi, M. et al. (2017). Green nanotechnology: biomimetic synthesis of metal nanoparticles using plants and their application in agriculture and forestry. In: *Nanotechnology* (eds. R. Prasad, M. Kumar and V. Kumar), 133–175. Singapore: Springer.

Gholami-Shabani, M., Gholami-Shabani, Z., Shams-Ghahfarokhi, M., and Razzaghi-Abyaneh, M. (2018). Application of nanotechnology in mycoremediation: current status and future prospects. In: *Fungal Nanobionics: Principles and Applications* (eds. R. Prasad, V. Kumar, M. Kumar and S. Wang), 89–116. Singapore: Springer.

Godin, B. and Touitou, E. (2003). Ethosomes: new prospects in transdermal delivery. *Critical Review in Therapeutic Drug Carrier Systems* 20 (1): 63–102.

Gonzalez-Rodriguez, R., Campbell, E., and Naumov, A. (2019). Multifunctional graphene oxide/iron oxide nanoparticles for magnetic targeted drug delivery dual magnetic resonance/fluorescence imaging and cancer sensing. *PLoS One* 14 (6): 0217072. https://doi.org/10.1371/journal.pone.0217072.

Guo, L., Wu, X., Liu, L. et al. (2018). Gold nanoparticle-based paper sensor for simultaneous detection of 11 benzimidazoles by one monoclonal antibody. *Small* 14: 1701782. https://doi.org/10.1002/smll.201701782.

Hsu, S.H. and Luo, P.W. (2019). From nanoarchitectonics to tissue architectonics: nanomaterials for tissue engineering. In: *Advanced Supramolecular Nanoarchitectonics: Micro and Nano Technologies* (eds. K. Ariga and M. Aono), 277–288. UK: Elsevier.

Hua, S., De Matos, M.B., Metselaar, J.M., and Storm, G. (2018). Current trends and challenges in the clinical translation of nanoparticulate nanomedicines: pathways for translational development and commercialization. *Frontiers in Pharmacology* 9 (790) https://doi.org/10.3389/fphar.2018.00790.

Huang, X., Qi, X., Boey, F., and Zhang, H. (2012). Graphene-based composites. *Chemical Society Reviews* 41: 666–686.

Hyafil, F., Cornily, J.C., Feig, J.E. et al. (2007). Noninvasive detection of macrophages using a nanoparticulate contrast agent for computed tomography. *Nature Medicine* 13: 636–641.

Inbaraj, B.S. and Chen, B.H. (2016). Nanomaterial-based sensors for detection of foodborne bacterial pathogens and toxins as well as pork adulteration in meat products. *Journal of Food and Drug Analysis* 24 (1): 15–28.

Jackson, T.C., Patani, B.O., and Ekpa, D.E. (2017). Nanotechnology in diagnosis: a review. *Advances in Nanoparticles* 6: 93–102.

Jafari, S., Derakhshankhah, H., Alaei, L. et al. (2019). Mesoporous silica nanoparticles for therapeutic/diagnostic applications. *Biomedicine & Pharmacotherapy* 109: 1100–1111.

Jain, K.K. (2003). Nanodiagnostics: application of nanotechnology in molecular diagnostics. *Expert Review of Molecular Diagnostics* 3: 153–161.

Jeong, H.H., Choi, E., Ellis, E., and Lee, T.C. (2019). Recent advances in gold nanoparticles for biomedical applications: from hybrid structures to multi-functionality. *Journal of Materials Chemistry B* 7: 3480–3496.

Kaur, R., Sharma, S.K., and Tripathy, S.K. (2019). Advantages and limitations of environmental nanosensors. In: *Advances in Nanosensors for Biological and Environmental Analysis* (eds. A. Deep and S. Kumar), 119–132. UK: Elsevier.

Kesharwani, P., Gorain, B., Low, S.Y. et al. (2018). Nanotechnology based approaches for anti-diabetic drugs delivery. *Diabetes Research and Clinical Practice* 136: 52–77.

Kievit, F.M., Stephen, Z.R., Veiseh, O. et al. (2012). Targeting of primary breast cancers and metastases in a transgenic mouse model using rationally designed multifunctional SPIONs. *ACS Nano* 6 (3): 2591–2601.

Kim, D., Park, S., Lee, J.H. et al. (2007). Antibiofouling polymer-coated gold nanoparticles as a contrast agent for *in vivo* X-ray computed tomography imaging. *Journal of the American Chemical Society* 129 (24): 7661–7665.

de Kraker, M.E.A., Stewardson, A.J., and Harbarth, S. (2016). Will 10 million people die a year due to antimicrobial resistance by 2050? *PLoS Medicine* 13 (11): e1002184. https://doi.org/10.1371/journal.pmed.1002184.

Kulthe, S.S., Choudhari, Y.M., Inamdar, N.N., and Mourya, V. (2012). Polymeric micelles: authoritative aspects for drug delivery. *Designed Monomers and Polymers* 15 (5): 465–521.

Kurbanoglu, S. and Ozkan, S.A. (2018). Electrochemical carbon based nanosensors: a promising tool in pharmaceutical and biomedical analysis. *Journal of Pharmaceutical and Biomedical Analysis* 147: 439–457.

Leary, J.F. (2010). Nanotechnology: what is it and why is small so big? *Canadian Journal of Ophthalmology* 45 (5): 449–456.

Lewinski, N., Colvin, V., and Drezek, R. (2008). Cytotoxicity of nanoparticles. *Small* 4: 26–49.

Li, W. and Chen, X. (2015). Gold nanoparticles for photoacoustic imaging. *Nanomedicine (London, England)* 10: 299–320.

Li, Z., Barnes, J.C., Bosoy, A. et al. (2012). Mesoporous silica nanoparticles in biomedical applications. *Chemical Society Reviews* 41: 2590–2605.

Li, Y., Wang, Z., Sun, L. et al. (2019). Nanoparticle-based sensors for food contaminants. *TrAC Trends in Analytical Chemistry* 113: 74–83.

Liang, R., Wei, M., Evans, D.G., and Duan, X. (2014). Inorganic nanomaterials for bioimaging, targeted drug delivery and therapeutics. *Chemistry Communication* 50: 14071–14081.

Liu, H.J. and Xu, P. (2019). Smart mesoporous silica nanoparticles for protein delivery. *Nanomaterials (Basel)* 9 (4): 511. https://doi.org/10.3390/nano9040511.

Lombardo, D., Kiselev, M.A., and Caccamo, M.T. (2019). Smart nanoparticles for drug delivery application: development of versatile nanocarrier platforms in biotechnology and nanomedicine. *Journal of Nanomaterials* 2019 (3702518) https://doi.org/10.1155/2019/3702518.

Lujan, H., Griffin, W.C., Taube, J.H., and Sayes, C.M. (2019). Synthesis and characterization of nanometer-sized liposomes for encapsulation and microRNA transfer to breast cancer cells. *International Journal of Nanomedicine* 14: 5159–5173.

Lyberopoulou, A., Efstathopoulos, E.P., and Gazouli, M. (2015). Nanodiagnostic and nanotherapeutic molecular platforms for cancer management. *Journal of Cancer Research Updates* 4: 153–162.

Ma, Q., Wang, Y., Jia, J., and Xiang, Y. (2018). Colorimetric aptasensors for determination of tobramycin in milk and chicken eggs based on DNA and gold nanoparticles. *Food Chemistry* 249: 98–103.

Medintz, I.L., Uyeda, H.T., Goldman, E.R., and Mattoussi, H. (2005). Quantum dot bioconjugates for imaging, labelling and sensing. *Nature Materials* 4 (6): 435–446.

Meng, J., Fan, J., Galiana, G. et al. (2009). LHRH-functionalized superparamagnetic iron oxide nanoparticles for breast cancer targeting and contrast enhancement in MRI. *Materials Science and Engineering C* 29 (4): 1467–1479.

Mitragotri, S. and Stayton, P. (2019). Organic nanoparticles for drug delivery and imaging. *MRS Bulletin* 39: 219–223.

Mukherjee, B., Dey, N., Maji, R. et al. (2014). Current status and future scope for nanomaterials in drug delivery. In: *Application of Nanotechnology in Drug Delivery* (ed. A. Sezer), 1–21. UK: Intech Open.

Mura, S., Nicolas, J., and Couvreur, P. (2013). Stimuli-responsive nanocarriers for drug delivery. *Nature Materials* 12 (11): 991–1003.

Muzzalupo, R. and Mazzotta, E. (2019). Do niosomes have a place in the field of drug delivery? *Expert Opinion on Drug Delivery* 16 (11): 1145–1147.

Novoselov, K.S., Geim, A.K., Morozov, S.V. et al. (2004). Electric field effect in atomically thin carbon films. *Science* 306: 666–669.

Núñez, C., Estévez, S.V., and Chantada, M.P. (2018). Inorganic nanoparticles in diagnosis and treatment of breast cancer. *Journal of Biological Inorganic Chemistry* 23 (3): 331–345.

Otles, S. and Yalcin, B. (2012). Review on the application of nanobiosensors in food analysis. *Acta Scientiarum Polonorum Technologia Alimentaria* 11: 7–18.

Pang, B., Zhao, Y., Luehmann, H. et al. (2016). ^{64}Cu-doped PdCu@Au tripods: a multifunctional nanomaterial for positron emission tomography and image-guided photothermal cancer treatment. *ACS Nano* 10 (3): 3121–3131.

Patel, S., Nanda, R., and Sahoo, S. (2015). Nanotechnology in healthcare: applications and challenges. *Medicinal Chemistry* 5: 528–533.

Patra, J.K., Das, G., Fraceto, L.F. et al. (2018). Nano based drug delivery systems: recent developments and future prospects. *Journal of Nanobiotechnology* 16: 71. https://doi.org/10.1186/s12951-018-0392-8.

Peiris, P.M., Toy, R., Doolittle, E. et al. (2012). Imaging metastasis using an integrin-targeting chain-shaped nanoparticle. *ACS Nano* 6 (10): 8783–8795.

Pérez-Medina, C., Tang, J., Abdel-Atti, D. et al. (2015). PET imaging of tumor-associated macrophages with ^{89}Zr-labeled high-density lipoprotein nanoparticles. *Journal of Nuclear Medicine* 56 (8): 1272–1277.

Petrie, J.R., Guzik, T.J., and Touyz, R.M. (2018). Diabetes, hypertension, and cardiovascular disease: clinical insights and vascular mechanisms. *Canadian Journal of Cardiology* 34 (5): 575–584.

Pink, D.L., Loruthai, O., Ziolek, R.M. et al. (2019). On the structure of solid lipid nanoparticles. *Small* 15 (45): 1903156. https://doi.org/10.1002/smll.201903156.

Pinto, M.F., Moura, C.C., Nunes, C. et al. (2014). A new topical formulation for psoriasis: development of methotrexate-loaded nanostructured lipid carriers. *International Journal of Pharmaceutics* 477: 519–526.

Prakitchaiwattana, C. and Detudom, R. (2017). Contaminant sensors: nanosensors, an efficient alarm for food pathogen detection. In: *Nanobiosensors* (ed. A.M. Grumezescu), 511–572. Academic Press.

Rabiee, N., Deljoo, S., and Rabiee, M. (2018). Curcumin-hybrid nanoparticles in drug delivery system. *Asian Journal of Nanoscience and Materials* 2 (1): 66–91.

Rai, M., Deshmukh, S., Ingle, A., and Gade, A. (2012). Silver nanoparticles: the powerful nano-weapon against multidrug resistant bacteria. *Journal of Applied Microbiology* 112 (5): 841–852.

Rai, M., Ingle, A.P., Yadav, A. et al. (2016). Strategic role of selected noble metal nanoparticles in medicine. *Critical Reviews in Microbiology* 42 (5): 696–719.

Rajasundari, K. and Hamurugu, K. (2011). Nanotechnology and its application in medical diagnosis. *Journal of Basic and Applied Chemistry* 1: 26–32.

Ramezani, M., Danesh, N.M., Lavaee, P. et al. (2015). A novel colorimetric triple-helix molecular switch aptasensor for ultrasensitive detection of tetracycline. *Biosensors and Bioelectronics* 70: 181–187.

Reimhult, E. and Höök, F. (2015). Design of surface modifications for nanoscale sensor applications. *Sensors* 15 (1): 1635–1675.

Ríos-Corripio, M.A., López-Díaz, A.S., Ramírez-Corona, N. et al. (2020). Metallic nanoparticles: development, applications, and future trends for alcoholic and nonalcoholic beverages. In: *Nanoengineering in the Beverage Industry* (eds. A.M. Grumezescu and A.M. Holban), 263–300. Academic Press.

Rizvi, S.A.A. and Saleh, A.M. (2018). Applications of nanoparticles systems in drug delivery technology. *Saudi Pharmaceutical Journal* 26: 64–70.

Senapati, S., Mahanta, A.K., Kumar, S., and Maiti, P. (2018). Controlled drug delivery vehicles for cancer treatment and their performance. *Signal Transduction and Targeted Therapy* 3 (1): 7. https://doi.org/10.1038/s41392-017-0004-3.

Sharma, M. (2019). Transdermal and intravenous nano drug delivery systems: present and future. In: *Applications of Targeted Nano Drugs and Delivery Systems* (eds. S. Mohapatra, S. Ranjan, N. Dasgupta, et al.), 499–550. UK: Elsevier.

Sharma, D., Sharma, N., Pathak, K. et al. (2018). Nanotechnology-based drug delivery systems: challenges and opportunities. In: *Drug Targeting and Stimuli Sensitive Drug Delivery Systems* (ed. A. Grumezescu), 39–79. UK: Elsevier.

Shehada, N., Brönstrup, G., Funka, K. et al. (2015). Ultrasensitive silicon nanowire for real-world gas sensing: noninvasive diagnosis of cancer from breath volatolome. *Nano Letters* 15 (2): 1288–1295.

Singh, A.P., Biswas, A., Shukla, A., and Maiti, P. (2019). Targeted therapy in chronic diseases using nanomaterial-based drug delivery vehicles. *Signal Transduction and Targeted Therapy* 4 (33) https://doi.org/10.1038/s41392-019-0068-3.

Srisuk, P., Thongnopnua, P., Raktanonchai, U., and Kanokpanont, S. (2012). Physico-chemical characteristics of methotrexate-entrapped oleic acid-containing deformable liposomes for *in vitro* transepidermal delivery targeting psoriasis treatment. *International Journal of Pharmaceutics* 427: 426–434.

Stark, D.D., Weissleder, R., Elizondo, G. et al. (1988). Superparamagnetic iron oxide: clinical application as a contrast agent for MR imaging of the liver. *Radiology* 168 (2): 297–301.

Sun, T., Gao, J., Han, D. et al. (2019). Fabrication and characterization of solid lipid nano-formulation of astraxanthin against DMBA-induced breast cancer via Nrf-2-Keap1 and NF-kB and mTOR/Maf-1/PTEN pathway. *Drug Delivery* 26 (1): 975–988.

Takahashi, S., Shiraishi, T., Miles, N. et al. (2015). Nanowire analysis of cancer-testis antigens as biomarkers of aggressive prostate cancer. *Urology* 85: 704.e1–704.e7.

Tang, W., Fan, W., Lau, J. et al. (2019). Emerging blood-brain-barrier-crossing nanotechnology for brain cancer theranostics. *Chemical Society Reviews* 48: 2967–3014.

Tong, R., Christian, D.A., Tang, L. et al. (2009). Nanopolymeric therapeutics. *MRS Bulletin* 34 (6): 422–431.

Vengurlekar, S. and Chaturvedi, S.C. (2019). Elevating toward a new innovation: carbon nanotubes (CNTs). In: *Biomedical Applications of Nanoparticles* (ed. A.M. Grumezescu), 271–294. UK: Elsevier.

Verma, P. and Pathak, K. (2010). Therapeutic and cosmeceutical potential of ethosomes: An overview. *Journal of Advanced Pharmaceutical Technology & Research* 1 (3): 274–282.

Vigneshvar, S., Sudhakumari, C.C., Senthilkumaran, B., and Prakash, H. (2016). Recent advances in biosensor technology for potential applications-an overview. *Frontiers in Bioengineering and Biotechnology* 4: 1–9.

Virlan, M.J.R., Miricescu, D., Radulescu, R. et al. (2016). Organic nanomaterials and their applications in the treatment of oral diseases. *Molecules* 21: 207. https://doi.org/10.3390/molecules21020207.

Wang, H., Liang, Y., Mirfakhrai, T. et al. (2011). Advanced asymmetrical supercapacitors based on graphene hybrid materials. *Nano Research* 4: 729–736.

Wang, X., Feng, Y., Dong, P., and Huang, J. (2019). A mini review on carbon quantum dots: preparation, properties, and electrocatalytic application. *Frontiers in Chemistry* 7 (671): 1–9. https://doi.org/10.3389/fchem.2019.00671.

Warriner, K., Reddy, S.M., Namvar, A., and Neethirajan, S. (2014). Developments in nanoparticles for use in biosensors to assess food safety and quality. *Trends in Food Science & Technology* 40 (2): 183–199.

Werner, M., Auth, T., Beales, P.A. et al. (2018). Nanomaterial interactions with biomembranes: bridging the gap between soft matter models and biological context. *Biointerphases* 13 (2): 028501.

Wright, P.F.A. (2016). Potential risks and benefits of nanotechnology: perceptions of risk in sunscreens. *Medical Journal of Australia* 204 (10): 369–370.

Yang, W., Deng, X., Huang, W. et al. (2019). The physicochemical properties of graphene nanocomposites influence the anticancer effect. *Journal of Oncology* 2019 (7254534): 1–10. https://doi.org/10.1155/2019/7254534.

Yavuz, M.S., Cheng, Y., Chen, J. et al. (2009). Gold nanocages covered by smart polymers for controlled release with near-infrared light. *Nature Materials* 8: 935–939.

Zhang, Z., Tsai, P.C., Ramezanli, T., and Michniak-Kohn, B. (2013a). Polymeric nanoparticles-based topical delivery systems for the treatment of dermatological diseases. *Wiley Interdisciplinary Reviews: Nanomedicine and Nanobiotechnology* 5 (3): 205–218.

Zhang, Z.J., Wang, J., and Chen, C.Y. (2013b). Near-infrared light-mediated nano-platforms for cancer thermo-chemotherapy and optical imaging. *Advanced Materials* 25 (28): 3869–3880.

Zhao, C.Y., Cheng, R., Yang, Z., and Tian, Z.M. (2018). Nanotechnology for cancer therapy based on chemotherapy. *Molecules* 23 (4): 826. https://doi.org/10.3390/molecules23040826.

2

Selenium Nanocomposites in Diagnosis, Drug Delivery, and Treatment

Irina A. Shurygina and Michael G. Shurygin

Irkutsk Scientific Center of Surgery and Traumatology, Bortsov Revolutsii st., Irkutsk, Russia

Nomenclature

AG	arabinogalactan
Cur-SeNPs	Curcumin-loaded SeNPs
DIC	differential interference contrast
DNA	deoxyribonucleic acid
Dox	doxorubicin
EGFR	epidermal growth factor receptor
ELISA	enzyme-linked immunosorbent assay
EPI	epirubicin
FA-SeNP	SeNPs loaded with ferulic acid
FU	5-fluorouracil
GE11 SeNPs	oridonin peptide-conjugated GE11 SeNPs
GRL-11372	human osteoblast-like cells
HeLa	immortal cell line
HepG2	human hepatoma cell
IC50	half maximal inhibitory concentration
IL	interleukin
MAPK	mitogen-activated protein kinase
MBC	minimum bactericidal concentration
MCP-1	monocyte chemoattractant protein 1
MDA-MB-231	human breast cancer cells
MIC	minimal inhibiting concentration
NF-κB	nuclear factor kappa-light-chain-enhancer of activated B cells
ProGRP	progastrin-releasing-peptide
PTX	paclitaxel
RNA	ribonucleic acid
ROS	reactive oxygen species
RPMI-1640	growth medium used in cell culture

Nanobiotechnology in Diagnosis, Drug Delivery, and Treatment, First Edition. Edited by Mahendra Rai, Mehdi Razzaghi-Abyaneh, and Avinash P. Ingle.

SA	Sialic acid
SeLPs	Se-functionalized liposomes
SeNP	selenium nanoparticle
siRNA	small interfering RNA
TNF-α	tumor necrosis factor alpha

2.1 Introduction

Nowadays, there is a growing interest in studies on various types of biological activity, toxicity, and usage of both nonorganic and organic forms of selenium. It seems clear that such high scientific interest exists because maintaining the physiological level of selenium in the body is vital. The main biological role of selenium involves being a cofactor unit of selenium-containing enzymes (Dumitrescu and Refetoff 2011). It needs to be noted that those enzymes are among the main ones in the functioning of the redox system of the cell and, thus, all the basic parameters of cell vital functions depend on their activity (Huang et al. 2012). It is found that action of selenium-dependent enzymes in tissues, deiodinase and glutathione peroxidase, directly depend on the selenium intake in the body (Villette et al. 1998).

Naturally, selenium enters the human and animal body mainly in the form of selenium-containing amino acids (Gammelgaard et al. 2011). At the same time, there are only a few reports available on biological activity of selenium in a nanosized form than on organic and nonorganic selenium compounds. Particularly, it is observed that red nanoselenium is less toxic and more biologically active than other nonorganic (Zhang et al. 2001; Sadeghian et al. 2012) and organic forms of selenium (Wang et al. 2007; Zhang et al. 2008).

In Caco-2 cell line model it is established that intracellular transport of selenium depends on its chemical form. The lowest speed was documented for sodium selenite while selenomethionine and nanoselenium did not differ for this indicator (Wang and Fu 2012). To date, there is no consensus about the effect of the size of selenium nanoparticles (SeNPs) on its biological activity. Thus, activation of selenium-dependent enzyme systems in mouse liver and human hepatoma cell line HepG2 does not depend on the size of SeNPs (Zhang et al. 2004). Meanwhile, Peng et al. (2007) showed that smaller size SeNPs (36 nm) had more biological activity than bigger ones (90 nm) (Peng et al. 2007).

Nowadays, different fields of application of selenium in nanoform are studied. A large number of studies are dedicated to the use of SeNPs for diagnosis and treatment of various diseases. In addition, their use as drug delivery systems has been also studied. Considering these facts, the present chapter is focused on the analysis of known data and the prospects of selenium nanocomposites used for biomedical purposes.

2.2 Nanoselenium: Application in Diagnosis

One of the promising fields for possible applications of nanoselenium is their use for diagnosis of different pathological conditions and evaluation of cell activity. The use of SeNPs for medical diagnosis and application as biosensors depends on the physical and chemical properties of the elemental selenium. It is most promising to use SeNPs to visualize cells and structures. Those prospects are associated primarily with the SeNPs' fluorescent abilities, which allow direct identification of nanoparticles in tissues (Figure 2.1) (Shurygina et al. 2015a; Khalid et al. 2016). Khalid et al. (2016)

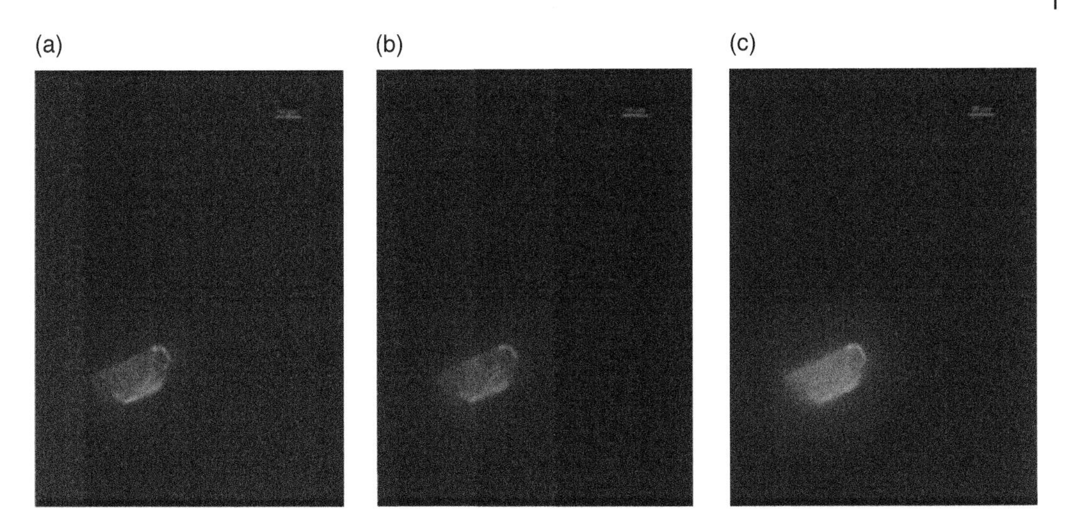

Figure 2.1 Nanocomposite of elemental selenium and arabinogalactan, fluorescence microscopy. (a) Nikon B-2A filter (excitation 450–490 nm, dichroic mirror 505 LP, emission ≥515 nm); (b) Nikon TRITC filter (excitation 528–553 nm, dichroic mirror 565 LP, emission 590–650 nm); (c) Nikon DAPI filter (excitation 325–375 nm, dichroic mirror 400 LP, emission 435–485 nm).

used SeNPs having photoluminescent spectra in the near-infrared region to study the behavior of SeNPs in fibroblasts. Here, the fluorescence of SeNPs was directly used for visualization and tracking *in vitro* in cells of fibroblasts without the need for any additional marks (Khalid et al. 2016).

Rodionova et al. (2015) synthesized arabinogalactan (AG) stabilized selenium nanocomposite, thus prepared selenium and arabinogalactan (Se-AG) nanocomposites were in the form of red-orange powder and highly soluble in water. The UV-Vis absorption spectrum of arabinogalactan (a substance used as a matrix in the synthesized nanocomposite) was characterized by peaks at 199 and 287 nm, which may be due to the presence of aldehyde groups. While studying the optical absorption spectrum of the Se-AG nanocomposite in the range of 190–1000 nm, a minimal rise (0.005 abs) was recorded in the region of 926 nm; as well, a gradual increase in the range of 600–190 nm to 0.772 abs with small plateaus in the region of 230–221 nm and 204–197 nm was also recorded which is typical for nanoselenium (0) (Singh et al. 2010). The electron microscopy analysis revealed that thus prepared composite consists of globules of arabinogalactan covered with SeNPs, the size of individual SeNPs is about only 2–3 nm. Considering the important role of selenium-containing proteins in cell energy supply, a study was performed evaluating the effect of selenium nanocomposites in the process of wound healing including bone tissue damage. Experimental studies were conducted on models of hole fracture of the tibia (Chinchilla male rabbits, N = 8), including five that followed the natural course of the reparative process (the control group) and three with the local intraoperative introduction to the fracture site of 50.33 $\mu g\,kg^{-1}$ of nanoselenium expressed in terms of selenium (the nanoselenium group).

The specimens were fixed in a FineFix solution (Milestone, Italy). The decalcification and subsequent preparation of bone tissue for a histological study were done according to the method proposed by Shurygina and Shurygin (2013, 2018). The specific fluorescence of the fluorescent labels was visualized using a Zeiss LSM 710 laser confocal microscope. It was found that the preparation had properties of quantum dots. The nanocomposite particles observed most often fluoresced over a wide range of the spectrum upon laser excitation at 405 nm with irradiation maximum at 480 nm, at 458 nm with maximum at 538 nm, and at 514 nm with maximum at 555 nm (Figure 2.2).

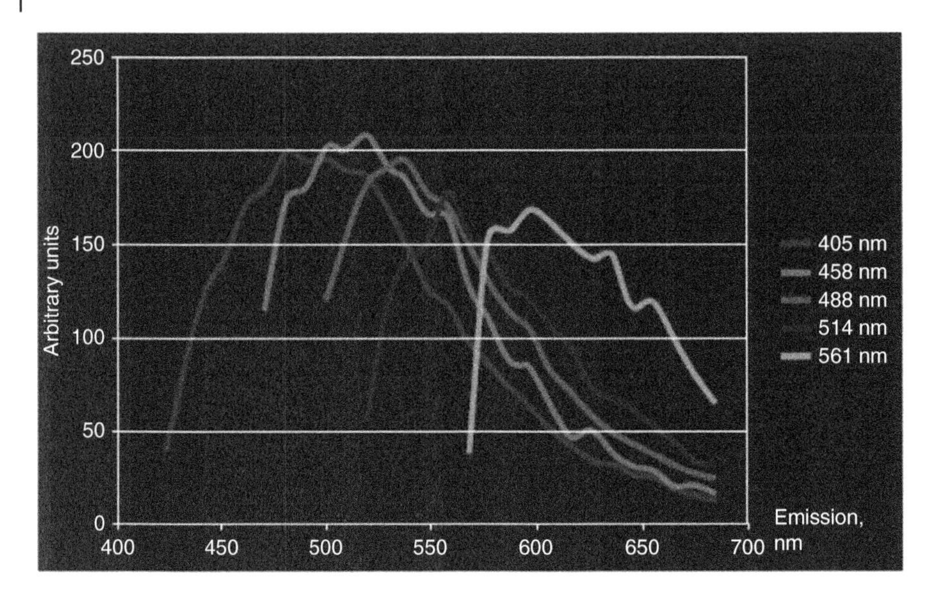

Figure 2.2 The emission spectra of the nanocomposite elemental selenium and arabinogalactan.

In animals that were administered with selenium-arabinogalactan nanocomposite, there was no complete bone tissue regeneration in the fracture site. Trabeculae of bones were thinned. A large amount of fluorescent amorphous masses outside the fracture site, as well as intense fluorescence marks in the Haversian canals, were noted. In confocal microscopy study these amorphous masses had the fluorescence spectrum closer to the selenium-arabinogalactan nanocomposites (from 430 to 630 nm with a maximum in the region of 490 nm) (Shurygina et al. 2015a).

In another study, it was reported that the local application of nanoselenium at the fracture sites significantly impairs the reparative processes, slowing down bone regeneration and impairing mineralization in calluses that have formed. No such deviations in the natural course of the reparative process were observed in studies performed earlier on the biological effects of the arabinogalactan matrix in altering tissue. This allows authors to exclude the development of the observed changes due to the toxic effect the matrix substance has on an organism's cells (Kostyro et al. 2013).

The local use of SeNPs associated with macromolecules of arabinogalactan on a fracture site seems to lead above all to significant diffusion impediments to the evacuation of the nanocomposite from this area and thus to pronounced prolonged local effects of nanoselenium on the site of the trauma (Shurygina et al. 2015a). Spherical forms of SeNPs for dot-blot immunoassay connected to multiple native antigens for rapid serodiagnosis of human lung cancer were developed. The sensitivity of dot immunoassay for the detection of progastrin-releasing peptide (ProGRP) was found to be 75 $pg\,ml^{-1}$. The detection time of the colloidal enzyme-linked immunosorbent assay (ELISA) tests of Se Dot for ProGRP was only five minutes (Zhao et al. 2018). Moreover, SeNPs were also studied as hydrogen peroxide (H_2O_2) biosensors. For example, Wang et al. (2010) synthesized semiconductor monoclinic SeNPs for accurate detection of H_2O_2. It was shown that H_2O_2 biosensor had high-speed response and affinity for H_2O_2 with detection limit of 8×10^{-8} M (Wang et al. 2010).

Rational design of multifunctional nanoplatforms with drugs is a promising strategy for simultaneous diagnosis, real-time monitoring, and cancer treatment (Huang et al. 2019). Multi-component complexes were synthesized targeting the epidermal growth factor receptor (EGFR) and the ability to respond to the tumor microenvironment (Se-5Fu-Gd-P [Cet/YI-12]) using EGFR as a target molecule, gadolinium chelate as a contrast agent for magnetic resonance tomography, 5-fluorouracil (5Fu) and cetuximab as drugs, and polyamidoamine and 3,3′-dithiobis (sulfosuccinimidyl propionate) as effectors to affect glutathione in the tumor cells for the diagnosis and treatment of nasopharyngeal cancer. This Se nanoplatform demonstrated excellent ability to visualize with the help of magnetic resonance tomography and had potential for clinical application as a diagnostic mean for investigating tumor tissue. In addition, *in vitro* experiments showed that due to administration of targeting drugs and peptides, intracellular saturation of Se nano-platform in tumor cells significantly increased. Also, it improved the directional delivery and anticancer efficacy of drugs included in the platform (Huang et al. 2019).

Thus, the applications of SeNPs and nanocomposites which have selenium in their composition is very promising for simultaneous diagnosis and treatment of different pathological conditions.

2.3 Nanoselenium and Antitumor Activity

SeNPs reported to have potential anticancer activity and hence they can be used in chemotherapy for cancer (Yang et al. 2012; Bao et al. 2015; Jia et al. 2015; Liao et al. 2015; Yanhua et al. 2016). Antitumor effects of SeNPs are usually mediated by their ability to inhibit the growth of cancer cells through induction of S phase arrest of cell cycle (Luo et al. 2012). Particularly, SeNPs induced mitochondria-mediated apoptosis in A375 human melanoma cells. Treatment of this Nano-Se cancer cell line resulted in a dose-dependent apoptosis of the cells manifested by DNA fragmentation and phosphatidylserine translocation (Chen et al. 2008). Apart from unique anti-cancer efficacy, SeNPs provide better selectivity between normal and cancer cells. It was demonstrated that SeNPs were not toxic to human osteoblast-like cells of the GRL-11372 line (Tran and Webster 2008); however, these nanoparticles were able to inhibit the growth of mouse osteosarcoma cells (Tran et al. 2010). In addition, it was shown that SeNPs, when given in conjugation to anastrozole, lower the bone toxicity caused by anastrozole and thus reduce the probable damage to the bone (Vekariya et al. 2013). SeNPs at a concentration of only 2 μg Se per ml effectively inhibited proliferation and induced caspase-independent apoptosis in adenocarcinoma cells of human prostate glands without any significant toxicity to human peripheral blood mononuclear cells (Sonkusre et al. 2014). SeNPs inhibit the growth of HeLa cells and human breast cancer cells MDA-MB-231 depending on the dosage. In this case, the dose of Nano-Se0 (10 $\mu mol\,l^{-1}$) happened to be effective (Luo et al. 2012).

Selenium nanocomposite and arabinogalactan had significant inhibitory effect on the cancerous cells lines such as A549, HepG-2, and MCF-7 in a dose-dependent manner. The nanocomposite induced apoptosis of these cancer cells (Tang et al. 2019). The studies performed showed that elemental selenium nanocomposites and arabinogalactan are nanoparticles of zero-valent selenium with particle size of 0.5–250 nm (depending on the production conditions) stabilized by nontoxic polysaccharide matrix-arabinogalactan. The selenium concentration in the obtained samples of nanocomposites is 0.5–60.0% (depending on the initial ratio of arabinogalactan/precursor of selenium and on other synthesis conditions). Nanocomposites have an antitumor effect with accumulation of Se in the nucleus of a tumor cell. Tests were conducted in a culture of Ehrlich's carcinoma cells. These cells were incubated with nanocomposite of elemental selenium and arabinogalactan at a dose of 2.5, 5, and 7.5 $mg\,l^{-1}$ (calculated as Se) in RPMI-1640 nutritional medium at 37 °C for 24 hours, and with no addition of nanocomposite to the control group (Sukhov et al. 2017).

The evaluation of effect on the culture of tumor cells and distribution of nanocomposite of elemental selenium and arabinogalactan was carried out using light microscopy in a mixed mode (differential interference contrast + fluorescence). It was observed that nanostructured selenium-containing compounds based on arabinogalactan have fluorescent abilities in a wide range of wavelengths from 405 to 514 nm (Shurygina et al. 2015a). Swabs were prepared, and the visualization of luminescence was recorded using a Nikon Eclipse 80i research microscope with a DIH-M epifluorescence device with a Nikon TRITC filter (excitation 528–553 nm, dichroic mirror 565 LP, emission 590–650 nm). It was found that there was no luminescence of Ehrlich's carcinoma cells in the control group after 24 hours of incubation (Figure 2.3a). On the contrary, during incubation of Ehrlich's carcinoma cells after incubation with nanocomposite of elemental selenium and arabinogalactan at a dose of 7.5 $mg\,l^{-1}$ calculated as Se after 24 hours of incubation, a bright luminescence of cell nuclei was observed (Figure 2.3b). Thus, selective accumulation of selenium nanocomposite in the nucleus of tumor cells is shown (Sukhov et al. 2017). Moreover, it was also reported that during incubation at concentrations of 20 and 10 $mg\,l^{-1}$ Ehrlich's carcinoma cells actively died on the first day of exposure. At the same time, at concentrations of 5 and 2.5 $mg\,l^{-1}$, cell death was more pronounced in the first hours and reached 50% faster than when exposed to higher concentrations. When incubating cells at low concentrations (1.25 $mg\,l^{-1}$), the effect of the test agent was minimal: death of carcinoma cells was observed only on day 1 and then their number stabilized and did not reach 50% mortality until the end of the experiment (Trukhan et al. 2018). The study was conducted using the equipment of the center of collective use of scientific equipment "Diagnostic images in surgery."

In *in vivo* experiments in the Ehrlich's carcinoma model, it was found that when the elemental selenium and arabinogalactan nanocomposite were administered intraperitoneally singly at a dose of 2.5, 5, and 7.5 $mg\,kg^{-1}$ of live weight (calculated as Se), a sharp increase in the number of cells with signs of degeneration was noted (Figure 2.4a,b) (Sukhov et al. 2017). Zeng et al. (2019) synthesized SeNPs covered in water-soluble polysaccharides extracted from various mushrooms having spherical shape and particle size of 91–102 nm. Further, they demonstrated significant *in vivo* antitumor activity, inducing caspases and mitochondria-mediated apoptosis, but did not show pronounced toxicity for normal cells. Similarly, Huang et al. (2018) studied SeNPs conjugated with *Pleurotus tuber-regium* for the treatment of colorectal cancer. It was found that these nanoparticles were absorbed by cancer cells through clathrin-mediated endocytosis into lysosomes and caveolae-mediated endocytosis into

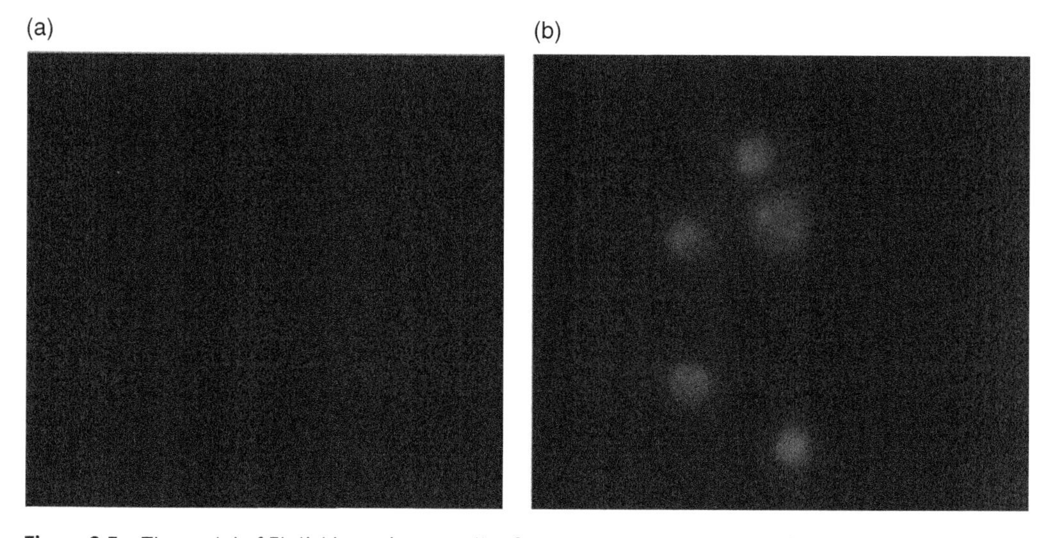

Figure 2.3 The nuclei of Ehrlich's carcinoma cells after exposure to nanocomposite elemental selenium and arabinogalactan, fluorescence microscopy: (a) control group, no light; (b) experimental group, bright glow of nuclei.

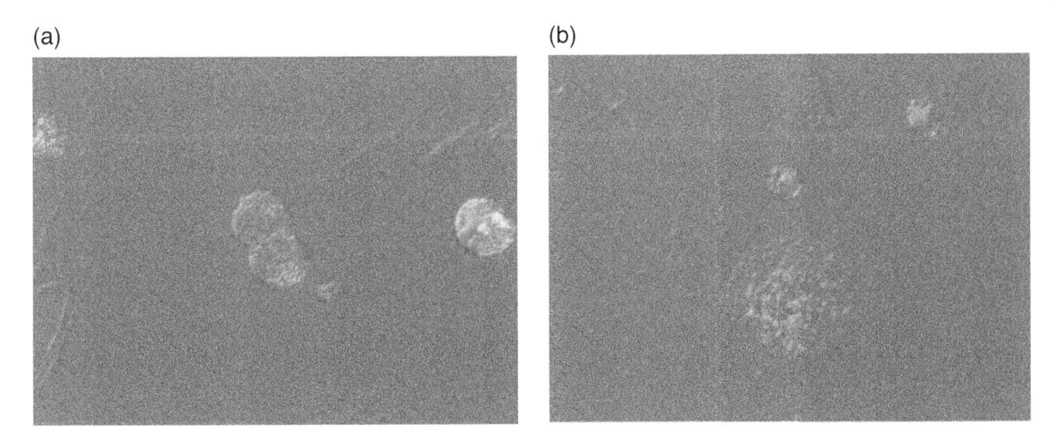

Figure 2.4 Ehrlich's carcinoma cells after exposure to nanocomposite elemental selenium and arabinogalactan, DIC (a) control group; (b) experimental group.

the Golgi apparatus. Nanocomposites stopped cell growth in the phase G2/M and started apoptosis depending on dosage and time through nanocomposite-activated autophagy (Huang et al. 2018). The mechanism of the antitumor effect of biogenic SeNPs obtained from *Bacillus licheniformis* on PC-3 cells is associated with the fact that SeNPs at a concentration of Se $2\,\mu g\,ml^{-1}$ induce cell death using reactive oxygen species (ROS)-mediated necroptosis activation (Sonkusre and Cameotra 2017).

In another study, the antitumor activity of SeNPs synthesized biologically using *Acinetobacter* sp. SW30 and chemically in breast cancer cells (4T1, MCF-7) was evaluated. The obtained results revealed that chemically synthesized SeNPs demonstrated higher anticancer activity than SeNPs synthesized by *Acinetobacter* sp. SW30. However, chemically synthesized SeNPs were also found to be toxic to non-cancerous cells (NIH/3T3, HEK293). On the contrary, biogenic SeNPs were found to be more selective for breast cancer cells (Wadhwani et al. 2017). Krug et al. (2019) synthesized SeNPs coated with sulforaphane. The *in vivo* studies in rats showed SeNPs administered intraperitoneally were mainly excreted with urine (and, to a lesser degree, with feces), however it was partially accumulated in the animal organism. On the other hand, modified SeNPs are mainly accumulated in liver. Moreover, SeNPs conjugated with sulforaphane showed significant anticancer effect *in vitro*. At the same time, the cytotoxic effect on normal cells is relatively low. High antitumor activity and selectivity of the conjugate toward sick and healthy cells are extremely promising from the point of view of cancer treatment (Krug et al. 2019). Considering these facts it is clear that modification of SeNPs can increases cellular uptake and anticancer efficacy (Yang et al. 2012; Wu et al. 2013). For example, decorating the surface of SeNPs with spirulina polysaccharides significantly increased the cellular uptake and cytotoxicity of SeNPs against several cancer cell lines (Yang et al. 2012).

SeNPs can be selectively internalized by cancer cells through endocytosis by means of affinity of membrane proteins to the components administered into the composition of nanoparticles, which leads to activation of the transmission pathway of apoptotic signal and induction apoptosis of cells (Pi et al. 2013; Wu et al. 2013; Zhang et al. 2013). Folate-coated SeNPs showed greater cytotoxicity and potential tumor growth inhibitory effect in mice for both *in vitro* and *in vivo* tests against breast cancer compared to naked SeNPs (without surface modification). Moreover, folate-modified SeNPs showed a significant anti-proliferative effect against 4T1 cells, significantly increased lifespan, and also prevented tumor growth (Shahverdi et al. 2018). In other study, SeNPs loaded with ferulic acid

(FA-SeNPs) were reported to cause damage of tumor cells as a result of apoptosis induction and direct interaction with DNA. Although the antitumor effect of both ferulic acid and SeNPs singly is relatively weak, the combination of these two biologically active ingredients exhibits high antitumor activity. It was shown that FA-SeNPs induced intracellular overproduction of ROS and destruction of mitochondrial membrane potential by activating caspase-3/9 to trigger HepG-2 cell apoptosis through the mitochondrial pathway. The antitumor activity of FA-SeNPs has also been associated with their binding to DNA (Cui et al. 2018).

SeNPs functionalized with walnut peptides and having average size diameter of 89.22 nm showed high antitumor activity. These modified SeNPs were also reported to show excellent selectivity between cancer cells and normal cells. Targeted induction of apoptosis in human mammary adenocarcinoma cells (MCF-7) was confirmed by cell-cycle arrest in the S-phase, nuclear condensation, and DNA disruption (Liao et al. 2016). Similarly, in another study chitosan-stabilized iron oxide nanoparticles decorated with selenium having size diameter of 5–9 nm, zeta potential 29.59 mV, and magnetic properties of 35.932 $emu\,g^{-1}$ were prepared. Further, the authors evaluated their anticancerous activity using breast cancer cells MB-231. After one day of incubation the viability of breast cancer cells was reduced to 40.5% in the presence of 1 $\mu g\,ml^{-1}$ of these composite nanoparticles without using chemotherapeutic pharmaceutical drugs (Hauksdóttir and Webster 2018). Chen et al. (2018) evaluated the possibility of SeNPs as new radio-sensitizers in MCF-7 breast cancer cells. Nano-Se enhanced the toxic effects of radiation which resulted in high tumor cell death compared to any separate treatment causing cell cycle arrest in G2/M phase and activation of autophagy, and increasing the formation of both endogenous and radiation-induced active oxygen forms (Chen et al. 2018). Similar findings were reported against lung cancer cell lines by Cruz et al. (2019).

Biocompatible crystalline nanoparticles which release antitumor non-organic elements are promising therapy for bone tumors. Selenium-doped hydroxyapatite nanoparticles were reported to cause apoptosis of bone cancer cells *in vitro* with the help of caspase-dependent apoptosis pathway and inhibit tumor growth *in vivo* while reducing systemic toxicity (Wang et al. 2016). Thus, all these studies performed in the last decade demonstrate the prospect of SeNPs for cancer therapy, which forms an actively progressing field for anticancer agent development.

2.4 Nanoselenium As a Part of Drug Delivery System

One of the important applications of nanotechnology in medicine is the delivery of active ingredients and diagnostic agents to certain cells or tissues using nanoparticles. Many reports available proposed that various complexes of SeNPs can be used as potential delivery systems. To date, many studies have been devoted to the use of SeNPs in drug delivery systems for cancer and diabetes treatment (Guan et al. 2018). For this purpose, SeNPs are modified by functional ligands to achieve specific affinity for certain cells or organelles, in particular targeting cancer cells and mitochondria of various cells. A lot of diseases are associated with mitochondria dysfunction including cancer, cardiovascular diseases, diabetes, and neurological disorders. Some of the organic cations can penetrate the mitochondrial membrane and deliver therapeutic agents to mitochondria (Hou et al. 2017).

Functionalized SeNPs loaded with various chemotherapeutic drugs offer new prospects for cancer treatment. Due to their own anticancer activity and good responsiveness in the formation of complex forms, SeNPs are widely used for the systemic delivery of different antitumor drugs. Bioactivity in combination with higher selectivity toward cancer cells promises stable delivery with reduced systemic toxicity and higher chemotherapeutic efficacy (Chen et al. 2008).

Nanomaterials tend to accumulate in cancer cells via the process of passive targeting (Yang et al. 2012) and often serve as "nanocarriers" for chemotherapy (Kano et al. 2007; Cho et al. 2010; Yang et al. 2010; Liu et al. 2012; Ramamurthy et al. 2013). It is well known that SeNPs have higher selectivity toward cancer cells compared to normal cells than selenite in similar concentrations (Chen et al. 2008). SeNPs can selectively penetrate cancer cells more than normal cells (Faghfuri et al. 2015) and they have rather low toxicity, high bio-accessibility, convenient routes of administration, and good passive targeting. Also, SeNPs can maintain a prolonged release of selenium and have the ability to target the tumor, thereby reducing the distribution of selenium in normal tissues and increasing the accumulation in tumor tissues. This provides favorable conditions for the use of the fine-line selenium drug (Menon et al. 2018).

SeNPs can be used as a carrier of 5-fluorouracil (FU) to achieve anticancer synergism (Liu et al. 2012). Thus, SeNPs functionalized with 5-fluorouracil showed anticancer activity against five human cancer cell lines (A375, MCF-7, HepG2, Colo201, and PC3) with IC50 values ranging from 6.2 to 14.4 μM. It is worthy to note that despite the activity, the compound has high selectivity between cancer and normal cells. SeNPs loaded with 5-fluorouracil in breast and colon cancer cell lines enhanced the chemosensitivity of FU-NPs in MCF7 and Caco-2 cells more than in MDA-MB-231 and HCT 116 cell lines. The effect was achieved by inhibiting the bioenergy of cancer cells by blocking the glucose uptake (Abd-Rabou et al. 2019). SeNPs and irinotecan in combination dramatically inhibit tumor growth and significantly induce apoptosis of tumor cells in the HCT-8 cell xenograft model; they also decrease systematic toxicity (Gao et al. 2014).

A combination of SeNPs and doxorubicin (Dox) demonstrated better antitumor effect than treatment with each of the component separately (Ramamurthy et al. 2013). Se-functionalized liposomes (SeLPs) for systemic Dox delivery were obtained by applying selenium in situ on liposomes. It was shown that Dox loaded Se-functionalized liposomes have a long-term release effect of Dox and they can increase cellular uptake of Dox compared to normal liposomes. Selenium cover increases the circulation time of liposomes in the body and, therefore, it prolongs the overall release of the drug *in vivo*. In addition, selenium attached to liposomes doubles the antitumor effect of liposomal Dox (Xie et al. 2018). Similarly, folic acid-modified SeNPs were loaded with Dox for targeting the surface of tumor cells that overexpress receptors to folic acid (for example, HeLa cells). These nanocomposites can be easily absorbed by HeLa cells (folate receptor overexpression cells) compared to A54 cells of lung cancer (folate receptor deficient cells) and entered HeLa cells mainly through the clathrin-mediated endocytosis pathway. The nanocomposite inhibited the proliferation of HeLa cells and induced cell apoptosis; it could specifically accumulate itself at the site of tumor which contributed to the significant antitumor efficacy of the nanocomposite *in vivo* (Xia et al. 2018a).

In another study conducted by Xia et al. (2018b), SeNPs modified by cyclic peptide (Arg-Gly-Asp-d-Phe-Cys [RGDfC]) and loaded with Dox were used in non-small cell lung cancer therapy. This structure demonstrated effective uptake by A549 cells mainly through the clathrin-mediated endocytosis pathway. Compared to free Dox, this compound showed more inhibiting proliferation and caused A549 cell apoptosis. This delivery system with active targeting showed high antitumor efficacy in *in vivo* studies (Xia et al. 2018b). Moreover, mesoporous SeNPs coated with human serum albumin, associated with Dox, showed the ability not only to target the tumor in mice but also to reduce the side effects associated with Dox, and they enhanced its antitumor activity (Zhao et al. 2017a). Cyclophosphamide is one of the most effective anticancer drugs, but it has serious toxic effects on normal host cells due to its nonspecific action. Coadministration of cyclophosphamide and SeNPs caused a significant decrease in tumor volume and number of viable tumor cells while providing increased survival in mice (Bhattacharjee et al. 2017).

Multiple drug resistance is one of the major challenges in cancer therapy. Liu et al. (2015) manufactured folate-conjugated SeNPs loaded with ruthenium polypyridyl as a new nanotherapeutic system. This nanosystem provided direct uptake visualization of nanoparticles by cells and it was able to prevent multidrug resistance effectively in liver cancer. The authors noted that it is possible to overcome the multidrug resistance in R-HepG2 cells using SeNPs by inhibiting the expression of the ABC family protein. Internalized SeNPs caused an overproduction of ROS in the tumor and induced apoptosis by activating the p53 and mitogen-activated protein kinase (MAPK) pathways (Liu et al. 2015). Curcumin-loaded SeNPs (Cur-SeNPs) were reported to enhance the antitumor effect. *In vitro* results showed that Cur-SeNPs were most effective against colorectal carcinoma cells (HCT116) and had pleiotropic anticancer effects primarily associated with increased levels of autophagy and apoptosis. On the other hand, *in vivo* studies on the Ehrlich's carcinoma model showed that Cur-SeNPs significantly reduced the tumor progression and increased average survival time in mice (Kumari et al. 2017).

Paclitaxel (PTX) represents one of the most effective natural anticancer drugs. Bidkar et al. (2017) developed SeNPs for PTX delivery and estimated its antiproliferative efficacy against cancer cells *in vitro*. SeNPs doped with antitumor agent PTX showed significant antiproliferative activity against cancer cells causing apoptosis associated with cell cycle arrest in G2/M phase. To increase anticancer activity of oridonin peptide – conjugated GE11 SeNPs (GE11-SeNPs) aimed at EGFR – overexpressing cancer cells were synthesized. It was found that GE11-SeNPs increased the cellular uptake of oridonin in cancer cells which leads to increased inhibition of tumor cell growth and reduction of toxicity toward normal cells (Pi et al. 2017). Targeted co-delivery of epirubicin (EPI, an anticancer agent) and NAS-24 aptamer (inducer of apoptosis) into cancer cells using SeNPs to enhance tumor response *in vitro* and *in vivo* was performed by Jalalian et al. (2018). A significant reduction in toxicity in non-target cells and inhibition of tumor growth in mice compared to the use of epirubicin was recorded.

The use of small interfering RNA (siRNA) for cancer therapy is one of the promising modern trends. However, the traditionally used viral carriers of siRNA are prone to have immunogenicity and the risk of mutagenesis. The creation of hyaluronic acid-coated SeNPs and polycationic polymers polyethylenimine was an innovative approach. SiRNA was loaded onto the surface of the nanoparticles through an electrostatic interaction between siRNA and polycationic polymers polyethylenimine. The resulting particles, due to the active effect on the tumor, mediated by hyaluronic acid, penetrated HepG2 cells mainly by clathrin-mediated endocytosis. HepG2 cell cycle arrest in the G0/G1 phase and apoptosis in the tumor were caused, and they were practically nontoxic to the key organs of mice (Xia et al. 2018c).

2.5 Nanoselenium for Alzheimer's Disease

The main causes of Alzheimer's disease are the accumulation of amyloid protein in the brain and the formation of extracellular amyloid plaques. Nowadays, a great deal of attention has been given on the promising use of selenium in nanoform for the treatment of Alzheimer's disease. Preliminarily, it was shown that small-sized SeNPs (5–15 nm) deplete the formation of amyloid β by reducing the production of ROS, which may be promising in Alzheimer's disease treatment (Nazıroğlu et al. 2017).

Huo et al. (2019) used SeNPs embedded into poly-lactide-co-glycolide composites for Alzheimer's disease treatment. This system was reported to reduce the load of amyloid-β in brain samples using transgenic mice (5XFAD) and significantly reduced the memory deficit in model mice. The authors visualized specific linking of curcumin-loaded nanospheres to amyloid plaques by fluorescence microscopy (Huo et al. 2019). Similarly, Sialic acid (SA) modified SeNPs coated with a high-penetration peptide for the hematoencephalic barrier, i.e. the peptide-B6 (B6-SA-SeNPs), showed high permeability for the hematoencephalic barrier, and can act as a new nanoplatform for Alzheimer's disease treatment. The inhibitory effect of B6-SA-SeNP on amyloid aggregation has been demonstrated in PC12 and bEnd3 cells (Yin et al. 2015). Further, Yang et al. (2018) demonstrated that inclusion of resveratrol antioxidant into SeNPs can be promisingly used for the management of Alzheimer's disease (Yang et al. 2018). Therefore, both using SeNPs themselves and using nanoselenium as a drug delivery system is a promising field for developing alternative approaches for Alzheimer's disease treatment.

2.6 Antibacterial Activity of Nanoselenium

Selenium nanocomposites have attracted considerable attention in terms of their antimicrobial activity. It was shown that SeNPs inhibited growth of a variety of bacteria such as *Pseudomonas aeruginosa*, *Streptococcus aureus*, and *Streptococcus pyogenes* in concentration of $100\,\mu g\,ml^{-1}$, but *Escherichia coli* was of $250\,\mu g\,ml^{-1}$. Moreover, it was found that SeNPs at concentrations of $500\,\mu g\,ml^{-1}$ inhibit growth of pathogenic fungi like *Aspergillus clavatus* (Srivastava and Mukhopadhyay 2015). In another study, it was found that SeNPs synthesized with *Enterococcus faecalis* can be effectively used to prevent and treat infections caused by *S. aureus* (Shoeibi and Mashreghi 2017).

It was demonstrated that antimicrobial activity of SeNPs depends on the method of their synthesis and also on their size. It was found that SeNPs synthesized by the biological (green synthesis) methods usually have greater antimicrobial activity compared to chemically synthesized nanoparticles (Cremonini et al. 2016; Piacenza et al. 2017). Selenium nanocomposites synthesized using *Aspergillus orayzae* with average size of 55 nm were found to be effective against *Acinetobacter calcoaceticus*, *S. aureus*, and *Candida albicans* (Mosallam et al. 2018). Similarly, SeNPs synthesized using gram-negative (*Stenotrophomonas maltophilia*) and gram-positive (*Bacillus mycoides*) bacteria were reported active at low minimal inhibitory concentrations (MICs) against *P. aeruginosa* clinical isolates, but they did not inhibit clinical fungi isolates such as *C. albicans* and *Candida parapsilosis*. These biogenic nanocomposites demonstrated a stronger antimicrobial effect than synthetic SeNPs (Cremonini et al. 2016, 2018). SeNPs stabilized with polyvinyl alcohol showed strong growth inhibition against *S. aureus* at a concentration of 1 ppm, but they did not inhibit growth of *E. coli* (Tran et al. 2016). Lara et al. (2018) demonstrated the antifungal effect of SeNPs and chitosan against *C. albicans*.

Moreover, SeNPs obtained through laser ablation in water have MIC of 50 ppm in case of *E. coli* and *S. aureus*. However, minimum bactericidal concentration (MBC) toward *E. coli* and *S. aureus* was found to be 107 ± 12 and 79 ± 4 ppm, respectively (Guisbiers et al. 2016), but for *C. albicans* MIC was recorded as 25 ppm (Guisbiers et al. 2017). Using electron microscopy, it was observed that SeNPs can easily stick to the biofilm and then penetrate into the pathogen and damage their cellular structure, replacing sulfur (Guisbiers et al. 2017). Therefore, it is strongly believed that selenium can be promisingly used as an effective antibacterial drug including multidrug-resistant organisms (Shurygina et al. 2011, 2015b, 2016; Fadeeva et al. 2015).

2.7 Nanoselenium in Diabetes Treatment

SeNPs not only counteract oxidative stress but also have hypoglycemic activity and hence can be used as hypoglycemic agents. Thus, both type 1 and type 2 diabetes can be treated with SeNPs by reducing the oxidative damage of macromolecules and increasing insulin sensitivity. The hypoglycemic effect of SeNPs in rats with diabetes induced by streptozotocin (a model of type 1 diabetes) was investigated. In rats with diabetes, there was a significant decrease in blood glucose levels in fasting state after treatment with SeNPs was recorded when SeNPs were given orally for 28 days. The concentration of insulin in the serum of these animals was also found to be higher than in non-treated rats. It confirmed that SeNPs were able to reduce hepatic cytolysis and renal dysfunction, total lipids, total cholesterols, triglyceride levels, and low-density lipoprotein cholesterol. In addition, SeNPs reduced the intensity of morphological disorders in liver and kidney tissues of rats. Further, the obtained findings showed that SeNPs may reduce the manifestations of hyperglycemia and hyperlipidemia in patients with diabetes, possibly causing an insulin-like effect (Al-Quraishy et al. 2015).

Liu et al. (2018) investigated the antidiabetic activity of SeNPs loaded with *Catathelasma ventricosum* polysaccharides in mice with diabetes induced by streptozotocin. These nanocomposites showed a potential antidiabetic efficacy which was established by studying the serum profiles of glucose levels and antioxidant enzymes. In addition, SeNPs had significantly higher antidiabetic activity than other drugs of organic and nonorganic selenium (Liu et al. 2018). This study was in accordance with the observation recorded by Zeng et al. (2018). The authors demonstrated that chitosan-stabilized SeNPs at a selenium dose of 2.0 mg kg^{-1} can achieve higher antidiabetic activity than other doses of SeNPs and other selenium drugs with the same dose of selenium.

It was proposed that the inclusion of SeNPs in liposomes was a highly effective form of treatment for streptozotocin-induced experimental diabetes. Liposomal forms of SeNPs made it possible to preserve the integrity of β-cells of the pancreas followed by an increase in insulin secretion and, thus, led to reduction of blood glucose levels, suppression of oxidative stress, increase of antioxidant defense system, and inhibiting of pancreatic inflammation (Ahmed et al. 2017). Selenium nanocomposites loaded with mulberry leaf and *Pueraria lobata* extracts showed a slow phytomedicines release and good physiological stability in the simulated digestive fluid. These nanocomposites exert pronounced hypoglycemic effects in both normal and diabetic rats after oral administration. Ex vivo intestinal visualization showed that the nanocomposite had good permeability into the intestinal wall and the ability of transepithelial transport. It was also found that the composite improved the function of the pancreas and promoted glucose utilization by adipocytes (Deng et al. 2019).

In recent years, SeNPs have commonly been used for the delivery of proteins and peptides. Proteins, being a kind of stabilizer, are often involved in the synthesis of SeNPs (Zhang et al. 2018). SeNPs was used for oral delivery of insulin. In this case insulin acts as a therapeutic agent and stabilizer of SeNPs and acquires the ability to avoid degradation by digestive enzymes in the gastrointestinal tract. The composite had intestinal permeability, and after oral administration it exerts significant hypoglycemic effect in both normal rats and rats with type 2 diabetes. The relative pharmacological bioavailability was up to 9.15% compared to subcutaneously injected insulin (Deng et al. 2017). SeNPs loaded with pituitary adenylate cyclase-activating peptide resulted in increased insulin secretion and sustained hypoglycemic effect with an injection dose of 20 nmol l^{-1} in mice with type 2 diabetes. Repeated administration for 12 weeks significantly improved glucose and lipids profile, insulin sensitivity, and histomorphology of pancreatic and adipose tissues (Zhao et al. 2017b). Therefore, diabetes treatment can be one of the potential applications for the use of selenium nanocomposites.

2.8 Other Applications of Nanoselenium

SeNPs coated with polysaccharides from *Ulva lactuca* were developed and investigated for their anti-inflammatory efficacy in a model of acute colitis by Zhu et al. (2017). The treatment with SeNPs reduced body weight loss and inflammatory damage of the colon in mice and reduced infiltration with macrophages. It was shown that anti-inflammatory effects of SeNPs were associated with a decrease in the level of pro-inflammatory cytokines, in particular IL-6 and tumor necrosis factor alpha (TNF-α), by inhibiting the activation of macrophages through suppressing nuclear translocation of NF-κB responsible for the transcription of these pro-inflammatory cytokines (Zhu et al. 2017). Atteia et al. (2018) investigated and described the potential protective mechanism of SeNPs against lead acetate-induced thyrotoxicity. It was found that SeNPs can prevent acetate-induced hypothyroidism and oxidative damage in the tissues of the thyroid gland. SeNPs can also be used as effective drug carriers based on nucleic acids, and their use may be promising in gene therapy (Li et al. 2016).

Ren et al. (2019) in a rat model of rheumatoid arthritis induced by Freund's adjuvant showed that SeNPs loaded with p-Coumaric acid are an effective therapeutic agent for inflammatory diseases. The data were confirmed by histological examination, the level of antioxidant enzyme activity, and inflammatory cytokines (TNF-α, IL-1β, IL-6, and MCP-1) (Ren et al. 2019). El-Ghazaly et al. (2017) showed that nano-Se has potential anti-inflammatory activity against radiation-induced inflammation in rats.

In another study, Nematollahi et al. (2018) in ex vivo experiments showed that SeNPs can play an important role in the treatment of hydatid cyst. Considering the significant biological role of selenium in the implementation of vital functions for the cell and in bone mineralization, Hoeg

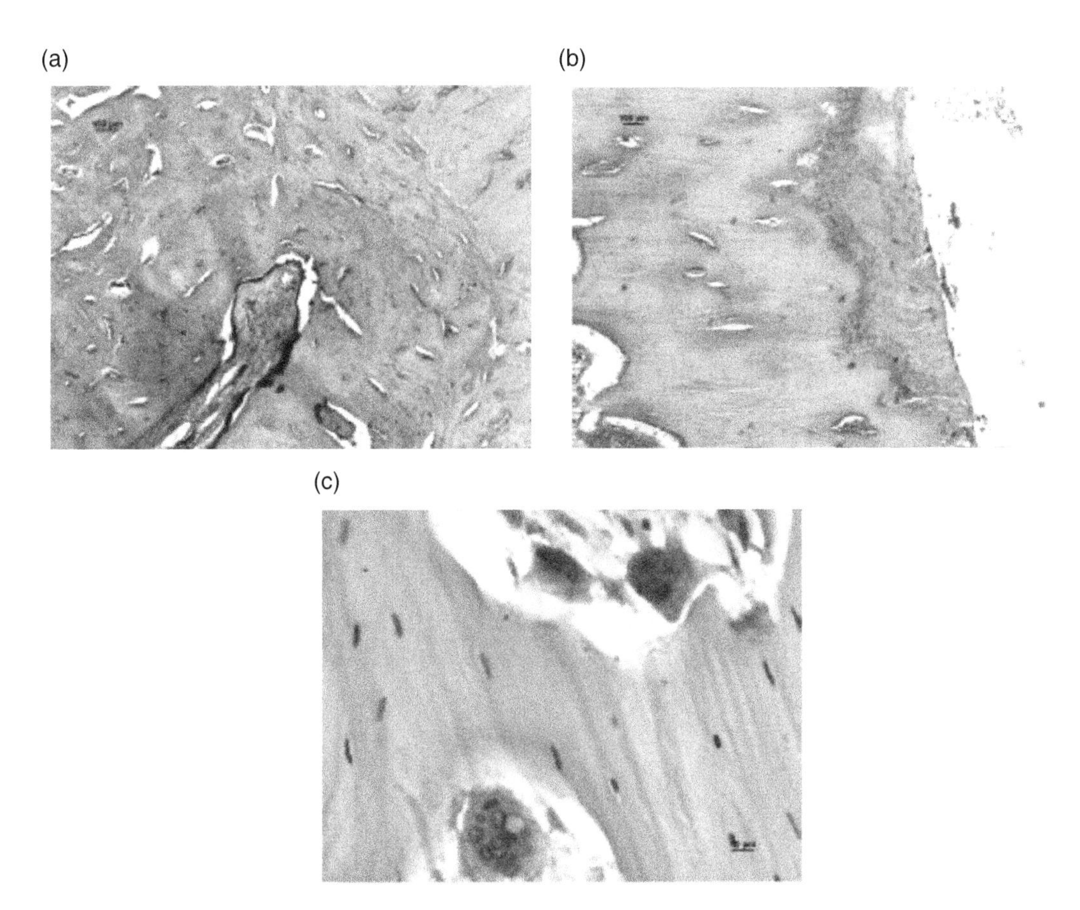

Figure 2.5 Regenerative process during fracture (a) The control group. Bone regenerates in the upper third of the perforated fracture of the tibia. Hematoxylin and eosin. (b) Experimental group. Lack of bone regeneration in perforated fracture. Hematoxylin and eosin. (c) Experimental group. Osteoclast activation in the area of the fracture. Hematoxylin and eosin.

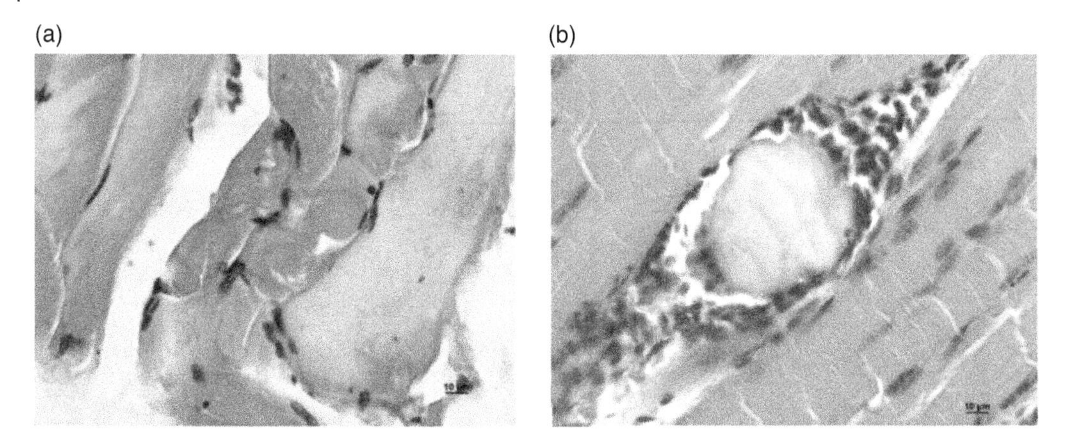

Figure 2.6 The effect of the nanocomposite of elemental selenium and arabinogalactan on the repair of muscle tissue when administered locally for 35 days. (a) Experimental group: Loss of cross-striations, disorientation of individual muscle fibers in the fracture zone. Hematoxylin and eosin. (b) Experimental group: Death of individual muscle fibers, infiltration of lymphocytes and macrophages in the area of the fracture. Hematoxylin and eosin.

et al. (2012) studied the possibility of use of selenium in the form of nanoparticles in the polysaccharide arabinogalactan with local application to modify the reparative process for bone and muscle tissue injuries. The effect of elemental selenium/arabinogalactan nanocomposite (local introduction) on reparation of bone and muscular tissue in a trauma area has been studied. In animals, to which elemental selenium/arabinogalactan nanocomposite was introduced, the regeneration of bone tissue was absent in the fracture area. Bone trabeculars were destroyed. Activation of osteoclasts both in the area of new bone formation and in the area of old bone was observed (Figure 2.5). In this group of animals, the muscular tissue in the area adjacent to trauma zone was stained unevenly. Many fibers have lost cross-striation. Necrosis of separate muscular fibers owing to infiltration with macrophages and lymphocytes was observed (Figure 2.6).

The electron microscope study of the muscular tissue from damaged area in animals after introduction of the selenium nanocomposite has shown that the muscular tissue is considerably changed compared to the control group. Areas of abnormal orientation of muscular fibers were revealed, Z-lines were considerably curved, I-bands were expanded, and contractile proteins were degradable. Nuclei of myocyte were of irregular shape with clearing zones. The biotesting data indicate that local application of the nanobiocomposite of elemental selenium in the reparation area at injury of bone and muscular tissue leads to a considerably impaired reparative process of both bone and muscular tissue. Impairments of osteoreparation are expressed in osteoresorption, and slowing of bone regeneration was also observed. In our opinion, significant damage to muscle fibers of a toxic nature is observed: loss of transverse striation of muscle fibers, death of some fibers, and infiltration by lymphocytes and macrophages.

2.9 Conclusion

Selenium nanocomposites are promising compounds with great potential for medicinal use. Nowadays, the main spectrum of beneficial use of these nanocomposites is associated with their use for the diagnosis and treatment of various diseases. In addition, these compounds can act as multifunctional platforms for targeted drug delivery and simultaneous detection of SeNPs in tissues due to their fluorescent ability. Prospects for the use of SeNPs are associated with the synergism of their antitumor activity and ability for target delivery of antitumor agents. Alzheimer's disease and diabetes are two quite common severe diseases in which the use of selenium nanoparticles may be promising.

The antibacterial activity of SeNPs is of particular interest. This area is particularly relevant because of the increase in the amount of antibacterial resistance in pathogens against main antibacterial drugs. At the same time, the reaction of damaged tissues to the introduction of selenium nanocomposites is ambiguous, and additional studies are needed to determine the effects of inflammation and wound healing. Rational design of multifunctional nano platforms with drugs is a promising strategy for simultaneous diagnosis, real-time monitoring, and treatment.

References

Abd-Rabou, A.A., Shalby, A.B., and Ahmed, H.H. (2019). Selenium nanoparticles induce the chemosensitivity of fluorouracil nanoparticles in breast and colon cancer cells. *Biological Trace Element Research* 187 (1): 80–91.

Ahmed, H.H., Abd El-Maksoud, M.D., Abdel Moneim, A.E., and Aglan, H.A. (2017). Pre-clinical study for the antidiabetic potential of selenium nanoparticles. *Biological Trace Element Research* 177 (2): 267–280.

Al-Quraishy, S., Dkhil, M.A., and Abdel Moneim, A.E. (2015). Anti-hyperglycemic activity of selenium nanoparticles in streptozotocin-induced diabetic rats. *International Journal of Nanomedicine* 10: 6741–6756.

Atteia, H.H., Arafa, M.H., and Prabahar, K. (2018). Selenium nanoparticles prevents lead acetate-induced hypothyroidism and oxidative damage of thyroid tissues in male rats through modulation of selenoenzymes and suppression of miR-224. *Biomedicine & Pharmacotherapy* 99: 486–491.

Bao, P., Chen, Z., Tai, R.Z. et al. (2015). Selenite-induced toxicity in cancer cells is mediated by metabolic generation of endogenous selenium nanoparticles. *Journal of Proteome Research* 14 (2): 1127–1136.

Bhattacharjee, A., Basu, A., Biswas, J. et al. (2017). Chemoprotective and chemosensitizing properties of selenium nanoparticle (Nano-Se) during adjuvant therapy with cyclophosphamide in tumor-bearing mice. *Molecular and Cellular Biochemistry* 424 (1–2): 13–33.

Bidkar, A.P., Sanpui, P., and Ghosh, S.S. (2017). Efficient induction of apoptosis in cancer cells by paclitaxel-loaded selenium nanoparticles. *Nanomedicine* 12 (21): 2641–2651.

Chen, T., Wong, Y.S., Zheng, W. et al. (2008). Selenium nanoparticles fabricated in *Undaria pinnatifida* polysaccharide solutions induce mitochondria-mediated apoptosis in A375 human melanoma cells. *Colloids and Surfaces B: Biointerfaces* 67 (1): 26–31.

Chen, F., Zhang, X.H., Hu, X.D. et al. (2018). The effects of combined selenium nanoparticles and radiation therapy on breast cancer cells in vitro. *Artificial Cells, Nanomedicine, and Biotechnology* 46 (5): 937–948.

Cho, H.S., Dong, Z., Pauletti, G.M. et al. (2010). Fluorescent, superparamagnetic nanospheres for drug storage, targeting, and imaging: a multifunctional nanocarrier system for cancer diagnosis and treatment. *ACS Nano* 4 (9): 5398–5404.

Cremonini, E., Zonaro, E., Donini, M. et al. (2016). Biogenic selenium nanoparticles: characterization, antimicrobial activity and effects on human dendritic cells and fibroblasts. *Microbial Biotechnology* 9 (6): 758–771.

Cremonini, E., Boaretti, M., Vandecandelaere, I. et al. (2018). Biogenic selenium nanoparticles synthesized by *Stenotrophomonas maltophilia* SeITE02 loose antibacterial and antibiofilm efficacy as a result of the progressive alteration of their organic coating layer. *Microbial Biotechnology* 11 (6): 1037–1047.

Cruz, L.Y., Wang, D., and Liu, J. (2019). Biosynthesis of selenium nanoparticles, characterization and X-ray induced radiotherapy for the treatment of lung cancer with interstitial lung disease. *Journal of Photochemistry and Photobiology B: Biology* 191: 123–127.

Cui, D., Yan, C., Miao, J. et al. (2018). Synthesis, characterization and antitumor properties of selenium nanoparticles coupling with ferulic acid. *Materials Science and Engineering: C* 90: 104–112.

Deng, W., Xie, Q., Wang, H. et al. (2017). Selenium nanoparticles as versatile carriers for oral delivery of insulin: insight into the synergic antidiabetic effect and mechanism. *Nanomedicine* 13 (6): 1965–1974.

Deng, W., Wang, H., Wu, B., and Zhang, X. (2019). Selenium-layered nanoparticles serving for oral delivery of phytomedicines with hypoglycemic activity to synergistically potentiate the antidiabetic effect. *Acta Pharmaceutica Sinica B* 9 (1): 74–86.

Dumitrescu, A.M. and Refetoff, S. (2011). Inherited defects of thyroid hormone metabolism. *Annales d'Endocrinologie* 72 (2): 95–98.

El-Ghazaly, M.A., Fadel, N., Rashed, E. et al. (2017). Anti-inflammatory effect of selenium nanoparticles on the inflammation induced in irradiated rats. *Canadian Journal of Physiology and Pharmacology* 95 (2): 101–110.

Fadeeva, T.V., Shurygina, I.A., Sukhov, B.G. et al. (2015). Relationship between the structures and antimicrobial activities of argentic nanocomposites. *Bulletin of the Russian Academy of Sciences: Physics* 79: 273–275.

Faghfuri, E., Yazdi, M.H., Mahdavi, M. et al. (2015). Dose-response relationship study of selenium nanoparticles as an immunostimulatory agent in cancer-bearing mice. *Archives of Medical Research* 46 (1): 31–37.

Gammelgaard, B., Jackson, M.I., and Gabel-Jensen, C. (2011). Surveying selenium speciation from soil to cell-forms and transformations. *Analytical and Bioanalytical Chemistry* 399 (5): 1743–1763.

Gao, F., Yuan, Q., Gao, L. et al. (2014). Cytotoxicity and therapeutic effect of irinotecan combined with selenium nanoparticles. *Biomaterials* 35 (31): 8854–8866.

Guan, B., Yan, R., Li, R., and Zhang, X. (2018). Selenium as a pleiotropic agent for medical discovery and drug delivery. *International Journal of Nanomedicine* 13: 7473–7490.

Guisbiers, G., Wang, Q., Khachatryan, E. et al. (2016). Inhibition of *E. coli* and *S. aureus* with selenium nanoparticles synthesized by pulsed laser ablation in deionized water. *International Journal of Nanomedicine* 11: 3731–3736.

Guisbiers, G., Lara, H.H., Mendoza-Cruz, R. et al. (2017). Inhibition of *Candida albicans* biofilm by pure selenium nanoparticles synthesized by pulsed laser ablation in liquids. *Nanomedicine* 13 (3): 1095–1103.

Hauksdóttir, H.L. and Webster, T.J. (2018). Selenium and iron oxide nanocomposites for magnetically-targeted anti-cancer applications. *Journal of Biomedical Nanotechnology* 14 (3): 510–525.

Hoeg, A., Gogakos, A., Murphy, E. et al. (2012). Bone turnover and bone mineral density are independently related to selenium status in healthy euthyroid postmenopausal women. *The Journal of Clinical Endocrinology & Metabolism* 97 (11): 4061–4070.

Hou, J., Yu, X., Shen, Y. et al. (2017). Triphenyl phosphine-functionalized chitosan nanoparticles enhanced antitumor efficiency through targeted delivery of doxorubicin to mitochondria. *Nanoscale Research Letters* 12 (1): 158.

Huang, Z., Rose, A.H., and Hoffmann, P.R. (2012). The role of selenium in inflammation and immunity: from molecular mechanisms to therapeutic opportunities. *Antioxidants & Redox Signaling* 16 (7): 705–743.

Huang, G., Liu, Z., He, L. et al. (2018). Autophagy is an important action mode for functionalized selenium nanoparticles to exhibit anti-colorectal cancer activity. *Biomaterials Science* 6 (9): 2508–2517.

Huang, J., Huang, W., Zhang, Z. et al. (2019). Highly uniform synthesis of selenium nanoparticles with EGFR targeting and tumor microenvironment-responsive ability for simultaneous diagnosis and therapy of nasopharyngeal carcinoma. *ACS Applied Materials & Interfaces* 11 (12): 11177–11193.

Huo, X., Zhang, Y., Jin, X. et al. (2019). A novel synthesis of selenium nanoparticles encapsulated PLGA nanospheres with curcumin molecules for the inhibition of amyloid β aggregation in Alzheimer's disease. *Journal of Photochemistry and Photobiology B: Biology* 190: 98–102.

Jalalian, S.H., Ramezani, M., Abnous, K., and Taghdisi, S.M. (2018). Targeted co-delivery of epirubicin and NAS-24 aptamer to cancer cells using selenium nanoparticles for enhancing tumor response in vitro and in vivo. *Cancer Letters* 416: 87–93.

Jia, X., Liu, Q., Zou, S. et al. (2015). Construction of selenium nanoparticles/β-glucan composites for enhancement of the antitumor activity. *Carbohydrate Polymers* 117: 434–442.

Kano, M.R., Bae, Y., Iwata, C. et al. (2007). Improvement of cancer-targeting therapy, using nanocarriers for intractable solid tumors by inhibition of TGF-β signaling. *Proceedings of the National Academy of Sciences of the United States of America* 104 (9): 3460–3465.

Khalid, A., Tran, P.A., Norello, R. et al. (2016). Intrinsic fluorescence of selenium nanoparticles for cellular imaging applications. *Nanoscale* 8 (6): 3376–3385.

Kostyro, V.V., Kostyro, Y.A., Lepekhova, S.A. et al. (2013). Research of vein-protective effect of the preparation "Agsular"® on the models of vascular pathology. *Acta Biomedica Scientifica* 1 (89): 106–110.

Krug, P., Mielczarek, L., Wiktorska, K. et al. (2019). Sulforaphane-conjugated selenium nanoparticles: towards a synergistic anticancer effect. *Nanotechnology* 30 (6): 065101.

Kumari, M., Ray, L., Purohit, M.P. et al. (2017). Curcumin loading potentiates the chemotherapeutic efficacy of selenium nanoparticles in HCT116 cells and Ehrlich's ascites carcinoma bearing mice. *European Journal of Pharmaceutics and Biopharmaceutics* 117: 346–362.

Lara, H.H., Guisbiers, G., Mendoza, J. et al. (2018). Synergistic antifungal effect of chitosan-stabilized selenium nanoparticles synthesized by pulsed laser ablation in liquids against *Candida albicans* biofilms. *International Journal of Nanomedicine* 13: 2697–2708.

Li, Y., Lin, Z., Zhao, M. et al. (2016). Multifunctional selenium nanoparticles as carriers of HSP70 siRNA to induce apoptosis of HepG2 cells. *International Journal of Nanomedicine* 11: 3065–3076.

Liao, W., Yu, Z., Lin, Z. et al. (2015). Biofunctionalization of selenium nanoparticle with *Dictyophora indusiata* polysaccharide and its antiproliferative activity through death-receptor and mitochondria-mediated apoptotic pathways. *Scientific Reports* 5: 18629.

Liao, W., Zhang, R., Dong, C. et al. (2016). Novel walnut peptide-selenium hybrids with enhanced anticancer synergism: facile synthesis and mechanistic investigation of anticancer activity. *International Journal of Nanomedicine* 11: 1305–1321.

Liu, W., Li, X., Wong, Y.S. et al. (2012). Selenium nanoparticles as a carrier of 5-fluorouracil to achieve anticancer synergism. *ACS Nano* 6 (8): 6578–6591.

Liu, T., Zeng, L., Jiang, W. et al. (2015). Rational design of cancer-targeted selenium nanoparticles to antagonize multidrug resistance in cancer cells. *Nanomedicine* 11 (4): 947–958.

Liu, Y., Zeng, S., Liu, Y. et al. (2018). Synthesis and antidiabetic activity of selenium nanoparticles in the presence of polysaccharides from *Catathelasma ventricosum*. *International Journal of Biological Macromolecules* 114: 632–639.

Luo, H., Wang, F., Bai, Y. et al. (2012). Selenium nanoparticles inhibit the growth of HeLa and MDA-MB-231 cells through induction of S phase arrest. *Colloids and Surfaces B: Biointerfaces* 94: 304–308.

Menon, S., Devi, S., Santhiya, R. et al. (2018). Selenium nanoparticles: a potent chemotherapeutic agent and an elucidation of its mechanism. *Colloids and Surfaces B: Biointerfaces* 170: 280–292.

Mosallam, F.M., El-Sayyad, G.S., Fathy, R.M., and El-Batal, A.I. (2018). Biomolecules-mediated synthesis of selenium nanoparticles using *Aspergillus oryzae* fermented Lupin extract and gamma radiation for hindering the growth of some multidrug-resistant bacteria and pathogenic fungi. *Microbial Pathogenesis* 122: 108–116.

Nazıroğlu, M., Muhamad, S., and Pecze, L. (2017). Nanoparticles as potential clinical therapeutic agents in Alzheimer's disease: focus on selenium nanoparticles. *Expert Review of Clinical Pharmacology* 10 (7): 773–782.

Nematollahi, A., Shahbazi, P., Rafat, A., and Ghanbarlu, M. (2018). Comparative survey on scolicidal effects of selenium and silver nanoparticles on protoscolices of hydatid cyst. *Open Veterinary Journal* 8 (4): 374–377.

Peng, D., Zhang, J., Liu, Q., and Taylor, E.W. (2007). Size effect of elemental selenium nanoparticles (Nano-Se) at supranutritional levels on selenium accumulation and glutathione S-transferase activity. *Journal of Inorganic Biochemistry* 101 (10): 1457–1463.

Pi, J., Jin, H., Liu, R. et al. (2013). Pathway of cytotoxicity induced by folic acid modified selenium nanoparticles in MCF-7 cells. *Applied Microbiology and Biotechnology* 97 (3): 1051–1062.

Pi, J., Jiang, J., Cai, H. et al. (2017). GE11 peptide conjugated selenium nanoparticles for EGFR targeted oridonin delivery to achieve enhanced anticancer efficacy by inhibiting EGFR-mediated PI3K/AKT and Ras/Raf/MEK/ERK pathways. *Drug Delivery* 24 (1): 1549–1564.

Piacenza, E., Presentato, A., Zonaro, E. et al. (2017). Antimicrobial activity of biogenically produced spherical Se-nanomaterials embedded in organic material against *Pseudomonas aeruginosa* and *Staphylococcus aureus* strains on hydroxyapatite-coated surfaces. *Microbial Biotechnology* 10 (4): 804–818.

Ramamurthy, C., Sampath, K.S., Arunkumar, P. et al. (2013). Green synthesis and characterization of selenium nanoparticles and its augmented cytotoxicity with doxorubicin on cancer cells. *Bioprocess and Biosystems Engineering* 36 (8): 1131–1139.

Ren, S.X., Zhan, B., Lin, Y. et al. (2019). Selenium nanoparticles dispersed in phytochemical exert anti-inflammatory activity by modulating catalase, GPx1, and COX-2 gene expression in a rheumatoid arthritis rat model. *Medical Science Monitor: International Medical Journal of Experimental and Clinical Research* 25: 991–1000.

Rodionova, L.V., Shurygina, I.A., Sukhov, B.G. et al. (2015). Nanobiocomposite based on selenium and arabinogalactan: synthesis, structure, and application. *Russian Journal of General Chemistry* 85 (1): 485–487.

Sadeghian, S., Kojouri, G.A., and Mohebbi, A. (2012). Nanoparticles of selenium as species with stronger physiological effects in sheep in comparison with sodium selenite. *Biological Trace Element Research* 146 (3): 302–308.

Shahverdi, A.R., Shahverdi, F., Faghfuri, E. et al. (2018). Characterization of folic acid surface-coated selenium nanoparticles and corresponding in vitro and in vivo effects against breast cancer. *Archives of Medical Research* 49 (1): 10–17.

Shoeibi, S. and Mashreghi, M. (2017). Biosynthesis of selenium nanoparticles using *Enterococcus faecalis* and evaluation of their antibacterial activities. *Journal of Trace Elements in Medicine and Biology* 39: 135–139.

Shurygina, IA, & Shurygin, MG 2013, Method of preparing bone tissue preparation and set for its realization, RU patent 2500104.

Shurygina, I.A. and Shurygin, M.G. (2018). Method of decalcination of bone tissue. *Journal of Clinical and Experimental Morphology* 4 (28): 34–37.

Shurygina, I.A., Sukhov, B.G., Fadeeva, T.V. et al. (2011). Bactericidal action of Ag(0)-antithrombotic sulfated arabinogalactan nanocomposite: coevolution of initial nanocomposite and living microbial cell to a novel nonliving nanocomposite. *Nanomedicine: Nanotechnology, Biology, and Medicine* 7 (6): 827–833.

Shurygina, I.A., Rodionova, L.V., Shurygin, M.G. et al. (2015a). Using confocal microscopy to study the effect of an original pro-enzyme Se/arabinogalactan nanocomposite on tissue regeneration in a skeletal system. *Bulletin of the Russian Academy of Sciences: Physics* 79 (2): 256–258.

Shurygina, I.A., Shurygin, M.G., Dmitrieva, L.A. et al. (2015b). Bacterio- and lymphocytotoxicity of silver nanocomposite with sulfated arabinogalactan. *Russian Chemical Bulletin* 64 (7): 1629–1632.

Shurygina, I.A., Shurygin, M.G., and Sukhov, B.G. (2016). Nanobiocomposites of metals as antimicrobial agents. In: *Antibiotic Resistance: Mechanisms and New Antimicrobial Approaches*, 1e (eds. M. Rai and K. Kon), 167–186. Academic Press.

Singh, S.C., Mishra, S.K., Srivastava, R.K., and Gopal, R. (2010). Optical properties of selenium quantum dots produced with laser irradiation of water suspended Se nanoparticles. *The Journal of Physical Chemistry C* 114 (41): 17374–17384.

Sonkusre, P. and Cameotra, S.S. (2017). Biogenic selenium nanoparticles induce ROS-mediated necroptosis in PC-3 cancer cells through TNF activation. *Journal of Nanobiotechnology* 15 (1): 43.

Sonkusre, P., Nanduri, R., Gupta, P., and Cameotra, S.S. (2014). Improved extraction of intracellular biogenic selenium nanoparticles and their specificity for cancer chemoprevention. *Journal of Nanomedicine & Nanotechnology* 5 (2): 194.

Srivastava, N. and Mukhopadhyay, M. (2015). Green synthesis and structural characterization of selenium nanoparticles and assessment of their antimicrobial property. *Bioprocess and Biosystems Engineering* 38 (9): 1723–1730.

Sukhov, BG, Ganenko, TV, Pogodaeva, NN, et al. 2017, Agent with antitumor activity based on arabinogalactan nanocomposites with selenium and methods for prepariation of such nanobiocomposites, RU patent 2614363.

Tang, S., Wang, T., Jiang, M. et al. (2019). Construction of arabinogalactans/selenium nanoparticles composites for enhancement of the antitumor activity. *International Journal of Biological Macromolecules* 128: 444–451.

Tran, P. and Webster, T.J. (2008). Enhanced osteoblast adhesion on nanostructured selenium compacts for anti-cancer orthopedic applications. *International Journal of Nanomedicine* 3 (3): 391–396.

Tran, P.A., Sarin, L., Hurt, R.H., and Webster, T.J. (2010). Differential effects of nanoselenium doping on healthy and cancerous osteoblasts in coculture on titanium. *International Journal of Nanomedicine* 5: 351–358.

Tran, P.A., O'Brien-Simpson, N., Reynolds, E.C. et al. (2016). Low cytotoxic trace element selenium nanoparticles and their differential antimicrobial properties against *S. aureus* and *E. coli*. *Nanotechnology* 27 (4): 045101.

Trukhan, I.S., Dremina, N.N., Lozovskaya, E.A., and Shurygina, I.A. (2018). Assessment of potential cytotoxicity during vital observation at the BioStation CT. *Acta Biomedica Scientifica* 3 (6): 48–53.

Vekariya, K.K., Kaur, J., and Tikoo, K. (2013). Alleviating anastrozole induced bone toxicity by selenium nanoparticles in SD rats. *Toxicology and Applied Pharmacology* 268 (2): 212–220.

Villette, S., Bermano, G., Arthur, J.R., and Hesketh, J.E. (1998). Thyroid stimulating hormone and selenium supply interact to regulate selenoenzyme gene expression in thyroid cells (FRTL-5) in culture. *FEBS Letters* 438 (1–2): 81–84.

Wadhwani, S.A., Gorain, M., Banerjee, P. et al. (2017). Green synthesis of selenium nanoparticles using *Acinetobacter* sp. SW30: optimization, characterization and its anticancer activity in breast cancer cells. *International Journal of Nanomedicine* 12: 6841–6855.

Wang, Y. and Fu, L. (2012). Forms of selenium affect its transport, uptake and glutathione peroxidase activity in the Caco-2 cell model. *Biological Trace Element Research* 149 (1): 110–116.

Wang, H., Zhang, J., and Yu, H. (2007). Elemental selenium at nano size possesses lower toxicity without compromising the fundamental effect on selenoenzymes: comparison with selenomethionine in mice. *Free Radical Biology & Medicine* 42 (10): 1524–1533.

Wang, T., Yang, L., Zhang, B., and Liu, J. (2010). Extracellular biosynthesis and transformation of selenium nanoparticles and application in H_2O_2 biosensor. *Colloids and Surfaces B: Biointerfaces* 80 (1): 94–102.

Wang, Y., Wang, J., Hao, H. et al. (2016). In vitro and in vivo mechanism of bone tumor inhibition by selenium-doped bone mineral nanoparticles. *ACS Nano* 10 (11): 9927–9937.

Wu, H., Zhu, H., Li, X. et al. (2013). Induction of apoptosis and cell cycle arrest in A549 human lung adenocarcinoma cells by surface-capping selenium nanoparticles: an effect enhanced by polysaccharide–protein complexes from *Polyporus rhinocerus*. *Journal of Agricultural and Food Chemistry* 61 (41): 9859–9866.

Xia, Y., Xu, T., Zhao, M. et al. (2018a). Delivery of doxorubicin for human cervical carcinoma targeting therapy by folic acid-modified selenium nanoparticles. *International Journal of Molecular Sciences* 19: E3582.

Xia, Y., Chen, Y., Hua, L. et al. (2018b). Functionalized selenium nanoparticles for targeted delivery of doxorubicin to improve non-small-cell lung cancer therapy. *International Journal of Nanomedicine* 13: 6929–6939.

Xia, Y., Guo, M., Xu, T. et al. (2018c). siRNA-loaded selenium nanoparticle modified with hyaluronic acid for enhanced hepatocellular carcinoma therapy. *International Journal of Nanomedicine* 13: 1539–1552.

Xie, Q., Deng, W., Yuan, X. et al. (2018). Selenium-functionalized liposomes for systemic delivery of doxorubicin with enhanced pharmacokinetics and anticancer effect. *European Journal of Pharmaceutics and Biopharmaceutics* 122: 87–95.

Yang, X., Grailer, J.J., Pilla, S. et al. (2010). Tumor-targeting, pH-responsive, and stable unimolecular micelles as drug nanocarriers for targeted cancer therapy. *Bioconjugate Chemistry* 21 (3): 496–504.

Yang, F., Tang, Q., Zhong, X. et al. (2012). Surface decoration by Spirulina polysaccharide enhances the cellular uptake and anticancer efficacy of selenium nanoparticles. *International Journal of Nanomedicine* 7: 835–844.

Yang, L., Wang, W., Chen, J. et al. (2018). A comparative study of resveratrol and resveratrol-functional selenium nanoparticles: inhibiting amyloid β aggregation and reactive oxygen species formation properties. *Journal of Biomedical Materials Research Part A* 106 (12): 3034–3041.

Yanhua, W., Hao, H., Li, Y., and Zhang, S. (2016). Selenium-substituted hydroxyapatite nanoparticles and their in vivo antitumor effect on hepatocellular carcinoma. *Colloids and Surfaces B: Biointerfaces* 140: 297–306.

Yin, T., Yang, L., Liu, Y. et al. (2015). Sialic acid (SA)-modified selenium nanoparticles coated with a high blood-brain barrier permeability peptide-B6 peptide for potential use in Alzheimer's disease. *Acta Biomaterialia* 25: 172–183.

Zeng, S., Ke, Y., Liu, Y. et al. (2018). Synthesis and antidiabetic properties of chitosan-stabilized selenium nanoparticles. *Colloids and Surfaces B: Biointerfaces* 170: 115–121.

Zeng, D., Zhao, J., Luk, K.H. et al. (2019). Potentiation of in vivo anticancer efficacy of selenium nanoparticles by mushroom polysaccharides surface decoration. *Journal of Agricultural and Food Chemistry* 67 (10): 2865–2876.

Zhang, J.S., Gao, X.Y., Zhang, L.D., and Bao, Y.P. (2001). Biological effects of a nano red elemental selenium. *BioFactors* 15 (1): 27–38.

Zhang, J., Wang, H., Bao, Y., and Zhang, L. (2004). Nano red elemental selenium has no size effect in the induction of seleno-enzymes in both cultured cells and mice. *Life Sciences* 75 (2): 237–244.

Zhang, J., Wang, X., and Xu, T. (2008). Elemental selenium at nano size (Nano-Se) as a potential chemopreventive agent with reduced risk of selenium toxicity: comparison with se-methylselenocysteine in mice. *Toxicological Sciences* 101 (1): 22–31.

Zhang, Y., Li, X., Huang, Z. et al. (2013). Enhancement of cell permeabilization apoptosis-inducing activity of selenium nanoparticles by ATP surface decoration. *Nanomedicine: Nanotechnology, Biology, and Medicine* 9 (1): 74–84.

Zhang, J., Teng, Z., Yuan, Y. et al. (2018). Development, physicochemical characterization and cytotoxicity of selenium nanoparticles stabilized by beta-lactoglobulin. *International Journal of Biological Macromolecules* 107 (Pt B): 1406–1413.

Zhao, S., Yu, Q., Pan, J. et al. (2017a). Redox-responsive mesoporous selenium delivery of doxorubicin targets MCF-7 cells and synergistically enhances its anti-tumor activity. *Acta Biomaterialia* 54: 294–306.

Zhao, S.J., Wang, D.H., Li, Y.W. et al. (2017b). A novel selective VPAC2 agonist peptide-conjugated chitosan modified selenium nanoparticles with enhanced anti-type 2 diabetes synergy effects. *International Journal of Nanomedicine* 12: 2143–2160.

Zhao, Y., Sun, Q., Zhang, X. et al. (2018). Self-assembled selenium nanoparticles and their application in the rapid diagnostic detection of small cell lung cancer biomarkers. *Soft Matter* 14 (4): 481–489.

Zhu, C., Zhang, S., Song, C. et al. (2017). Selenium nanoparticles decorated with *Ulva lactuca* polysaccharide potentially attenuate colitis by inhibiting NF-κB mediated hyper inflammation. *Journal of Nanobiotechnology* 15 (1): 20.

3

Emerging Applications of Nanomaterials in the Diagnosis and Treatment of Gastrointestinal Disorders

Patrycja Golińska and Magdalena Wypij

Department of Microbiology, Nicolaus Copernicus University, Torun, Poland

3.1 Introduction

Nanotechnology provides different types of nanomaterials (NMs) which are built from various materials and show varied shapes, sizes, and chemical and surface properties (Laroui et al. 2013). Moreover, all such nanomaterials have been reported to have a broad spectrum of applications in industry, environmental protection and medicine. Several kinds of nanomaterials, namely metallic nanoparticles, quantum dots (QDs), silica nanospheres, magnetic nanoparticles, carbon nanotubes, graphene nanostructured surfaces, etc., were found to have attractive applications in diagnostic tests such as genotyping techniques, immunohistochemistry assays, detection of biomarkers, early cancer detection, and many others (Lyberopoulou et al. 2016). Nanomaterials have been also used as drug carriers in bioimaging and cancer treatment (Zottel et al. 2019). However, the balance between physical properties of nanomaterials, their biocompatibility, and the evidence of no cytotoxic effects is the key to their successful use in clinical applications. Nanomaterials can offer interesting interactions with biomolecules present on cell surfaces or inside the cell (Laroui et al. 2013). A particularly important feature is the configuration of the ligands and their interaction with the atoms present on the particle surface which play a significant role in determining the physiochemical properties of the nanomaterials and thus nanoparticle interaction with the human body and biological material (Bayford et al. 2017).

The application of nanotechnology in cancer research has provided hope within the scientific community for the development of novel cancer therapeutic strategies. Therefore, nanomaterials are being advanced as novel and more targeted treatments for diseases such as cancers which are difficult to manage (Zottel et al. 2019). Gastrointestinal (GI) diseases affect the GI tract, from the esophagus to the rectum, and the accessory digestive organs. These diseases include acute, chronic, recurrent, or functional disorders while covering a broad range of diseases, including the most common ones, namely acute and chronic inflammatory bowel diseases (IBDs) (Riasat et al. 2016). The GI cancers are especially dangerous as they contribute to more than 55% of deaths associated with cancer. Therefore, tremendous efforts have been made to develop the novel diagnostic and therapeutic methods for improving quality and life span of patient's (Laroui et al. 2013).

The GI tract is an attractive target system for nanotechnology applications. The GI tract is the site of adsorption of various compounds, including water, nutrients, or therapeutics. The behavior of

Nanobiotechnology in Diagnosis, Drug Delivery, and Treatment, First Edition. Edited by Mahendra Rai, Mehdi Razzaghi-Abyaneh, and Avinash P. Ingle.

nanomaterials used for diagnosis or therapy of GI diseases can be regulated during transport through the digestive tract according to conditions of varying pH, transit time, pressure, and bacterial content (Laroui et al. 2011). Of all nanomaterials, nanoparticles have shown great promise in gastroenterology because their interactions with intestinal epithelial cells, macrophages, immune cells, and M cells are tunable, suggesting their potential as a vehicle for vaccinations (Laroui et al. 2011).

In this chapter we present applications of nanomaterials in the diagnosis and treatment of GI disorders. The major types of nanoparticles that have potential use both in gastroenterology and general medicine are discussed; moreover, nanoparticle behavior in the GI tract is also discussed. The application of nanotechnology in medicine is a rapidly developing area of investigation. It is believed that nanotechnology will play an important role in the assessment and treatment of gastroenterological diseases. Some of the nanomaterial-based therapies and diagnostics presented here outperform conventional materials in terms of efficacy, reliability, and practicality.

3.2 Properties of Nanomaterials Affecting Their Potential Use in Medicine

Nanomaterials are characterized by their small size, commonly defined to be of diameter in the range of 1–100 nm and large surface area to volume ratio. However, in principle, NMs are described as materials with a length of 1–1000 nm in at least one dimension. Size is an important feature of nanomaterials as it affects their cellular uptake, physical properties, and interactions with biomolecules. It is observed that the smaller the size the easier the penetration of nanoparticles through the cell envelope (Jeevanandam et al. 2018). Kumar et al. (2016) reported that nanoparticles in the range of 1–10 nm have the capacity to diffuse into tumor cells. This helps to overcome limitations related to chemotherapy using free drugs such as poor *in vivo*/*in vitro* correlation and other possible resistances exhibited by tumors.

Powers et al. (2007) demonstrated that decrease in the size of any materials leads to an exponential increase in surface area to volume ratio, thereby making the nanomaterial surface more reactive to itself and to its contiguous environments. Moreover, it is suggested that size-dependent toxicity of nanoparticles can be attributed to its ability to enter into the biological systems and then modify the structure of various macromolecules, thereby interfering with critical biological functions (Lovrić et al. 2005; Aggarwal et al. 2009). Small particles in the size range of 5–110 nm can be used as potential carriers of anticancer drugs via intracellular drug delivery (Laroui et al. 2011). However, evaluation of other physicochemical properties of nanomaterials including surface area, solubility, chemical composition, shape, agglomeration state, crystal structure, surface energy, surface charge, surface morphology, and surface coating are essential for their safe use in clinical applications. Therefore, the role of individual, characteristic properties of nanomaterials in imparting toxic manifestations is so important (Gatoo et al. 2014).

Nanomaterials possess good stability and much longer shelf life compared with molecular carriers (Laroui et al. 2011). The drugs can be loaded into nanoparticles at a specific concentration, and such nanoconjugates may avoid digestive processes in the GI tract, which ultimately helps in efficient drug delivery at targeted sites. Moreover, the kinetics of drug release can be modulated, and nanomaterial surfaces may be modified with ligands to affect site-specific drug delivery (Laroui et al. 2011). Similarly, nanostructures can be conjugated to biological molecules, including hormones and antibodies, which enable their targeting to tissues expressing their cognate receptors (Fortina et al. 2007).

These capabilities of nanomaterials allow design and use of nanostructures in various fields of medicine including gastroenterology that help in diagnosis, bioimaging, and treatment processes, and can favorably compete with conventional methods (Laroui et al. 2011).

3.3 Nanomaterials Used in Diagnosis and Treatment of Gastrointestinal Disorders

3.3.1 Liposomes

Liposomes are nanostructures comprised of a lipid bilayer membrane surrounding an aqueous interior (Gaur et al. 2008). They may carry hydrophilic drugs inside the capsule or lipophilic drugs inserted into the phospholipid bilayer (Figure 3.1). Liposomes possess good biocompatibility because the raw materials that compose them are natural phospholipids, sterols, or glycerolipids, thus they may interfere with the cell membrane (Laroui et al. 2011; He et al. 2019). Liposome-based nanoparticles are commonly used nanoparticles for delivering small peptides, nucleic acids, and proteins in nanoplatform drug delivery (Huynh et al. 2009). They behave as a modified release system (Laroui et al. 2011). To date, several nanoliposomes have been developed for therapy of colorectal cancer (CRC), namely Doxorubicin (Doxil®) or Marqibo® which are examples of Food and Drug Administration (FDA)-approved nanoliposomes for chemotherapy of CRC (Barenholz 2012; Stang et al. 2012; Allen and Cullis 2013). Thermo-sensitive liposome doxorubicin (Thermodox®) is another promising nanoliposomal drug for colorectal liver metastases in combination with radiofrequency ablation. This nanoliposome with doxorubicin formulation releases the drug upon a mild hyperthermic trigger and can deliver 25-fold more doxorubicin into tumors than IV doxorubicin does (Stang et al. 2012).

3.3.2 Polymers

The polymers used as particulate vectors may be natural or synthetic, synthesized by standard polymerization chemical methods. The polymers used for diagnosis and treatment of diseases must be biocompatible, nontoxic, nonimmunogenic, and noncarcinogenic. They must also be (bio) degraded in the body, and their degradation products must be well tolerated and quickly eliminated (Laroui et al. 2011). Variable polymer chains can be used to form swollen nanosized structures called nano-size hydrogels or nanogels. These polymer chains are usually formed by

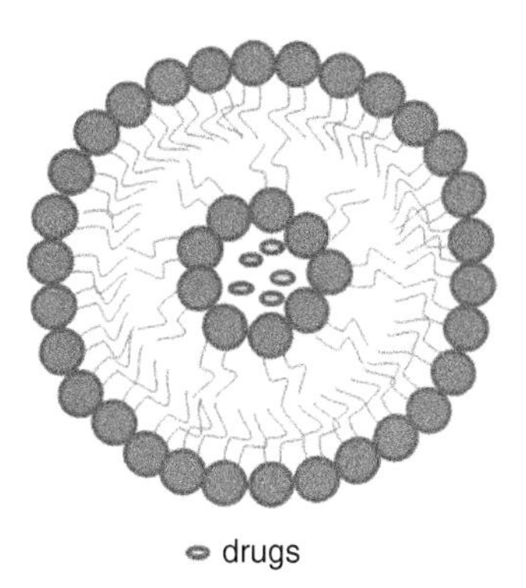

Figure 3.1 Nanoliposome as a carrier for drug delivery.

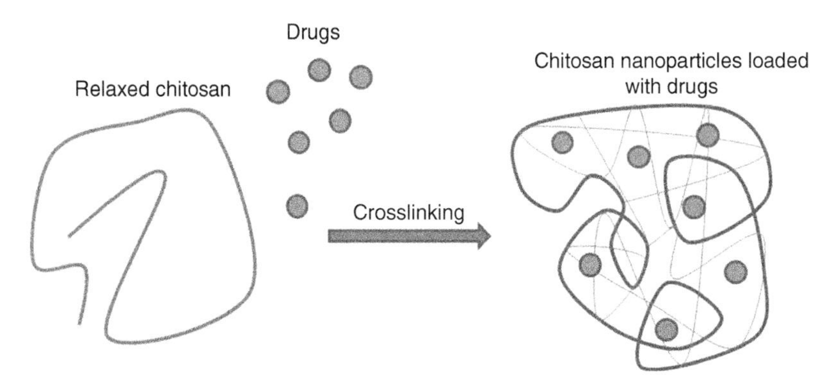

Figure 3.2 Chitosan nanoparticles loaded with drugs.

non-covalent interactions or covalent crosslinking. Nanogels, in addition to typical nanomaterial properties, e.g. large surface area to volume ratio, are also characterized by size tunability, controlled drug release profile, excellent drug loading capacity, and responsiveness to environmental stimuli. In nanomedicine, they have attracted significant attention as imaging labels and targeted drug delivery, while reducing systemic side effects (Arunraj et al. 2014). Nanogels may be tailored to exhibit exceptional stability, low cytotoxicity, and higher blood compatibility (Chacko et al. 2012; Kong et al. 2015). They showed potential to be an oral drug delivery system (Senanayake et al. 2013). The polysaccharide chitosan nanoparticles have been also used as drug delivery systems (Figure 3.2). Polymers can be conjugated with proteins, and such nanoconstructs display reduced immunogenicity, enhanced stability, and prolong plasma half-life (Ekladious et al. 2018). The conjugation of therapeutic agents to polymeric carriers offers improved drug solubilization, prolonged circulation, controlled release, and enhanced safety (Ekladious et al. 2019). The nano-sized multi-structural constructs of polymers with drugs displayed potential to improve pharmacological therapy of a variety of solid tumors. Polymer-drug nanoconstructs promote tumor targeting throughout the increased permeability and retention effect, and in the cells, following endocytic capture, allow lysosomotropic drug delivery (Lee 2006; Gaur et al. 2008; Greco and Vicent 2009).

3.3.3 Core-Shell Nanoparticles

The core-shell nanoparticles are composed of two or more materials which can be synthesized with different combinations of inorganic and organic materials. Their functionality and stability can be increased by coating. Superparamagnetic iron oxide nanoparticles (SPIONs) are one of the most common core-shell nanoparticles that are used in medical imaging and therapy (Damascelli et al. 2001; Gholami and Engel 2018). Their biocompatible polymer coating and core surface modification enable their use in nanomedicine and nuclear medicine applications. SPIONs exhibit magnetization only in an applied magnetic field and are able to load drugs and medical radioisotopes (due to their highly active surface) (Gholami and Engel 2018). There are key advantages of SPION drug delivery including longer circulation half-lives, improved pharmacokinetics, capability of carrying a large amount of drugs, reduction in side effects, and targeting the drug to a specific location in the body (Laroui et al. 2011).

3.3.4 Quantum Dots (QDs)

QDs are fluorescent nanocrystals produced from semiconductor materials with unique optical and electrical properties (Matea et al. 2017). QDs have drawn a lot of attention for their simplicity of synthesis and abundance of the raw material in nature. They are rich in carboxyl groups on their surface, therefore QDs can absorb a lot of single-strand carcinoembryonic antigen (CEA) aptamer through π-π stacking interactions, leading to effective fluorescence quenching (Zhu and Gao 2019). The utility of fluorescent properties of QDs for cancer targeting and imaging applications has been suggested in many studies (Laroui et al. 2013; Gao et al. 2014). Semiconductor nanoparticles can accumulate at a target site due to their enhanced permeability and retention at a tumor site. For example, fluorescent QDs conjugated to various peptides specifically target either the vasculature of normal tissues or, alternatively, cancer cells (Fortina et al. 2007). QDs were found to be useful in diagnosis of leishmaniasis, a parasitic disease caused by parasites of the *Leishmania* type (Andreadou et al. 2016). Authors developed a *Leishmania*-specific surface antigen and DNA detection methods based on a combination of magnetic beads and CdSe QDs with a test specificity of 100% and a low limit of detection of $3125\,\text{ng}\,\mu\text{l}^{-1}$ for *Leishmania* DNA and 10^3 cells ml^{-1} for *Leishmania* protein. Based on obtained results the authors concluded that this method showed considerable potential for clinical application in human and veterinary medicine.

3.4 Nanoparticle Uptake in the Gastrointestinal Tract

The GI tract is one of the portals for nanoparticles to get across the human body. However, inhaled nanoparticles can also be ingested by the GI tract once they are cleared through the respiratory tract (Hoet et al. 2004; Gaur et al. 2008). The kinetics of particle uptake in the GI tract depends on diffusion through the mucus layer, initial contact with enterocytes, cellular trafficking, and post-translocation events (Medina et al. 2007). Once ingested, nanoparticles readily penetrate the mucus layer and come into contact with enterocytes of the intestinal lining. The smaller the particle diameter is, the faster they can diffuse through the mucus layer and reach the colonic enterocytes. However, nanoparticles may escape from active uptake by enterocytes as they are scavenged by M-cells overlying the intestinal mucosa. Due to cellular transposition they can reach the bloodstream and distribute all over the body (Szentkuti 1997; Gaur et al. 2008).

It is suggested that similar to the lungs, the GI tract is also easily exposed to stimuli that can induce an inflammatory response. IBD, which is a group of inflammatory chronic disorders of the gut, can result from a combination of genetic predisposition and environmental factors (Podolsky 2002; Gaur et al. 2008). However, none of the published studies have reported direct toxicological effects of nanoparticles in the GI tract (Gaur et al. 2008).

On the one hand, the successful action of nanomaterials used for diagnosis or therapy of GI diseases depends heavily on their size, size distribution, morphology, hydrophilic–hydrophobic balance, and surface functionalization (Laroui et al. 2011). However, this action of nanomaterials also depends on conditions in each part of the digestive tract. These distinct conditions introduce many challenges to the application of therapeutics to GI tract. It should be highlighted that physicochemical properties and aggregation of nanomaterials will be also affected by co-ingested material present in the gut, namely food matrices, proteins, mucus, and bile acids secreted within the gut (Walczak et al. 2015; Bouwmeester et al. 2018). Studies by Peters et al. (2012) and Walczak et al. (2013) reported that properties of 60 nm silver nanomaterials and nanometer-sized silica were affected by the food matrix during transit before they were available for uptake in the small intestine.

The knowledge about the properties of nanomaterials is required for successful application of nanotechnology in diagnosis and therapy of GI tract disorders. The nanomaterials can be designed and their behavior regulated according to conditions of changing pH, transport time, pressure, enzyme-catalyzed degradation, and content of bacterial population to reach the target site (Laroui et al. 2011).

3.5 Gastrointestinal Disorders and Their Treatment with Nanomaterials

The digestive system consists of the GI tract, liver, pancreas, and gallbladder (Giau et al. 2019). This system allows the body to digest and break down the food into nutrients, which are subsequently used for energy, growth, and cell repair (Angsantikul et al. 2018). The digestive system diseases and disorders can be acute and last for a short time (e.g. various bacterial or viral infections), while others are chronic or long-lasting (e.g. cancers, *Helicobacter pylori* infection, etc.) (Giau et al. 2019). Irritable bowel syndrome (IBS) belongs to the functional GI disorder group as it shows a group of symptoms such as abdominal pain and changes in the pattern of bowel movements but without any evidence of underlying damage (Lacy et al. 2016). Some of the reports available proposed the use of nanodelivery systems as carriers for the delivery of active compounds in the treatment of IBS (Collnot et al. 2012; Xiao and Merlin 2012). These reports demonstrate promising results showing physiological changes in IBS, after application of such nanoconjugates with drugs and exploiting these differences to enhance specific delivery of drugs to affected tissue.

GI tract disorders are statistically noticed in 5 to 50% of patients with primary immunodeficiencies (PIDs) which include rare, chronic, and serious disorders of the immune system. Patients suffering from PIDs cannot mount a sufficiently protective immune response, leading to an increased susceptibility to infections. The gut is the largest lymphoid organ in the body, containing the majority of lymphocytes and producing large amounts of immunoglobulin (Ig), therefore patients with PIDs might suffer from various GI disorders caused by microorganisms or parasites. Dysfunction of the regulatory mechanisms responsible for the balance between active immunity and tolerance in the gut may lead to mucosal inflammation and damage, and also lead to GI diseases (Agarwal and Mayer 2013).

Overall, nanoparticle and nanomolecule drug-delivery mechanisms can be classified into active and passive targeting. Active targeting highly depends on the interaction between the target cell receptors and nanoparticles, whereas passive targeting relies on a number of factors such as longer biological half-life, long-circulating time at tumor locations, and the flow rate of nanoparticles to the impaired lymphatic system (Hong et al. 1999; Kaasgaard et al. 2001; Mishra et al. 2004; Gryparis et al. 2007; Romberg et al. 2008; Gholami and Engel 2018). Moreover, the effectiveness of the nanoplatform drug-delivery system is determined by the enhanced permeability, retention effects, and nanoparticle clearance by the mononuclear phagocyte system (Maeda et al. 2000; Greish 2007). The reticuloendothelial system (RES) effect is one of the most common problems among all different types of nanoparticles used for diagnosis and therapy of various diseases. The RES effect refers to the quick absorption of nanoparticles by macrophages which usually results in clearing nanoparticles from the circulation *in vivo* (Owens and Peppas 2006; Torchilin 2007; Howard et al. 2008). Therefore, modification of nanomaterials, e.g. specific types of nanoparticle coating, may prevent and minimize the RES effect (Gholami and Engel 2018). Kovacevic et al. (2011) reported that nanoparticles with surfactants or covalent linkage of polyoxyethylene have shown to effectively minimize the RES effect. Similarly, the size of nanoparticles was found to affect the delivery of conventional therapeutics to solid tumors. Nanoparticles larger than 500 nm are shown to be rapidly removed from the circulation *in vivo* (Maeda et al. 2000; Cho et al. 2009).

Although the application of diagnostics and therapeutics to the targeted sites along the GI tract and endoscopy are sometimes complicated, especially into the distal small intestine, it has been tried in many types of diseases, with varied success. The targeted delivery of therapeutic agents to the terminal ileum and colon was performed in the case of IBD and to the stomach in gastrin ulcers, while the theranostic probe was tried in the diagnostics of pancreatic, gastric and colonic cancers, or genes in gene therapy of gastric and colonic cancers (Jha et al. 2012). It suggests that application of nanomaterials in diagnosis and therapy of GI disorders is a very promising approach in nanomedicine.

3.6 Nanomaterials: Potential Treatment for Gastric Bacterial Infections

Gastric diseases are defined as diseases that affect the stomach. Inflammation of the stomach resulting from infections caused by any infectious agent is called gastritis. Moreover, when this condition affects various other parts of the GI tract, it is called gastroenteritis. Long-lasting (chronic) state of gastritis is associated with several diseases, including atrophic gastritis, pyloric stenosis, and gastric cancer. Another common disorder of the GI tract includes gastric ulceration or peptic ulcers. Ulceration damages the gastric mucosa, responsible for the protection of stomach tissues from the acids present inside this organ. The *H. pylori* infection is a common chronic infectious disease which is considered to be a main agent causing peptic ulcers (College et al. 2010). Besides peptic ulcer disease, it can lead to atrophic gastritis, gastric adenocarcinoma, and mucosa-associated lymphoid tissue lymphoma (Angsantikul et al. 2018). The eradication rates of *H. pylori* are far from desirable for infectious diseases (Lopes-De-Campos et al. 2019) as the resistance of *H. pylori* to antibiotics has reached alarming levels worldwide (Angsantikul et al. 2018). Nowadays many researchers are focused on developing rapid detection and efficient drug delivery systems to meet the challenge of antibiotic resistance. Therefore, conjugation of antibiotics with micro- or nanodelivery materials is considered one of the most promising strategies to improve the efficacy of conventionally used drugs (Giau et al. 2019). Moreover, such carriers acting as encapsulating agents can protect antibiotics from enzyme deactivation, resulting in an increase of the therapeutic effectiveness of the drug (Lopes-De-Campos et al. 2019). Development of nanoparticles that encapsulate multiple antibiotics for concurrent delivery has been reported by Angsantikul et al. (2018). For example, amoxicillin (AMX) antibiotic has been encapsulated in several delivery systems such as polymeric nanoparticles, gastro-retentive tablets, and liposomes (Lopes et al. 2014; Arif et al. 2018; Lopes-De-Campos et al. 2019).

Overall, several types of nanomaterials such as micelles (Ahmad et al. 2014), carbon nanomaterials (Al-Jumaili et al. 2017), magnetic nanoparticles (Tokajuk et al. 2017), mesoporous silica nanoparticles (Martinez-Carmona et al. 2018), polymer-based nanomaterials (Álvarez-Paino et al. 2017), and dendrimers (Mintzer et al. 2012) have been used as vehicles to carry antimicrobial drugs in various type of diseases. Similarly, the micro- and nanosized materials such as liposomes, dendrimers, peptides, polymer, and inorganic materials were found to be compatible with and enhanced the sensitivities of conventional diagnostic tests by several orders of magnitude (College et al. 2010). It is also suggested that drug-free nanomaterials that do not kill the pathogen but affect its virulent factors such as adhesins, toxins, or secretion systems can be used to decrease resistance of the pathogen and severity of the caused infection (Giau et al. 2019; Lopes-De-Campos et al. 2019).

Recently, *H. pylori* have been successfully eradicated using a variety of antibiotic-loaded chitosan micro- and nanoparticles (Giau et al. 2019). The proposed mechanism of bacterial cell inactivation is presented in Figure 3.3. Westmeier and coauthors (2018) studied the behavior of a panel of model nanoparticles, varying in size, shape, surface functionalization, and material on *H. pylori*. They noticed that conjugation of nanoparticles with bacteria did not require specific functionalization or negative surface charge and that assembly in interfering medium was significantly influenced by low pH which effected on the bacterial surface. Moreover, it was also reported that silica nanoparticles (25% surface coverage) did not show significant bactericidal activity but it was observed that these nanoparticles have the ability to attenuate the pathogenesis of *H. pylori* by inhibition of the type IV secretion system (Giau et al. 2019). Beside all of these reports of nanotechnology in combating bacterial gastric infections, still there is a major challenge for the scientific community to develop such nanomaterials that should be effective in the changing pH of the GI tract (Lopes et al. 2015). One of the such promising materials is the ternary copolymer poly (ethylene glycol)-block-poly (ami-nolated glycidyl methacrylate)-block-poly(2-(diisopropyl amino) ethyl methacrylate) (PEG-b-PAGA-b-PDPA) which was recently used as a pH-responsive micelle to target metastatic breast cancer (Giau et al. 2019).

Similarly, the efficacy of lipid nanoparticles (LNPs) against *H. pylori* was reported (Seabra et al. 2017). This type of nanomaterial is generally considered cost-effective, easily scaled up, and also biocompatible and biodegradable, which enhances its interest for commercial purposes (Giau et al. 2019). Thamphiwatana et al. (2013) studied the novel pH-responsive gold nanoparticle-stabilized liposome system for gastric antimicrobial delivery. The anionic phospholipid liposomes were stabilized by positively charged small chitosan-modified gold nanoparticles deposited on their surface. These liposome nanomaterials were stable in gastric acid while at neutral pH released gold stabilizers from their surfaces and gained the capability of fusing with bacterial membranes. The liposomes loaded with antibiotic (doxycycline) were used as a targeted delivery system to treat

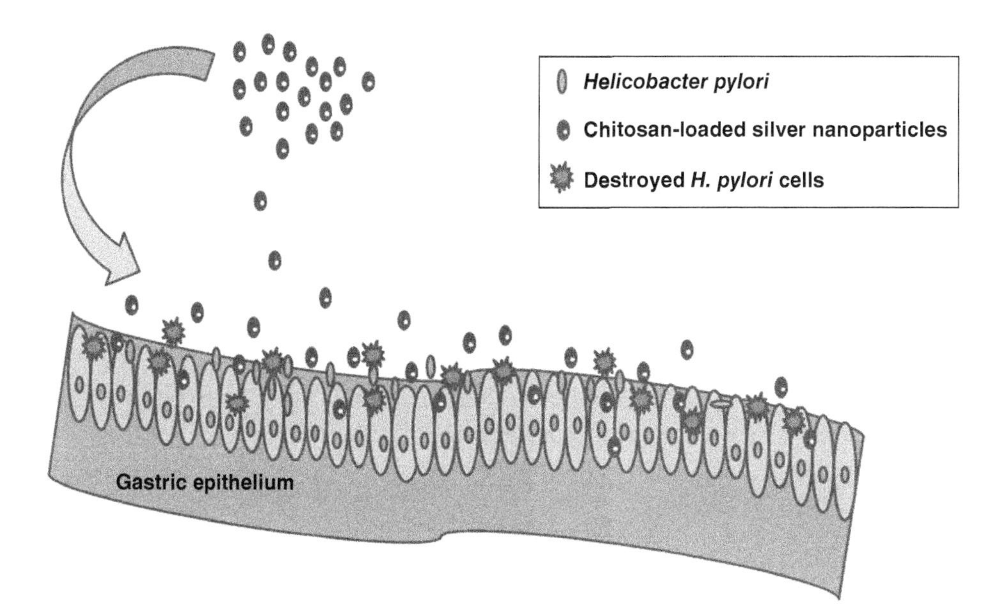

Figure 3.3 Mechanism of *Helicobacter pylori* inactivation in epithelium cells by chitosan nanoparticles loaded with antibiotic.

gastric pathogens such as *H. pylori*. Authors reported that the use of doxycycline-loaded liposomes, which rapidly fused with bacterial membranes, showed superior bactericidal efficacy compared to the use of free doxycycline. Based on obtained results the authors concluded that the liposome system has the potential for gastric drug delivery (Thamphiwatana et al. 2013).

The nanoparticles have been also conjugated with ligands such as mannose- or fucose-specific lectins in order to target the carbohydrate receptors on *H. pylori* cells (Umamaheshwari and Jain 2003). *H. pylori* is able to change the rheological properties of the mucus layer due to urease secretion (Celli et al. 2009), resulting in an increase of the pH and subsequently in reduction of the viscosity of the surrounding environment. Therefore, mucoadhesiveness of the gastric mucosa is considered as a target in treatment of *H. pylori* infections. Changes in mucosa viscosity may improve the time of contact between the drug and bacterial cells, and consequently, the efficacy against these bacterial infections (Lopes et al. 2015).

3.7 Nanostructures for Colon Cancer Diagnostics and Therapeutics

Colorectal cancer (CRC) is a severe health problem and has the third highest incidence of tumors in both males and females. The global incidence of CRC is about 1.3 million cases per year (Viswanath et al. 2016). However, according to estimates from the International Agency for Research on Cancer in 2018, globally, CRC constituted approximately 1.8 million new cases and 900 000 deaths annually (Keum and Giovannucci 2019). Thus, CRC is a common health problem due to its high prevalence and high mortality rate. Adjuvant and neo-adjuvant strategies, chemotherapy, and radiotherapy alone or in combination, have substantially improved survival and local recurrence rates. Their effectiveness remains limited due to the intrinsic build-up of resistance of cancer cells to chemotherapy drugs, dose-limiting toxicities and other major side effects (Gholami and Engel 2018). Colon cancer occurs due to certain routine factors and increasing age, with only some cases resulting from fundamental genetic disorders (Gulbake et al. 2016). Over the last few years, nanotechnology has improved available and developed novel methods for the detection and treatment of CRCs that cannot be achieved using the existing conventional technologies (Laroui et al. 2013; Viswanath et al. 2016). Moreover, targeted nanostructures offer potential solutions against the restrictions of standard chemo- and radiotherapy that may cause damages to normal tissues in proximity to and distant from tumors and clinical toxicities (Viswanath et al. 2016). Nanomaterials of different shapes, size, and compositions are considered as promising and important novel tools for CRC diagnosis, staging, and therapeutics (Dong et al. 2015).

There are some examples of the application of nanotechnology that directly address the *in vitro* diagnosis of CRC. Nanostructures, particularly nanoparticles, might be used as a label for direct visual detection. They offer multiplexing capabilities for detecting proteins or nucleic acids (Seydack 2004). Nanostructures may be also used for modifications of traditional contrast or imaging agents such as gadolinium or iron oxide, respectively, and thus enhance the diagnostic power of clinical imaging by using magnetic resonance (Laroui et al. 2013). Among the developing technologies that have potential in imaging GI diseases such as CRC, the use of near-infrared fluorescence (NIRF) imaging can be highlighted, especially if the current clinical evaluation of CRC already uses fiber optic examination of luminal surfaces (Weissleder et al. 2005). Application of NIRF imaging agents such as tunable QDs during endoscopic visualization can enhance standard imaging techniques used for diagnostics of GI tumors, including CRC (Viswanath et al. 2016). Vigor et al. (2010) reported that SPIONs could be used to target and image cancer cells if functionalized with recombinant single-chain Fv antibody fragments (scFv). Authors generated antibody-functionalized (abf) SPIONs using a scFv specific for CEA, an oncofoetal cell surface protein present in human tumor cells. Obtained results demonstrated that abf-SPIONs bound specifically to CEA-expressing human tumor cells, generated selective image contrast on magnetic resonance imaging (MRI) (Vigor et al. 2010). Similarly, Tiernan and coauthors (2015) studied binding of antibody-targeted fluorescent nanoparticles (CEA-targeted nanoparticles) to colorectal cancer cells *in vitro* and *in vivo* after systemic delivery to murine xenografts. They found that CEA-targeted, polyamidoamine dendrimer-conjugated nanoparticles allowed strong tumor-specific imaging and concluded that these nanoparticles have the potential to allow intra-operative fluorescent visualization of tumor cells (Tiernan et al. 2015).

Beyond diagnostics, targeted nanostructures offer opportunities to develop novel treatment approaches of tumor diseases (e.g. colorectal tumor). Indeed, nanostructures have been assembled with antibodies, proteins, and small-molecule ligands targeted to specific tumor-associated receptors to deliver chemotherapeutic agents. The targeted use of chemotherapeutics in animal tumor-xenograft models showed greater pharmacological and clinical efficacy and decreased adverse events as antibody-targeted liposomes effectively accumulate in CRC cells (Fortina et al. 2007). Nanogels containing 5-fluorouracil (5-FU) were assembled to be used as new colon-targeting drug carrier systems. This delivery system was characterized by its excellent pH sensitive release property and effectively reduced toxicity (Ashwanikumar et al. 2012). Moreover, potential application of activated nucleoside analogs for the treatment of drug-resistant tumors by oral delivery of nanogel drug was shown in studies of Senanayake et al. (2013).

Targeted nanoparticles as a drug delivery system based on monoclonal antibodies are currently one of the main approaches for CRC therapy under preclinical development (Brennan et al. 2004; Weinberg et al. 2005).

An enhanced antitumor activity of the photosensitizer *meso*-Tetra(N-methyl-4-pyridyl) porphine tetra tosylate (TMP) through encapsulation in antibody-targeted chitosan/alginate nanoparticles was reported by Abdelghany et al. (2013). TMP is a photosensitizer that can be used in photodynamic therapy to induce cell death through generation of reactive oxygen species in targeted tumor cells. However, TMP is highly hydrophilic, and therefore, its ability to accumulate intracellularly is limited. Thus, its encapsulation in chitosan/alginate nanoparticles improved TMP uptake into human colorectal carcinoma HCT116 cells, unlike TMP or nanoparticles used alone. Moreover, these nanoparticles were further conjugated with antibodies targeting death receptor 5 (DR5), the cell surface apoptosis-inducing receptor, upregulated in various types of cancer and found on HCT116 cells. This conjugation of nanoparticles had antibody-enhanced uptake and cytotoxic potency of the generated nanoparticles (Abdelghany et al. 2013). Similarly, Fay et al. (2011) demonstrated the induction of apoptosis in colorectal HCT116 cancer cells using Poly(lactic-co-glycolic acid) (PLGA) nanoparticles coated with Conatumumab (AMG 655) death receptor 5-specific antibodies (DR5-NP). This conjugate preferentially targeted DR5-expressing cells and presented a sufficient density of antibody paratopes to induce apoptosis via DR5, unlike free AMG 655 or non-targeted control nanoparticles. They also demonstrated that DR5-targeted nanoparticles encapsulating the cytotoxic drug camptothecin were effectively targeted to the tumor cells, affecting enhanced cytotoxicity through simultaneous drug delivery and apoptosis induction. The authors concluded that antibodies on nanoparticulate surfaces can be exploited for dual modes of action to enhance the therapeutic utility of the modality (Fay et al. 2011).

3.8 Conclusion and Future Perspectives

Nanotechnology is a multidisciplinary research field that integrates a broad and diverse array of equipments derived from chemistry, engineering, biology, and medicine. Nanomaterials have a wider range of potential applications for the detection and treatment of various diseases, including GI disorders, due to the suitable physical and chemical properties of nanoparticles for *in vivo* applications.

Nanostructures and nanotechnology-based devices are still under active development of the design of diagnostic and therapeutic tools and devices. It is particularly important in case of cancer diseases as the effectiveness of traditional anticancer therapies (chemotherapy, radiotherapy) or other adjuvant and neo-adjuvant strategies used alone and in combinations remain limited due to the intrinsic build-up of resistance of cancer cells to chemotherapy drugs, dose-limiting toxicities, and other major side effects. Therefore, new strategies to overcome these issues are being developed, one of which is cancer nanomedicine, a rapidly developing interdisciplinary research field.

The last few decades have seen a rapid growth of interest in utilizing nanoparticles and nanotechnology in cancer medicine, mainly for targeted drug delivery and imaging. Nanoscale imaging technology significantly improves the precision and accuracy of tumor diagnoses, while nanomaterial-based chemotherapeutic drug delivery's accuracy of dose reduces the toxic side effects. Overall, the use of nanomaterials in diagnosis and treatment of various diseases has been widely investigated both preclinically and clinically. However, the diagnosis and treatment of GI disorders is mainly preclinically related to *in vitro* (colorectal cell lines; e.g. HCT116 cells) or *in vitro* (mouse xenograft models) tests. Nanomedicine has been particularly considered as a novel solution to enhance CRC diagnosis and treatment, both separately and in combination with theranostic techniques. To date, the diagnostic and therapeutic potential of GI disorders has been reposted for liposome and polymer nanomaterials as well as QDs.

Although nanotechnology application could considerably improve GI disease detection and therapy, there are still many challenges that need to be addressed before it is accepted in routine clinical use, e.g. improvement of delivery and targeting in the body to provide effective treatment for specific disease conditions. The introduction of nanomaterial into the human body must be controlled, as there are many issues with possible toxicity and long-term effects which should be considered. However, it is expected that nanotechnology will continue bringing improvements to diagnostics and therapies of GI diseases.

References

Abdelghany, S.M., Schmid, D., Deacon, J. et al. (2013). Enhanced antitumor activity of the photosensitizer meso-tetra(N-methyl- 4-pyridyl) porphine tetra tosylate through encapsulation in antibody-targeted chitosan/alginate nanoparticles. *Biomacromolecules* 14: 302–310.

Agarwal, S. and Mayer, L. (2013). Diagnosis and treatment of gastrointestinal disorders in patients with primary immunodeficiency. *Clinical Gastroenterology and Hepatology* 11: 1050–1063.

Aggarwal, P., Hall, J.B., McLeland, C.B. et al. (2009). Nanoparticle interaction with plasma proteins as it relates to particle biodistribution, biocompatibility and therapeutic efficacy. *Advanced Drug Delivery Reviews* 61: 428–437.

Ahmad, Z., Shah, A., Siddiq, M., and Kraatz, H.B. (2014). Polymeric micelles as drug delivery vehicles. *RSC Advances* 4: 17028–17038.

Al-Jumaili, A., Alancherry, S., Bazaka, K., and Jacob, M.V. (2017). Review on the antimicrobial properties of carbon nanostructures. *Materials* 10 (1066): 1–26.

Allen, T.M. and Cullis, P.R. (2013). Liposomal drug delivery systems: from concept to clinical applications. *Advanced Drug Delivery Reviews* 65: 36–48.

Álvarez-Paino, M., Muñoz-Bonilla, A., and Fernández-García, M. (2017). Antimicrobial polymers in the nano-world. *Nanomaterials* 7: 48.

Andreadou, M., Liandris, E., Gazouli, M. et al. (2016). Detection of *Leishmania*-specific DNA and surface antigens using a combination of functionalized magnetic beads and cadmium selenite quantum dots. *Journal of Microbiological Methods* 123: 62–67.

Angsantikul, P., Thamphiwatana, S., Zhang, Q. et al. (2018). Coating nanoparticles with gastric epithelial cell membrane for targeted antibiotic delivery against *Helicobacter pylori* infection. *Advanced Therapeutics* 1: 1–9.

Arif, M., Dong, Q.J., Raja, M.A. et al. (2018). Development of novel pH-sensitive thiolated chitosan/PMLA nanoparticles for amoxicillin delivery to treat *Helicobacter pylori*. *Materials Science & Engineering, C: Materials for Biological Applications* 83: 17–24.

Arunraj, T.R., Sanoj, R.N., Kumar, A.N., and Jayakumar, R. (2014). Bio-responsive chitin-poly(L-lactic acid) composite nanogels for liver cancer. *Colloids and Surfaces B: Biointerfaces* 113: 394–402.

Ashwanikumar, N., Kumar, N.A., Nair, S.A., and Kumar, G.V. (2012). Methacrylic-based nanogels for the pH-sensitive delivery of 5-fluorouracil in the colon. *International Journal of Nanomedicine* 7: 5769–5779.

Barenholz, Y. (2012). Doxil®-the first FDA-approved nano-drug: lessons learned. *Journal of Controlled Release* 160: 117–134.

Bayford, R., Rademacher, T., Roitt, I., and Wang, S.X. (2017). Emerging applications of nanotechnology for diagnosis and therapy of disease: a review. *Physiological Measurement* 38: R183–R203.

Bouwmeester, H., van der Zande, M., and Jepson, M.A. (2018). Effects of food-borne nanomaterials on gastrointestinal tissues and microbiota. *Nanomedicine and Nanobiotechnology* 10: 1–12.

Brennan, F.R., Shaw, L., Wing, M.G., and Robinson, C. (2004). Preclinical safety testing of biotechnology-derived pharmaceuticals: understanding the issues and addressing the challenges. *Molecular Biotechnology* 27: 59–74.

Celli, J.P., Turner, B.S., Afdhal, N.H. et al. (2009). *Helicobacter pylori* moves through mucus by reducing mucin viscoelasticity. *Proceedings of the National Academy of Science* 106: 1421–1426.

Chacko, R.T., Ventura, J., Zhuang, J., and Thayumanavan, S. (2012). Polymer nanogels: a versatile nanoscopic drug delivery platform. *Advanced Drug Delivery Reviews* 64: 836–851.

Cho, M., Cho, W.S., Choi, M. et al. (2009). The impact of size on tissue distribution and elimination by single intravenous injection of silica nanoparticles. *Toxicology Letters* 189: 177–183.

College, N.R., Walker, B.R., and Ralston, S.H. (2010). *Davidson's Principles and Practice of Medicine*, 21e. Edinburgh: Churchill Livingstone/Elsevier.

Collnot, E.M., Ali, H., and Lehr, C.M. (2012). Nano and microparticulate drug carriers for targeting of the inflamed intestinal mucosa. *Journal of Controlled Release* 161: 235–246.

Damascelli, B., Cantù, G., Mattavelli, F. et al. (2001). Intraarterial chemotherapy with polyoxyethylated castor oil free paclitaxel, incorporated in albumin nanoparticles (ABI-007): phase I study of patients with squamous cell carcinoma of the head and neck and anal canal: preliminary evidence of clinical activity. *Cancer* 92: 2592–2602.

Dong, Z., Cui, M.Y., and Peng, Z. (2015). Nanoparticles for colorectal cancer-targeted drug delivery and MR imaging: current situation and perspectives. *Current Cancer Drug Targets* 30: 536–550.

Ekladious, I., Liu, R., Varongchayakul, N. et al. (2018). Reinforcement of polymeric nanoassemblies for ultra-high drug loadings, modulation of stiffness and release kinetics, and sustained therapeutic efficacy. *Nanoscale* 10: 8360–8366.

Ekladious, I., Colson, Y.L., and Grinstaff, M.W. (2019). Polymer–drug conjugate therapeutics: advances, insights and prospects. *Nature Reviews Drug Discovery* 18: 273–294.

Fay, F., McLaughlin, K.M., Small, D.M. et al. (2011). Conatumumab (AMG 655) coated nanoparticles for targeted proapoptotic drug delivery. *Biomaterials* 32: 8645–8653.

Fortina, P., Kricka, L.J., Graves, D.J. et al. (2007). Applications of nanoparticles to diagnostics and therapeutics in colorectal cancer. *Trends in Biotechnology* 25: 145–152.

Gao, W., Thamphiwatana, S., Angsantikul, P., and Zhang, L. (2014). Nanoparticle approaches against bacterial infections. *Wiley Interdisciplinary Reviews: Nanomedicine and Nanobiotechnology* 6: 532–547.

Gatoo, M.A., Naseem, S., Arfat, M.Y. et al. (2014). Physicochemical properties of nanomaterials: implication in associated toxic manifestations. *BioMed Research International* 2014: 1–8.

Gaur, A., Midha, A., and Bhatia, A.L. (2008). Significance of nanotechnology in medical sciences. *Asian Journal of Pharmaceutics* 2008: 80–85.

Gholami, Y.H. and Engel, A. (2018). Theranostic nanoplatforms for treatment and diagnosis of rectal and colon cancer: a brief review. *Mini-invasive Surgery* 2 (44): 1–9.

Giau, V.V., An, S.S.A., and Hulme, J. (2019). Recent advances in the treatment of pathogenic infections using antibiotics and nano-drug delivery vehicles. *Drug Design, Development and Therapy* 13: 327–343.

Greco, F. and Vicent, M.J. (2009). Combination therapy: opportunities and challenges for polymer-drug conjugates as anticancer nanomedicines. *Advanced Drug Delivery Reviews* 61: 1203–1213.

Greish, K. (2007). Enhanced permeability and retention of macromolecular drugs in solid tumors: a royal gate for targeted anticancer nanomedicines. *Journal of Drug Targeting* 15: 457–464.

Gryparis, E.C., Hatziapostolou, M., Papadimitriou, E., and Avgoustakis, K. (2007). Anticancer activity of cisplatin-loaded PLGA-mPEG nanoparticles on LNCaP prostate cancer cells. *European Journal of Pharmaceutics and Biopharmaceutics* 67: 1–8.

Gulbake, A., Jain, A., Jain, A. et al. (2016). Insight to drug delivery aspects for colorectal cancer. *World Journal of Gastroenterology* 22: 582–599.

He, H., Lu, Y., Qia, J. et al. (2019). Adapting liposomes for oral drug delivery. *Acta Pharmaceutica Sinica B* 9: 36–48.

Hoet, P., Bruske-Hohlfeld, I., and Salata, O. (2004). Nanoparticles-known and unknown health risks. *Journal of Nanobiotechnology* 2: 1–15.

Hong, R.L., Huang, C.J., Tseng, Y.L. et al. (1999). Direct comparison of liposomal doxorubicin with or without polyethylene glycol coating in C-26 tumor-bearing mice: is surface coating with polyethylene glycol beneficial? *Clinical Cancer Research* 5: 3645–3652.

Howard, M.D., Jay, M., Dziubla, T.D., and Lu, X. (2008). PEGylation of nanocarrier drug delivery systems: state of the art. *Journal of Biomedical Nanotechnology* 4: 133–148.

Huynh, N.T., Passirani, C., Saulnier, P., and Benoit, J.P. (2009). Lipid nanocapsules: a new platform for nanomedicine. *International Journal of Pharmaceutics* 379: 201–209.

Jeevanandam, J., Barhoum, A., Chan, J.S. et al. (2018). Review on nanoparticles and nanostructured materials: history, sources, toxicity and regulations. *Beilstein Journal of Nanotechnology* 9: 1050–1074.

Jha, A.K., Goenka, M.K., Nijhawan, S. et al. (2012). Nanotechnology in gastrointestinal endoscopy: a primer. *Journal of Digestive Endoscopy* 3: S77–S80.

Kaasgaard, T., Mouritsen, O.G., and Jørgensen, K. (2001). Screening effect of PEG on avidin binding to liposome surface receptors. *International Journal of Pharmaceutics* 214: 63–65.

Keum, N. and Giovannucci, E. (2019). Global burden of colorectal cancer: emerging trends, risk factors and prevention strategies. *Nature Reviews Gastroenterology & Hepatology* https://doi.org/10.1038/s41575-019-0189-8.

Kong, S.H., Noh, Y.W., Suh, Y.S. et al. (2015). Evaluation of the novel near-infrared fluorescence tracers pullulan polymer nanogel and indocyanine green/gamma-glutamic acid complex for sentinel lymph node navigation surgery in large animal models. *Gastric Cancer* 18: 55–64.

Kovacevic, A., Savic, S., Vuleta, G. et al. (2011). Polyhydroxy surfactants for the formulation of lipid nanoparticles (SLN and NLC): effects on size, physical stability and particle matrix structure. *International Journal of Pharmaceutics* 406: 163–172.

Kumar, P.S., Datta, M.S., Kumar, D.M. et al. (2016). Potential application of dendrimers in drug delivery: a concise review and update. *Journal of Drug Delivery and Therapeutics* 6: 71–88.

Lacy, B.E., Mearin, F., Chang, L. et al. (2016). Bowel disorders. *Gastroenterology* 150: 1393–1407.

Laroui, H., Wilson, D.S., Dalmasso, G. et al. (2011). Nanomedicine in GI. *American Journal of Physiology - Gastrointestinal and Liver Physiology* 300: G371–G383.

Laroui, H., Rakhya, P., Xiao, B. et al. (2013). Nanotechnology in diagnostics and therapeutics for gastrointestinal disorders. *Digestive and Liver Disease* 45: 995–1002.

Lee, L.J. (2006). Polymer nano-engineering for biomedical applications. *Annals of Biomedical Engineering* 34: 75–88.

Lopes, D., Nunes, C., Martins, M.C. et al. (2014). Eradication of *Helicobacter pylori*: past, present and future. *Journal of Controlled Release* 189: 169–186.

Lopes, D., Nunes, C., Martins, M.C.L. et al. (2015). Targeting strategies for the treatment of *Helicobacter pylori* infections. In: *Nano Based Drug Delivery* (ed. J. Naik), 339–366. Zagreb, Croatia: IAPC Publishing.

Lopes-de-Campos, D., Pinto, R.M., Lima, S.A. et al. (2019). Delivering amoxicillin at the infection site – a rational design through lipid nanoparticles. *International Journal of Nanomedicine* 14: 2781–2795.

Lovrić, J., Bazzi, H.S., Cuie, Y. et al. (2005). Differences in subcellular distribution and toxicity of green and red emitting CdTe quantum dots. *Journal of Molecular Medicine* 83: 377–385.

Lyberopoulou, A., Efstathopoulos, E.P., and Gazouli, M. (2016). Nanotechnology-based rapid diagnostic tests. In: *Proof and Concepts in Rapid Diagnostic Tests and Technologies* (ed. K.S. Saxena), 89–105. London: IntechOpen.

Maeda, H., Wu, J., Sawa, T. et al. (2000). Tumor vascular permeability and the EPR effect in macromolecular therapeutics: a review. *Journal of Controlled Release* 65: 271–284.

Martinez-Carmona, M., Gunko, Y.K., and Vallet-Regi, M. (2018). Mesoporous silica materials as drug delivery: the nightmare of bacterial infection. *Pharmaceutics* 10 (279): 1–29.

Matea, C.T., Mocan, T., Tabaran, F. et al. (2017). Quantum dots in imaging, drug delivery and sensor applications. *International Journal of Nanomedicine* 12: 5421–5431.

Medina, C., Santos-Martinez, M.J., Radomski, A. et al. (2007). Nanoparticles: pharmacological and toxicological significance. *British Journal of Pharmacology* 150: 552–558.

Mintzer, M.A., Dane, E.L., O'Toole, G.A., and Grinstaff, M.W. (2012). Exploiting dendrimer multivalency to combat emerging and re-emerging infectious diseases. *Molecular Pharmaceutics* 9: 342–354.

Mishra, S., Webster, P., and Davis, M.E. (2004). PEGylation significantly affects cellular uptake and intracellular trafficking of non-viral gene delivery particles. *European Journal of Cell Biology* 83: 97–111.

Owens, D.E. and Peppas, N.A. (2006). Opsonization, biodistribution, and pharmacokinetics of polymeric nanoparticles. *International Journal of Pharmaceutics* 307: 93–102.

Peters, R., Kramer, E., Oomen, A.G. et al. (2012). Presence of nano-sized silica during *in vitro* digestion of foods containing silica as a food additive. *ACS Nano* 6: 2441–2451.

Podolsky, D.K. (2002). Inflammatory bowel disease. *The New England Journal of Medicine* 347: 417–429.

Powers, K.W., Palazuelos, M., Moudgil, B.M., and Roberts, S.M. (2007). Characterization of the size, shape, and state of dispersion of nanoparticles for toxicological studies. *Nanotoxicology* 1: 42–51.

Riasat, R., Guangjun, N., Riasat, N. et al. (2016). Effects of nanoparticles on gastrointestinal disorders and therapy. *Journal of Clinical Toxicology* 6 (313): 1–10.

Romberg, B., Hennink, W.E., and Storm, G. (2008). Sheddable coatings for long-circulating nanoparticles. *Pharmaceutical Research* 25: 55–71.

Seabra, C.L., Nunes, C., and Gomez-Lazaro, M. (2017). Docosahexaenoic acid loaded lipid nanoparticles with bactericidal activity against *Helicobacter pylori*. *International Journal of Pharmaceutics* 519: 128–137.

Senanayake, T.H., Warren, G., Wei, X., and Vinogradov, S.V. (2013). Application of activated nucleoside analogs for the treatment of drug-resistant tumors by oral delivery of nanogel-drug conjugates. *Journal of Controlled Release* 167: 200–209.

Seydack, M. (2004). Nanoparticle labels in immunosensing using optical detection methods. *Biosensors and Bioelectronics* 20: 2454–2469.

Stang, J., Haynes, M., Carson, P., and Moghaddam, M. (2012). A preclinical system prototype for focused microwave thermal therapy of the breast. *IEEE Transactions on Biomedical Engineering* 59: 2431–2438.

Szentkuti, L. (1997). Light microscopical observations on luminally administered dyes, dextrans, nanospheres and microspheres in the pre-epithelial mucus gel layer of the rat distal colon. *Journal of Controlled Release* 46: 233–242.

Thamphiwatana, S., Fu, V., Zhu, J. et al. (2013). Nanoparticle-stabilized liposomes for pH-responsive gastric drug delivery. *Langmuir* 29: 12228–12233.

Tiernan, J.P., Ingram, N., Marston, G. et al. (2015). CEA-targeted nanoparticles allow specific in vivo fluorescent imaging of colorectal cancer models. *Nanomedicine (London, England)* 10: 1223–1231.

Tokajuk, G., Niemirowicz, K., Deptuła, P. et al. (2017). Use of magnetic nanoparticles as a drug delivery system to improve chlorhexidine antimicrobial activity. *International Journal of Nanomedicine* 12: 7833–7846.

Torchilin, V.P. (2007). Targeted pharmaceutical nanocarriers for cancer therapy and imaging. *The AAPS Journal* 9: E128–E147.

Umamaheshwari, R.B. and Jain, N.K. (2003). Receptor mediated targeting of lectin conjugated gliadin nanoparticles in the treatment of *Helicobacter pylori*. *Journal of Drug Targeting* 11: 415–424.

Vigor, K.L., Kyrtatos, P.G., Minogue, S. et al. (2010). Nanoparticles functionalized with recombinant single chain Fv antibody fragments (scFv) for the magnetic resonance imaging of cancer cells. *Biomaterials* 31: 1307–1315.

Viswanath, B., Kim, S., and Lee, K. (2016). Recent insights into nanotechnology development for detection and treatment of colorectal cancer. *International Journal of Nanomedicine* 11: 2491–2504.

Walczak, A.P., Fokkink, R., Peters, R. et al. (2013). Behaviour of silver nanoparticles and silver ions in an *in vitro* human gastrointestinal digestion model. *Nanotoxicology* 7: 1198–1210.

Walczak, A.P., Kramer, E., Hendriksen, P.J. et al. (2015). *In vitro* gastrointestinal digestion increases the translocation of polystyrene nanoparticles in an in vitro intestinal co-culture model. *Nanotoxicology* 9: 886–894.

Weinberg, W.C., Frazier-Jessen, M.R., Wu, W.J. et al. (2005). Development and regulation of monoclonal antibody products: challenges and opportunities. *Cancer Metastasis Reviews* 24: 569–584.

Weissleder, R., Kelly, K., Sun, E.Y. et al. (2005). Cell-specific targeting of nanoparticles by multivalent attachment of small molecules. *Nature Biotechnology* 23: 1418–1423.

Westmeier, D., Posselt, G., Hahlbrock, A. et al. (2018). Nanoparticle binding attenuates the pathobiology of gastric cancer-associated *Helicobacter pylori*. *Nanoscale* 10: 1453–1463.

Xiao, B. and Merlin, D. (2012). Oral colon-specific therapeutic approaches toward treatment of inflammatory bowel disease. *Expert Opinion in Drug Delivery* 9: 1393–1407.

Zhu, X. and Gao, T. (2019). Spectrometry. In: *Nano-Inspired Biosensors for Protein Assay with Clinical Applications* (ed. G. Li), 237–264. Amsterdam, Netherlands: Elsevier.

Zottel, A., Videtič, A., and Jovcevska, A. (2019). Nanotechnology meets oncology: nanomaterials in brain cancer research, diagnosis and therapy. *Materials* 12: 1–28.

4

Nanotheranostics: Novel Materials for Targeted Therapy and Diagnosis

Sougata Ghosh[1] *and Rohini Kitture*[2]

[1] *Department of Microbiology, School of Science, RK University, Rajkot, Gujarat, India*
[2] *Springer Nature Scientific Technology and Publishing Solutions, Magarpatta City, Pune, Maharashtra, India*

4.1 Introduction

Nanotechnology has promising applications in medicine, which has inspired the global scientific community to explore novel nanomedicine. Various exotic nanostructures are multifunctionalized by a rational combinatorial approach to make them applicable in both therapy and diagnosis, popularly termed as nanotheranostics. Poor target tissue specificity and severe adverse effects are the major drawbacks of conventional chemotherapy that have encouraged fabrication of clinically significant theranostics via systematic integration of imaging probes and chemotherapeutic drugs to a single nanostructure-based cargo (Wang et al. 2015; Coates et al. 1983; Liu et al. 2014a,b).

The notable advantage of such nanotheranostics is the spatial control of the therapeutic effect only in the site of interest by external stimuli such as light, magnetic field, X-ray, radiofrequency, and ultrasound (Guan et al. 2015; Chen et al. 2015a; Cheng et al. 2014; Mikhaylov et al. 2011; Sasidharan et al. 2015; Chen and Du 2013). Novel nanomaterials such as nanoliposomes, fullerenes, nanotubes, magnetic nanomaterials, polymeric micelles, dendrimers, nanoshells, and polymeric microspheres are fabricated for targeted drug delivery, triggered release, enhanced tissue localization, and bioimaging for treatment of cancer and diseases related to heart, lung, and blood (Partha and Conyers 2009).

Nanocarriers are able to carry therapeutic payloads to biogenic sites which can also be exploited for gene therapy, photothermal, and photodynamic therapy (PDT) in addition to biosensing (Partha et al. 2008). Preclinical evaluations of several nanocomposites are underway to study biodistribution, excretion, histocompatibility, hemocompatibility, cytotoxicity, and biocompatibility to define the effective dosage, appropriate route of administration, and the period of treatment (Kale et al. 2017; Ghosh 2018). Biocompatible inorganic mesoporous materials possess large surface area, high pore volume, and tunable pore sizes that serve as reservoirs for guest molecules and exhibit sustained drug-release profiles (Ghosh 2019). Hybrid nanotheranostics are composed of a targeting agent such as antibody (Ab) or aptamer (Ap), imaging agent (e.g. radionuclide, fluorophore), and drugs for simultaneous targeted detection and therapy. Novel theranostic agents with tunable biodegradability can be useful to various diagnostic modalities that include single-photon emission computed tomography (SPECT), positron emission tomography (PET), computed tomography

Nanobiotechnology in Diagnosis, Drug Delivery, and Treatment, First Edition. Edited by Mahendra Rai, Mehdi Razzaghi-Abyaneh, and Avinash P. Ingle.

(CT), magnetic resonance imaging (MRI), and optical imaging (Kitture and Ghosh 2019). Imaging probes labeled nanohybrids are often used for selective macrophage recognition.

Nanoparticles are rapidly engulfed by circulating monocytes, splenic macrophages, and disease-associated macrophages owing to their intrinsic phagocytic ability. Dextran-coated gold nanorods (AuNR) or single-walled carbon nanotubes are reported for selective killing of plaque macrophages by PDT and photosensitizer-mediated photothermal therapy (PTT) (Kosuge et al. 2012). More recently, macrophages of the M1 inflammatory phenotype have been selectively targeted by model hybrid lipid–latex (LiLa) nanoparticles functionalized with hydrophobic entities such as drug cargos, signaling lipids, and imaging agents. Theranostic LiLa formulation with gadolinium, fluorescein, and "eat-me" phagocytic signals (Gd-FITC-LiLa) has exhibited high MRI sensitivity, efficient drug loading percentage, and selective targeting of inflammatory M1 macrophages concomitant with phospholipase A2 activity-dependent controlled release of anti-inflammatory drug.

These particles allow noninvasive imaging of atherosclerotic plaque by MRI (Bagalkot et al. 2015). Numerous bioactive metals like gold, silver, copper, platinum, palladium, and their bimetallic nanoalloys with antimicrobial, antifungal, anticancer, antidiabetic, antioxidant, anti-inflammatory, antiprotozoal, and anti-hyperlipidemia activity are considered potential nanomaterials which can be used as key constituents of engineered nanotheranostics (Shinde et al. 2018; Shende et al. 2018; Bhagwat et al. 2018; Rokade et al. 2017; Shende et al. 2017; Ghosh 2018; Ghosh et al. 2015a; Adersh et al. 2015; Ghosh et al. 2015b; Jamdade et al. 2019; Kitture et al. 2015b).

This chapter focuses on the theranostic application of hybrid nanomaterials of various metallic and non-metallic origins with application in targeted drug delivery, sustained drug release and multimodal imaging.

4.2 Magnetic Nanostructures

Magnetic nanoparticles have been one of the favorite nanoscale materials owing to their cost-effectiveness, inherent magnetic properties, and their biocompatibility, enabling their application not only as MRI contrast agents but also for targeted drug delivery and hyperthermia applications (Ghosh et al. 2015c; Kitture et al. 2012). Moreover, magnetic nanoparticle-assisted theranostics have been advantageous over the conventional optical-dependent therapies which suffer from penetration depth issues. The deeper tissue penetration and stability make the MRI based magnetic nanoparticles favorable compared to the conventional counterparts such as organic dyes. Carr et al. (1984) reported the first MRI agent in human volunteers. The material used as MRI agent was gadolinium-diethylenetriamine penta-acetic acid, also known as Gd-DTPA, one of the most commonly used T_1 contrast agents. The chemical interaction between the water molecules and the paramagnetic molecules (Gd^{3+}) gives rise to the T_1-weighted signal, wherein the unpaired seven electrons from Gd^{3+} contribute to the strong magnetic field, affecting the water protons in close proximity which causes modulations in the relaxation times along the longitudinal axis (spin-lattice relaxation). Herein, it is interesting to note that the seven unpaired electrons play a crucial role in generating longer signaling time. Thus Gd^{3+} has been one of the most conventional MRI contrast agents. The hydrophilic nature of the Gd^{3+} ions help them not to pass through blood-brain barrier, thus beneficial for enhancing the contrast of brain vessels and tumors. Moreover, the potential gadolinium leakage can be avoided by tuning the physicochemical properties, offering easy excretion through renal clearance. Nevertheless, stability of the ions needs to be improved for promising applications. Table 4.1 represents the list of commercially available Gd^{3+} chelate T_1-weighted MRI contrast agents and their in vivo targets.

Table 4.1 Examples of commercially available Gd^{3+} chelate T_1-Weighted MRI contrast agents and their *in vivo* targets.

Sr. no.	Commercial gadolinium-based contrast agents	Organs or tissues
1	Gadopentetate dimeglumine (Magnevist®) Gadoterate meglumine (Dotarem®)	Central nervous system (CNS) (for blood-brain barrier, tumor, or spine imaging), whole body
2	Gadoteridol (ProHance®) Gadodiamide (Omniscan®)	CNS, abdominal cavities
3	Gadobutrol (Gadovist®)	CNS
4	Gadobenate dimeglumine (MultiHance®)	CNS and liver
5	Gadoversetamide (OptiMark®) Gadoxetic acid (Primovist® or Eovist®)	Liver
6	Gadofosveset (Vasovist®)	Abdominal cavities, limb vessels, vascularization

Source: Reprinted with permission from Wallyn et al. (2019).

Besides gadolinium phosphates ($GdPO_4$), gadolinium oxides (Gd_2O_3), GdF_3:$CeFn_3$ and some other non-lanthanides such as Mn^{2+}, Fe^{3+}, Cu^{2+} have also been extensively studied for T_1 contrast imaging, offering brighter images (Smith and Gambhir 2017). Wallyn et al. (2019) have reviewed current developments in the major imaging techniques including MRI. On the other hand, iron oxide nanoparticles (IONPs) have favored for their T_2 weighted negative contrast. With the advancements in nanoscience, the properties of these nanoparticles could be tuned to archive expected results. Table 4.2 lists the commercially available IONP-based (along with their sizes) T_2-weighted MRI contrast agents, and their *in vivo* targets. This list also includes some other inorganic nanoparticles viz. cobalt (Co), Au@Fe, Au@Pt, Pt@Fe, Au@Ag, Au@Fe_2O_3, Au@Co, Au@TiO_2, etc.(Blasiak et al. 2013).

The past few decades have witnessed tremendous developments in the nanomaterials for MRI and image-guided therapeutics, which is not limited to only cancer theranostics. These developments offered tailoring the shape and size of the nanoparticles, functionalizing them to offer multifunctional sophisticated structures that overcome the limitations of the conventional systems, e.g. limited specificity, resistance by the immune system, and high toxicity, and further offer controlled, external stimuli-assisted drug delivery at the desired site. For example, liposomes help the contrast agents to overcome their rapid clearance, non-specific cellular interaction which usually leads to low-contrast images. The contrast agents can either be entrapped within the core of the liposome or dispersed within the bilayer. When entrapped within the core, the hindered exchange of bulk water protons with the contrast agents causes reduced relativities, thus hampering the efficacy of the system. This limited exchange is addressed by incorporating hydrophobic contrast agents into the lipid bilayer. To enhance the blood circulation, micelles have been preferred over the liposomes, wherein the efficacy of the system could be tuned through surface modification. Besides these, dendrimers, polymer-modified nanoparticles, have also been in focus for their application in MRI. Mody et al. (2009) have reviewed modification of the nanoparticles for their enhanced imaging used for gynecological malignancies (Mody et al. 2009). Several studies reported the use of nanoparticles for imaging of atherosclerotic plaques, cardiovascular angiography, etc., (Schalla et al. 2002).

Magnetic field-assisted drug targeting has been extensively studied owing to its advantages over conventional drug delivery. For example, magnetic fields can enhance the dye delivery efficacy

Table 4.2 Examples of commercially available IONP-based T_2-weighted MRI contrast agents, their *in vivo* targets, and the size of the IONPs included within each contrast agents' formulation.

Sr. no.	Commercial iron-oxide based contrast agents	Target	IONPs size (Ø)
1	Ferumoxsil (Lumirem® or GastroMark®) – Silicon coating Ferristene (Abdoscan®) – Sulfonated styrene-divinylbenzene copolymer citrate coating	Bowel, lumen organs	300 nm > Ø > 3.5 μm
2	Ferumoxide (Endorem®) – Dextran coated-Fe_3O_4	Liver, spleen	80 nm < Ø < 180 nm
3	Ferucarbotran (Resovist®) – Carboxyldextran coating		Ø = 60 nm
4	Ferumoxtran-10 (Sinerem® or Combinex®) – Dextran coating	Lymph node	20 nm < Ø < 40 nm
5	Feruglose (Clariscan®) – PEG starch coating	Bone marrow, perfusion, vessel	Ø = 20 nm

Source: Reprinted with permission from Wallyn et al. (2019).

from 0.14 to ~0.22 ng per 1 g of tumor tissue, in the nude mouse xenografted with MCF-7 breast cancer cells, as reported by Foyet al. (2010). The study was based on encapsulation of the commercial dye molecules and a magnetite nanoparticle into a pluronic F127 polymer. This drug-delivery vehicle was injected into the nude mouse bearing MCF-7 breast cancer cells. In another interesting study, Hayashi and group reported the synthesis of smart magnetic nanoparticles showing hyperthermia in the alternating magnetic field followed by release of doxorubicin (DOX) anti-cancer drugs (Hayashi et al. 2014). The IONPs have been modified with polypyrrole (PPy), polyethylene glycol (PEG), and folic acid (FA) and loaded with DOX. On the application of alternating current magnetic field, the Fe_3O_4/DOX/PPy-PEG-FA produces heat and sequentially releases DOX, thus curing the tumor. Figure 4.1 shows the synergistic effect of magnetic hyperthermia and drug release on female CB17/Icr-*Prkdcscid* mice. Moreover, there were no reports of tumor malignancy. The synergistic effect was accompanied by no toxicity, indicating the potential use of nanoparticles for effective therapeutics. Readers are encouraged to refer to recent reviews on the magnetic nanoparticles for theranostics applications for further details (Sun et al. 2019; Jahangirian et al. 2019; Hayashi et al. 2014; Lisjak and Mertelj 2018).

Figure 4.1 (a) Photograph and thermal image of a mouse exposed to ACMF for 20 minutes after injection with Fe_3O_4/DOX/PPy-PEG-FA NPs. (b) Average change of the tumor temperature of the mice injected with Fe_3O_4/DOX/PPy-PEG-FA NPs, Fe_3O_4/PPy-PEG-FA NPs, and no NPs with respect to ACMF exposure time (n = 5). (c) Follow-up photographs of mouse exposed to ACMF for 20 minutes after injection with Fe_3O_4/DOX/PPy-PEG-FA NPs. Change of (d) tumor volume, (e) survival rate, and (f) body weight: non-treated mice (black), mice treated with chemotherapy (yellow), mice exposed to ACMF (green), mice injected with Fe_3O_4/DOX/PPy-PEG-FA NPs intratumorally (purple), mice treated with MHT (blue), and mice treated with the combination of MHT and chemotherapy (red). The inset in Fig. 3D shows the magnified view for the first 12 days after treatment. (g) Photographs of non-treated mice, mice treated with chemotherapy, mice exposed to ACMF, mice injected with Fe_3O_4/DOX/PPy-PEG-FA NPs intratumorally, mice treated with MHT, and mice treated with the combination of MHT and chemotherapy 45 days after treatment. *Source:* Reprinted with permission from Hayashi et al. (2014); Copyright © 2014 Ivyspring International Publisher.

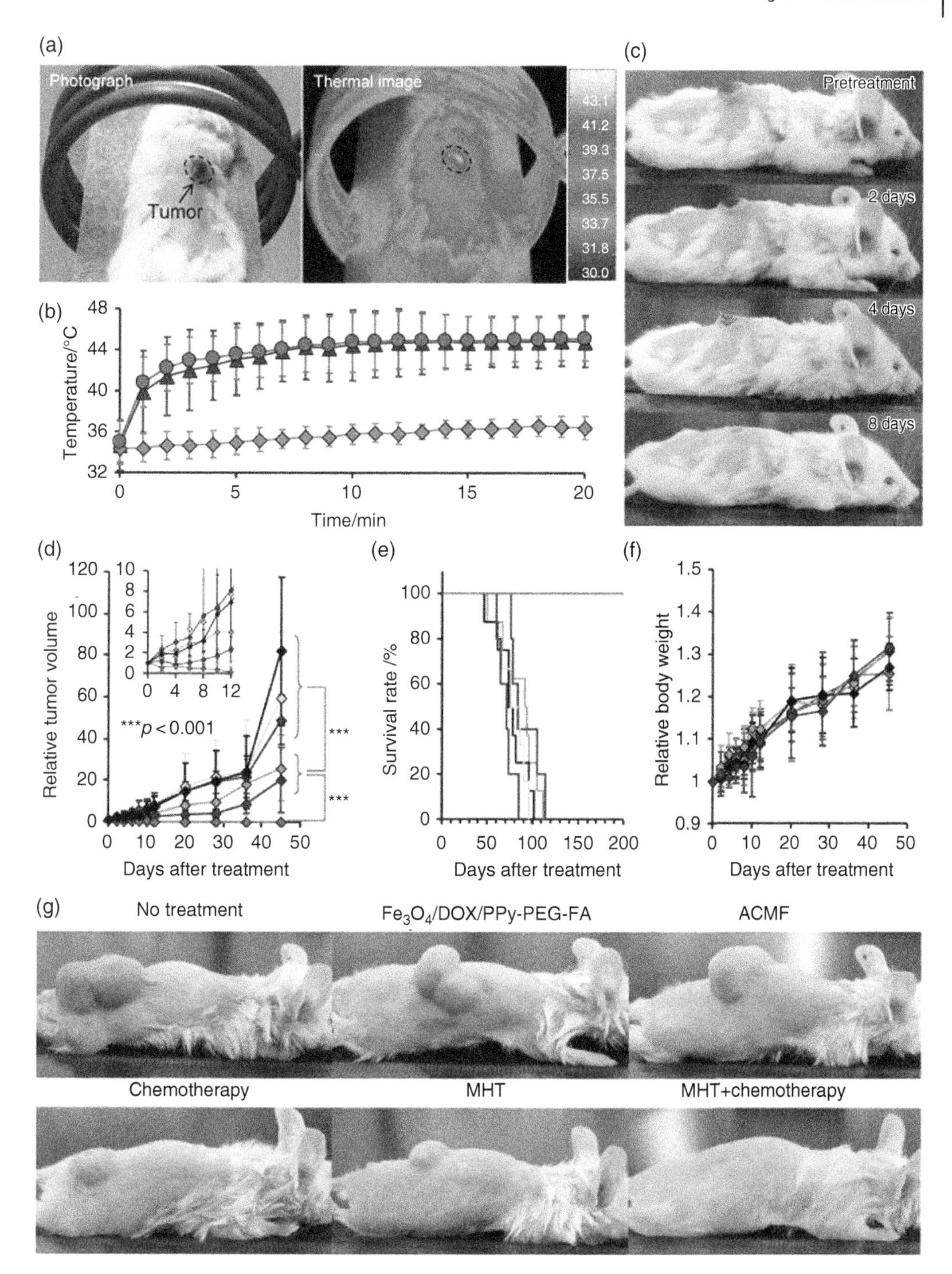
(a)
Photograph
Tumor
Thermal image
43.1
41.2
39.3
37.5
35.5
33.7
31.8
30.0
(b)
Temperature/°C
Time/min
(c)
Pretreatment
2 days
4 days
8 days
(d)
Relative tumor volume
***p < 0.001
Days after treatment
(e)
Survival rate /%
Days after treatment
(f)
Relative body weight
Days after treatment
(g)
No treatment
Fe3O4/DOX/PPy-PEG-FA
ACMF
Chemotherapy
MHT
MHT+chemotherapy

4.3 Gold/Silver-Based Nanomaterials

Besides biocompatibility, ease of synthesis of various morphologies, and their functionalization, gold nanoparticles (AuNPs) and silver nanoparticles (AgNPs) have been extensively studied in the domain of theranostics (Sant et al. 2013; Ghosh et al. 2016a; Ghosh et al. 2016b). Excellent plasmonic properties, thermal conductivity, chemical stability, and antibacterial properties of AuNPs and AgNPs have been utilized to develop new diagnostic markers and nanocarriers with enhanced therapeutic efficacy (Salunke et al. 2014; Ghosh et al. 2016c; Ghosh et al. 2016d). The localized surface plasmon resonance (LSPR) can be utilized for imaging while the photo-responsive properties can be well harvested for therapeutic applications. Moreover, the localized heat generated on illumination with LASER source, the photothermal effect, and external stimuli-assisted drug release can be combined with the LSP- assisted imaging to offer simultaneous diagnostics and therapeutics.

The first synthesis of colloidal AuNPs was done by Faraday in 1857, wherein the gold chloride was reduced by phosphorus and stabilized by carbon disulphide. Later, with the advances in nanomaterial synthesis and characterization techniques, various forms of AuNPs have been developed. In addition to the spherical AuNPs, AuNR, nanocages, nanotriangles, and various other nanostructures have been reported so far (Ghosh et al. 2011; Ghosh et al. 2012a). The shape of the nanomaterials plays a key role in tuning the surface plasmon resonance (SPR). On changing the shape from sphere to rod, the SPR splits into two bands. Among these, one corresponding to the electron oscillations results in a strong band in the near infrared (NIR) region, while the other is a weak band in the visible region. Thus, by tuning the aspect ratio of the nanorods, desired optical properties can be achieved. The higher atomic number and X-ray absorption coefficient, AuNPs have been extensively studied as contrast agents in CT imaging and are excellent radiotherapy sensitizers. Moreover, the feasibility of functionalizing has been beneficial for cladding the nanoparticles for site-specific drug delivery.

Since tissues show low absorption for the wavelength between 650 to 2000 nm, the NIR light has been preferred for imaging tumors deep within the body. Herein, the inherent optical properties of AuNPs are beneficial for their use as contrast agents for tumor imaging. Moreover, as AuNPs are known for enhancing the Raman scattering signal of adjacent molecules, the AuNP-assisted surface-enhanced Raman spectroscopy (SERS) imaging has been extensively studied for imaging of viruses and cancer cells. The light stimuli-assisted therapy that comprises PTT and PDT has been harvested for therapeutics. To address stability, toxicity, and lesser circulation issues, Au nanostructures have been conjugated with various inorganic and organic materials, making them multifunctional theranostic candidates. PEG and other stabilizing ligands have been used to improve the pharmacokinetics of nanoparticles. One study has summarized various Au nanocomplexes for systemic cancer imaging in papilloma, colon carcinoma, melanoma, lung cancer, glioblastoma, squamous carcinoma, glioma, ovarian cancer, pancreatic cancer, hepatoma, breast cancer, and prostate cancer (Guo et al. 2017).

Recently, mesoporous silica-coated AuNR ($Au@SiO_2$) have been developed for cancer theranostics. The mesoporous silica shell protected the inner AuNR core, without hampering its LSPR, and also offered a larger surface area for higher drug payload. Moreover, the functionality of the nanostructure was improved by adding a well-known chemotherapy drug, DOX. The theranostic principle based on the unique properties of the $Au@SiO_2$ is illustrated in Figure 4.2. In addition, Au nanomaterials have been combined with various other components to extend their applications beyond conventional imaging and therapies. Zhang et al. (2013) have reviewed and summarized typical combined applications of AuNRs (Table 4.3).

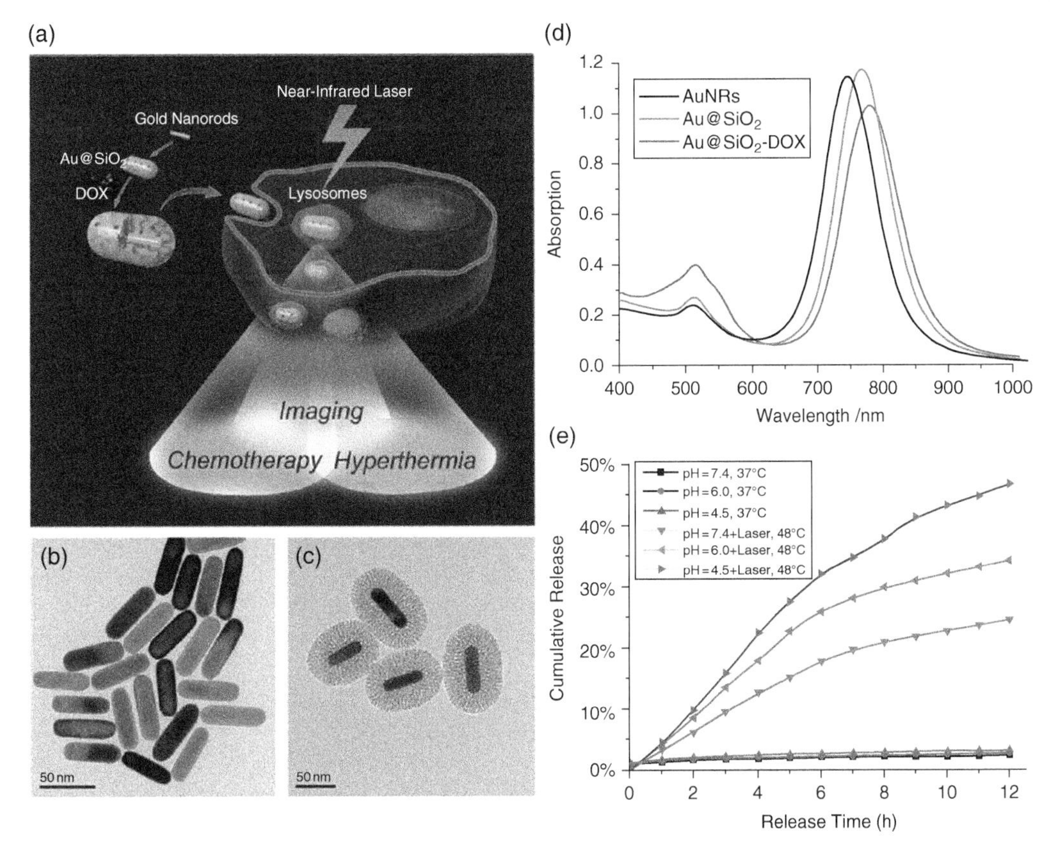

Figure 4.2 Gold-based nanotheranostic agents. (a) Schematic illustration of mesoporous silica-coated gold nanorods (Au@SiO_2) as a novel multifunctional theranostic platform for cancer treatment. TEM images of (b) AuNRs and (c) Au@SiO_2, (d) extinction spectra of AuNRs, Au@SiO_2, and Au@SiO_2–DOX, and (e) DOX release profiles from Au@SiO_2-DOX with and without NIR laser irradiation at different pHs. *Source:* Reprinted with permission from Zhang et al. (2012); Copyright © 2012 WILEY-VCH Verlag GmbH and Co. KGaA, Weinheim.

AuNPs have also been studied for enhancing therapeutic efficacy of conventional antibiotics, which suffer from partial drug adsorption, drug excretion in urine, gastrointestinal complications, and poor solubility of drugs in lipid. Various antibiotics including polymyxins, lincosamides, aminoglycosides, aminopenicillins, ansamycins, etc., have been conjugated for enhanced therapeutic efficacy. The study has also been extended in the domain of multidrug-resistant pathogenic microbial strains. For example, AuNPs conjugated with antimicrobial peptide (CACWQVSRRRRG) showed promising antimicrobial activities against *Staphylococcus aureus* while that conjugated with peptide VG16KRKP (VARGWKRKCPLFGKGG) showed encouraging results against *Salmonella typhi* (Dykman and Khlebtsov 2019).

Another interesting domain where AuNPs have been used is in diagnosis of Alzheimer's disease (AD). AD and tumors have some similarities, like changes in the redox balance disorders in the diseased location. This is characterized by appearance of more hydrogen peroxide, NADH, free radicals, and also accompanied by lack of oxygen. Considering these facts, Lai et al. (2016) have reported *in vivo* fluorescence imaging for AD. Herein, in situ formation of gold nanoclusters at the affected site, upon injection of gold salt in the form of chloroauric acid, has been utilized for imaging the AD affected site. The results are shown in Figure 4.3. It is interesting to note the

Table 4.3 Typical combined applications of Au NRs.

Sr. no.	Surface group	Targeting ligand	Imaging modalities	Drug delivered	Therapeutic characteristics
1	PSS	epidermal growth factor receptor (EGFR) antibodies	light scattering of AuNRs	none	half laser energy needed for malignant cells to be photothermally destroyed than nonmalignant cells
2	PEG	folic acids	TPL	none	extensive cell membrane blebbing by hyperthermia
3	PEG	Deltorphin	AuNRs luminescence	none	selective destruction of receptor-expressing cells while sparing receptor-free cells
4	PSMA	none	ICG luminescence	none	hyperthermia combined with PDT with improved photo-destruction efficacy
5	Cy5.5-peptide	MMP substrate peptide	Cy5.5 fluorescence	none	simultaneous *in vivo* diagnosis and photothermal therapy
6	PEG	CET	NIR absorption of AuNRs	none	extensive pyknosis and cell vacuolization in tumor tissue by hyperthermia
7	PSS, PDDAC	none	light scattering of AuNRs	siRNA	significant impact on suppression of GAPDH gene expression (70% gene silencing, >10 days post-injection)
8	PDDAC	none	TEM	cisplatin	with AuNRs heating, drug dosage lowered to roughly 33% of the unheated amount to achieve comparable cytotoxicity
9	PEG, dDNA	folate	Cy3luminescence	DOX	selective drug delivery to target cells, effective inhibition of tumor growth through thermo-chemotherapy
10	mesoporous silica	none	TPL	DOX	two therapeutic modes of chemotherapy and hyperthermia, switchable by changing laser power density

Source: Partially reprinted with permission from Zhang et al. (2013).

increased fluorescence signal with longer circulation time after injection of $HAuCl_4$ solution in the case of AD mice, whereas there is almost negligible fluorescence signal in the case of mice with no AD or the control group.

Liao et al. (2012) predicted the potential use of AuNPs with negative surface potential to inhibit and redirect amyloid-β fibrillization, contributing to the application of AD. Similarly, AgNPs represent one of the most common man-made nanoparticles. The therapeutic effects of AgNPs have been well documented. The metallic nanoparticles are found to be as efficient as their ionic counterparts. These nanoparticles have proven to be excellent antimicrobial, antifungal, antiprotozoal, antifungal, antiviral agents, hence are one of the most favored candidates for therapeutics

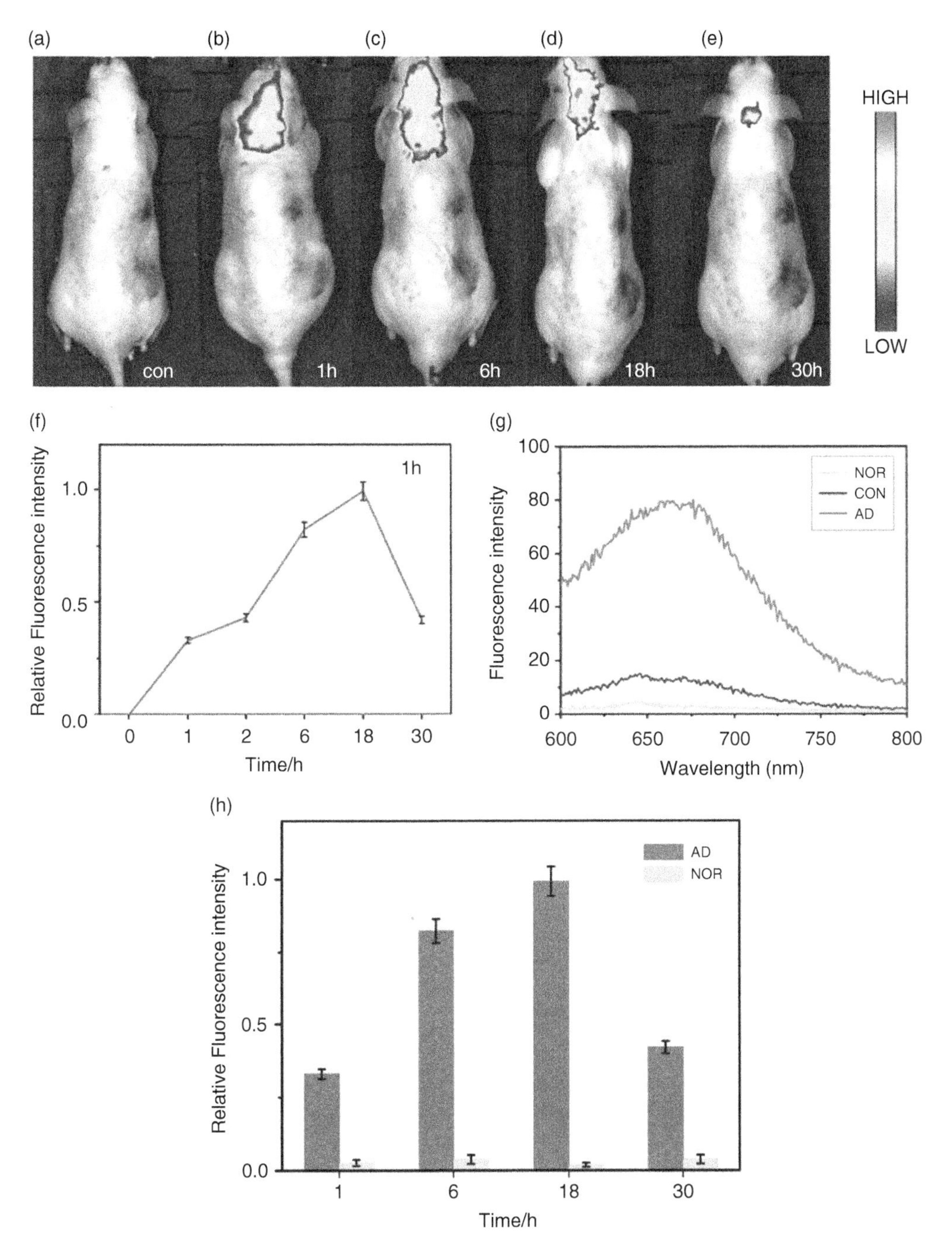

Figure 4.3 Nanotheranostics against Alzheimer's. (a) Fluorescence imaging of the blank control group of Alzheimer's mice (without given $HAuCl_4$). (b–e) Fluorescence imaging of the same Alzheimer's model mice via tail-vein injection of 10 mmol l^{-1}/L $HAuCl_4$ solution at different time points (1, 6, 18, 30 hours). (f) The variations of mean fluorescence intensity of the Alzheimer's model mice at different time points via tail-vein injection of 10 mmol l^{-1} HAuCl4 solution (CON, 1, 2, 6, 18, 30 hours). (g) Emission spectra of the normal control group of mice (NOR), control group of mice (CON) and AD's brain tissue which was grinded and then diluted by ultrapure water. Excitation wavelength: 420 nm. Excitation/emission slit: 5/5 nm. AD has a strong emission peak at 660 nm, the CON has a low emission peak at 650 nm, and the NOR has a very weak fluorescence at 635 nm (same concentration and identical experimental conditions for AD, CON, and NOR). (h) The variations of mean fluorescence intensity of Alzheimer's model mice (AD), and normal control group of mice (NOR) at various time points (1, 6, 18, 30 hours). *Source:* Reprinted with permission from Lai et al. (2016); Copyright © 2016 Royal Society of Chemistry.

and infectious disease control (Yamada et al. 2015). Chemical synthesis has been one of the most convenient methods for synthesizing AgNPs of desired and physicochemical characteristics; until their potential hazard including carcinogenicity, genotoxicity, and cytotoxicity have been noticed. In this context, green synthesis of AgNPs is preferred by researchers. Ghosh et al. (2012b) have reported on phytogenic AgNPs, using *Dioscorea bulbifera* tuber extract, which is rich in flavonoid, phenolics, reducing sugars, starch, diosgenin, ascorbic acid, and citric acid. By tuning the precursor ratio and other synthesis parameters, various morphologies including spheres, triangles, and hexagons have been developed. The antimicrobial activity of the nanoparticles was studied against a library of Gram-negative and Gram-positive bacteria. Moreover, promising activity was seen on multidrug-resistant bacteria, *Acinetobacter baumannii*. These and similar promising results have opened a wide scope of therapeutic applications of AgNPs. Along similar lines, the diagnostic and therapeutic applications of AgNPs have shown promising results over pathogens. Gopinath et al. (2016) have reported sunlight irradiated synthesis of AgNPs using an actinobacterial cell-filtrate-*Streptomyces* sp.- GRD cell-filtrate. The optical properties of the as synthesized nanoparticles have been used for imaging. The biologically derived nano silver has shown pronounced inhibition of the endospores of *Bacillus subtilis*, *Bacillus cereus*, *Bacillus amyloliquefaciens*, *Clostridium perfringens*, and *Clostridium difficile* even at a very low concentration, as compared to the conventional sporicides. The vital organs of mice exposed to these nanoparticles showed no signs of pathological lesions, proving the complete disinfection of the spore; thus the multifunctionality of the nanoparticles and the theranostics applications of the same. Besides their topical applications, AgNPs have been extensively used against various other diseases (Yamada et al. 2015). Although earlier, Ag was used only for therapy, emerging supramolecular assembly using rational nanoscale engineering has enabled its simultaneous application for targeted therapy and diagnostic modalities (Hada et al. 2019). Chitosan-based AgNPs, polyvinylpyrrolidone (PVP)-coated AgNPs, starch-coated AgNPs, AgNPs- Temozolomide (TMZ) combination, and Ag sulfide nanoparticles/nanorods/nano-wires were reported to inhibit an array of cancer cells like human colon cancer cells (HT 29), acute myeloid leukemia cells (AML), human lung fibroblast cells (IMR-90), human glioblastoma cells (U251), C6 glioma cells, and human hepatocellular carcinoma Bel-7402 cells (Sanpui et al. 2011; Guo et al. 2013; Asharani et al. 2009; Liang et al. 2017). Apart from acting as anticancer agents, AgNPs can also work as drug delivery vehicles and imaging facilitators, which was confirmed by functionalization of AgNPs with PVP and the antitumoral drug doxorubicin (DOX) that can be tracked rapidly for tumor tissue localization using ^{125}I isotope (Farrag et al. 2017).

Chitosan-coated Ag nanotriangles (Chit-Ag NTs) with strong resonances in NIR are potent photothermal agents lung cancer cells (NCI-H460) (Matteis et al. 2018). Previously we have reported that $Au_{core}Ag_{shell}$ nanoparticles synthesized using *D. bulbifera* could effectively inhibit bacterial biofilms against *A. baumannii*, *Escherichia coli*, *Pseudomonas aeruginosa,* and *S. aureus*. Further, it could inhibit and kill, causing remarkable deformation in *Leishmania donovani* which is a leading cause of kala-azar (Ghosh et al. 2015d). Novel gold–silver core-shell nanoparticles labeled with para-mercaptobenzoic acid (4MBA) molecules can be used as SERS-nanotags with ultra-bright traceability inside cells. These novel nanohybrids can convey spectrally coded information about the intracellular pH of human ovarian adenocarcinoma cells (NIH:OVCAR-3) by means of SERS (Hada et al. 2019). However, AgNPs associated toxicological responses on circulatory, respiratory, central nervous and hepatic systems cannot be overruled which are supposed to be major drawbacks (Gopinath et al. 2016; Li et al. 2018; Yeşilot and AydınAcar 2019; Yamada et al. 2015; Stensberg et al. 2011). Efforts are being made to reduce such cytotoxic and genotoxic effect of AgNPs by careful manipulation of their concentration, morphology, surface-functionalization, exposure time, route of administration, and physiological environment (Tran et al. 2013).

4.4 Quantum Dot-Based

Luminescent semiconductor nanocrystals, called quantum dots (QDs), are typically comprised of elements from periodic groups of II–VI (e.g. CdSe and CdTe) or III–V (e.g. InP and InAs) semiconductor materials. These nanomaterials exhibit superior optical properties, such as narrow, symmetric, and size-tunable emission spectra, and broad excitation spectra, rendering them particularly valuable for multicolor fluorescent applications. Compared to organic fluorophores or fluorescent proteins, QDs show stronger fluorescence (10–100 times brighter) and higher fluorescence stability against photobleaching (100–1000 times more stable), ideal for long-term monitoring of intermolecular and intramolecular interactions in live cells and organisms (Ho and Leong 2010). These properties have made QDs popular fluorescent cellular probes for light microscopy. Electron-dense semiconductor cores can be directly imaged by electron microscopy even without any contrasting treatment and thus distinct size, shape, and elemental fingerprint of QDs facilitate multi-labeling for correlative microscopy. QD-mediated Förster resonance energy transfer (FRET) is the underlying mechanism for QD-based fluorescence biosensing. QD–aptamer (Apt)–doxorubicin (Dox) conjugate, abbreviated as QD–Apt(Dox) is a promising nanostructure for simultaneous targeted cancer imaging, therapy, and sensing. Functionalization of A10 RNA aptamer onto the surface of fluorescent CdSe/ZnS core-shell QD490 helps to recognize the extracellular domain of a model aptamer-targeting molecule, prostate-specific membrane antigen (PSMA) on the surface of LNCaP cells (prostate cancer cell lines). This novel nanoconjugate consists of three components where QDs act as fluorescent imaging vehicles, covalently conjugated RNA aptamers function as targeting molecules as well as drug-carrying vehicles, and doxorubicin (Dox), an anthracycline drug that intercalates within the double-stranded coarse-grained (CG) sequences of RNA and DNA to act as a cancer-inhibiting potent therapeutic agent. The A10 PSMA aptamer is a 57 base pair nuclease-stabilized 2′-flouropyrimidine RNA molecule with a single 5′-CG-3′ sequence in its predicted double-stranded stem region that binds with Dox as a reversible physical conjugate due to the ability of Dox to intercalate into a single CG sequence present in this aptamer. Dox with remarkable fluorescence properties can be maximally excited by absorbing light with wavelength of 480 nm, resulting in the emission of light in the range of 520–640 nm and thus act as the photon acceptor of CdSe/ZnS QD490, which emits light in the range of 470–530 nm (Bagalkot et al. 2007). This Bi-FRET complex consists of two closely associated models. A donor-acceptor model FRET is comprised of QD and Dox where fluorescence of QD is quenched as a result of Dox absorbance. On the other hand, the donor-quencher model FRET consists of Dox and aptamer, where Dox is quenched by a double-stranded RNA aptamer. Loading of Dox on QD-Apt [QD-Apt(Dox)] is represented as the fluorescence "OFF" state for both QD and Dox. Internalization of the nanoconjugate by the targeted cancer cells leads to gradual release of Dox, and eventually activation of QD and Dox fluorescence is considered as the "ON" state (Figure 4.4a). It is important to note that this multifunctional nanomaterial has the potential for both delivering Dox to the targeted cells as well as effective sensing of Dox delivery with facilitation of concurrent imaging of the cancer cells (Figure 4.4b).

Recently, QD–liposome (QD–L) systems have gained wide attention as a novel group of theranostic agents. Functionalized-quantum-dot–liposome (f-QD-L) hybrid nanoparticles of 80–100 nm in average diameter can be achieved by encapsulating COOH–PEG–QD (PEG: poly[ethylene glycol]) into small unilamellar vesicles (SUV) of various lipid compositions. Such f-QD–liposome hybrid nanoparticles (f-QD-L) exhibit enhanced internalization into tumor cells, achieving a high degree of penetration through the tumor interstitium, and increasing retention within the tumor mass *in vivo*, which allow for efficient labeling of cancer cells and dramatic reduction in the dose of QD needed. Moreover, such hybrid systems are designed to take advantage of the

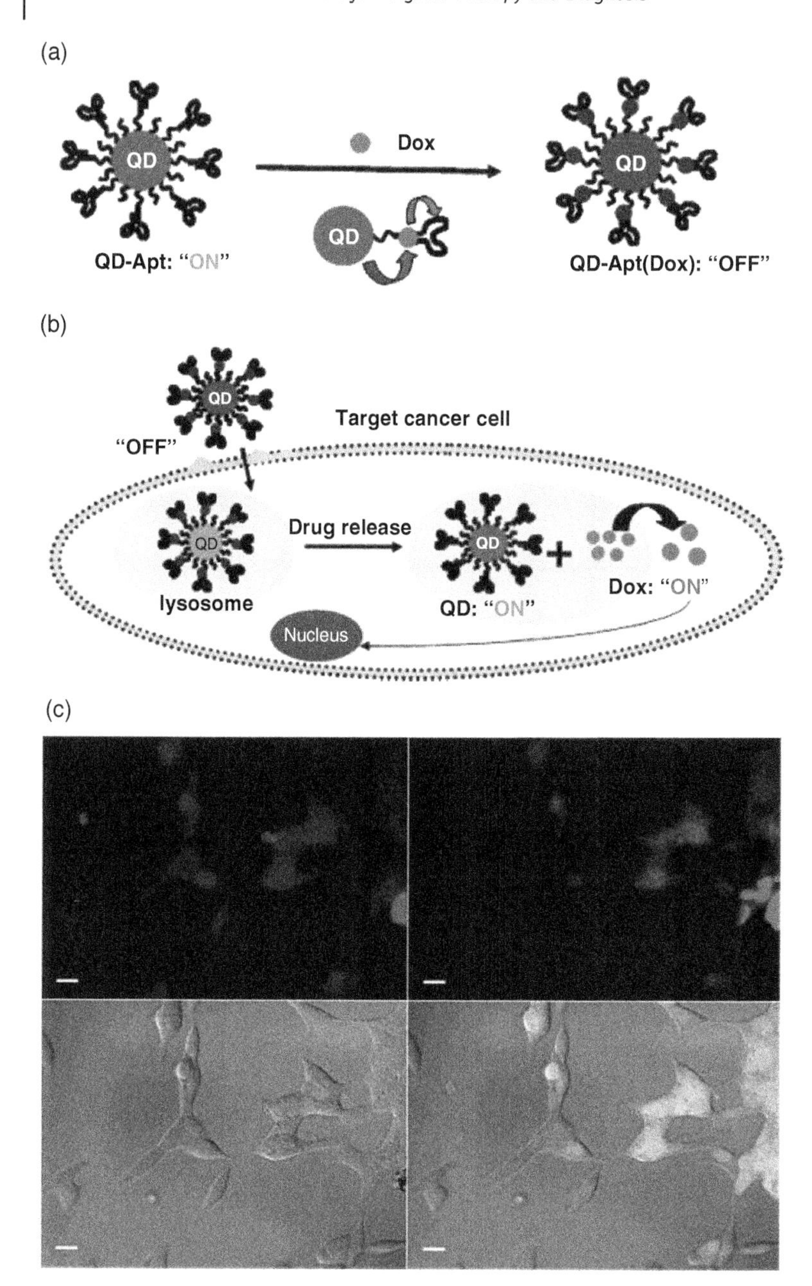

Figure 4.4 QD-Apt(Dox) Bi-FRET system as novel theranostic agent against cancer. (a) In the first step, the CdSe/ZnS core-shell QD are surface functionalized with the A10 PSMA aptamer. The intercalation of Dox within the A10 PSMA aptamer on the surface of QDs results in the formation of the QD-Apt(Dox) and quenching of both QD and Dox fluorescence through a Bi-FRET mechanism: the fluorescence of the QD is quenched by Dox while simultaneously the fluorescence of Dox is quenched by intercalation within the A10 PSMA aptamer resulting in the "OFF" state. (b) Schematic illustration of specific uptake of QD-Apt(Dox) conjugates into the target cancer cell through PSMA mediate endocytosis. The release of Dox from the QD-Apt(Dox) conjugates induces the recovery of fluorescence from both QD and Dox ("ON" state), thereby sensing the intracellular delivery of Dox and enabling the synchronous fluorescent localization and killing of cancer cells. (c) Confocal laser scanning microscopy images of PSMA expressing LNCaP cells after incubation with 100 nM QD-Apt-(Dox) conjugates for 0.5 hours at 37 °C, washed two times with PBS buffer, and further incubated at 37 °C for 1.5 hours. Dox and QD are shown in red and green, respectively, and the lower right images of each panel represents the overlay of Dox and QD fluorescent. The scale bar is 20 μm. *Source:* Reprinted with permission from Bagalkot et al. 2007; Copyright © 2007, American Chemical Society.

physicochemical and pharmacodynamic versatility offered by the liposome structure combined with the wide range of photochemical characteristics of the different available types of f-QD (Al-Jamal et al. 2008). Similarly, functionalized PEG-lipid coated QD (f-QD) encapsulated into the aqueous core of 100 nm cationic (DOPC:Chol:DOTAP), sterically stabilized fluid-phase (DOPC:Chol:DSPE-PEG2000) and sterically stabilized gelphase (DSPC:Chol:DSPE-PEG2000) liposome vesicles are stable in blood, showing excellent tissue distribution in B16F10 melanoma tumor-bearing mice, after intravenous administration. These sterically stabilized f-QD-L hybrid vesicles have high serum stability. Rigid PEGylated *f*-QD-L hybrids accumulate in the tumor lesions much more rapidly compared to f-QD which emphasizes many advantages for the development of simultaneous therapeutic and imaging modalities by incorporating both drug molecules and QD within the different compartments of a single vesicle (Al-Jamal et al. 2009; Ho and Leong 2010).

4.5 Polymer-Based Nanomaterials

Clinically approved polymers used for macroformulations include PEG, poly(D,L-lactic acid), poly(D,L-glycolic acid), and poly(ε-caprolactone), as they have numerous advantages like enhancement of drug efficacy compared with free drugs due to improved drug encapsulation and delivery, prolonged circulation half-life, and sustained or triggered drug release. Several targeting ligands can be effectively functionalized onto polymeric nanoparticles, which may eventually lead to enhanced cellular uptake and retention due to passive targeting and tissue-specific drug release and accumulation. These polymeric nanomaterials ideally consist at the region of interest. In general, a polymer-based theranostic material is comprised of at least three main components: (i) a polymer component which gives stability and biocompatibility; (ii) a therapeutic agent (i.e. small-molecule drug, siRNA, etc.); and (iii) an imaging agent (i.e. MRI contrast agent, radionuclide, fluorophore, etc.). Sometimes, the therapeutic agents with inherent fluorescence may also help in simultaneous imaging. Such polymeric nanoparticles are now widely explored for drug delivery, gene delivery, and PDT, combined with imaging agents for MRI, radionuclide imaging, and fluorescence imaging, making them promising theranostics (Luk and Zhang 2014). High hydrophobicity of efficient photosensitizers like zinc phthalocyanine and zinc naphthalocyanine is a major drawback that can be circumvented using an advanced polymer-based nano-delivery system having a PEG-coated polymeric core. These theranostic probes show efficient *in vitro* fluorescence and singlet oxygen quantum yields upon irradiation with 620–750 nm (30 mW cm^{-2}) light (Figure 4.5). Clathrin-mediated endocytosis (CME) may enhance cellular internalization and thus improve accumulation and higher cytotoxicity in cancer tissue. Such polymeric nanotheranostics do not show hepato/nephrotoxicity indicating that passive targeting mechanism renders it clinically safe (Thakur et al. 2019).

UV–NIR wavelength range fluorescent probes are incorporated into hyperbranched polyhydroxyl polymeric nanoparticles simultaneously with apoptosis-inducing protein cytochrome c, which can be effectively targeted using folic acid ligand-facilitating cellular uptake and enhance the therapeutic effect against cancer. These nanomaterials emit photons via excitation of encapsulated indocyanine green (ICG) which is advantageous for imaging. Further, core-multishell nanoparticles (CMS NPs) composed of hyperbranched polyethylene imine (PEI) core functionalized by alkyl diacids connected to monomethyl poly(ethylene glycol) are used to encapsulate and transport the antitumor drugs doxorubicin hydrochloride (Dox), methotrexate (Mtx), and sodium ibandronate (Ibn), as well as dye molecules like tetrasulfonated indotricarbocyanine (ITCC) and nilered (Figure 4.6). These CMS NPs demonstrated a strong contrast within the tumor tissues compared to free dye even after six hours of injection into F9 teratocarcinoma-bearing mice. (Quadir et al. 2008). Various bio-templates like plant-derived polymeric matrix are being used to ensure stability, biocompatibility, and sustained release of the bioactive agents. A self-assembled *Aloe vera* template

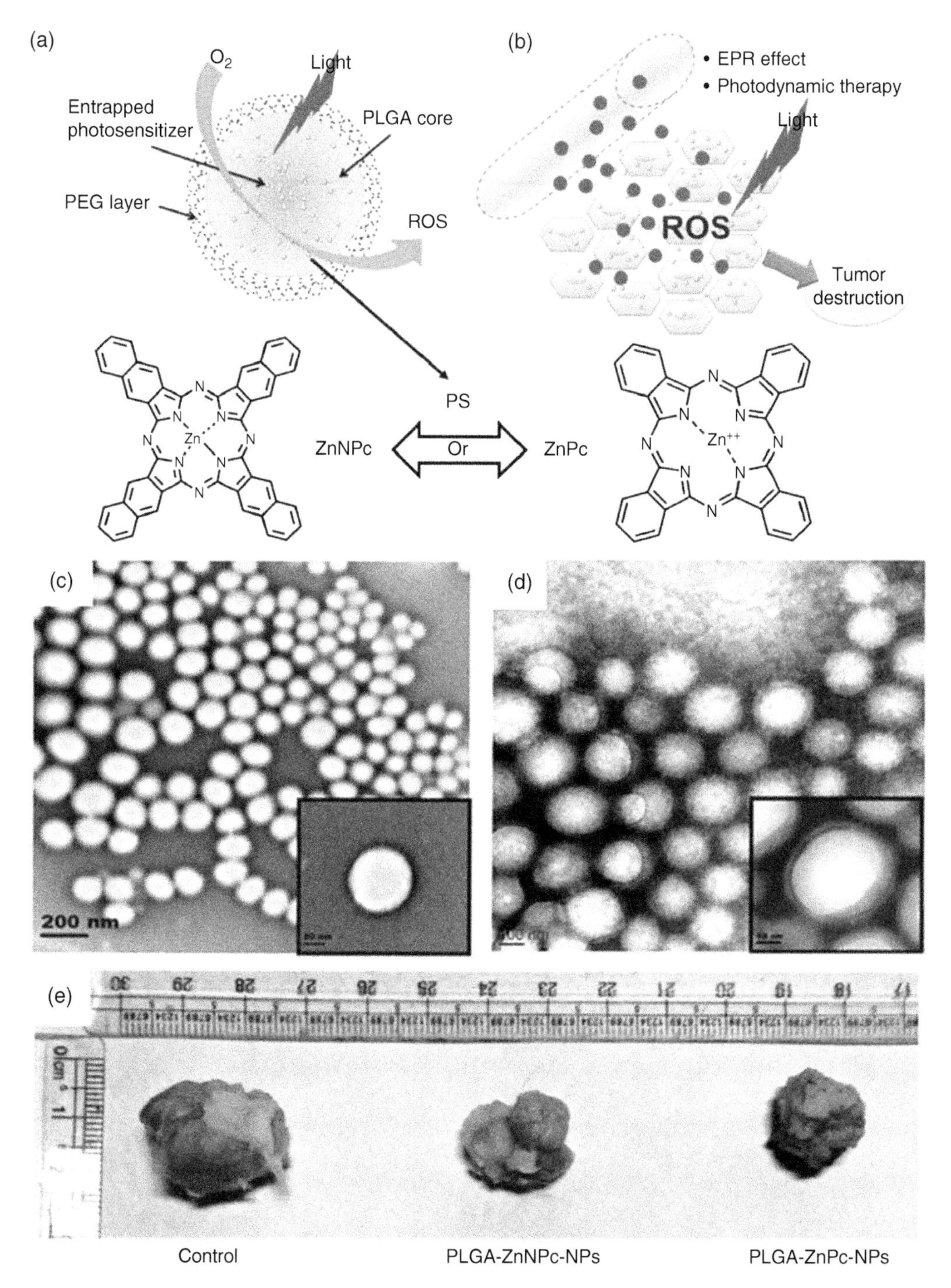

Figure 4.5 Schemes for the generation of reactive oxygen species, EPR effect, and simultaneous tumor destruction. Characterization of NPs using dynamic light scattering (DLS), transmission electron microscopy (TEM), and powder X-ray diffraction spectroscopy (PXRD). (a) Schematic diagram of PS loaded PLGA nanoparticle, showing the reactive oxygen generation when irradiated with light; (b) Schematic presentation of rationale and the prospective strategy of current work; (c) TEM images of synthesized ZnNPc loaded PLGA nanoparticles (at 200 nm scale and inset at 50 nm scale); (d) TEM images of synthesized ZnPc loaded PLGA nanoparticles (at scale 100 nm, inset at 50 nm scale); (e) Excised tumor tissue images from DMBA (7,12-dimethylbenzanthracene) induced breast tumor-bearing SD rats treated with different groups of prepared phototheranostic nanoagents. Here, the volumes of control, PLGA-ZnNPc-NPs, and PLGAZnPc-NPs are ~5820, 2760, and 1800 mm^3, respectively. *Source:* Reprinted with permission from Thakur et al. (2019); Copyright © 2019 Elsevier B.V.

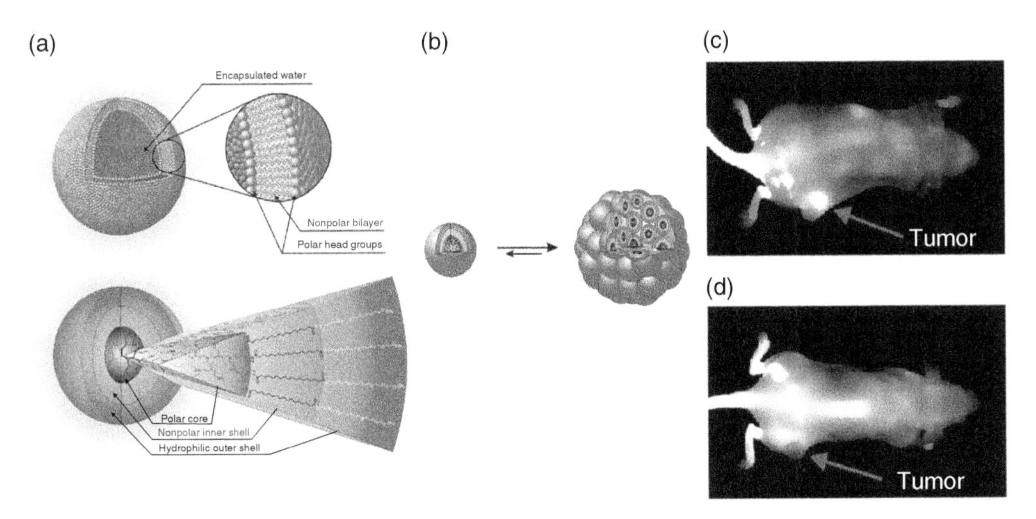

Figure 4.6 Polymeric nanoparticles against teratocarcinoma: (a) Dendritic core-multi shell architecture at the bottom panel with PEI core, inner hydrophobic segment, and terminal mPEG chain. On the top panel, the typical liposome structure to reveal the structural resemblance to CMS NPs. (b) Dynamics between CMS NP unimers and their aggregates those are responsible for encapsulation. (c) Strong contrast was observed after 6 hours of administration of ITCC dye loaded CMS NP to F9 teratocarcinoma bearing mice; (d) contrast achieved with free dye after the same time period is not very prominent. *Source:* Reprinted with permission from Quadir et al. (2008); Copyright © 2008 Elsevier B.V.

was used for loading of curcumin for trans-membrane drug release and superior antioxidants. These particles may be used as complementary and alternative nanomedicine against oxidative stress and associated diseases (Kitture et al. 2015a). More recently, we have successfully engineered sub 200 nm dual drug conjugates by direct attachment of a mitochondria-damaging drug (α-TOS) and clinically approved DNA-damaging drugs (cisplatin and doxorubicin) or a microtubule-binding drug (paclitaxel) without any additional linker strategy. These nanovehicles were internalized into the lysosomal compartments in a time-dependent manner through endocytosis in HeLa cervical cancer cells and the dual drugs were released in a slow and sustained manner over three days at $pH = 5.5$, mimicking the lysosomal compartments inside the tumor cells. These novel nanoparticles can induce apoptosis, arrest cell cycle, damage mitochondrial morphology, and release cytochrome c, damaging the nucleus and tubulin of the HeLa cells. Rational modification and functionalization of these highly effective nanocomposites may lead to novel next-generation multiple organelle-damaging cancer theranostics which can combat both toxic side effects of anticancer drugs and increasing tendency toward drug resistance (Mallick et al. 2015). Multifunctional micelles with fluorescent imaging and drug delivery capabilities are prepared via the co-assembly of DOX-conjugated monomethoxy lPEGblock-poly(L-lactide-co-mercaptoethanol) copolymer, Rhodamine B-conjugated mPEG-b-p(LA-co-ME), and folic acid-conjugated PEG-b-PLA copolymer which exhibits enhanced antitumor efficacy compared to free DOX in hepatocarcinomas (Luk and Zhang 2014).

4.6 Silica-Based Nanomaterials

Among various theranostic nanomaterials, mesoporous silica nanoparticles (MSNs) are a most-preferred agent due to their unique pore structure, tunable surface, and biocompatibility. MSNs can be chemically modified for imaging and drug delivery in a living system, rationalizing their role in simultaneous diagnosis and therapy (Ghosh et al. 2018). Functionalization of nanoparticles onto/within MSNs is carried out by various methods like covalent bonding for AuNPs to MSN surface. AuNPs are linked with adenosine triphosphate (ATP) aptamer through Au–S bonds and further derivative of the ATP molecules (adenosine-5-carboxylic acid) which is immobilized on the external surface of the amine-functionalized MSN through amidation reaction. Dye-doped MSNs are also used to assemble the IONPs for optical and MRI diagnostics. Alternatively, sonochemical synthesis is used to cover the silica surface with ZnS NPs where ultrasound physical irradiation of a slurry (made of amorphous silica, zinc acetate, and thioacetamide) is used. Other surface functionalization methods include layer-by-layer (LBL), self-assembly, surface chemical reaction, in situ formation, sol-gel and refluxing, and redox reaction (Singh et al. 2017). Multi-functionalized MSNs are associated with various imaging modalities that include MRI, CT, nuclear imaging (PET and SPECT), optical imaging, and ultrasound. The porous nature of MSNs results in high specific surface area, and also makes them ideal for incorporation of therapeutic agents, targeting proteins and ligands in large quantities. Similarly, even the release patterns of drugs can be controlled by changing mesopore properties like pore channel geometry. Silanol groups, present on the surface of fully hydroxylated MSN can be of different forms like free silanol, silanol with physically adsorbed water, siloxane bond, germinal silanol groups (≡Si–OH), and hydrogen-bonded silanol groups which determine the type and quantity of biomolecule functionalization. Although silanol groups render a highly hydrophilic surface to the MSNs facilitating incorporation of hydrophilic drugs, other functional groups, like trimethylsilyl, can be used to modify the surface for incorporation of hydrophobic drugs as well. This rational tailoring of the MSN surface is very critical for successful functionalization of imaging agents, drugs, and targeting ligands for potential theranostic applications including targeted delivery and triggered release (Singh et al. 2017). Radionuclides are introduced to MSNs for PET imaging owing to their high penetration and sensitivity toward diagnosis of tumors at an early stage. Commonly used PET agents include heavy metal ^{64}Cu, ^{68}Ga, ^{99}mTc, and ^{111}In, as well as radioisotopes ^{11}C, ^{18}F, and ^{15}O labeling with targeted molecules. Recently, copper sulfide (CuS) nanoparticles encapsulated in biocompatible mesoporous silica shells are fabricated to develop novel theranostic agents using multistep-surface engineering methodology. In short, water-soluble CuS nanoparticles with strong NIR optical absorption and high molar extinction coefficient are selected as the core while MSNs act as a protective shell and drug reservoir for potential targeted drug delivery and thermal-chemotherapy against cancer. These CuS@MSNs are further surface engineered with amino groups ($-NH_2$), NOTA (1,4,7-triazacyclononane triacetic acid, a well-known chelator for copper-64 [^{64}Cu] labeling), polyethylene glycol (PEG, for improved *in vivo* biocompatibility and stability), TRC105 (a human/murine chimeric IgG1monoclonal antibody, which binds to both human and murine CD105 on tumor neovasculature), and finally the radioisotope ^{64}Cu (a positron emitter with a 12.7 hours half-life, for PET imaging and biodistribution studies). These ^{64}Cu-CuS@MSN-TRC105 theranostic nanoparticles are effectively taken up by the breast tumor tissue in a time-dependent manner as evident from a PET scan (Figure 4.7). However,

Figure 4.7 *In vivo* photothermal therapeutic evaluation using silica-based theranostic agent. Photothermal images of mice after laser treatment. (a) (CuS@MSN + 980 nm laser) group, (b) (980 nm laser only) group, and (c) (CuS@MSN only) group. Digital photos of mice with 4T1 tumors on Day 30 after treatment. (d) (CuS@MSN + 980 nm laser) group, (e) (980 nm laser only) group, and (f) (CuS@MSNonly) group. (g) A digital photo of mouse from (CuS@MSN + 980 nm laser) group on Day 67 after treatment. (h) Changes of tumor size of mice from three different groups after treatment (n = 5). Laser dose: 4 W cm^{-2}, 15 minutes. CuS@MSN dose: 33 mg kg^{-1}. Tumors were marked with red arrows. The differences between the treatment group and two control groups were statistically significant (*$P < 0.05$ on Day 4, **$P < 0.01$ at later time points). (i) *In vivo* CD105 targeted PET imaging, tumor uptake comparison, and histology studies. *In vivo* serial coronal PET images of ^{64}Cu-CuS@MSN-TRC105 nanoconjugates in targeted group; (j) ^{64}Cu-CuS@MSN in nontargeted group and (k) ^{64}Cu-CuS@MSN-TRC105 with a large dose of free TRC105 in blocking group in 4T1 murine breast tumor-bearing mice at different time points post-injection. *Source:* Reprinted with permission from Chen et al. (2015b); Copyright © 2015 American Chemical Society.

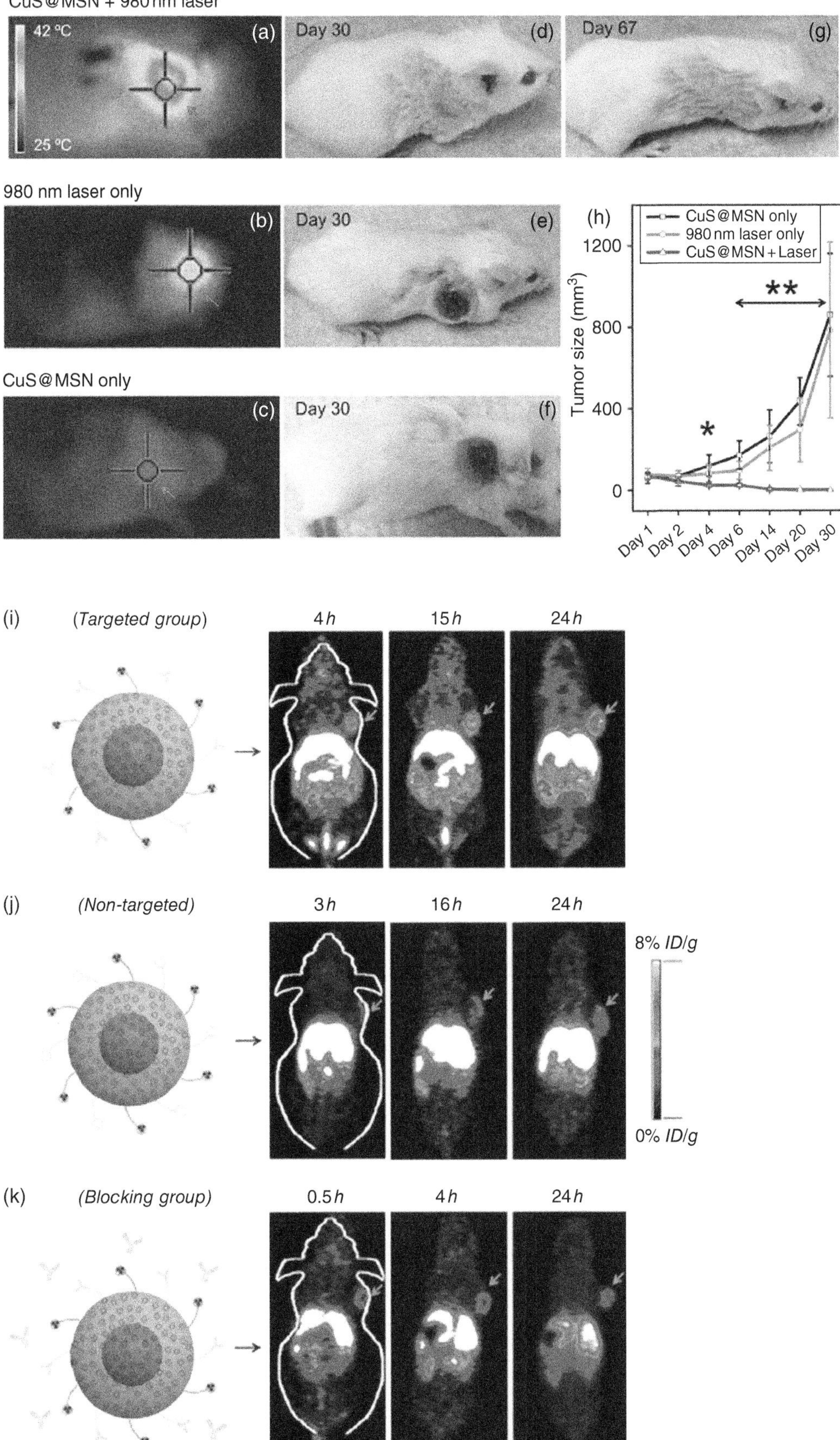

CuS@MSN + 980 nm laser
42 °C
25 °C
(a)
Day 30
(d)
Day 67
(g)
980 nm laser only
(b)
Day 30
(e)
CuS@MSN only
(c)
Day 30
(f)
(h)
CuS@MSN only
980 nm laser only
CuS@MSN + Laser
Tumor size (mm3)
1200
800
400
0
*
**
Day 1
Day 2
Day 4
Day 6
Day 14
Day 20
Day 30
(i)
(Targeted group)
4 h
15 h
24 h
(j)
(Non-targeted)
3 h
16 h
24 h
8% ID/g
0% ID/g
(k)
(Blocking group)
0.5 h
4 h
24 h

without the conjugation of TRC105, the 4T1 tumor uptake of 64Cu-CuS@MSN is comparatively low, indicating that TRC105 conjugation is a critical factor for enhanced tumor accumulation. The serial PET scans reflect higher uptake of CuS@MSN-TRC105 than CuS@MSN at all time points (Singh et al. 2017; Chen et al. 2015b).

Various organic groups like 3-aminopropyl (AP), *N*-(2-aminoethyl)-3-aminopropyl (AEAP), *N*-floate-3-aminopropyl (FAP), guanidinopropyl (GP), and 3-{N-(2-guanidinoethyl)guanidino}propyl (GUGP) play a significant role in selective cellular internalization (Slowing et al. 2006). FITC (fluorescin isothiocyanate)-MSN and FAP-MSN exhibit clathrin-mediated intracellular trafficking while the AP-MSN and GP-MSN show caveolae-mediated internalization. The size of MSNs determines efficacy, pathways of cellular internalization, and interactions with cell membrane and vesicles. MSNs show no tissue toxicity with minimal inflammatory signs in all tissues in animal model studies. However, the biodistribution pattern of MSNs are remarkably size dependent. Larger particles are found in higher amounts in urine after administration and are more easily captured by the reticuloendothelial system (RES) than smaller ones. Intravenous administration of MSNs shows their presence mainly in the liver, spleen, and lungs. Particles with smaller dimensions are accumulated in the liver whereas larger particles are accumulated in the spleen. It is important to note that the clearance rate of MSNs is particle-shape dependent (Singh et al. 2017). For optical imaging, fluorescent MSNs are fabricated by combining cyanine dyes such as trimethinecyanines (Cy3), pentamethinecyanines (Cy5), heptamethine cyanines (Cy7), and ICG available in the NIR range of the biological window (700–900 nm). The ICG NIR contrast agent can be incorporated within pores of trimethylammonium-modified MSNs by an electrostatic attraction which is able to show an effective bio-distribution profile by the high fluorescent intensity. Negatively charged and water-soluble cyanine NIR dye like IR-783 can be electrostatically adsorbed onto the silica shell which is able to be revealed in fluorescence and X-ray CT imaging even four hours after injection of the probe into nude mice (Song et al. 2015). Similarly, NIR dye ZW800 attached onto MSN is used for optical imaging of tumor tissue without any quenching effects *in vivo* which is advantageous over traditional organic dyes that are susceptible to bio-erosion and degrade quickly. PEGylated liposome-coated QDs with mesoporous silica shell are used for molecular imaging due to their superior optical imaging efficiency in labeling cancer cells. These particles are more biocompatible and stable, unlike traditional Cd-based QDs which are more cytotoxic. Recently, rare-earth based upconversion NPs exhibit fluorescence upon the 980 nm laser excitation which is characterized by efficient tissue penetration and is free of tissue auto-fluorescence. Monodispersed upconversion NPs, $NaYF_4$:Tm/Yb/Gd, layered with a thin mesoporous silica, show excellent fluorescent imaging performance after the intratumor administration. Similarly, rattle-structured upconversion mesoporous nanotheranostics (UCMSN) like $NaYF_4$:Yb/Er/Tm@$NaGdF_4$ (UCNP [upconversion nanoparticles]) as the core and mesoporous silica as the shell show strong NIR upconversion luminescent (UCL) signal in the tumor tissue by an intravenous injection. These sub-80 nm multifunctional Gd-UCNPs core/mesoporous silica shell nanotheranostics (UCMSNs) are constructed for the codelivery of a radio-/photosensitizer hematoporphyrin (HP) and a radiosensitizer/chemo drug docetaxel (Dtxl). After being subjected to NIR excitation and X-ray irradiation, the complete tumor elimination can be achieved by the synergetic chemo-/radio-/photodynamic tri-modal therapy under the assistance of simultaneous magnetic/upconversion luminescent (MR/UCL) bimodal imaging (Figure 4.8). Thus, such nanotheranostic systems may serve as immensely useful platforms for simultaneous diagnosis and efficient treatment of cancer (Fan et al. 2014).

Oxygen defects in the MSN structure is also exploited in optical imaging. Porous MSNs achieved by electrochemical etching possess luminescence property, being attributed to quantum confinement effects and defects localized at the Si–SiO_2 interface. Porous silicon nanostructures with intrinsic NIR

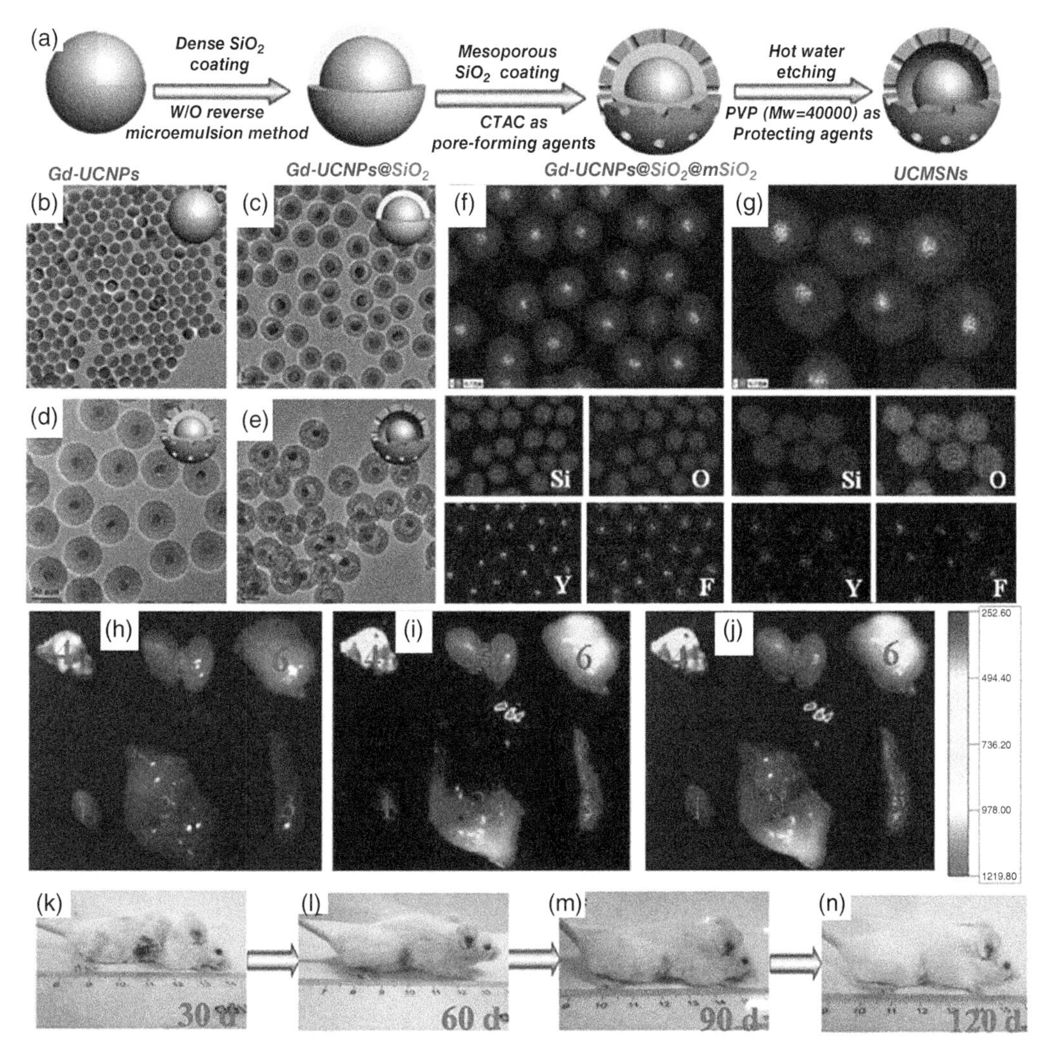

Figure 4.8 Synthesis and application of UCMSNs in imaging. (a) Schematic illustration of the synthetic procedure of UCMSNs. Gd-UCNPs were prepared by epitaxial growth $NaGdF_4$ layer on $NaYF_4$:Yb/Er/Tm through a typical thermal decomposition process. Then a dense silica layer was coated on Gd-UCNPs by a reverse microemulsion method, designed as Gd-UCNPs@SiO_2. Subsequently, a mesoporous silica shell was deposited on Gd-UCNPs@SiO_2 via the template of CTAC (Cetyltrimethylammonium chloride), designed as Gd-UCNPs@SiO_2@mSiO_2. Finally, UCMSNs were successfully fabricated based on a "surface-protected hot water etching" strategy. (b–e) Transmission electron microscopic (TEM) images of (b) Gd-UCNPs ($NaYF_4$:Yb/Er/Tm@$NaGdF_4$), (c) Gd-UCNPs@SiO_2, (d) Gd-UCNPs@SiO_2@mSiO_2, and (e) UCMSNs (scale bar ¼ 50 nm). (f and g) STEM (scanning transmission electron microscopy) image and the corresponding element (Si, O, Y, F) mappings of (f) Gd-UCNPs@SiO2@mSiO_2 and (g) UCMSNs. (h–j) Ex vivo NIReNIR upconversion luminescent imaging (UCL) of dissected organs from a 4T1-tumor bearing mouse after the intravenous injection of UCMSNs: 1, heart; 2, liver; 3, spleen; 4, lung; 5, kidney; 6, tumor. (k–n) *In vivo* evaluation of synergetic chemo-/radio-/photodynamic therapy on 4T1-tumor bearing mice after intratumoral injection of UCMSNs-HP-Dtxl. Digital photos of mice at (k) 30, (l) 60, (m) 90, (n) 120 days after treatment with UCMSNs-HP-Dtxl + RT + NIR. *Source:* Reprinted with permission from Fan et al. (2014); Copyright © 2014 Elsevier Ltd.

luminescence can be used for *in vivo* monitoring. Apart from loading with therapeutics, they can be engineered to degrade *in vivo* into benign components that can be cleared renally within specific timescales. Injected luminescent porous Si nanoparticles (LPSiNPs), on degradation, show no histopathological alterations in kidney, liver and spleen tissues of mice. Histopathologically, no significant toxicity was observed in these tissues relative to the controls. Hepatocytes in the liver samples appeared unremarkable, and there were no inflammatory infiltrates. Intrinsic photoluminescent properties of LPSiNPs helps in cellular imaging like HeLa cells using excitation wavelengths of 370488 and 750nm (two-photon excitation) two hours after incubation. These nanomaterials are also useful for vivo imaging by subcutaneous and intramuscular injections of LPSiNP dispersions as evident in experimental models using nude mouse. Imaging is achieved using fluorescence mode (green fluorescent protein [GFP] excitation filter, 445–490nm and ICG emission filter, 810–875nm) which is clearly visible without any skin autofluorescence. The fluorescence spectrum of LPSiNPs enables imaging in the near-infrared-emission range. Further, passive accumulation of D-LPSiNP is observed in a nude mouse bearing an MDA-MB-435 tumor as indicated by NIR fluorescence imaging. However, these accumulated nanoparticles exhibit rapid clearance within four weeks indicating ensuring biological safety (Park et al. 2009). Figure 4.9 shows the characterization of LPSiNPs and their applications in imaging.

4.7 Carbon-Based Nanomaterials

The unique structure of carbon-based nanomaterials with attractive physicochemical properties and biological behavior like chemical inertness, superior electrical conductivity, mechanical stability, high drug-loading capacity, and excellent biocompatibility have recently emerged as potential theranostic agents (Chen and Shi 2015). Carbon (sp^2)-based nanostructures exploited for drug delivery, biosensing, photothermal/photodynamic therapy, gene transfection, and tissue engineering/regeneration are comprised of fullerene (Da Ros and Prato 1999; Jensen et al. 1996), carbon quantum dots (CQDs) (Lim et al. 2015; Zheng et al. 2014), carbon nanotubes (CNTs) (Li et al. 2010; Liu et al. 2014a,b), and graphene (Morales-Narváez et al. 2017; Bai et al. 2018).

4.7.1 Fullerene

Buckminsterfullerene (C_{60}) is a truncated icosahedron containing 60 carbon atoms, with C_5–C_5 single bonds forming pentagons and C_5–C_6 double bonds forming hexagons. The diameter of a C_{60} fullerene molecule being 0.7nm, it is considered as a significant base for rational designing of nanomedicine. The poor solubility of C_{60} in coupled aqueous solvents and the tendency to aggregate in aqueous solutions can be well handled by various chemical and supramolecular approaches to functionalize fullerenes (Wilson 2000). Various fullerene-based nanomedicine consists of amphiphilic fullerene (AF-1, buckysomes, PEB), dendrofullerene (DF-1), fullerene-paclitaxel, fullerene polyamine, amino-fullerene adducts, fullerene-based amino acids and peptide, fullerene lipidosome, gadofullerenes, and bisphosphonate fullerene that are used for drug/gene delivery, radioprotection, cancer therapy, antiviral activity, MRI contrast agents, oxidative stress reduction, and bone-related therapy (Partha and Conyers 2009). Buckminsterfullerenes can add multiple radicals in fullerene molecule. This radical scavenging feature could be well exploited using water-soluble malonic acid derivatives of C_{60} with a defined 3D structure. Further, regioisomers of carboxyfullerenes are reported as neuroprotective antioxidants *in vitro* and *in vivo* (Murthy et al. 2002). Amphiphilic fullerenes (AF-1) which are ideally composed of an anticancer drug, paclitaxel, embedded in the hydrophobic pockets of buckysomes (PEBs), are spherical nanostructures in a size range of 100–200nm. This advance material facilitates the delivery of increased amount

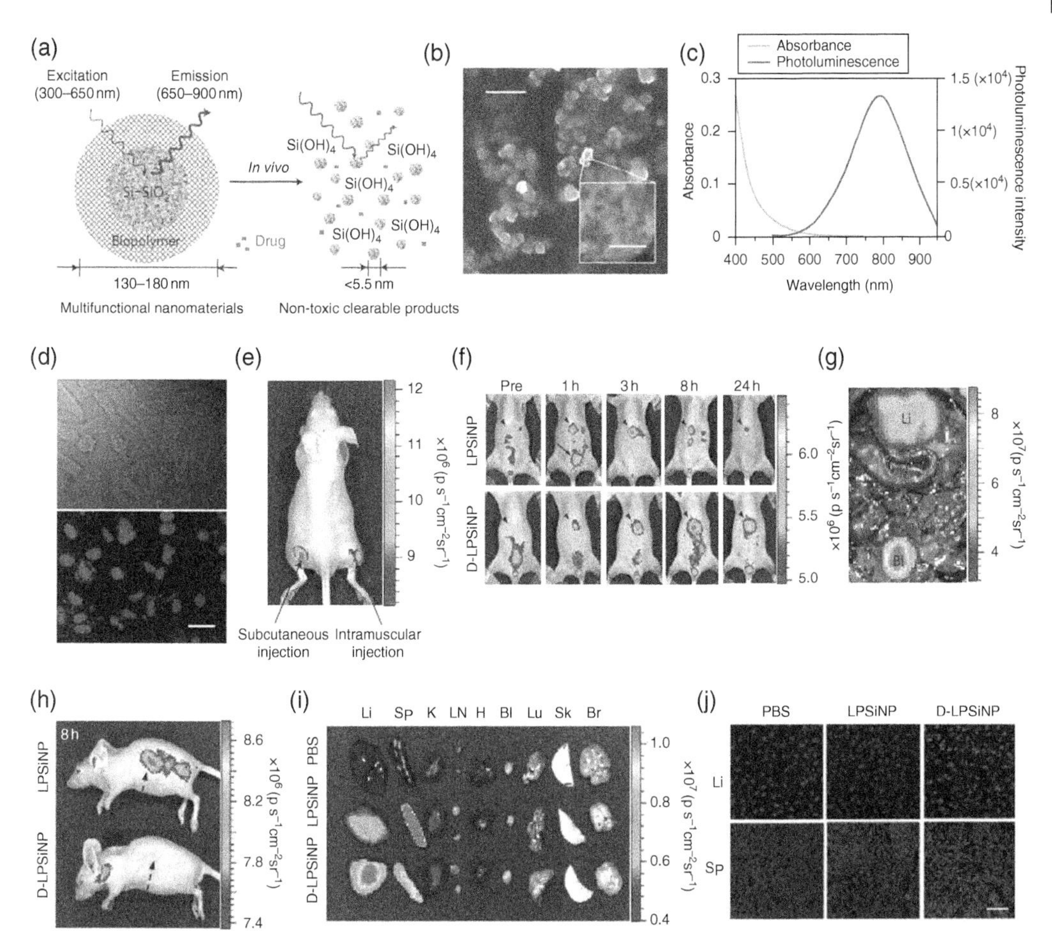

Figure 4.9 Characterization of LPSiNPs and their applications in imaging. (a) Schematic diagram depicting the structure and *in vivo* degradation process for the (biopolymer-coated) nanoparticles used in this study. (b) SEM image of LPSiNPs (the inset shows the porous nanostructure of one of the nanoparticles). The scale bar is 500 nm (50 nm for the inset). (c) Photoluminescence emission and absorbance spectra of LPSiNPs. Photoluminescence is measured using ultraviolet excitation (λ = 370 nm). (d) *In vitro* cellular imaging with LPSiNPs. HeLa cells were treated with LPSiNPs for 2 hours and then imaged. Red and blue indicate LPSiNPs and cell nuclei, respectively. The scale bar is 20 μm. (e)*In vivo* fluorescence image of LPSiNPs (20 μl of 0.1 mg ml^{-1}) injected subcutaneously and intramuscularly on each flank of a mouse. (f) *In vivo* images of LPSiNPs and D-LPSiNPs. The mice were imaged at multiple time points after intravenous injection of LPSiNPs and D-LPSiNPs (20 mg kg^{-1}). Arrowheads and arrows with solid lines indicate liver and bladder, respectively. (g) *In vivo* image showing the clearance of a portion of the injected dose of LPSiNPs into the bladder, 1 hour post-injection. Li and Bl indicate liver and bladder, respectively. (h) Lateral image of the same mice shown in (f), 8 hours after LPSiNP or D-LPSiNP injection. Arrows with dashed lines indicate spleen. (i) Fluorescence images showing the ex vivo biodistribution of LPSiNPs and D-LPSiNPs in a mouse. Organs were collected from the animals shown in f, 24 hours after injection. Li, Sp, K, LN, H, Bl, Lu, Sk, and Br indicate liver, spleen, kidney, lymph nodes, heart, bladder, lung, skin, and brain, respectively. (j) Fluorescence histology images of livers and spleens from the mice shown in f and i, 24 hours after injection. Red and blue indicate (D-) LPSiNPs and cellnuclei, respectively. The scale bar is 50 μm for all images. *Source:* Reprinted with permission from Park et al. (2009); Copyright © 2009 Springer Nature.

of paclitaxel, reduction in infusion times, higher uptake by tumor tissue and hence better anticancer efficacy (Partha et al. 2008). Fullerene-based delivery vectors of nanoscale dimensions facilitate passive targeting leading to accumulation and drug delivery specifically at tumor sites, as the endothelial cells of the tumor tissue create a leaky vasculature. The major advantages of this nanomaterial include prolonging circulation blood, prevention of enzymatic degradation, and reduction in nonspecific uptake by the RES. It is important to note that the dendritic groups present on the outer surface provide stealth function, reducing the clearance. Carboxyfullerene is another very significant nanomaterial which is highly potent against excitotoxic necrosis and protects against neuronal apoptosis. Amphiphilicity of such nanoparticles facilitate intercalation into brain membranes and enhances neuroprotective efficacy (Dugan et al. 1997). Thus, C_{60} derivatives are now widely explored as candidate neuroprotective drugs as they suppress iron-induced lipid peroxidation, and thus protect against degeneration of the nigrostriatal dopaminergic system. Moreover, intranigral infusion of carboxyfullerene is nontoxic to the nigrostriatal dopaminergic system, as per *in vivo* experiments in rats (Monti et al. 2000).

4.7.2 Carbon Quantum Dots

Fluorescent CQDs are a new class of carbon-based nanostructures with low toxicity, environmental friendliness, low cost, and simple synthetic routes. Their physicochemical properties can be tuned by rational surface passivation and functionalization, which has attributed to diverse applications in photocatalysis, electrocatalysis, chemical sensing, biosensing, bioimaging, and drug delivery (Lim et al. 2015). Being highly biocompatible, CQDs are being used for PDT against superficial tumors which requires localization and accumulation of photosensitizers in the tumor tissue followed by irradiation at specific wavelengths, triggering the formation of singlet oxygen species, leading to cell death as evident from experimental studies on MCF-7 and MDA-MB-231 cancer cells (Hsu et al. 2013; Bechet et al. 2008). Distribution, clearance, and tumor uptake of CQDs are dependent on surface functionalizing ligands and the route of administration (Yang et al. 2013). Moreover, they have the high tumor-to-background fluorescence contrast and low fluorescence levels in other tissues and organs, which make them more suitable as candidate photosensitizers. CQDs functionalized with PPEI-EI (PPEI-EI-CQDs) exhibit efficient photodynamic effect in Du145 and PC3 cells after irradiation with UV light which is attributed chiefly to photo-induced generation of singlet oxygen (Type II mechanism) and other reactive oxygen species (ROS) and radicals (Type I mechanism) (Juzenas et al. 2013). Attachment of chlorin e6, which is a potent photosensitizer, onto CQDs leads to a synergistic PDT where CQDs can indirectly excite the photosensitizer by the FRET mechanism (Huang et al. 2012). Additionally, CQDs can also be used for radiotherapy, where PEG-CQDs coated with a silver shell (C-Ag-PEG CQDs) could effectively and selectively damage Du145 cancer cells due to generation of free radicals by electrons ejected from C-Ag-PEG CQDs on irradiation with low-energy X-rays (Kleinauskas et al. 2013). bPEI-coated CQDs (bPEI-CQDs) with large numbers of amino groups on their surface which could condense DNA, display great potential in the application of gene delivery (Hu et al. 2014). Transfection experiments using enhanced green fluorescent protein (EGFP) reporter genes exhibited spontaneous attraction and condensation of the gene (polyanionic DNA strands) by bPEI with positive charge density and proton-sponge effect to form toroidal complexes which can easily be subjected to endocytosis-mediated cellular uptake. It is interesting to note that unprotonated amine moieties associated to the nanoconjugate buffer endolysosomal pH help in the cytoplasmic release of gene (Godbey et al. 1999). Organic dye-conjugated CQDs can be used as promising fluorescent probes to track H_2S level change in live cells, as it was reported to display the fluorescence images of HeLa and L929 cells, which on exposure to H_2S for 30 minutes at 37 °C turned green (Yu et al. 2013). Different functionalization strategies like conjugation of platinum(VI)-based anticancer pro-drug (oxaliplatin), photosensitive molecule (quinolone), targeting agent (folic acid), anticancer drug (doxorubicin), and PEG oligomers have enabled CQDs to serve as drug carriers and fluorescent tracers and thereby powerful theranostic agents (Lim et al. 2015; Zheng et al. 2014).

4.7.3 Carbon Nanotubes

Functionalized water-soluble CNTs are another type of carbon-based nanomaterials that are biocompatible and nontoxic and are being explored as cancer theranostics. Multidrug resistance (MDR) is a major challenge to cancer therapy as it exhibits increased efflux of anticancer drugs by overexpression of P-glycoprotein (P-gp). Antibody of P-gp (anti-P-gp) functionalized water-soluble single-walled carbon nanotubes (Ap-SWNTs) loaded with doxorubicin (Dox), Dox/Ap-SWNTs, are reported to specifically recognize the multidrug-resistant human leukemia cells (K562R), as well as exhibit effective loading and controllable near-infrared radiation (NIR), triggered release of Dox at the target K562R cells. These nanoparticles have 23-fold higher binding affinity toward drug-resistant K562R cells compared to drug-sensitive K562S cells. These Ap-SWNTs localize on the cell membrane and the fluorescence of Dox in K562R cells is reported to be significantly enhanced. Dox/Ap-SWNTs composite exhibited higher cytotoxicity and inhibition of cell proliferation compared to free Dox against K562R cells showing their potential to destroy the tumor stem cells, and inhibit the metastasis of tumor (Li et al. 2010). Modified CNTs can be very useful for imaging and diagnosis. Rational combination of superparamagnetic iron oxide (SPIO) nanoparticles with multiwalled carbon nanotubes (MWCNTs) led to fabrication of novel and efficient MRI T_2-weighted contrast agents with potential liver-targeting functionality. A coating of poly(diallyldimethylammonium chloride) (PDDA) was applied onto the surface of acid-treated MWCNTs via electrostatic interactions and SPIO nanoparticles were modified with a potential targeting agent, lactose-glycine adduct (Lac-Gly), which was eventually immobilized on the surface of the PDDA-MWCNTs (Figure 4.10). These novel CNT-based theranostic agents display superparamagnetism at room temperature. Their biocompatibility was indicated by low cytotoxicity against HEK293 and Huh7 cell lines. Enhanced T_2 relaxivities were observed for the hybrid material and could be effectively administered to an *in vivo* liver cancer model in mice showing remarkable enhancement in tumor to liver contrast ratio (277%) in T2-weighted magnetic resonance images (Figure 4.10) (Liu et al. 2014b).

4.7.4 Graphene

Graphene-based materials (GBMs) are used for encapsulation of drugs, nanoparticles, polymers, oxides, and cells resulting in excellent hybrid theranostics that have diversified applications in micro/nanomotors, power generation, super-capacitors, biosensors, environmental remediation, bioimaging, drug delivery, and tumor targeting. GBMs are basically composed of one-atom thick sheets containing sp^2 bonded carbon atoms patterned in a two-dimensional honeycomb lattice. GBMs include pristine graphene, polycrystalline graphene, graphene oxide (GO), reduced graphene oxide, and graphene quantum dots which are synthesized using various methods, namely electrostatic self-assembly, layer-by-layer self-assembly (LBL), aerosol-phase method, hydrothermal method, emulsification method, covalent bonding, chemical vapor deposition (CVD), co-pyrolysis method, and microwave-assisted solvothermal synthesis. These exotic GBMs have seen immense use in gene/drug delivery, imaging, toxicology, and regenerative medicine (Morales-Narváez et al. 2017). The multimodal bioimaging probe in both MRI and X-ray CT has been engineered based on multifunctional material composed of graphene oxide (GO), gold, and magnetite nanoparticles. This nanomaterial synthesized using aerosol-phase graphene encapsulation is magnetically responsive, exhibits excellent contrast enhancement, and can be used as cargo-filled graphene nanosacks for sustained release of nanoparticles in an aqueous environment for improved tumor-associated vascular penetration (Chen et al. 2013). Further, GO-encapsulated AuNPs into poly(lactic acid) microcapsules [Au@PLA-$(PAH/GO)_2$] are immensely effective in multimodal imaging (diagnostics) guided cancer photothermal therapies (Jin et al. 2013). GO-coated poly(lactic acid) microcapsules

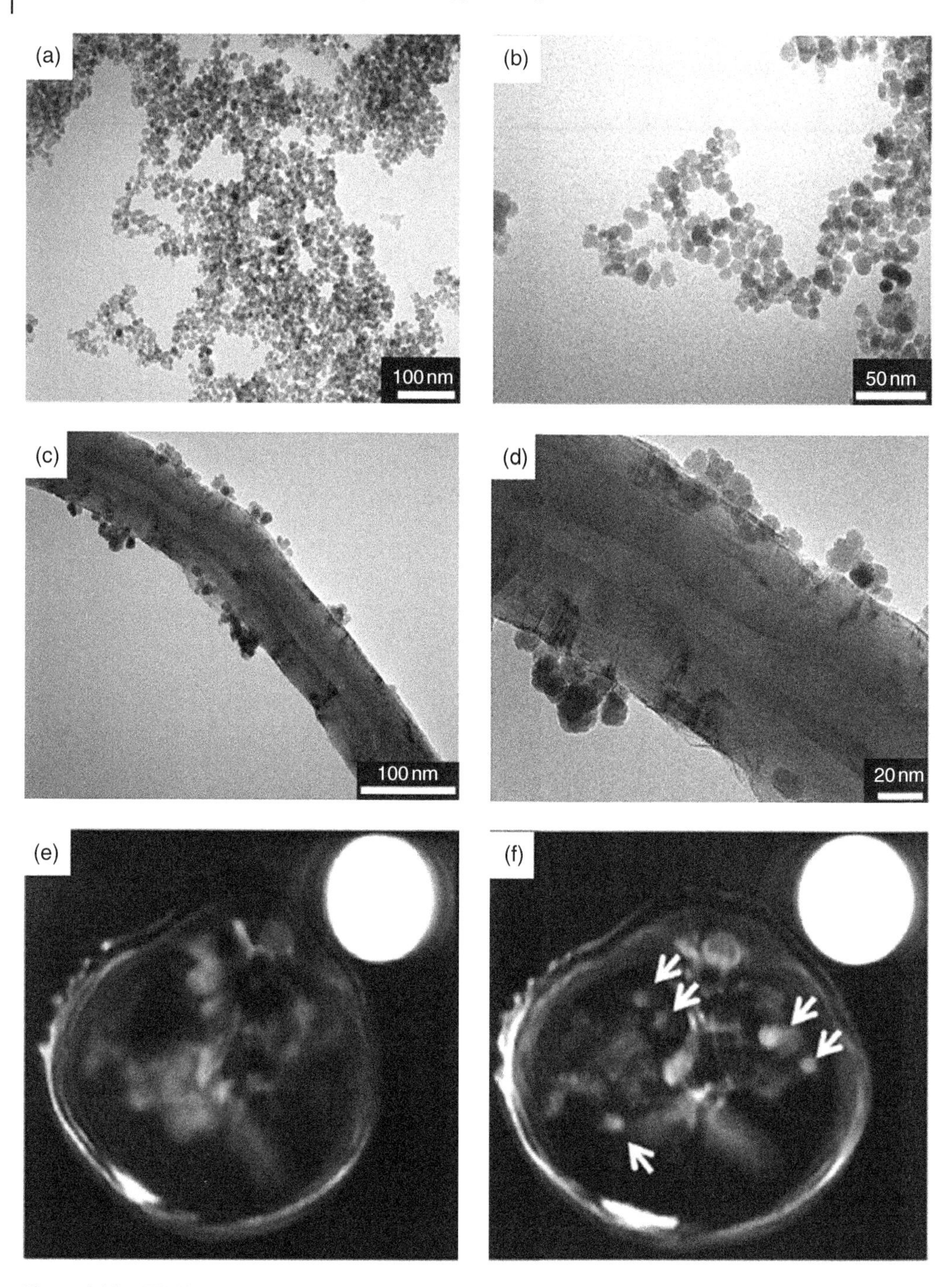

Figure 4.10 TEM images of SPIO@Lac-Gly (a, b) and CNT-PDDA-SPIO@Lac-Gly (c, d). *In vivo* MRI images of mouse liver (e) pre- and (f) post-injection of CNT-PDDA-SPIO@Lac-Gly at a dose of $10\,\mathrm{mg\,kg^{-1}}$ (white arrows indicate tumors) compared to internal standard (water, top right). *Source:* Reprinted with permission from Liu et al. (2014b); Copyright © 2013 Elsevier Ltd.

doped with IONPs synthesized by a double emulsion evaporation process with electrostatic self-assembly mediated GO coating can operate as contrast agents and simultaneously enhance ultrasound, magnetic resonance, and photoacoustic imaging, not only *in vitro* but also in vivo. These novel nanostructures are biocompatible and kill cancer cells upon NIR laser irradiation that can be further enhanced by using an external magnetic field (Li et al. 2014). GO/poly(allylamine hydrochloride) capsules synthesized using LBL assembly could be effectively loaded with an anticancer drug (e.g. doxorubicin) via incubation. These capsules with a "core–shell" loading property are biocompatible in nature (Kurapati and Raichur 2012). This novel nanohybrid shows NIR triggered drug release (Figure 4.11a) (Kurapati and Raichur 2013). Rationally fabricated GO-encapsulated AuNPs provide the double enhancement effect of GO and AuNPs on a Raman signature which can

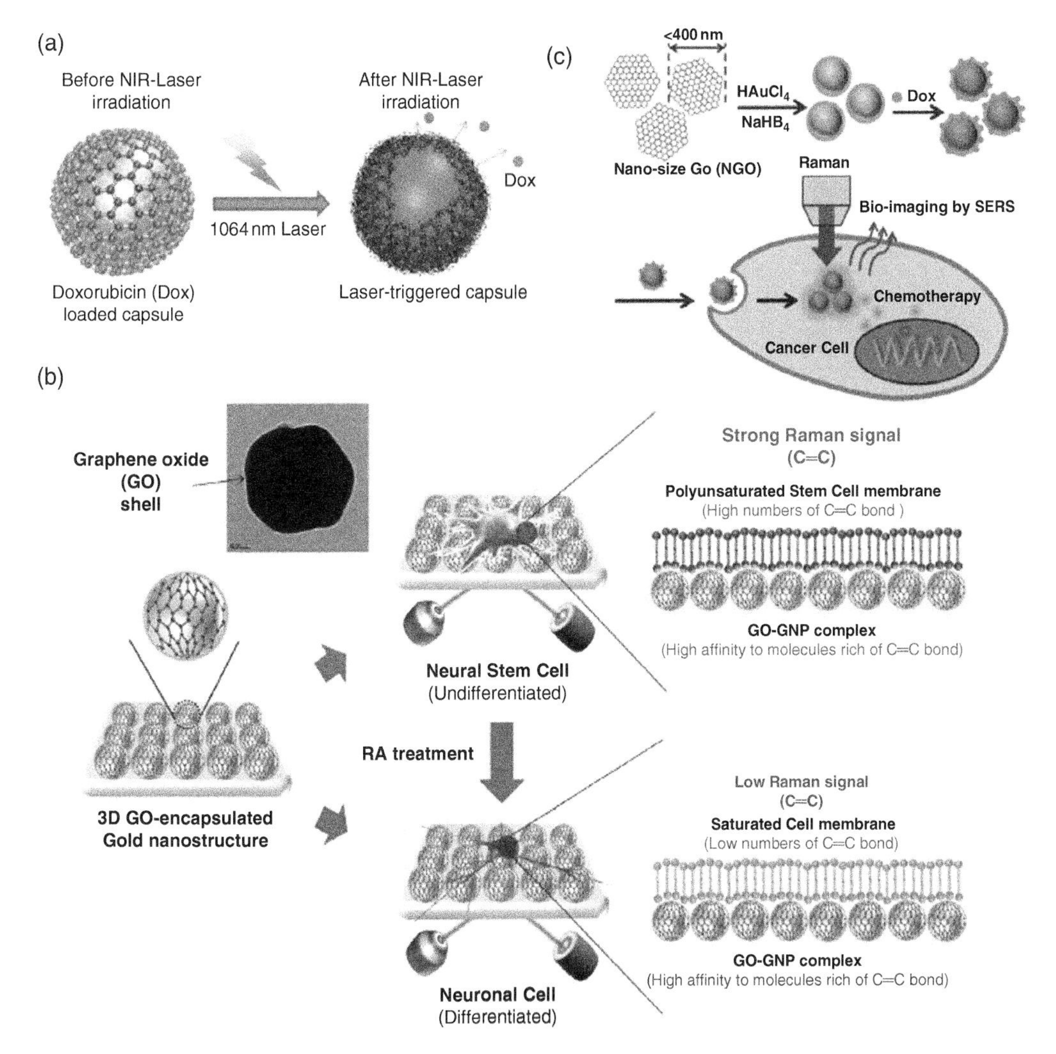

Figure 4.11 Graphene-encapsulated particles in biomedical applications (ii). (a) (NIR)-light controlled release of drugs using GO/poly(allylamine hydrochloride) capsules. (b) Monitoring of mouse neural stem cells differentiation using GO-encapsulated AuNPs. (c) GO-wrapped AuNPs as SERS (bioimaging) agents and drug delivery systems. *Source:* Reprinted with permission from Morales-Narváez et al. (2017); Copyright © 2017 Elsevier Ltd.

be observed specifically in undifferentiated neural stem cells (Figure 4.11b). This nanocomposite can be used to study the differentiation state of mesenchymal, hematopoietic, and single neural stem cells using electrochemical and electrical techniques (Kim et al. 2013). GO-wrapped AuNPs are not only useful for intracellular Raman imaging using in cancer cells (HeLa) but also can be loaded with the drug doxorubicin via non-covalent interactions for effective chemotherapy with low non-target cytotoxicity and high delivery efficacy (Figure 4.11c) (Ma et al. 2013). Apart from the aforementioned applications, various multi-functionalized GO/gold combinations with exotic shape, size, and physicochemical properties may be exploited for gene delivery, optical labeling, photothermal treatment, and both *in vitro* or *in vivo* fluorescent cellular imaging (Morales-Narváez et al. 2017).

4.8 Conclusion and Future Perspectives

This chapter provides an elaborate account of recent efforts for developing nanotheranostics composed of magnetic, gold, silver, quantum dots, polymer, silica, and carbon-based materials. Further research on mechanistic aspects involving efficient characterization of the materials, rational fabrication of the nanomaterials, and route of synthesis to avoid hazardous and toxic solvents to ensure biocompatibility are key points which are needed to be focused. Similarly, detailed toxicity and dosage evaluation studies coupled with pharmacokinetic and pharmacodynamic evaluation must be conducted before claiming the clinical suitability of the theranostic agents. Some of these nanotheranostics may provide great improvements over pre-existing medications and improve the life expectancy of the patients with severe infections, cancer, or neurodegenerative diseases. Although there have been encouraging and promising progress in the development of nanotheranostics, the toxicity, undesirable side effects, limited bioavailability, and obstacles in clinical transformations need to be primarily addressed for their widespread acceptance and applications in medical science. Moreover, certain other drawbacks such as the cost of AuNPs, low sensitivity of magnetic nanostructures acting as MRI contrast agents, complexity and larger dimension of silica-associated composite carrier, and non-biodegradable nature of carbon-based materials necessitate substantial attention and thus create scope for more advanced research. In recent years, functionalized nanomaterials have drastically altered and extended the application of nanotechnology in biomedical science.

Thus, proper knowledge of tailor-made synthesis of nanoparticles, chemical surface modification, and biological applications is a prerequisite for successful diagnosis and therapeutic outcomes in disease. In this chapter, we have provided an overview and most detailed ever-expanding research studies on functionalized nanostructures for diagnosis and therapy. Many of these novel nanotheranostic agents can deliver drugs to target tissues, ensure sustained release, enhanced bioavailability, quench ROS, function as MRI contrast agents, and enhance the quality of noninvasive multimodal imaging. Current studies are mostly based on *in vitro* and *in vivo* preclinical data and probably the major reasons behind the inability to get clinical approval include lack of consistent data on the cytotoxicity, genotoxicity, and immunotoxicity profiles on animal models. However, with the rapid technological advancement in the field of nanomedicine, soon this scenario is expected to change and the ultimate goal of theranostics meeting the demand for personalized medicine will become a reality.

References

Adersh, A., Kulkarni, A.R., Ghosh, S. et al. (2015). Surface defect rich ZnO quantum dots as antioxidant inhibiting α-amylase and α-glucosidase: a potential anti-diabetic nanomedicine. *Journal of Materials Chemistry B3* 3: 4597–4606.

Al-Jamal, W.T., Al-Jamal, K.T., Bomans, P.H. et al. (2008). Functionalized-quantum-dot–liposome hybrids as multimodal nanoparticles for cancer. *Small* 4: 1406–1415.

Al-Jamal, W.T., Al-Jamal, K.T., Tian, B. et al. (2009). Tumor targeting of functionalized quantum dot-liposome hybrids by intravenous administration. *Molecular Pharmaceutics* 6: 520–530.

AshaRani, P.V., Low Kah Mun, G., Hande, M.P., and Valiyaveettil, S. (2009). Cytotoxicity and genotoxicity of silver nanoparticles in human cells. *ACS Nano* 3: 279–290.

Bagalkot, V., Zhang, L., Levy-Nissenbaum, E. et al. (2007). Quantum dotaptamer conjugates for synchronous cancer imaging, therapy, and sensing of drug delivery based on bi-fluorescence resonance energy transfer. *Nano Letters* 7: 3065–3070.

Bagalkot, V., Badgeley, M.A., Kampfrath, T. et al. (2015). Hybrid nanoparticles improve targeting to inflammatory macrophages through phagocytic signals. *Journal of Controlled Release* 217: 243–255.

Bai, R.G., Ninan, N., Muthoosamy, K., and Manickam, S. (2018). Graphene: a versatile platform for nanotheranostics and tissue engineering. *Progress in Materials Science* 91: 24–69.

Bechet, D., Couleaud, P., Frochot, C. et al. (2008). Nanoparticles as vehicles for delivery of photodynamic therapy agents. *Trends in Biotechnology* 26: 612–621.

Bhagwat, T.R., Joshi, K.A., Parihar, V.S. et al. (2018). Biogenic copper nanoparticles from medicinal plants as novel antidiabetic nanomedicine. *World Journal of Pharmaceutical Research* 7: 183–196.

Blasiak, B., van Veggel, F.C.J.M., and Tomanek, B. (2013). Applications of nanoparticles for MRI cancer diagnosis and therapy. *Journal of Nanomaterials* 2013: 148578.

Carr, D.H., Brown, J., Bydder, G.M. et al. (1984). Gadolinium-DTPA as a contrast agent in MRI: initial clinical experience in 20 patients. *AJR. American Journal of Roentgenology* 143: 215–224.

Chen, W.Q. and Du, J.Z. (2013). Ultrasound and pH dually responsive polymer vesicles for anticancer drug delivery. *Scientific Reports* 3: 2162.

Chen, Y. and Shi, J. (2015). Mesoporous carbon biomaterials. *Science China Materials* 58: 241–257.

Chen, Y., Guo, F., Qiu, Y. et al. (2013). Encapsulation of particle ensembles in graphene nanosacks as a new route to multifunctional materials. *ACS Nano* 7: 3744–3753.

Chen, Q., Ke, H.T., Dai, Z.F., and Liu, Z. (2015a). Nanoscale theranostics for physical stimulus-responsive cancer therapies. *Biomaterials* 73: 214–230.

Chen, F., Hong, H., Goel, S. et al. (2015b). In vivo tumor vasculature targeting of CuS@MSN based theranostic nanomedicine. *ACS Nano* 9: 3926–3934.

Cheng, L., Wang, C., Feng, L.Z. et al. (2014). Functional nanomaterials for phototherapies of cancer. *Chemical Reviews* 114: 10869–10939.

Coates, A., Abraham, S., Kaye, S.B. et al. (1983). On the receiving end–patient perception of the side-effects of cancer chemotherapy. *European Journal of Cancer and Clinical Oncology* 19: 203–208.

Da Ros, T. and Prato, M. (1999). Medicinal chemistry with fullerenes and fullerene derivatives. *Chemical Communications* 8: 663–669.

Dugan, L.L., Turetsky, D.M., Du, C. et al. (1997). Carboxyfullerenes as neuroprotective agents. *Proceedings of the National Academy of Sciences of the United States of America* 94: 9434–9439. Erratum in: Proceedings of the National Academy of Sciences of the United States of America, 1997, vol. 94, pp. 12241.

Dykman, L.A. and Khlebtsov, N.G. (2019). Gold nanoparticles in chemo-, immuno-, and combined therapy: review [invited]. *Biomedical Optics Express* 10: 3152–3182.

Fan, W., Shen, B., Bu, W. et al. (2014). A smart upconversion-based mesoporous silica nanotheranostic system for synergetic chemo−/radio−/photodynamic therapy and simultaneous MR/UCL imaging. *Biomaterials* 35: 8992–9002.

Farrag, N.S., El-Sabagh, H.A., Al-mahallawi, A.M. et al. (2017). Comparative study on radiolabeling and biodistribution of core-shell silver/polymeric nanoparticles-based theranostics for tumor targeting. *International Journal of Pharmaceutics* 529: 123–133.

Foy, S.P., Manthe, R.L., Foy, S.T. et al. (2010). Optical imaging and magnetic field targeting of magnetic nanoparticles in tumors. *ACS Nano* 4: 5217–5224.

Ghosh, S. (2018). Copper and palladium nanostructures: a bacteriogenic approach. *Applied Microbiology and Biotechnology* 102: 7693–7701.

Ghosh, S. (2019). Mesoporous silica-based nano drug-delivery system synthesis, characterization, and applications. In: *Nanocarriers for Drug Delivery, Nanoscience and Nanotechnology in Drug Delivery Micro and Nano Technologies* (eds. S.S. Mohapatra, S. Ranjan, N. Dasgupta, et al.), 285–317. Elsevier.

Ghosh, S., Patil, S., Ahire, M. et al. (2011). Synthesis of gold nano-anisotrops using *Dioscorea bulbifera* tuber extract. *Journal of Nanomaterials* 2011 https://doi.org/10.1155/2011/354793.

Ghosh, S., Patil, S., Ahire, M. et al. (2012a). *Gnidia glauca* flower extractmediated synthesis of gold nanoparticles and evaluation of its chemocatalytic potential. *Journal of Nanobiotechnology* 10: 17.

Ghosh, S., Patil, S., Ahire, M. et al. (2012b). Synthesis of silver nanoparticles using *Dioscorea bulbifera* tuber extract and evaluation of its synergistic potential in combination with antimicrobial agents. *International Journal of Nanomedicine* 7: 483–496.

Ghosh, S., Nitnavare, R., Dewle, A. et al. (2015a). Novel platinum-palladium bimetallic nanoparticles synthesized by *Dioscorea bulbifera*: anticancer and antioxidant activities. *International Journal of Nanomedicine* 10: 7477–7490.

Ghosh, S., More, P., Nitnavare, R. et al. (2015b). Antidiabetic and antioxidant properties of copper nanoparticles synthesized by medicinal plant *Dioscorea bulbifera*. *Journal of Nanomedicine and Nanotechnology* S6: 007. https://doi.org/10.4172/2157-7439.S6-007.

Ghosh, S., More, P., Derle, A. et al. (2015c). Diosgenin functionalized iron oxide nanoparticles as novel nanomaterial against breast cancer. *Journal of Nanoscience and Nanotechnology* 15: 9464–9472.

Ghosh, S., Jagtap, S., More, P. et al. (2015d). *Dioscorea bulbifera* mediated synthesis of novel $Au_{core}Ag_{shell}$ nanoparticles with potent antibiofilm and antileishmanial activity. *Journal of Nanomaterials* 2015: 562938.

Ghosh, S., Gurav, S.P., Harke, A.N. et al. (2016a). *Dioscorea oppositifolia* mediated synthesis of gold and silver nanoparticles with catalytic activity. *Journal of Nanomedicine and Nanotechnology* 7: 5.

Ghosh, S., Chacko, M.J., Harke, A.N. et al. (2016b). *Barleria prionitis* leaf mediated synthesis of silver and gold nanocatalysts. *Journal of Nanomedicine and Nanotechnology* 7: 4.

Ghosh, S., Harke, A.N., Chacko, M.J. et al. (2016c). *Gloriosa superba* mediated synthesis of silver and gold nanoparticles for anticancer applications. *Journal of Nanomedicine and Nanotechnology* 7: 4.

Ghosh, S., Patil, S., Chopade, N.B. et al. (2016d). *Gnidia glauca* leaf and stem extract mediated synthesis of gold nanocatalysts with free radical scavenging potential. *Journal of Nanomedicine and Nanotechnology* 7: 358.

Ghosh, S., Sanghavi, S., and Sancheti, P. (2018). Metallic biomaterial for bone support and replacement. In: *Fundamental Biomaterials: Metals, Woodhead Publishing Series in Biomaterials* (eds. P. Balakrishnan, M.S. Sreekala and S. Thomas), 139–165. Elsevier.

Godbey, W.T., Wu, K.K., and Mikos, A.G. (1999). Poly(ethylenimine) and its role in gene delivery. *Journal of Controlled Release* 60: 149–160.

Gopinath, P.M., Ranjani, A., Dhanasekaran, D. et al. (2016). Multi-functional nano silver: a novel disruptive and theranostic agent for pathogenic organisms in real-time. *Scientific Reports* 6: 34058.

Guan, M., Dong, H., Ge, J. et al. (2015). Multifunctional upconversion-nanoparticles-trismethylpyridylporphyrin-fullerene nanocomposite: a near-infrared light-triggered theranostic platform for imaging-guided photodynamic therapy. *NPG Asia Materials* 7: e205.

Guo, D., Zhu, L., Huang, Z. et al. (2013). Anti-leukemia activity of PVP-coated silver nanoparticles via generation of reactive oxygen species and release of silver ions. *Biomaterials* 34: 7884–7894.

Guo, J., Rahme, K., He, Y. et al. (2017). Gold nanoparticles enlighten the future of cancer theranostics. *International Journal of Nanomedicine* 12: 6131–6152.

Hada, A.M., Potara, M., Suarasan, S. et al. (2019). Fabrication of gold-silver core-shell nanoparticles for performing as ultrabright SERS-nanotags inside human ovarian cancer cells. *Nanotechnology* 30: 315701.

Hayashi, K., Nakamura, M., Miki, H. et al. (2014). Magnetically responsive smart nanoparticles for cancer treatment with a combination of magnetic hyperthermia and remote-control drug release. *Theranostics* 4: 834–844.

Ho, Y.P. and Leong, K.W. (2010). Quantum dot-based theranostics. *Nanoscale* 2: 60–68.

Hsu, P.C., Chen, P.C., Ou, C.M. et al. (2013). Extremely high inhibition activity of photoluminescent carbon nanodots toward cancer cells. *Journal of Materials Chemistry B* 1: 1774–1781.

Hu, L.M., Sun, Y., Li, S.L. et al. (2014). Multifunctional carbon dots with high quantum yield for imaging and gene delivery. *Carbon* 67: 508–513.

Huang, P., Lin, J., Wang, X.S. et al. (2012). Light-triggered theranostics based on photosensitizer-conjugated carbon dots for simultaneous enhanced-fluorescence imaging and photodynamic therapy. *Advanced Materials* 24: 5104–5110.

Jahangirian, H., Kalantari, K., Izadiyan, Z. et al. (2019). A review of small molecules and drug delivery applications using gold and iron nanoparticles. *International Journal of Nanomedicine* 14: 1633–1657.

Jamdade, D.A., Rajpali, D., Joshi, K.A. et al. (2019). *Gnidia glauca* and *Plumbago zeylanica* mediated synthesis of novel copper nanoparticles as promising antidiabetic agents. *Advances in Pharmacological Sciences* 2019: 9080279.

Jensen, A.W., Wilson, S.R., and Schuster, D.I. (1996). Biological applications of fullerenes – a review. *Bioorganic and Medicinal Chemistry* 4: 767–779.

Jin, Y., Wang, J., Ke, H. et al. (2013). Graphene oxide modified PLA microcapsules containing gold nanoparticles for ultrasonic/CT bimodal imaging guided photothermal tumor therapy. *Biomaterials* 34: 4794–4802.

Juzenas, P., Kleinauskas, A., Luo, P.G., and Sun, Y.P. (2013). Photoactivatable carbon nanodots for cancer therapy. *Applied Physics Letters* 103: 063701.

Kale, S.N., Kitture, R., Ghosh, S. et al. (2017). Nanomaterials as enhanced antimicrobial agent/activity-enhancer for transdermal applications: a review. In: *Antimicrobial Nanoarchitectonics* (ed. A.M. Grumezescu), 279–321. Elsevier.

Kim, T.H., Lee, K.B., and Choi, J.W. (2013). 3D graphene oxide-encapsulated gold nanoparticles to detect neural stem cell differentiation. *Biomaterials* 34: 8660–8670.

Kitture, R. and Ghosh, S. (2019). Hybrid nanostructures for in vivo imaging. In: *Hybrid Nanostructures for Cancer Theranostics*, vol. 2019 (eds. R.A. Bohara and N. Thorat), 173–208. Elsevier.

Kitture, R., Ghosh, S., Kulkarni, P. et al. (2012). Fe_3O_4-citrate-curcumin: promising conjugates for superoxide scavenging, tumor suppression and cancer hyperthermia. *Journal of Applied Physics* 111: 064702–064707.

Kitture, R., Chordiya, K., Gaware, S. et al. (2015a). ZnO nanoparticles-red sandalwood conjugate: a promising anti-diabetic agent. *Journal of Nanoscience and Nanotechnology* 15: 4046–4051.

Kitture, R., Ghosh, S., More, P.A. et al. (2015b). Curcumin-loaded, self-assembled *Aloe vera* template for superior antioxidant activity and trans-membrane drug release. *Journal of Nanoscience and Nanotechnology* 15: 4039–4045.

Kleinauskas, A., Rocha, S., Sahu, S. et al. (2013). Carbon-core silver-shell nanodots as sensitizers for phototherapy and radiotherapy. *Nanotechnology* 24: 325103–325112.

Kosuge, H., Sherlock, S.P., Kitagawa, T. et al. (2012). Near infrared imaging and photothermal ablation of vascular inflammation using single-walled carbon nanotubes. *Journal of the American Heart Association* 1: e002568.

Kurapati, R. and Raichur, A.M. (2012). Graphene oxide based multilayer capsules with unique permeability properties: facile encapsulation of multiple drugs. *Chemical Communications* 48: 6013–6015.

Kurapati, R. and Raichur, A.M. (2013). Near-infrared light-responsive graphene oxide composite multilayer capsules: a novel route for remote controlled drug delivery. *Chemical Communications* 49: 734–736.

Lai, L., Zhao, C., Li, X. et al. (2016). Fluorescent gold nanoclusters for in vivo target imaging of Alzheimer's disease. *RSC Advances* 6: 30081–30088.

Li, R., Wu, R., Zhao, L. et al. (2010). P-glycoprotein antibody functionalized carbon nanotube overcomes the multidrug resistance of human leukemia cells. *ACS Nano* 4: 1399–1408.

Li, X.D., Liang, X.L., Yue, X.L. et al. (2014). Imaging guided photothermal therapy using iron oxide loaded poly(lactic acid) microcapsules coated with graphene oxide. *Journal of Materials Chemistry B* 2: 217–223.

Li, Y., Chang, Y., Lian, X. et al. (2018). Silver nanoparticles for enhanced cancer theranostics: in vitro and in vivo perspectives. *Journal of Biomedical Nanotechnology* 14: 1515–1542.

Liang, P., Shi, H., Zhu, W. et al. (2017). Silver nanoparticles enhance the sensitivity of temozolomide on human glioma cells. *Oncotarget* 8: 7533–7539.

Liao, Y.H., Chang, Y.J., Yoshiike, Y. et al. (2012). Negatively charged gold nanoparticles inhibit Alzheimer's amyloid-β fibrillization, induce fibril dissociation, and mitigate neurotoxicity. *Small* 8: 3631–3639.

Lim, S.Y., Shen, W., and Gao, Z.Q. (2015). Carbon quantum dots and their applications. *Chemical Society Reviews* 44: 362–381.

Lisjak, D. and Mertelj, A. (2018). Anisotropic magnetic nanoparticles: a review of their properties, syntheses and potential applications. *Progress in Materials Science* 95: 286–328.

Liu, Q., Jin, C., Wang, Y. et al. (2014a). Aptamer-conjugated nanomaterials for specific cancer cell recognition and targeted cancer therapy. *NPG Asia Materials* 6: e95.

Liu, Y., Hughes, T.C., Muir, B.W. et al. (2014b). Water-dispersible magnetic carbon nanotubes as T_2-weighted MRI contrast agents. *Biomaterials* 35: 378–386.

Luk, B.T. and Zhang, L. (2014). Current advances in polymer-based nanotheranostics for cancer treatment and diagnosis. *ACS Applied Materials and Interfaces* 6: 21859–21873.

Ma, X., Qu, Q., Zhao, Y. et al. (2013). Graphene oxide wrapped gold nanoparticles for intracellular Raman imaging and drug delivery. *Journal of Materials Chemistry B* 1: 6495–6500.

Mallick, A., More, P., Ghosh, S. et al. (2015). Dual drug conjugated nanoparticle for simultaneous targeting of mitochondria and nucleus in cancer cells. *ACS Applied Materials and Interfaces* 7: 7584–7598.

Matteis, V.D., Cascione, M., Toma, C.C., and Leporatti, S. (2018). Silver nanoparticles: synthetic routes, in vitro toxicity and theranostic applications for cancer disease. *Nanomaterials* 8: 319.

Mikhaylov, G., Mikac, U., Magaeva, A.A. et al. (2011). Ferri-liposomes as an MRI-visible drug-delivery system for targeting tumours and their microenvironment. *Nature Nanotechnology* 6: 594–602.

Mody, V.V., Nounou, M.I., and Bikram, M. (2009). Novel nanomedicine-based MRI contrast agents for gynecological malignancies. *Advanced Drug Delivery Reviews* 61: 795–807.

Monti, D., Moretti, L., Salvioli, S. et al. (2000). C60 carboxyfullerene exerts a protective activity against oxidative stress-induced apoptosis in human peripheral blood mononuclear cells. *Biochemical and Biophysical Research Communications* 277: 711–717.

Morales-Narváez, E., Sgobbi, L.F., Machado, S.A.S., and Merkoçi, A. (2017). Graphene-encapsulated materials: synthesis, applications and trends. *Progress in Materials Science* 86: 1–24.

Murthy, C.N., Choi, S.J., and Geckeler, K.E. (2002). Nanoencapsulation of [60] fullerene by a novel sugar-based polymer. *Journal of Nanoscience and Nanotechnology* 2: 129–132.

Park, J.H., Gu, L., von Maltzahn, G. et al. (2009). Biodegradable luminescent porous silicon nanoparticles for in vivo applications. *Nature Materials* 8: 331–336.

Partha, R. and Conyers, J.L. (2009). Biomedical applications of functionalized fullerene-based nanomaterials. *International Journal of Nanomedicine* 4: 261–275.

Partha, R., Mitchell, L.R., Lyon, J.L. et al. (2008). Buckysomes: fullerene-based nanocarriers for hydrophobic molecule delivery. *ACS Nano* 2: 1950–1958.

Quadir, M.A., Radowski, M.R., Kratz, F. et al. (2008). Dendritic multishell architectures for drug and dye transport. *Journal of Controlled Release* 132: 289–294.

Rokade, S.S., Joshi, K.A., Mahajan, K. et al. (2017). Novel anticancer platinum and palladium nanoparticles from *Barleria prionitis*. *Global Journal of Nanomedicine* 2: 555600.

Salunke, G.R., Ghosh, S., Santosh, R.J. et al. (2014). Rapid efficient synthesis and characterization of AgNPs, AuNPs and AgAuNPs from a medicinal plant, *Plumbago zeylanica* and their application in biofilm control. *International Journal of Nanomedicine* 9: 2635–2653.

Sanpui, P., Chattopadhyay, A., and Ghosh, S.S. (2011). Induction of apoptosis in cancer cells at low silver nanoparticle concentrations using chitosan nanocarrier. *ACS Applied Materials and Interfaces* 3: 218–228.

Sant, D.G., Gujarathi, T.R., Harne, S.R. et al. (2013). *Adiantum philippense* L. frond assisted rapid green synthesis of gold and silver nanoparticles. *Journal of Nanoparticles* 2013: 1–9.

Sasidharan, A., Sivaram, A.J., Retnakumari, A.P. et al. (2015). Radiofrequency ablation of drug-resistant cancer cells using molecularly targeted carboxyl-functionalized biodegradable graphene. *Advanced Healthcare Materials* 4: 679–684.

Schalla, S., Higgins, C.B., and Saeed, M. (2002). Contrast agents for cardiovascular magnetic resonance imaging: current status and future directions. *Drugs in R and D* 3: 285–302.

Shende, S., Joshi, K.A., Kulkarni, A.S. et al. (2017). *Litchi chinensis* peel: a novel source for synthesis of gold and silver nanocatalysts. *Global Journal of Nanomedicine* 3: 555603.

Shende, S., Joshi, K.A., Kulkarni, A.S. et al. (2018). *Platanus orientalis* leaf mediated rapid synthesis of catalytic gold and silver nanoparticles. *Journal of Nanomedicine and Nanotechnology* 9: 2.

Shinde, S.S., Joshi, K.A., Patil, S. et al. (2018). Green synthesis of silver nanoparticles using *Gnidia glauca* and computational evaluation of synergistic potential with antimicrobial drugs. *World Journal of Pharmaceutical Research* 7: 156–171.

Singh, R.K., Patel, K.D., Leong, K.W., and Kim, H.W. (2017). Progress in nanotheranostics based on mesoporous silica nanomaterial platforms. *ACS Applied Materials and Interfaces* 9: 10309–10337.

Slowing, I., Trewyn, B.G., and Lin, V.S.Y. (2006). Effect of surface functionalization of MCM-41-type mesoporous silica nanoparticles on the endocytosis by human cancer cells. *Journal of the American Chemical Society* 128: 14792–14793.

Smith, B.R. and Gambhir, S.S. (2017). Nanomaterials for in vivo imaging. *Chemical Reviews* 117: 901–986.

Song, J.T., Yang, X.Q., Zhang, X.S. et al. (2015). Facile synthesis of gold nanospheres modified by positively charged mesoporous silica, loaded with near-infrared fluorescent dye, for in vivo x-ray computed tomography and fluorescence dual mode imaging. *ACS Applied Materials and Interfaces* 7: 17287–17297.

Stensberg, M.C., Wei, Q., McLamore, E.S. et al. (2011). Toxicological studies on silver nanoparticles: challenges and opportunities in assessment, monitoring and imaging. *Nanomedicine* 6: 879–898.

Sun, Q., He, F., Bi, H. et al. (2019). An intelligent nanoplatform for simultaneously controlled chemo-, photothermal, and photodynamic therapies mediated by a single NIR light. *Chemical Engineering Journal* 362: 679–691.

Thakur, N.S., Patel, G., Kushwah, V. et al. (2019). Facile development of biodegradable polymer-based nanotheranostics: hydrophobic photosensitizers delivery, fluorescence imaging and photodynamic therapy. *Journal of Photochemistry and Photobiology, B: Biology* 193: 39–50.

Tran, Q.H., Nguyen, V.Q., and Le, A.T. (2013). Silver nanoparticles: synthesis, properties, toxicology, applications and perspectives. *Advances in Natural Sciences: Nanoscience and Nanotechnology* 4: 033001.

Wallyn, J., Anton, N., Akram, S., and Vandamme, T.F. (2019). Biomedical imaging: principles, technologies, clinical aspects, contrast agents, limitations and future trends in nanomedicines. *Pharmaceutical Research* 36: 78.

Wang, H., Agarwal, P., Zhao, S. et al. (2015). A biomimetic hybrid nanoplatform for encapsulation and precisely controlled delivery of theranostic agents. *Nature Communications* 6: 10081.

Wilson, S.R. (2000). Biological aspects of fullerenes. In: *Fullerenes: Chemistry, Physics and Technology* (eds. K.M. Kadish and R.S. Ruoff), 431–436. New York, NY: Wiley.

Yamada, M., Foote, M., and Prow, T.W. (2015). Therapeutic gold, silver, and platinum nanoparticles: therapeutic metal nanoparticles. *Wiley Interdisciplinary Reviews. Nanomedicine and Nanobiotechnology* 7: 428–445.

Yang, K., Gong, H., Shi, X.Z. et al. (2013). In vivo biodistribution and toxicology of functionalized nano-graphene oxide in mice after oral and intraperitoneal administration. *Biomaterials* 34: 2787–2795.

Yeşilot, Ş. and AydınAcar, Ç. (2019). Silver nanoparticles; a new hope in cancer therapy? *Eastern Journal of Medicine* 24: 111–116.

Yu, M., Li, X.Z., Zeng, F. et al. (2013). Carbon-dot-based ratiometric fluorescent sensor for detecting hydrogen sulfide in aqueous media and inside live cells. *Chemical Communications* 49: 403–405.

Zhang, Z., Wang, L., Wang, J. et al. (2012). Mesoporous silica-coated gold nanorods as a light-mediated multifunctional theranostic platform for cancer treatment. *Advanced Materials* 24: 1418–1423.

Zhang, Z., Wang, J., and Chen, C. (2013). Gold nanorods based platforms for light-mediated theranostics. *Theranostics* 3: 223–238.

Zheng, M., Liu, S., Li, J. et al. (2014). Integrating oxaliplatin with highly luminescent carbon dots: an unprecedented theranostic agent for personalized medicine. *Advanced Materials* 26: 3554–3560.

5

Aptamer-Incorporated Nanoparticle Systems for Drug Delivery

Fahimeh Charbgoo[1,2], Seyed Mohammad Taghdisi[3,4], Rezvan Yazdian-Robati[5], Khalil Abnous[1,6], Mohammad Ramezani[1], and Mona Alibolandi[1]

[1] *Pharmaceutical Research Center, Pharmaceutical Technology Institute, Mashhad University of Medical Sciences, Mashhad, Iran*
[2] *DWI – Leibniz Institute for Interactive Materials, Aachen, Germany*
[3] *Targeted Drug Delivery Research Center, Pharmaceutical Technology Institute, Mashhad University of Medical Sciences, Mashhad, Iran*
[4] *Department of Pharmaceutical Biotechnology, School of Pharmacy, Mashhad University of Medical Sciences, Mashhad, Iran*
[5] *Molecular and Cell Biology Research Center, Faculty of Medicine, Mazandaran University of Medical Sciences, Sari, Iran*
[6] *Department of Medicinal Chemistry, School of Pharmacy, Mashhad University of Medical Sciences, Mashhad, Iran*

5.1 Introduction

Aptamers are usually single-stranded sequences of DNA (ssDNA) or RNA capable of folding into a three-dimensional structure. They can bind to specific target molecules with high affinity and specificity, providing a wide range of applications (Munzar et al. 2019). They range from 20 to 80 bases in length with a molecular mass of 6–26 kDa. Since their discovery in 1990 by Ellington and Szostak, the researchers have been wondering about what else oligonucleotides might be doing that we have not yet understood. Aptamers are very similar to antibodies in function, but they have promising properties comparing to them. Aptamers have improved reproducibility, rare immunogenicity, long self-life, and ease of modification. Thes biomolecules are more thermostable and can fold reversibly, making them more tolerant to environmental changes. Due to their identity, aptamers are sensitive to degradation by nucleases (Gotrik et al. 2016). This issue limits their wide clinical uses.

Walking at the interface of nanotechnology, aptamers looked to find solutions to reduce their limitations. A combination of aptamers with nanoparticles can be used in different fields including analytical chemistry, biosensing, purification, imaging, and diagnosis (Munzar et al. 2019). They are also used in disease treatment and drug delivery (Sun and Zu 2015). Nanotechnology-based aptameric systems have introduced magic results in biomedicine since they can solve the challenges related to each of them, separately. Aptamers lead the nanosystems to the target site and provide therapeutic effects. Nanoparticles increase the loading capacity, resistance to nucleases, and cell penetration level. Since aptamers are very small, their use in drug delivery systems has attracted great attention, especially in terms of reaching tumors through the enhanced permeability and retention (EPR) effect. Moreover, conjugating aptamers onto nanoparticle surfaces increases the aptamer density on the material due to the higher surface area of nanomaterials, and thus raises the possibility of interaction between aptamer and its target molecule (Urmann et al. 2017).

Nanobiotechnology in Diagnosis, Drug Delivery, and Treatment, First Edition. Edited by Mahendra Rai, Mehdi Razzaghi-Abyaneh, and Avinash P. Ingle.

There are two different approaches for attachment of aptamers onto nanoparticles: non-covalent and covalent bonding. For the first approach, surface charge of nanoparticle is pivotally important. Non-covalent attachment of aptamers on cationic nanoparticles is commonly used, since aptamers as negatively charged oligonucleotides will be attached on the positively charged nanoparticle. For covalent attachment, the main point is that the linkage should not affect the functionality and affinity of aptamer (Urmann et al. 2017). In this chapter, different types of aptamers and different types of nanoparticles conjugated with aptamer and their properties for drug delivery are discussed.

5.2 Different Types of Aptamers for Drug Delivery

Aptamers can be ssDNA or RNA oligonucleotides, while DNA aptamers are more stable than RNA aptamers. Both ssDNA or RNA aptamers can be used as targeting or therapeutic agents. As a targeting agent, the aptamer leads the nanosystem to the desired site, reducing side effects of treatment. As a therapeutic agent, the aptamer can attach to its target that makes some improvement in disease treatment. There are some aptamers that are used for design of smart nanosystems which lead to the cargo release of nanoparticles just in the presence of their targets. The following section implies some widely used targeting, therapeutic, and gating aptamers.

5.2.1 Aptamers for Targeting

5.2.1.1 Mucin 1

The first aptamer, mucin 1 (MUC1), was introduced by Ferreira and group in 2006. MUC1 is present on the apical surface of most normal secretory epithelial cells. In most adenocarcinomas, MUC1 is overexpressed, so it is found over the entire cell surface and also shed as MUC1 fragments into the blood (Nabavinia et al. 2017). Accordingly, MUC1 is a potential candidate as a molecular target for overcoming cancer therapy challenges. The most common MUC1 aptamers with highest affinity are S2.2, MA3, 5TR1, 5TRG2, and GalNAc3. The first MUC1 aptamer for targeting drug delivery was used in conjugation to poly(ethylene glycol) (PEG), which enhances its nuclease resistance and blood circulation half-life (Tan et al. 2011). After that, MUC1 aptamers were used for targeting nanosystems to achieve more effective drug delivery systems. A detailed discussion is given in Section 5.3.

5.2.1.2 AS1411

AS1411 is a 26-mer guanine-rich ssDNA that binds to nucleolin as its target. Nucleolin is a nucleolar phosphorprotein that is overexpressed on the surface of certain cancer cells. It is now well-established that AS1411 has very effective anticancer properties. AS1411 can be used for both targeting and therapy since nucleolin is present on the surface of cells, as well as the cytoplasm and nucleus membrane. Surface nucleolin serves as the receptor for AS1411, leading to selective uptake in cancer cells (Abnous et al. 2018). Ko et al. (2009) reported the first conjugation of AS1411 aptamer with nanoparticles (quantum dots) for imaging of cancer. A further wide range of studies were performed using AS1411 targeted nanosystems for drug delivery.

5.2.1.3 Prostate-Specific Membrane Antigen (PSMA)

Prostate-specific membrane antigen (PSMA) aptamer is specific against a membrane protein especially expressed by prostate cancer cells. In 2002, Lupold et al. (2002) synthesized a RNA aptamer as the first aptamer for PSMA. However, this aptamer has 79 nucleotides and a very high molecular weight, which severely restricts its function as a targeting molecule (Lupold et al. 2002). Further, an RNA-based aptamer with 37 nucleotides was introduced which showed significant reduction in molecular weight and increased affinity for PSMA (Rockey et al. 2011). It has been demonstrated that RNA aptamers show higher affinity than DNA-based aptamers. However, considering the instability of RNA aptamers and their high costs, these are only used in situations where DNA aptamers with significant affinity do not exist.

5.2.1.4 EGFR

A peptide aptamer against epidermal growth factor receptor (EGFR) was introduced by Buerger and Groner, which has strong ability to inhibit the growth factor signaling (Buerger and Groner 2003). EGFR plays an important role in survival, proliferation, differentiation, and migration, as well as progression of tumors. A RNA aptamer against EGFR was developed in 2009 (Liu et al. 2009) and its first conjugation with nanoparticles was reported in 2010 using gold nanoparticles (AuNPs) (Li et al. 2010).

5.2.1.5 Sgc8c

Acute lymphoblastic leukemia (CCRF-CEM) cells are targets of the sgc8 DNA aptamer introduced by Shangguan in 2008 (2008). Sgc8c is a truncated form of sgc8 which has the affinity to bind a transmembrane receptor named protein tyrosine kinase 7 (PTK7) (Kd ~1 nM). PTK7 is highly expressed on CCRF-CEM cells. Sgc8c can identify between target leukemia cells and normal human bone marrow aspirate, as well as distinguishing cancer cells nearly related to the target cell line in clinical specimens due to its high specificity (Leitner et al. 2017). The first application of sgc8 for targeted drug delivery was reported in 2009 through its conjugation to viral capsid protein (MS2) in order to produce multivalent cell-targeting vehicles (Ray and White 2010). The high binding affinity of sgc8 and sgc8c led to more improvements in targeted delivery which is discussed in section3.

5.2.1.6 EpCAM Aptamer

The epithelial cell adhesion molecule (EpCAM) aptamer is a 19-mer RNA aptamer which selectively binds to EpCAM. EpCAM is a glycosylated transmembrane protein that has an important role in cell growth through the persistent expression of the activated c-myc oncogene. EpCAM is overexpressed in the surface of many cancer cells like breast cancer, while it has low level of expression in normal cells (Alibolandi et al. 2015).

5.2.2 Therapeutic Aptamers

The main therapeutic strategy of aptamers is inhibiting receptor–ligand or protein–protein interactions as an antagonist. There are limited numbers of aptamers with confirmed therapeutic effects *in vivo* which are discussed in the following sections.

5.2.2.1 AS1411

AS1411 is a DNA aptamer with quadruplex-forming guanine-rich (G-rich) structure and the first aptamer entered clinical trials as a cancer therapeutic agent. AS1411 binds to nucleolin, which is overexpressed in cancer cells, leading to its inhibition with results in apoptosis induction. It binds to nucleolin and internalizes into target cells, thereby preventing nucleolin from attaching and stabilizing anti-apoptotic BCL2 to mRNA. Destabilization of this mRNA leads to the reduction of

BCL2 protein production, causing apoptosis and cell death. There are two types of anticancer drugs based on the AS1411 aptamer in clinical trial phase II. AS1411 is currently in clinical trials for renal cell carcinoma and leukemia treatment and is in preclinical investigation for other hematologic and solid malignancies.

5.2.2.2 Proliferating Cell Nuclear Antigen (α-PCNA) Aptamer

Proliferating cell nuclear antigen (α-PCNA) is a protein in the eukaryotic cell's nuclei, which functions as a component of DNA replication and repair machinery. The anti-PCNA aptamer, in the presence of its target, inhibits the activity of human DNA polymerase ε and δ at nM range of concentrations. A complex of PCNA- α-PCNA aptamer provides resistance against the exonuclease role of the DNA polymerases (Kowalska et al. 2018). There is no report on attachment of this aptamer to nanoparticles it may because this aptamer functions in nucleus. Accordingly, nanosystems should be able to enter the nucleus to ensure the functionality of the conjugated α-PCNA aptamer that makes the designs complicated.

5.2.2.3 Forkhead Box M1 (FOXM1)

FOXM1 is a transcription factor, which binds to DNA promoters to trigger expression of proliferation-associated genes. This protein is a candidate of antitumor agents since its overexpression is correlated with tumorigenesis and the progression of many cancers. The aptamer against FOXM1 was produced in 2017 (Xiang et al. 2017) and its first use in nano-particulate systems was introduced in 2018 (Abnous et al. 2018).

5.2.2.4 NOX-A12

NOX-A12, an RNA aptamer in L-configuration from the Spiegelmer company, binds and inactivates CXC chemokine ligand (CXCL12), which has a critical role in homing and retention of chronic lymphocytic leukemia (CLL) cells (Hoellenriegel et al. 2014). This aptamer was used to treat non-Hodgkin's lymphoma and multiple myeloma. There is no report on NOX-A12 conjugation to nanoparticle systems.

5.2.2.5 Vimentin

Vimentin is the main intermediate filament in non-muscle cells that is also named fibroblast intermediate filament. This protein is overexpressed in cancer cells, enhancing immigration and invasion (Hayat 2005). Zamay et al. (2014) introduced a DNA aptamer capable of attaching to vimentin protein and inducing apoptosis in cancer cells named as NAS-24. After that, to enhance its tumor response and penetration into cancer cells, it was attached to selenium nanoparticles by our research team (Figure 5.1) (Jalalian et al. 2018). Then, NAS-24 application in nanotechnology-based approaches was developed significantly (Bahreyni et al. 2017a).

5.2.2.6 Vascular Endothelial Growth Factor (VEGF)

Vascular endothelial growth factor (VEGF) glycoprotein plays an important role in increasing vascular permeability, promoting endothelial cell growth and angiogenesis, and inducing cell migration, as well as inhibiting apoptosis. Blocking of its function with aptamers was first reported by M.C. Willis. The inhibitory effect of the aptamer toward VEGF decreases endothelial cell proliferation *in vitro* and also reduces vascular permeability and angiogenesis *in vivo*. In this report, the VEGF aptamer was anchored to liposome, which is a common nanosystem for therapeutic agents' delivery (Willis et al. 1998).

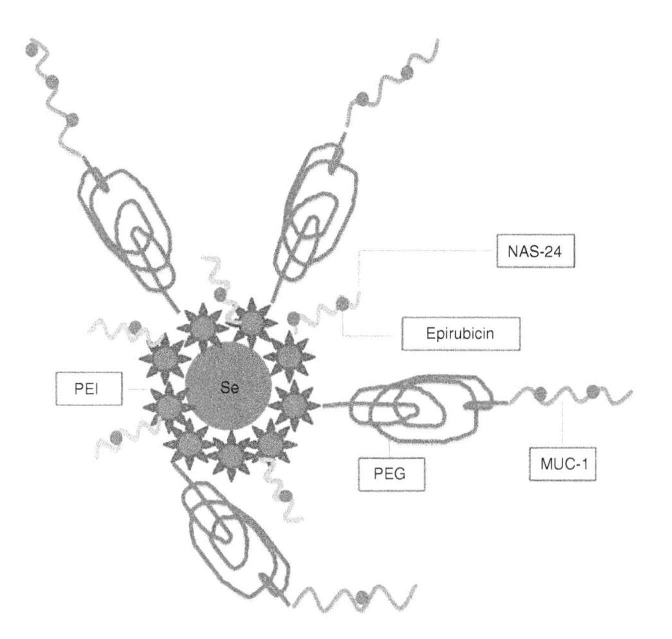

Figure 5.1 Schematic description of Epi-loaded-NAS-24-functionalized PEI-PEG-MUC1 aptamer-coated selenium nanoparticles. *Source:* Adapted from Jalalian et al. (2018); with copyright permission from Elsevier.

5.2.3 Gating/Sensing Aptamers

In addition to targeting and therapy, aptamers play other interesting roles such as gating and sensing. For example, the adenosine triphosphate (ATP) aptamer is widely used for both mentioned applications (Bahreyni et al. 2017b). The level of ATP molecule increases significantly in the cancer cell cytoplasm. So, accurate detection of it will help to diagnose tumor cells. Moreover, gating nanosystems carrying therapeutics or imaging agents will ensure the release of cargo in the cytoplasm of cancer cells, providing controlled release systems.

5.3 Aptamer-Conjugated Nanosystems for Targeted Delivery Platforms

5.3.1 Aptamer-Based Polymeric Nanoparticles

Polymers have dedicated specific properties for nanoscale delivery systems, including superior storage stability and sustained drug release. These types of materials are almost biodegradable and biocompatible. Therefore, they have attracted a great deal of attention from researchers for their use in drug delivery. The most widely used polymers for drug delivery are polyethylene glycol (PEG), chitosan, poly ethylene imine (PEI), and poly (lactic-co-glycolic acid) (PLGA) (Liechty et al. 2010; Li et al. 2018).

PEG is a polymer composed of $O(CH_2)_2$ monomers, which protects other elements of the delivery system in biological media. PEG is soluble in different types of solvents including both polar and nonpolar ones. Thus, PEG is widely applied as a carrier for hydrophobic therapeutics to enhance their solubility in biological media or increase their dissolution characteristics (Bunker

2012). For example, a PEGylated carboxymethyl cellulose was designed for the delivery of SN38, which is a hydrophobic and highly toxic drug used in cancer therapy. Further, a micelle-like structure was produced which targeted cancer cells by attaching CD133 aptamers to the exposed PEG moieties to reduce side effects of the drug. The data showed that aptamer targeted self-assembled nanoparticles could selectively enter and induce apoptosis in HT29 cells overexpressing CD133 (Alibolandi et al. 2018a). PEG can be also used as the linker for conjugation of targeting agents or drugs into other elements of the delivery system. For example, AS1411 and Wy5a aptamers were attached into PEG to be conjugated into PEI, which is a positively charged polymer useful for gene and DNA delivery (Lee et al. 2019). Conjugation of PEG with PEI is a logical design since PEG is a polymer electrolyte and associates strongly with cations (Bunker 2012).

The cationic charge of PEI is generally used to deliver nucleic acids (gene delivery) that are negatively charged and compensate for the PEI identity that makes some drawbacks for its usage *in vivo*. High positively charged polymers are toxic and attract more levels of protein corona in the bloodstream (Vinogradov et al. 1999). However, there are different reports on using this polymer for dual therapy in hybrid systems. In 2018, a complex composed of AS1411-chitosan-ss-polyethylenimine-urocanic acid was fabricated to deliver doxorubicin (Dox) and TLR4 siRNA selectively. The results showed that TLR4 expression was suppressed and reduced migration and invasion, and improved the Dox antitumor effect (Yang et al. 2018b). In another study, an interesting pH-responsive complex composed of single-walled carbon nanotube (SWCNT) and PEI was synthesized for co-delivery of Dox and surviving siRNA. In this report, PEI was covalently conjugated to betaine and the resulting polyethylene imine–betaine (PB) complex was further attached to SWCNTs to provide an excellent pH-responsive lysosomal scape (Cao et al. 2019).

Chitosan is a cationic polysaccharide formed from chitin with random β-(1-4)-linked N-acetyl-D-glucosamine (GlcNAc) and D-glucosamine (GlcN), and is commonly used for gene and drug delivery. This polymer is biocompatible and biodegradable with a high capability of cell penetrance. It also protects oligonucleotides against serum endonucleases. Additionally, its modification is easy and can be administered through versatile routes. It can be used for dual delivery of drugs like a report that delivered docetaxel and siRNA to cancer cells using MUC1 aptamers as targeting agents (Jafari et al. 2019). Chitosan can also be used in form of hydrogel by applying hydroxyl and amino groups of glucosamine units as reactive sites for attaching crosslink groups. The chitosan-based nanogels could deliver Dox into LNCaP cells in a targeted manner using a ssDNA aptamer against LNCaP cells (Atabi et al. 2017). They also have some drawbacks such as fast renal exertion, low solubility in water, and low stability, which can be addressed by conjugating them to other nanomaterials providing hybrid systems. For example, introducing a hydroxyl group significantly increases chitosan water solubility. Using an "all in one" approach, a complex of chitosan-AuNPs-AS1411 aptamer was synthesized for methotrexate delivery to promising effect in inducing apoptosis in cancer cells and tissues. The presence of Au provided the ability to track the chitosan-drug complex even *in vivo* (Guo et al. 2018). Chitosan was also combined with PLGA to overcome the limitation of PLGA for delivery of paclitaxel (PTX) to cancer cells (Lu et al. 2019).

PLGA or poly (lactic-co-glycolic acid) is a Food and Drug Administration (FDA) approved polymer with great biodegradability and biocompatibility. The most important limitation of PLGA is that it cannot specifically interact with cells or proteins, which results in a reduced level of drug accumulation in target tissues. Conjugation of chitosan with PLGA, increased the cell penetrance ability, encapsulation efficiency, and burst release property of PLGA (Lu et al. 2019). Conjugation of PEG into PLGA nanoparticles significantly reduced systemic clearance compared to the PLGA particles without PEG. Accordingly, AS1411-conjugated PEG-PLGA nanoparticles could effectively deliver PTX to C6 glioma cells (Guo et al. 2011). In another study, conjugation of PLGA to just CD133 aptamers could effectively deliver propranolol to hemangioma (Guo et al. 2017). In another work, AS1411-attached PLGA was also reported to successfully interact with cancer cells and release the PTX (Aravind et al. 2012b). The development of this system by inserting lecithin-PEG unites could remarkably increase loading capacity and sustained release profile (Aravind et al. 2012a). Furthermore, our group designed a targeted delivery system for transferring both epirubicin (Epi) and anti-miRNA 21 to cancer cells based on conjugation of MUC1 aptamer to nanoparticles composed of two biocompatible polymers of poly (β amino ester) (PβAE) and PLGA. PβAE was the core and PLGA acted as the shell of the complex. The results showed that the presented complex efficiently inhibited tumor growth in tumor-bearing mice compared with Epi alone (Figure 5.2) (Bahreyni et al. 2019).

Moreover, using different protocols, PLGA nanoparticles with specific morphology were produced for targeted delivery. For example, PLGA nano-bubbles were synthesized by applying a water-in-oil-in-water (water/oil/water) double emulsion and carbodiimide chemistry method for targeted delivery of PTX into prostate cancer cells using A10-3.2 aptamer (Wu et al. 2017). The available report proposed that PLGA is more appropriate for delivery of hydrophobic drugs such as PTX due to its unique identity, described previously.

5.3.2 Aptamer-Based Lipid Nanoparticles

The most common type of lipid-based nanoparticles are liposomes that can be formed using natural or synthetic amphipathic lipids in water. They have widely been used as delivery carriers owing to their biocompatibility and safety. Their surface can be modified by attaching different moieties, molecules, and polymers such as PEG which can prolong their circulation half-life (Mukherjee et al. 2019). Dox-loaded liposomes received FDA approval in 1995 for the first time (Barenholz 2012). To date, about 16 liposomal drugs have been clinically approved and a few of those are

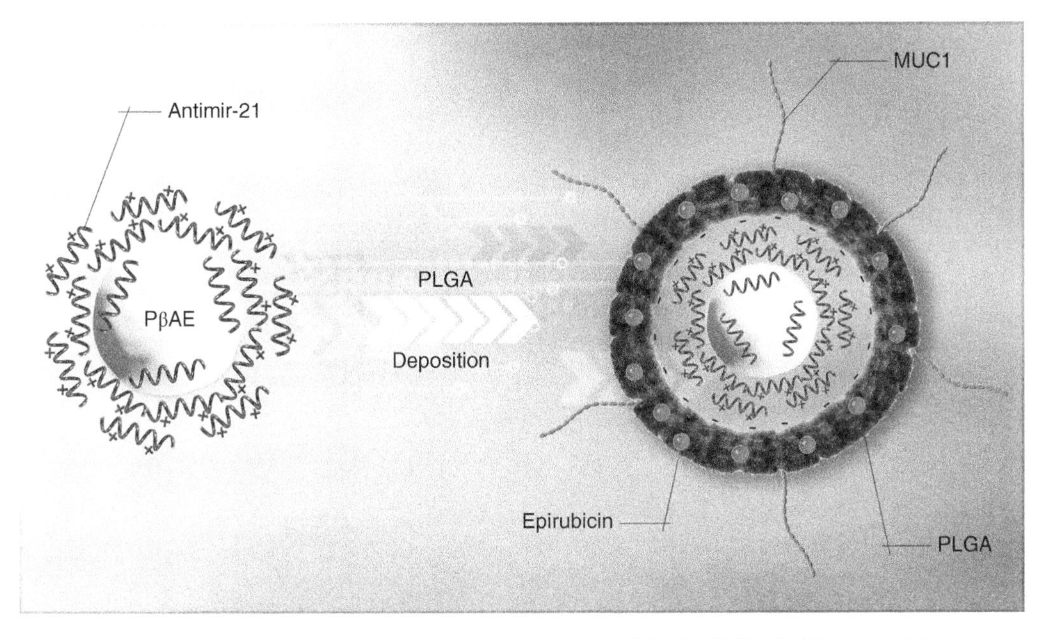

Figure 5.2 Schematic representation of the MUC1 aptamer-modified PLGA-Epi-PβAE-antimir-21 nanocomplex. *Source:* Adapted from Bahreyni et al. (2019); with copyright permission from Elsevier.

commercialized (Barenholz 2012). Despite clinical approval, most of the FDA-approved liposomal drugs do not show significant overall survival (OS) improvement over the traditional drugs (Petersen et al. 2016). In 2017, it was reported that a phase III study of a liposomal containing dual drugs (cytarabine and daunorubicin) revealed enhanced OS compared to their individual drugs (Chen et al. 2018).

Conjugating lipids to aptamers generally produces micellar structures with a hydrophobic core that is composed of lipid tail, and the hydrophilic part includes nucleic acid aptamers (Wu et al. 2010). Aptamer-based liposomes have been mostly used for targeted delivery of drugs. For example, a complex of HER3 aptamer liposome could deliver Dox to cancer cells and reduce its cardiotoxicity (Dou et al. 2018). During development of lipid-based delivery systems, they are usually combined to polymers to solve their limitations like low loading capacity, spontaneous degradation, and low stability. The conjugation of lipids to polymers helps to control the size of final nanoparticles by changing the ratio of lipid : polymer. MUC1-attached PEG-lipid nanoparticles effectively delivered vinorelbine (VRL) to the cancer cells (Liu et al. 2015). Also, lipid-PLGA nanoparticles modified with CD133 aptamer remarkably increased the delivery of salinomycin into osteosarcoma and cancer stem cells (Yu et al. 2018). In 2019, lipid-PLGA containing CD133 aptamer was applied to deliver all-trans retinoic acid (ATRA) into osteosarcoma-initiating cells. This was the first report on increasing ATRA delivery with nanoparticles to osteosarcoma initiating cells applying CD133 aptamers (Gui et al. 2019).

5.3.3 Aptamer-Based DNA Nanostructures

Targeting DNA nanostructures with aptamers is largely reported due to their nucleic acid structure. Based on Watson-Crick base pairing, a wide variety of predictable DNA nanostructures were proposed and conjugated to aptamers for targeted drug delivery. Among them, tetrahedral DNA nanostructures (TDNs) are the most common type because of the high capacity of drug loading, increased intracellular uptake capability, and outstanding stability (Li et al. 2017). Li et al. showed that AS1411-attached TDNs could remarkably enter target cells compared to TDNs alone. They used L929 cells as negative control (nucleolin^{-}) and MCF-7 cells as positive ones, demonstrating that AS1411-TDNs inhibited MCF-7 growth and promoted L929 growth while TDNs promoted growth of both MCF-7 and L929 without any selection (Li et al. 2017). In 2018, a MUC1 aptamer-modified TDN was introduced for remarkable delivery of Dox to cancer tissues. The presence of MUC1 aptamer significantly increased the cellular uptake level of Dox in target cells and reduced its entrance into normal cells. The level of Dox penetration into target cells interestingly depended on the level of MUC1 aptamer on the TDNs. The efficacy of tumor inhibition in mice treated with MUC1 Apt TDNs was also improved compared to the Dox-treated group (Han et al. 2018).

In another study, AS1411 aptamer was incorporated into the DNA pyramid and the results demonstrated that the attachment of aptamer into DNA nanostructures could increase resistance to nucleases compared to free aptamer (Charoenphol and Bermudez 2014). DNA icosahedrons were also used for targeted Dox delivery using MUC1 aptamers as targeting elements on their surface. Applying five DNA strands based on a five-point-star motif plus one other strand with a specific aptamer sequence could easily provide the targeted icosahedron structure (Chang et al. 2011). In another study, the authors introduced a diamond DNA structure carrying MUC1 aptamers on its edges for targeted delivery of Epi which was incorporated within the double-stranded regions of the DNA nanostructure. The Epi was released at the target site using a pH-sensitive approach (Abnous et al. 2017). In a similar design, our research team fabricated a three-way junction pocket DNA structure and also a polyvalent aptamer-based structure for anthracycline delivery using AS1411 and MUC1 aptamers (Figure 5.3), respectively (Yazdian-Robati et al. 2016; Taghdisi et al. 2018).

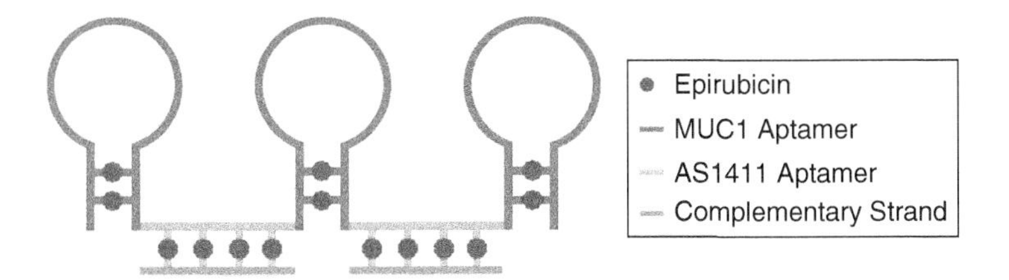

Figure 5.3 Schematic illustration of the polyvalent aptamer-based structure for Epi delivery using AS1411 and MUC1 aptamers. *Source:* Adapted from Yazdian-Robati et al. (2016); with copyright permission from Springer Nature.

Further, cruciform DNA nanostructures composed of AS1411 and FOXM1 aptamers were introduced as targeting and therapeutic agents, respectively. The structure was stable *in vivo* due to its structure and conjugation of PEG polymer to the 3'-end of aptamers. The nanostructure entered the target cells selectively and induced a significant level of cancer cell death without any remarkable effect on non-target cells (Abnous et al. 2018). All these structures are synthesized by providing different lengths of fragments capable of aligning with each other from specific parts. Other approaches can also be applied for preparing oligonucleotide nanostructures like rolling circle amplification (RCA). A DNA nanoparticle capable of carrying high levels of PMSA (prostate membrane-specific antigen) aptamers was developed for prostate cancer therapy using RCA. The template included a primer-binding site, complementary sequence to PSMA aptamer, drug-loading domain, and pH sensitive spacer. Accordingly, the produced DNA nanoparticle could selectively carry aptamer and Dox to the prostate cancer cells. Also, the pH-sensitive spacer enhanced the release of drugs in the target site which had acidic pH (Zhang et al. 2017a).

In order to improve the stability of DNA-based nanostructures, further they combined with other nanomaterials such as lipids, peptides, and polymers. In one of the studies, a targeted DNA micelle was prepared for dual therapy by conjugation of one strand of a dsDNA to cholesterol moieties. The complex formed a micelle in an aqueous environment to deliver both Dox and KLA peptide. KLA is a pro-apoptotic peptide that conjugated to outer side of one strand and the other outer strand was attached to a MUC1 aptamer as the targeting agent. The complex showed significant stability in serum due to KLA attachment and reduced adverse effects on normal cells due to the presence of MUC1 aptamer (Charbgoo et al. 2018a).

5.3.4 Aptamer-Based Peptide Nanoparticles

Application of peptide-based carriers for drug delivery is really a promising platform since it is more stable than DNA-based structures *in vivo*. It is made of bio-blocks and can be used in negative, positive, and neutral forms. Moreover, it can be synthesized easily and provides a high drug loading capacity (Tesauro et al. 2019). Peptides can form self-assembled structures such as nanovesicles, nanoparticles, and hydrogels. A methionine bis-alkylated peptide Wpc with only nine amino acid residues formed a nanoparticle capable of encapsulating AS1411 aptamers. The nanoparticles were disassembled in cytoplasm of cancer cells due to reduction of their disulfide bands (Ma et al. 2019). Some peptides also have therapeutic effects, so they can be used as carrier as well as therapeutic agents. For instance, melittin is a peptide composed of 26 amino acids with high anticancer efficacy was conjugated to AS1411 aptamer to be transferred to just target cancer cells (Rajabnejad et al. 2018).

Moreover, peptides have been used to solve the limitations of liposomes instead of polymers such as PEG as some peptides are not immunogenic. The peptide coating also reduces protein corona formation, outperforming PEGylated liposomes. For example, a lipo-peptide was provided by inserting peptide units within the liposomal membrane, resulting in more thermally and chemically stable structures. Aptamers were further attached on their surfaces as the targeting agents, ensuring delivery of encapsulated Dox into the target site with minimal nonspecific uptake (Ranalli et al. 2017).

5.3.5 Aptamer-Based Inorganic Nanoparticles

Inorganic nanoparticles such as gold, silver, superparamagnetic iron oxide nanoparticles (SPIONs), selenium, and silica nanoparticles have attracted great attention to be used as carriers in delivery systems due to their diversity, high loading capacity, and providing imaging properties besides carrying drugs. AuNPs can be easily attached to aptamers and other therapeutics like peptides. A His-tag DNA aptamer-attached to AuNPs was used for the delivery of antimicrobial peptides to the target site (Yeom et al. 2016). Recently, AuNPs have been widely used in hybrid systems. For example, AuNPs were attached to dendrimers to increase loading capacity and provide a platform for conjugation of MUC1 aptamer as a targeting agent and selectively deliver curcumin into the colon adenocarcinoma (Alibolandi et al. 2018b). In another design, a magnetite-gold nanocluster-aptamer nanocomposite, having a size around 35 nm, was developed for Epi delivery. It showed great results for delivering the cargo to the desired site (Binaymotlagh et al. 2019).

In addition, an acetylated carboxymethyl cellulose-coated hollow mesoporous silica hybrid nanosystem was fabricated for the targeted delivery of Dox to colon adenocarcinoma using AS1411 aptamers. *In vitro* and *in vivo* studies indicated that AS1411 aptamer-attached hybrid nanoparticles demonstrated a remarkable therapeutic efficiency over non-targeted complex and free Dox (Nejabat et al. 2018). Mesoporous silica nanoparticles (MSNPs) have been utilized widely due to their high drug-loading capacity and ease of gating their surfaces. For example, MSNPs were filled by Epi and their surfaces were covered by covalent attachment of MUC1 aptamers using disulfide bonds. After entering to the reductive cytoplasm of cancer cells, the S–S bond was opened and the drug was released at the target site (Hanafi-Bojd et al. 2018). Similarly, another research team has fabricated Epi-loaded-NAS-24-functionalized PEI-PEG-5TR1 aptamer-coated selenium nanoparticles (SeNPs) for targeted dual therapy of cancer cells *in vitro* and *in vivo*. As mentioned before, NAS-24-aptamer is a therapeutic aptamer with strong anticancer effects (Jalalian et al. 2018). SeNPs are biocompatible but they are quite unstable *in vivo* and should be stabilized by conjugating them to the polymers like chitosan (Hosnedlova et al. 2018). In another study, a MUC1 aptamer-attached SPION was developed for Epi delivery in adenocarcinomas. The results obtained revealed that SPION considerably accumulated in target tissues and transferred the drug to the target site by means of aptamer (Jalalian et al. 2013).

There are many reports of using different aptamer-modified nanoparticles for drug delivery. Here, we discussed those newly designed, as well as the ones that contain a conceptual approach. Nowadays, some more complicated platforms have been introduced that release the drugs at the desired site at the desired time; these are called smart delivery systems and are discussed in Section 5.4.

5.4 Aptamer-Conjugated Nanosystems for Smart Delivery Platforms

Stimuli-responsive aptamer-conjugated nanosystems are smart drug nanocarriers that mimic biological response behavior and release drugs in response to various endogenous stimuli, such as pH, temperature, redox, etc., or physical exogenous stimuli like light, X-ray, ultrasound, magnetic field, etc. (Kim et al. 2018). They are used for highly accurate treatment of cancers. Compare to internal stimuli, external triggers have less variability and are easier to be controlled. Stimuli-responsive systems are able to encapsulate drugs, protect them from the outer environment, and react on a given stimulus to improve therapeutic efficacy (Sahle et al. 2018).

5.4.1 Endogenous Stimuli-Responsive Aptamer-Conjugated Nanosystems

5.4.1.1 pH-Responsive Aptamer-Conjugated Nanosystems

pH variations have been employed to control the release of drugs in pathological situations or in specific organs. Tumor cells grow rapidly and generate enormous amounts of lactic acid. This is the main consideration for drug delivery from the points of pH-sensitive drug release and cellular internalization (Belleperche and DeRosa 2018). There are many pH-responsive aptamer-conjugated nanosystems that have taken benefit of the pH difference between normal tissues and the environment of tumors (Mura et al. 2013). For example, a dual-targeted (folate and the AS1411 aptamer) pH-sensitive biocompatible polymeric nanosystem was fabricated using atom transfer radical polymerization (ATRP)-based biodegradable tri-block copolymer, poly(poly(ethylene glycol) methacrylate)-poly(caprolactone)-poly(poly(ethylene glycol) methacrylate) (pPEGMA-PCL-pPEGMA). Dox was conjugated to the tri-block polymer by an acid-labile hydrazone linkage. A higher cumulative Dox release at pH 5.0 (~70%) in comparison to pH 7.4 (~25%) was detected (Lale et al. 2014).

In 2016, a delivery platform to carry daunorubicin to acute lymphoblastic leukemia T-cells using AuNPs conjugated with sgc8c aptamers was designed. Daunorubicin was bound to the AuNPs through electrostatic interactions. Therefore, protonation of NH_2- group of daunorubicin at acidic pH in the tumor cell microenvironment led to more and faster release of this anticancer drug (Taghdisi et al. 2016a). In another study, porous calcium carbonate nanoparticles coated with avidin were constructed to treat breast cancer cells. Calcium carbonate nanoparticles were loaded with Epi or melittin and conjugated to biotin-labeled dimer MUC1 aptamers. The findings obtained showed that under the acidic conditions of the lysosome, the structure of the calcium carbonate nanoparticle was destructed, leading to release of Epi or melittin (Yazdian-Robati et al. 2019).

In another pH-responsive and ATP-responsive system, a DNA dendrimer was fabricated for the delivery of Epi using three kinds of aptamers including MUC1 and AS1411 for targeting and internalization; the third aptamer was an ATP-binding aptamer which was integrated into the structure of the dendrimer (Figure 5.4). Within the lysosomes of target cells, the ATP aptamer was bound to ATP and destabilized the dendrimer structure, leading to the release of Epi as an intercalating agent. Moreover, protonation of the Epi at the acidic pH of lysosomes enhanced the rate of drug release (Taghdisi et al. 2016b). Taghdisi et al. (2018) designed a DNA nanostructure consisting of three strands of the AS1411 aptamer containing several double-helix sites where Dox drug was intercalated. Using AS1411 aptamer, the DNA structure was capable to attach to nucleolin and internalized into target cells. Then, protonation of Dox inside the acidic environment of the cells caused the anticancer drug to be released (Taghdisi et al. 2018).

5.4.1.2 Redox-Responsive Aptamer-Conjugated Nanosystems

Disulfide bonds are susceptible to quick cleavage by glutathione (GSH). This phenomenon has been exploited in the design of redox-responsive aptamer-conjugated nanosystems. Different amounts of GSH exist in extracellular (~2–10 μM) and intracellular (~2–10 mM) parts and in tumor tissues relative to normal cells, leading to cytosolic release of drugs in redox-responsive delivery systems (Mura et al. 2013).

Zhuang and coworkers engineered a novel redox-responsive hyper-branched polymer with a high cancer cell proliferation inhibition rate. The polymer backbone was successfully prepared by reversible addition-fragmentation chain transfer (RAFT) polymerization as well as self-condensing vinyl

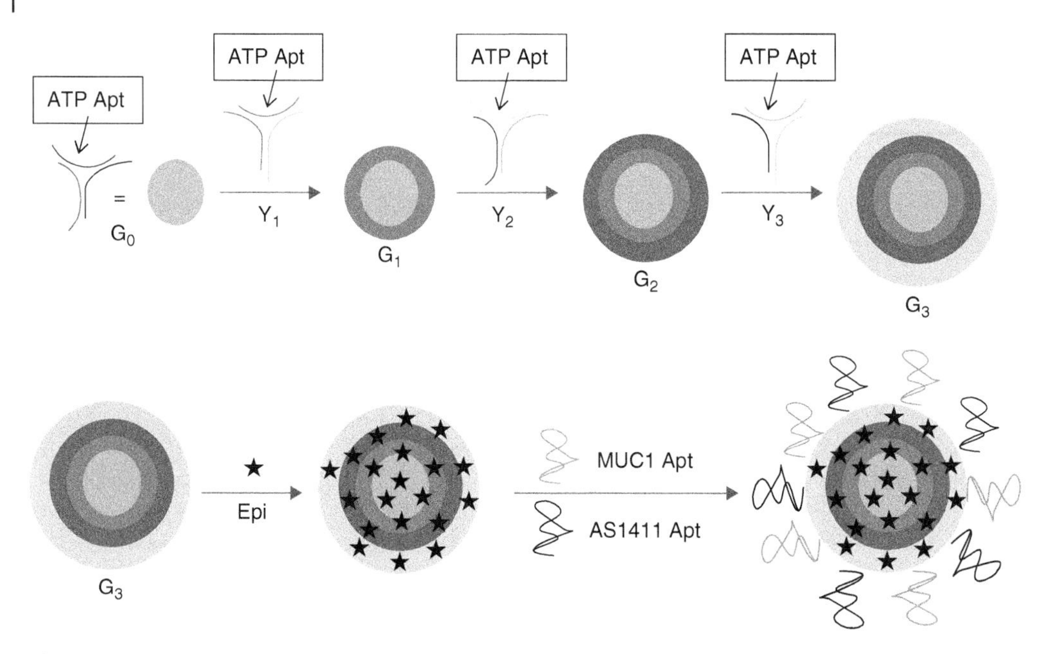

Figure 5.4 Schematic illustration of Apts-Dendrimer (G3)-Epi complex formation. *Source:* Adapted from Taghdisi et al. (2018); with copyright permission from Elsevier.

polymerization (SCVP). Subsequently, terminal vinyl groups were modified with AS1411 to mediate the endocytosis process. This amphiphilic polymer was self-assembled into nanoparticles, creating a hydrophobic inner core to encapsulated Dox. After internalization of hyper-branched polymer-AS1411 into cancer cells, the disulfide bonds in the backbone of polymer were cleaved by GSH in the cytoplasm, discharging Dox rapidly in the cytoplasm (Zhuang et al. 2016).

Mesoporous materials have been also used for the development of redox-sensitive systems. For instance, Zhang et al. (2017b) exploited Dox-loaded mesoporous carbon nanoparticles (MCNs) containing disulfide bonds, wherein polyacrylic acid (PAA) and PEI were encapsulated in MCNs to prevent the leakage of Dox. MUC1 aptamer was immobilized onto the surface of nanocarrier for specific recognition of human lung adenocarcinoma and breast cancer cells. Under reductive intracellular conditions, the MCNs-Dox-PAA/PEI-MUC1 aptamer complex was quickly disassembled, leading to the release of Dox inside target cells (Zhang et al. 2017b).

Incorporation of disulfide bond between the two polymer blocks is another route for design of a redox-responsive system. Using this approach, chitosan and PEI were linked together through disulfide bonds and urocanic acid was attached to PEI covalently by amide bonds. Then, thus formed micelles were functionalized with AS1411 aptamer. Toll-like receptor-4 siRNA (TLR4–siRNA) and Dox were loaded into the micelle as therapeutic agents. The results confirmed the excellent performance of the designed delivery system for tumor therapy *in vitro* and *in vivo* experiments (Yang et al. 2018b).

5.4.2 Physical Exogenous Stimuli-Responsive Aptamer-Conjugated Nanosystems

5.4.2.1 Light and Temperature-Responsive Aptamer-Conjugated Nanosystems

In recent years, numerous photo-based therapeutic modalities have been introduced which mainly includes photodynamic therapy (PDT), photothermal therapy (PTT), photo-induced chemotherapy, and photoimmunotherapy (PIT). Non-invasiveness in addition to the possibility of remote spatiotemporal control are two main features of this approach (Guo and You 2017). Yang and group designed a light-sensitive drug delivery platform that released its cargo under UV irradiation (365 nm) using a UV-sensitive highly branched polymer modified by sgc8 aptamer. Under UV irradiation, this biocompatible DNA-grafted polymer induced significantly more cancer cell death in comparison to the condition without UV irradiation (Yang et al. 2018a). In another study, researchers exploited a light strategy that used a complementary DNA sequence to hybridize and mask sgc8 aptamer grafted on the surface of two different photothermal agents including gold nanorods and SWCNTs. Upon exposure to near-infrared (NIR) light, localized heat generated on the surface of nano-agents led to increase of temperature. This phenomenon dehybridized the dsDNA and released the aptamer which could distinguish specific tumor cells. Overall, with external NIR laser irradiation and aptamer targeting, antitumor efficiency significantly enhanced (Yang et al. 2016).

A thermoresponsive ammonium-bicarbonate bubble-generating liposomal system was developed by Chuang et al. (2016) for encapsulations of gold nanocages and Dox. MUC-1 aptamer was grafted on the surfaces of the liposome to enhance drug delivery at the targeted site. It also acts as a molecular beacon signaling, following the systemic administration of the liposomes. Upon irradiation, the gold nanocages effectively converted the NIR into localized heat, making the decomposition of the loaded ammonium bicarbonate to generate CO_2 bubbles. Therefore, permeable defects on the lipid membrane were created and quickly triggered Dox release. Administration of the loaded thermoresponsive liposomal system in tumorigenic rat models decreased the relative tumor volume to ~25% and ~60% with and without NIR, respectively (Chuang et al. 2016).

5.4.2.2 Ultrasound-Responsive Aptamer-Conjugated Nanosystems

Application of ultrasound waves in smart targeted delivery platforms has gained great attention in clinical research because of some advantages including non-invasiveness, the lack of ionizing radiations, and the easy regulation of tissue penetration depth. In this context, Wang et al. (2012) incorporated Dox into acoustic droplets consisting of liquid perfluoropentane in core and lipid-based shell materials. The conjugation of sgc8c aptamer to the nanodroplets enhances the targeting CCRF-CEM cells for both imaging and therapeutic use. Upon high-intensity focused ultrasound insonation, the perfluorocarbon droplets were transited from an instant phase into a gas bubble leading to a fivefold extension in diameter and so, much higher contrast enhancement.

5.5 Clinical Applications of Aptamers

Currently, many aptamers having therapeutic potential are in clinical trial phases which mainly include AS1411, NOX-A12, and several others (Zhou and Rossi 2017). However, there are abundant *in vitro* results for aptamer-incorporated nanoparticle systems, only a modest amount of data in animal models, and to the best of our knowledge no human data using aptamer-conjugated nanosystems are available. The fact is that there are many parameters, which should be considered for introducing clinically usable systems. The main challenge of applying nanoparticle-based systems *in vivo* is the formation of corona on their surfaces that changes their identity and fate (Charbgoo et al. 2018b).

5.6 Conclusion

Targeted delivery of drugs with aptamer-conjugated nanoparticles is really a promising platform due to reduction of drug side effects, raising therapeutic index, and providing control over required dose. However, there are several areas that need further extensive studies to achieve these goals.

The most important thing is that aptamers are vulnerable to degradation by the ubiquitous nucleases present in biological environments; this is especially true for RNA-based aptamers. In addition, applying aptamers as targeting agents in nanosystems is not cost-effective and may not be usable for aptamers longer than 40–50 nucleotides, particularly if the application requires repeated systemic delivery. However, the methods for the chemical synthesis of oligonucleotides have been promoted; the yields and costs of synthesis have been improved but this approach still need different and expensive instruments. Moreover, as mentioned before, the efficacy of hybrid nanosystems are actually higher, but demonstrating the biocompatibility and safety of these complexes are time-consuming processes and need more investigations.

References

Abnous, K., Danesh, N.M., Ramezani, M. et al. (2017). A novel aptamer-based DNA diamond nanostructure for in vivo targeted delivery of epirubicin to cancer cells. *RSC Advances* 7 (25): 15181–15188.

Abnous, K., Danesh, N.M., Ramezani, M. et al. (2018). Targeted delivery of doxorubicin to cancer cells by a cruciform DNA nanostructure composed of AS1411 and FOXM1 aptamers. *Expert Opinion on Drug Delivery* 15 (11): 1045–1052.

Alibolandi, M., Ramezani, M., Sadeghi, F. et al. (2015). Epithelial cell adhesion molecule aptamer conjugated PEG–PLGA nanopolymersomes for targeted delivery of doxorubicin to human breast adenocarcinoma cell line in vitro. *International Journal of Pharmaceutics* 479 (1): 241–251.

Alibolandi, M., Abnous, K., Anvari, S. et al. (2018a). CD133-targeted delivery of self-assembled PEGylated carboxymethylcellulose-SN38 nanoparticles to colorectal cancer. *Artificial Cells, Nanomedicine, and Biotechnology* 46 (suppl. 1): 1159–1169.

Alibolandi, M., Hoseini, F., Mohammadi, M. et al. (2018b). Curcumin-entrapped MUC-1 aptamer targeted dendrimer-gold hybrid nanostructure as a theranostic system for colon adenocarcinoma. *International Journal of Pharmaceutics* 549 (1–2): 67–75.

Aravind, A., Jeyamohan, P., Nair, R. et al. (2012a). AS1411 aptamer tagged PLGA-lecithin-PEG nanoparticles for tumor cell targeting and drug delivery. *Biotechnology and Bioengineering* 109 (11): 2920–2931.

Aravind, A., Varghese, S.H., Veeranarayanan, S. et al. (2012b). Aptamer-labeled PLGA nanoparticles for targeting cancer cells. *Cancer Nanotechnology* 3 (1–6): 1–12.

Atabi, F., Mousavi Gargari, S.L., Hashemi, M., and Yaghmaei, P. (2017). Doxorubicin loaded DNA aptamer linked myristilated chitosan nanogel for targeted drug delivery to prostate cancer. *Iranian Journal of Pharmaceutical Research: IJPR* 16 (1): 35–49.

Bahreyni, A., Yazdian-Robati, R., Hashemitabar, S. et al. (2017a). A new chemotherapy agent-free theranostic system composed of graphene oxide nano-complex and aptamers for treatment of cancer cells. *International Journal of Pharmaceutics* 526 (1–2): 391–399.

Bahreyni, A., Yazdian-Robati, R., Ramezani, M. et al. (2017b). Identification and imaging of leukemia cells using dual-aptamer-functionalized graphene oxide complex. *Journal of Biomaterials Applications* 32 (1): 74–81.

Bahreyni, A., Alibolandi, M., Ramezani, M. et al. (2019). A novel MUC1 aptamer-modified PLGA-epirubicin-PβAE-antimir-21 nanocomplex platform for targeted co-delivery of anticancer agents in vitro and in vivo. *Colloids and Surfaces B: Biointerfaces* 175: 231–238.

Barenholz, Y.C. (2012). Doxil®—the first FDA-approved nano-drug: lessons learned. *Journal of Controlled Release* 160 (2): 117–134.

Belleperche, M. and DeRosa, M. (2018). pH-control in aptamer-based diagnostics, therapeutics, and analytical applications. *Pharmaceuticals* 11 (3): 80.

Binaymotlagh, R., Hajareh Haghighi, F., Aboutalebi, F. et al. (2019). Selective chemotherapy and imaging of colorectal and breast cancer cells by a modified MUC-1 aptamer conjugated to a poly(ethylene glycol)-dimethacrylate coated Fe_3O_4–AuNCs nanocomposite. *New Journal of Chemistry* 43 (1): 238–248.

Buerger, C. and Groner, B. (2003). Bifunctional recombinant proteins in cancer therapy: cell penetrating peptide aptamers as inhibitors of growth factor signaling. *Journal of Cancer Research and Clinical Oncology* 129 (12): 669–675.

Bunker, A. (2012). Poly (ethylene glycol) in drug delivery, why does it work, and can we do better? All atom molecular dynamics simulation provides some answers. *Physics Procedia* 34: 24–33.

Cao, Y., Huang, H.Y., Chen, L.Q. et al. (2019). Enhanced lysosomal escape of pH-responsive polyethylenimine-betaine functionalized carbon nanotube for the codelivery of survivin small interfering RNA and doxorubicin. *ACS Applied Materials & Interfaces* 11 (10): 9763–9776.

Chang, M., Yang, C.S., and Huang, D.M. (2011). Aptamer-conjugated DNA icosahedral nanoparticles as a carrier of doxorubicin for cancer therapy. *ACS Nano* 5 (8): 6156–6163.

Charbgoo, F., Alibolandi, M., Taghdisi, S.M. et al. (2018a). MUC1 aptamer-targeted DNA micelles for dual tumor therapy using doxorubicin and KLA peptide. *Nanomedicine* 14 (3): 685–697.

Charbgoo, F., Nejabat, M., Abnous, K. et al. (2018b). Gold nanoparticle should understand protein corona for being a clinical nanomaterial. *Journal of Controlled Release* 272: 39–53.

Charoenphol, P. and Bermudez, H. (2014). Aptamer-targeted DNA nanostructures for therapeutic delivery. *Molecular Pharmaceutics* 11 (5): 1721–1725.

Chen, E.C., Fathi, A.T., and Brunner, A.M. (2018). Reformulating acute myeloid leukemia: liposomal cytarabine and daunorubicin (CPX-351) as an emerging therapy for secondary AML. *Oncotargets and Therapy* 11: 3425.

Chuang, E.-Y., Lin, C.-C., Chen, K.-J. et al. (2016). A FRET-guided, NIR-responsive bubble-generating liposomal system for in vivo targeted therapy with spatially and temporally precise controlled release. *Biomaterials* 93: 48–59.

Dou, X.-Q., Wang, H., Zhang, J. et al. (2018). Aptamer-drug conjugate: targeted delivery of doxorubicin in a HER3 aptamer-functionalized liposomal delivery system reduces cardiotoxicity. *International Journal of Nanomedicine* 13: 763–776.

Gotrik, M.R., Feagin, T.A., Csordas, A.T. et al. (2016). Advancements in aptamer discovery technologies. *Accounts of Chemical Research* 49 (9): 1903–1910.

Gui, K., Zhang, X., Chen, F. et al. (2019). Lipid-polymer nanoparticles with CD133 aptamers for targeted delivery of all-trans retinoic acid to osteosarcoma initiating cells. *Biomedicine & Pharmacotherapy* 111: 751–764.

Guo, X. and You, J. (2017). Near infrared light-controlled therapeutic molecules release of nanocarriers in cancer therapy. *Journal of Pharmaceutical Investigation* 47 (4): 297–316.

Guo, J., Gao, X., Su, L. et al. (2011). Aptamer-functionalized PEG–PLGA nanoparticles for enhanced anti-glioma drug delivery. *Biomaterials* 32 (31): 8010–8020.

Guo, X., Zhu, X., Gao, J. et al. (2017). PLGA nanoparticles with CD133 aptamers for targeted delivery and sustained release of propranolol to hemangioma. *Nanomedicine* 12 (21): 2611–2624.

Guo, X., Zhuang, Q., Ji, T. et al. (2018). Multi-functionalized chitosan nanoparticles for enhanced chemotherapy in lung cancer. *Carbohydrate Polymers* 195: 311–320.

Han, X., Jiang, Y., Li, S. et al. (2018). Multivalent aptamer-modified tetrahedral DNA nanocage demonstrates high selectivity and safety for anti-tumor therapy. *Nanoscale* 11 (1): 339–347.

Hanafi-Bojd, M.Y., Moosavian Kalat, S.A., Taghdisi, S.M. et al. (2018). MUC1 aptamer-conjugated mesoporous silica nanoparticles effectively target breast cancer cells. *Drug Development and Industrial Pharmacy* 44 (1): 13–18.

Hayat, M.A. (2005). Handbook of Immunohistochemistry and In Situ Hybridization of Human Carcinomas: Molecular Pathology, Colorectal Carcinoma, and Prostate Carcinoma. Elsevier.

Hoellenriegel, J., Zboralski, D., Maasch, C. et al. (2014). The Spiegelmer NOX-A12, a novel CXCL12 inhibitor, interferes with chronic lymphocytic leukemia cell motility and causes chemosensitization. *Blood* 123 (7): 1032–1039.

Hosnedlova, B., Kepinska, M., Skalickova, S. et al. (2018). Nano-selenium and its nanomedicine applications: a critical review. *International Journal of Nanomedicine* 13: 2107–2128.

Jafari, R., Zolbanin, N.M., Majidi, J. et al. (2019). Anti-mucin1 aptamer-conjugated chitosan nanoparticles for targeted co-delivery of docetaxel and IGF-1R siRNA to SKBR3 metastatic breast cancer cells. *Iranian Biomedical Journal* 23 (1): 21–33.

Jalalian, S.H., Taghdisi, S.M., Shahidi Hamedani, N. et al. (2013). Epirubicin loaded super paramagnetic iron oxide nanoparticle-aptamer bioconjugate for combined colon cancer therapy and imaging in vivo. *European Journal of Pharmaceutical Sciences* 50 (2): 191–197.

Jalalian, S.H., Ramezani, M., Abnous, K., and Taghdisi, S.M. (2018). Targeted co-delivery of epirubicin and NAS-24 aptamer to cancer cells using selenium nanoparticles for enhancing tumor response in vitro and in vivo. *Cancer Letters* 416: 87–93.

Kim, M., Kwon, S.-H., Choi, J., and Lee, A. (2018). A promising biocompatible platform: lipid-based and bio-inspired smart drug delivery systems for cancer therapy. *International Journal of Molecular Sciences* 19 (12): 3859.

Ko, M.H., Kim, S., Kang, W.J. et al. (2009). In vitro derby imaging of cancer biomarkers using quantum dots. *Small* 5 (10): 1207–1212.

Kowalska, E., Bartnicki, F., Fujisawa, R. et al. (2018). Inhibition of DNA replication by an anti-PCNA aptamer/PCNA complex. *Nucleic Acids Research* 46 (1): 25–41.

Lale, S.V., Aravind, A., Kumar, D.S., and Koul, V. (2014). AS1411 aptamer and folic acid functionalized pH-responsive ATRP fabricated pPEGMA–PCL–pPEGMA polymeric nanoparticles for targeted drug delivery in cancer therapy. *Biomacromolecules* 15 (5): 1737–1752.

Lee, J., Oh, J., Lee, E.-s. et al. (2019). Conjugation of prostate cancer-specific aptamers to polyethylene glycol-grafted polyethylenimine for enhanced gene delivery to prostate cancer cells. *Journal of Industrial and Engineering Chemistry* 73: 182–191.

Leitner, M., Poturnayova, A., Lamprecht, C. et al. (2017). Characterization of the specific interaction between the DNA aptamer sgc8c and protein tyrosine kinase-7 receptors at the surface of T-cells by biosensing AFM. *Analytical and Bioanalytical Chemistry* 409 (11): 2767–2776.

Li, N., Larson, T., Nguyen, H.H. et al. (2010). Directed evolution of gold nanoparticle delivery to cells. *Chemical Communications (Camb)* 46 (3): 392–394.

Li, Q., Zhao, D., Shao, X. et al. (2017). Aptamer-modified tetrahedral DNA nanostructure for tumor-targeted drug delivery. *ACS Applied Materials & Interfaces* 9 (42): 36695–36701.

Li, J., Cai, C., Li, J. et al. (2018). Chitosan-based nanomaterials for drug delivery. *Molecules* 23 (10): 2661.

Liechty, W.B., Kryscio, D.R., Slaughter, B.V., and Peppas, N.A. (2010). Polymers for drug delivery systems. *Annual Review of Chemical and Biomolecular Engineering* 1: 149–173.

Liu, Y., Kuan, C.T., Mi, J. et al. (2009). Aptamers selected against the unglycosylated EGFRvIII ectodomain and delivered intracellularly reduce membrane-bound EGFRvIII and induce apoptosis. *Biological Chemistry* 390 (2): 137–144.

Liu, Z., Zhao, H., He, L. et al. (2015). Aptamer density dependent cellular uptake of lipid-capped polymer nanoparticles for polyvalent targeted delivery of vinorelbine to cancer cells. *RSC Advances* 5 (22): 16931–16939.

Lu, B., Lv, X., and Le, Y. (2019). Chitosan-modified PLGA nanoparticles for control-released drug delivery. *Polymers* 11 (2): 304.

Lupold, S.E., Hicke, B.J., Lin, Y., and Coffey, D.S. (2002). Identification and characterization of nuclease-stabilized RNA molecules that bind human prostate cancer cells via the prostate-specific membrane antigen. *Cancer Research* 62 (14): 4029–4033.

Ma, Y., Li, W., Zhou, Z. et al. (2019). Peptide-aptamer coassembly nanocarrier for cancer therapy. *Bioconjugate Chemistry* 30 (3): 536–540.

Mukherjee, A., Waters, A.K., Kalyan, P. et al. (2019). Lipid–polymer hybrid nanoparticles as a next-generation drug delivery platform: state of the art, emerging technologies, and perspectives. *International Journal of Nanomedicine* 14: 1937.

Munzar, J.D., Ng, A., and Juncker, D. (2019). Duplexed aptamers: history, design, theory, and application to biosensing. *Chemical Society Reviews* 48 (5): 1390–1419.

Mura, S., Nicolas, J., and Couvreur, P. (2013). Stimuli-responsive nanocarriers for drug delivery. *Nature Materials* 12 (11): 991.

Nabavinia, M.S., Gholoobi, A., Charbgoo, F. et al. (2017). Anti-MUC1 aptamer: a potential opportunity for cancer treatment. *Medicinal Research Reviews* 37 (6): 1518–1539.

Nejabat, M., Mohammadi, M., Abnous, K. et al. (2018). Fabrication of acetylated carboxymethylcellulose coated hollow mesoporous silica hybrid nanoparticles for nucleolin targeted delivery to colon adenocarcinoma. *Carbohydrate Polymers* 197: 157–166.

Petersen, G.H., Alzghari, S.K., Chee, W. et al. (2016). Meta-analysis of clinical and preclinical studies comparing the anticancer efficacy of liposomal versus conventional non-liposomal doxorubicin. *Journal of Controlled Release* 232: 255–264.

Rajabnejad, S.H., Mokhtarzadeh, A., Abnous, K. et al. (2018). Targeted delivery of melittin to cancer cells by AS1411 anti-nucleolin aptamer. *Drug Development and Industrial Pharmacy* 44 (6): 982–987.

Ranalli, A., Santi, M., Capriotti, L. et al. (2017). Peptide-based stealth nanoparticles for targeted and pH-triggered delivery. *Bioconjugate Chemistry* 28 (2): 627–635.

Ray, P. and White, R.R. (2010). Aptamers for targeted drug delivery. *Pharmaceuticals (Basel, Switzerland)* 3 (6): 1761–1778.

Rockey, W.M., Hernandez, F.J., Huang, S.Y. et al. (2011). Rational truncation of an RNA aptamer to prostate-specific membrane antigen using computational structural modeling. *Nucleic Acid Therapeutics* 21 (5): 299–314.

Sahle, F.F., Gulfam, M., and Lowe, T.L. (2018). Design strategies for physical-stimuli-responsive programmable nanotherapeutics. *Drug Discovery Today* 23 (5): 992–1006.

Shangguan, D., Cao, Z., Meng, L. et al. (2008). Cell-specific aptamer probes for membrane protein elucidation in cancer cells. *Journal of Proteome Research* 7 (5): 2133–2139.

Sun, H. and Zu, Y. (2015). Aptamers and their applications in nanomedicine. *Small (Weinheim an der Bergstrasse, Germany)* 11 (20): 2352–2364.

Taghdisi, S.M., Danesh, N.M., Lavaee, P. et al. (2016a). Double targeting, controlled release and reversible delivery of daunorubicin to cancer cells by polyvalent aptamers-modified gold nanoparticles. *Materials Science and Engineering: C* 61: 753–761.

Taghdisi, S.M., Danesh, N.M., Ramezani, M. et al. (2016b). Double targeting and aptamer-assisted controlled release delivery of epirubicin to cancer cells by aptamers-based dendrimer in vitro and in vivo. *European Journal of Pharmaceutics and Biopharmaceutics* 102: 152–158.

Taghdisi, S.M., Danesh, N.M., Ramezani, M. et al. (2018). A novel AS1411 aptamer-based three-way junction pocket DNA nanostructure loaded with doxorubicin for targeting cancer cells in vitro and in vivo. *Molecular Pharmaceutics* 15 (5): 1972–1978.

Tan, L., Neoh, K.G., Kang, E.T. et al. (2011). PEGylated anti-MUC1 aptamer-doxorubicin complex for targeted drug delivery to MCF7 breast cancer cells. *Macromolecular Bioscience* 11 (10): 1331–1335.

Tesauro, D., Accardo, A., Diaferia, C. et al. (2019). Peptide-based drug-delivery systems in biotechnological applications: recent advances and perspectives. *Molecules (Basel, Switzerland)* 24 (2): 351.

Urmann, K., Modrejewski, J., Scheper, T., and Walter Johanna, G. (2017). Aptamer-modified nanomaterials: principles and applications. *BioNanoMaterials* 18: 20160012.

Vinogradov, S., Batrakova, E., and Kabanov, A. (1999). Poly(ethylene glycol)–polyethyleneimine NanoGel™ particles: novel drug delivery systems for antisense oligonucleotides. *Colloids and Surfaces B: Biointerfaces* 16 (1): 291–304.

Wang, C.-H., Kang, S.-T., Lee, Y.-H. et al. (2012). Aptamer-conjugated and drug-loaded acoustic droplets for ultrasound theranosis. *Biomaterials* 33 (6): 1939–1947.

Willis, M.C., Collins, B.D., Zhang, T. et al. (1998). Liposome-anchored vascular endothelial growth factor aptamers. *Bioconjugate Chemistry* 9 (5): 573–582.

Wu, Y., Sefah, K., Liu, H. et al. (2010). DNA aptamer–micelle as an efficient detection/delivery vehicle toward cancer cells. *Proceedings of the National Academy of Sciences* 107 (1): 5.

Wu, M., Wang, Y., Wang, Y. et al. (2017). Paclitaxel-loaded and A10-3.2 aptamer-targeted poly (lactide-co-glycolic acid) nanobubbles for ultrasound imaging and therapy of prostate cancer. *International Journal of Nanomedicine* 12: 5313.

Xiang, Q., Tan, G., Jiang, X. et al. (2017). Suppression of FOXM1 transcriptional activities via a single-stranded DNA aptamer generated by SELEX. *Scientific Reports* 7: 45377.

Yang, Y., Liu, J., Sun, X. et al. (2016). Near-infrared light-activated cancer cell targeting and drug delivery with aptamer-modified nanostructures. *Nano Research* 9 (1): 139–148.

Yang, L., Sun, H., Liu, Y. et al. (2018a). Self-assembled aptamer-grafted hyperbranched polymer nanocarrier for targeted and photoresponsive drug delivery. *Angewandte Chemie International Edition* 57 (52): 17048–17052.

Yang, S., Ren, Z., Chen, M. et al. (2018b). Nucleolin-targeting AS1411-aptamer-modified graft polymeric micelle with dual pH/redox sensitivity designed to enhance tumor therapy through the codelivery of doxorubicin/TLR4 siRNA and suppression of invasion. *Molecular Pharmaceutics* 15 (1): 314–325.

Yazdian-Robati, R., Ramezani, M., Jalalian, S.H. et al. (2016). Targeted delivery of epirubicin to cancer cells by polyvalent aptamer system in vitro and in vivo. *Pharmaceutical Research* 33 (9): 2289–2297.

Yazdian-Robati, R., Arab, A., Ramezani, M. et al. (2019). Smart aptamer-modified calcium carbonate nanoparticles for controlled release and targeted delivery of epirubicin and melittin into cancer cells in vitro and in vivo. *Drug Development and Industrial Pharmacy* 45 (4): 603–610.

Yeom, J.-H., Lee, B., Kim, D. et al. (2016). Gold nanoparticle-DNA aptamer conjugate-assisted delivery of antimicrobial peptide effectively eliminates intracellular salmonella enterica serovar typhimurium. *Biomaterials* 104: 43–51.

Yu, Z., Chen, F., Qi, X. et al. (2018). Epidermal growth factor receptor aptamer-conjugated polymer-lipid hybrid nanoparticles enhance salinomycin delivery to osteosarcoma and cancer stem cells. *Experimental and Therapeutic Medicine* 15 (2): 1247–1256.

Zamay, T.N., Kolovskaya, O.S., Glazyrin, Y.E. et al. (2014). DNA-aptamer targeting vimentin for tumor therapy in vivo. *Nucleic Acid Therapeutics* 24 (2): 160–170.

Zhang, P., Ye, J., Liu, E. et al. (2017a). Aptamer-coded DNA nanoparticles for targeted doxorubicin delivery using pH-sensitive spacer. *Frontiers of Chemical Science and Engineering* 11 (4): 529–536.

Zhang, Y., Chang, Y.-Q., Han, L. et al. (2017b). Aptamer-anchored di-polymer shell-capped mesoporous carbon as a drug carrier for bi-trigger targeted drug delivery. *Journal of Materials Chemistry B* 5 (33): 6882–6889.

Zhou, J. and Rossi, J. (2017). Aptamers as targeted therapeutics: current potential and challenges. Nature reviews. *Drug Discovery* 16 (3): 181–202.

Zhuang, Y., Deng, H., Su, Y. et al. (2016). Aptamer-functionalized and backbone redox-responsive hyperbranched polymer for targeted drug delivery in cancer therapy. *Biomacromolecules* 17 (6): 2050–2062.

6

Application of Nanotechnology in Transdermal Drug Delivery

Dilesh Jagdish Singhavi and Shagufta Khan

Department of Pharmaceutics, Institute of Pharmaceutical Education and Research, Borgaon (Meghe), Wardha, Maharashtra, India

6.1 Introduction

Transdermal drug delivery (TDD) systems are extensively accepted since they can be used to inject drugs at specific sites, without causing any damage to the skin membrane. The drug can penetrate through the skin portal to the circulating system at a pre-programmed rate by maintaining the effective concentration of the drug for a more extended period (Ansel et al. 2002; Singh et al. 2011; Patel and Baria 2011).

Skin acts as a barrier, and it delivers protection from impacts of various factors which mainly include mechanical factors, microorganisms, radiations, chemicals, etc. It consists of receptors for touch, pain, heat, and cold. The skin consists of four layers: (i) stratum corneum; (ii) epidermis; (iii) dermis; and (iv) subcutaneous layer. Apart from this, there are several connected appendages like hair follicles, sweat ducts, glands, and nails. It also contains dead, flattened keratin-rich cells called corneocytes. Stratum corneum structure possesses a unique feature of bilayer arrays. The role and functions of the skin serve as an essential factor for the survival of humans and mammals. In general, it exists as a protective barrier, upholding homeostasis. The skin can protect us in various ways in the form of acting as physical barrier protection, thermal regulation, ultraviolet light protection, and water retention. The major barrier for transdermal permeation of drug through the skin is the stratum corneum. Various strategies have been studied to breach this barrier which include passive and active penetration enhancement strategies. In current years, the applications of nanotechnology in the domain of skin drug delivery are in the core of attention for local and systemic effects (Escobar-Chávez et al. 2012a; Ng and Lau 2015).

Nanotechnology can be defined as "the manipulation of material on an atomic scale such as 1–100 nm in size, with 1 nm being equivalent to one billionth part of a meter (1×10^{9} m)." In other words, it is a study of extremely small things measured in terms of nanoscale and its applications (Dowling et al. 2004). It can also be used in many fields of science, like chemistry, physics, material science, engineering, etc. Nanotechnology has various applications in areas such as drug development, water decontamination, and information and communication technologies, and for the making of more durable and agile (lighter) materials (Saini et al. 2010; Alvarez-Roman et al. 2004a). Nanotechnology implicates the foundation and manipulation of materials in measurements of nanosize. In recent years, increased interest has been paid to nanocarriers for drug delivery. These carriers

Nanobiotechnology in Diagnosis, Drug Delivery, and Treatment, First Edition. Edited by Mahendra Rai, Mehdi Razzaghi-Abyaneh, and Avinash P. Ingle.

can potentially: (i) guard labile drugs from early degradation; (ii) offer specific drug release rates by altering polymer composition; (iii) to achieve site-specific drug delivery, hence reducing systemic absorption; and (iv) trim down irritation (Alvarez-Roman et al. 2004a; Kuchler et al. 2009; Wu et al. 2009) The major goals in designing nanocarriers are to modify drug carrier size, surface engineering, and delivery of therapeutic agents in order to enhance transdermal delivery of the drug. The aim of the chapter is to provide detail insight on recent nanocarriers as transdermal formulation for TDD, the effect of physicochemical properties of nanocarriers on drug delivery, and future perspectives of nanotechnology on TDD.

6.2 What Is the Stratum Corneum (SC)?

It is a Latin word for "horny layer." It is an external layer of the epidermis which acts as an obstruction layer for the skin and performs rate controlling for diffusion. The SC measures about 10–20 μm thick and acts as a heterogeneous layer of the epidermis. The corneum has a brick-type structure called corneocytes. It acts as a dry layer of dead cells which prevents penetration of microbes. Each corneum consists of 16–20 layers of flattened cells called corneocytes which do not have cell organelles and nuclei, and their cytoplasm is filled with keratin filaments. Corneum cells are surrounded by a matrix of lipid composed of ceramides, fatty acids, and cholesterol (Escobar-Chávez et al. 2012a; Ng and Lau 2015).

6.2.1 SC as a Barrier

The epidermis is the outer layer of skin and is responsible for both physical protections from external damage including pathogens, micron-sized particulates, and several large, hydrophilic compounds, and water-retaining capacity. The stratum corneum acts as a barrier to guard the internal tissues from microbial infection, dryness, chemicals, and mechanical damage. The progression of cell molting from the surface of the SC is called desquamation. It balances the proliferating keratinocytes forms in the stratum basal region. It is a protein network involving keratin, filaggrin, and loricrin which is lined by a fat layer comprising of ceramides, cholesterol, fatty acids. Under normal conditions, the corneum constitutes the epidermis layer and it acts as an adequate barrier for many environmental exposures which can interrupt the drugs, and nanocarriers deep piercing into the skin (Brandner 2009; Matsui and Amagai 2015; Palmer and DeLouise 2016).

6.3 Nanocarriers

In recent years, nanomedicine has become very significant as there has been a large portion of novel research and patenting evolves in it. Nanoparticles for the delivery of drugs have gained much attention in overall pharmacological properties of the commonly used drug in chemotherapy.

6.3.1 Human Skin

Skin acts as the largest tissue in the human body and accounts for more than 10% of body mass. It enables the body to interact more closely with its surroundings. It is a stratified structure with different layers. A dead, flattened, keratin-rich cell, the corneocytes were present in human skin. Cells like dense cells are enclosed by a compound mixture of intercellular lipids, i.e. free fatty acids, ceramides, cholesterol sulfate, and cholesterol (Escobar-Chávez et al. 2012a; Ng and Lau 2015).

6.3.2 Interaction Between Nanocarriers and Skin

In recent years, the products which have nanosized material seem to have increased due to their unique properties. With the aid of nanocarriers, absorption, half-life, skin penetration, stability, bioavailability, etc., of drugs have been improved (Mishra et al. 2010; How et al. 2013). Nanocarriers are too tiny to be identified by the immune system, and supply the drug in the organ as a target with minimum drug doses to decrease the particular side effects (Soica et al. 2016). Nanocarriers can be controlled in organisms through all routes; the dermal path is one such route. The nanocarriers are used for topical/transdermal delivery of therapeutic agents in the field of pharmaceuticals and can also be studied. A nanocarrier can act as an advantageous material for the delivery of drugs through the skin. For optimum drug delivery into the skin, it is necessary that the encapsulated drug released by the carrier into the skin should be easily absorbed by the skin layer, which is involved in the illness (Desai et al. 2010).

The mechanism of interaction of the nanoparticulate carrier systems, skin, and the transport pathways within the membrane for the drug or the carrier is required to establish the possibility of using such methods to optimize the process of drug transport. Transdermal delivery has various benefits when compared to the oral route. The transdermal route of drug organization has unique advantages; the drug can easily bypass the metabolism of the first pass and reaches the universal circulation. Drug diffusion through the skin is a slow process which occurs due to the difference in concentration gradient between the drug delivery system and the skin (Gupta and Rai 2018; Lissarrague et al. 2013). As well, transdermal nanocarriers can get to specific organs by binding with antibodies, antigens, vitamins, and other molecules (Escobar-Chávez et al. 2012b).

SC organization, structural integrity, and an acidic environment with pH gradient are all pivotal for the maintenance of skin functions (Schmid-Wendtner and Korting 2006; Del Rosso and Levin 2011). Its nanoporous nature allows only the penetration of molecules smaller than 500 Da and its structural composition and arrangement disallow ionized molecule penetration. Metabolic enzymes with epidermis and dermis will enable the ingredient neutralization and degradation with the cells of immunological presence. Together, all these skin features significantly hinder the permeation of drugs into the skin and can be overwhelmed by a transfer system. So to overcome the above circumstances, the drug needs to surpass the SC. The penetration of drugs over the layer is the slowest infusion procedure, and also it has limitations in preventing elements to enter into the skin. The critical mechanism of substance penetration in the skin is passive diffusion, which involves partitioning and diffusion of drugs by various skin layers, following a concentration gradient, until they reach the blood and lymphatic vessels (Lemos et al. 2018).

Ethanol has been reported to be a skin penetration enhancer (Haq and Michniak-Kohn 2018). More studies have been done to find the mechanism of penetration of nanocarriers into the skin. In some research studies, phospholipid and ethanol are used for penetration of nanocarriers into murine skin layers over 24 hours of exposition to ethosomes. This study gives as a brief discussion as to why the enhancer for the permeation process causes some disturbances in the skin barrier and disturbs the organization of the SC lipids by enhancement in the fluidity of lipids and reducing the amount of density in intercellular lipid domains. Vesicles interact with the disturbed SC, which in turn alters intercellular lipid lamella and hence creates their pathways across the SC layers to deep skin layers (Touitou et al. 2000). The permeation enhancer will increase the vesicle fluidity and flexibility, causing an increase in the mobility of the polar lipid head of lipid molecules (Sharma et al. 2014; Haque and Talukder 2018). The penetration of drugs from ethosomes will be enhanced when compared to the other achieved conventional liposomes. The study came to the conclusion that the permeation enhancer will influence the permeation flux in ethosomes. Therefore, if the usage of permeation enhancer was more common, then it causes the increase of permeation flux through the skin (Verma and Utreja 2019).

6.4 Properties of Nanocarriers

The biological passage of a nanocarrier is based on its physical and chemical properties. To accomplish pharmacological targeting after transdermal drug administration, the nanocarrier has to preserve its structure. Improvement of its properties may be stimulating. However, nanocarriers must possess several resistant properties and be able to evade various stress factors.

6.4.1 Physicochemical Properties of Nanocarriers for TDD

The major physicochemical attributes that influence transdermal permeation and penetration of drug are size, shape, firmness, and surface charge of the nanocarriers (Lopez et al. 2011).

6.4.1.1 Size and Surface of the Particle

Size and surface area play a significant role in the interaction between nanocarriers and the biological system. Reduction in size of the material will dramatically lead to drastic enhancement in surface area; particle size and surface area dictate how the system responds to distribution (Powers et al. 2007). It has been recognized that different biological processes including endocytosis, cellular uptake, and efficiency of nanocarriers in the endocytic pathway depend on its size (Nel et al. 2006; Aillon et al. 2009). Moreover, the size of the nanoparticles also affects their transdermal toxicity (Lovrić et al. 2005).

The size of the particle is also a vital parameter that aids the access of the nanocarriers in the cell and to other subcellular domains (Behzadi et al. 2017). Typically, nanoparticles possess comparatively greater cellular uptake than microparticles (Singh and Lillard Jr 2009). Nanoparticles with a size of 10 nm exhibited a greater degree of organ distribution as compared to larger nanoparticles after intravenous administration. It shows the relation between size and transport of the particles (De Jong et al. 2008). The size of nanoparticles can also be modulated, and it depends on the lipid, the surfactant nature and proportion, and its production method (Üner and Yener 2007).

The suspended nanoparticle surfaces are electrically charged, and the counterions get absorbed onto the surface more or less to repay the electrical charges (Pfeiffer et al. 2014). Nanoparticles constituting of hydrophilic polymers, polyethylene glycol (PEG), form a sterically alleviating crown on the surface of the nanocarriers that prevents the particle degradation (Salmaso and Caliceti 2013). To balance out liposomes or some polymer nanoparticle formulations that are precarious, PEG can be covalently connected to the surface during the preparation process of formulation or by a post-insertion strategy (Nag and Awasthi 2013; Suk et al. 2016). The nanoparticles' surface and their properties play a substantial role in its toxicity, as they play a significant role in defining the consequence of their contact with the cells and other biological entities (Navya and Daima 2016).

Polymer nanoparticles, the arrangement procedure, the assembly of the monomer, and the subatomic weight of the polymer (higher than 10 000 Da as a rule) are important variables used to get the required size. For instance, poly alkyl cyanoacrylate nanoparticles created by anionic polymerization had sizes running from 20 to 770 nm with dextran as a stabilizing agent (the size was 193 and 585 nm with dextran 70 and dextran 10, separately). Whenever prepared by interfacial polymerization, poly alkyl cyanoacrylate nanoparticles of a size under 500 nm were obtained (Roger et al. 2010; Douglas et al. 1985).

6.4.2 Targeting of Nanocarriers

One of the considerable prerequisites of a system of drug delivery is to send a therapeutic agent efficiently to the pathological site as soon as it can. It is believed that active TDD can increase the efficacy by reducing side effects of drugs. Nanoparticles act as carriers and drug molecules attached on the surface of nanoparticles by physiochemical interaction. A targeting therapeutic moiety is bound to the drug-nanoparticle conjugate, like ligands, which direct to the receptor on the tumor cell surface (Friedman et al. 2013). Nanocarriers improve bioavailability and acceptability with very low cytotoxicity by increasing circulation time of the drug with high drug loading, site-specific, controlled release, and targeted approach (Singh and Lillard Jr 2009; Hsu et al. 2019).

Specific ligands can be added to the nanocarrier's surface to target explicit surface receptors on enterocytes (des Rieux et al. 2006). It is widely reported that few proteins are communicated in caveolae, such as the folic acid receptor, the glycophosphatidylinositol anchor, 60-kDa sialoglycoprotein, the autocrine motility factor receptor, the interleukin-2 receptor, monosialotetrahexosylganglioside, sialic acid, α2β1-integrin, platelet-derived growth factor (PDGF), epidermal growth factor (EGF), bradykinin, and the cholecystokinin (CCK) receptor. Thus, to advance this endocytosis pathway, nanocarriers which get conjugated to explicit ligands could be defined (for instance with folic acid, albumin, IRQ) (Mudhakir et al. 2008). Nevertheless, peptides that associate with receptors are commonly precarious in the gastrointestinal environment, bringing about the degradation of the compound before absorption (Ponchel and Irache 1998).

6.5 Drug Delivery Systems

Various nano-based drug delivery systems have been commonly used in targeted delivery of different drugs. Some of the important nanobased systems are discussed here.

6.5.1 TDD

In 1979, transdermal delivery of drugs, such as the scopolamine patch for motion sickness, was approved by the Food and Drug Administration (FDA) (Pastore et al. 2015). The TDD system is a dynamic area of research, and there are some FDA-approved transdermal drug formulations. The first approach of TDD is to allow the drug to enter the outermost layer of skin, called dermis, at the specific target site so that it can diffuse into the circulatory system. To design a TDD system, it is essential to recognize the anatomy of the skin as well as methods of drug delivery through the skin. The TDD system possesses many benefits in a system of drug delivery. However, it has become challenging to discover the minute and lipophilic drug molecules effectively to penetrate through the skin barrier. Most of the TDD systems have improved patient acceptance compared to other alternate invasive routes.

TDD system allows the delivery of a drug through the skin in a controlled manner. In TDD, various studies used skin parts like sweat glands, hair follicles, or broken skin lining to improve the drug penetration through the skin as well as nanocarrier retention. Studies on using penetration garnishes or physical abrasion to disrupt or damage the SC to raise the penetration of nanocarriers have been carried out. Nanocarrier formulation strategy is mainly affected by the structure of SC (Ehdaie 2011). A nanometric-sized drug delivery system provides favorable results for TDD, and this study has been widely used in research (Zhang et al. 2017). There are diverse technological solutions that have been proposed to expand the bioavailability of the drug.

6.5.1.1 Liposomes

Liposomes are sphere-shaped structures comprising of a hydrophilic core and a hydrophobic covering, which facilitate the carrying of hydrophilic and lipophilic drug molecules (Bulbake et al. 2017). It is demonstrated that liposomes can improve the transdermal penetration of drugs dependent on the molecular mass of drug (Peralta et al. 2018). Their physicochemical properties depend on the materials that were used for the fabrication and the process they performed. Liposomes are one of the best alternatives for drug delivery, as they are nontoxic and remain in the bloodstream for a long time (Yadav et al. 2017). Due to their lipophilic nature, they can undergo fusion with skin surface and lead to enhanced permeation of the drug (El Maghraby et al. 2000). Figure 6.1 represents the schematic illustration of possible mechanisms for the penetration of the drug from different nanocarriers through SC.

The size of liposomes relies upon their structure, which is identified by their readiness procedure: multilamellar (>500 nm), small unilamellar (10–50 nm), and large unilamellar (>100 nm) liposomes were obtained. It was tentatively seen that more smaller liposomes (<100 nm) were eliminated compared to larger liposomes (>100 nm) (Escobar-Chávez et al. 2012b; Bozzuto and Molinari 2015; Nkanga et al. 2019).

6.5.1.2 Transfersomes

Deformable liposomes can easily pass through the SC or get accumulated in the channel-like network in the SC, based on their composition. Moreover, their driving force is osmotic pressure; these liposomes are called transfersomes or transformable liposomes. These are supposed to infuse through the skin layers to the systemic circulation. Transfersomes as intact vesicles or deformable vesicles are stated to expand *in vitro* skin delivery of a range of drugs and *in vivo* permeation to attain therapeutic amounts that are comparable with subcutaneous injection. Transfersomes have been used successfully as carriers for a range of drugs. Transfersomes and deformable vesicles act as carrier systems by penetrating intact deep under the skin. Due to their flexible nature, they are able to squeeze through the pores of the skin (approximately 20 nm in diameter) and lead to the release of the drug (Cevc and Blume 1992; Benson 2006; Morrow et al. 2007; Roger et al. 2010) (Figure 6.1).

6.5.1.3 Ethosomes

Ethosomes are said to be the vesicular carriers that are made up of hydroalcoholic or hydro/alcoholic/glycolic acid. Ethosomes serve as a promising nanocarrier for dermal/TDD. These systems contain soft vesicles and are largely composed of phospholipid and ethanol at comparatively high amounts along with water molecules (Pandey et al. 2015). They can penetrate deep into the skin and also allow the speedy delivery of drugs to the deep layer of the skin or the systemic circulation. The advantage of ethosomes is that they have an extreme nature of high flexibility in their membrane. Their ethanol content causes disruption of the SC barrier which aids in the penetration of drugs through it (Verma and Pathak 2010) (Figure 6.1).

6.5.1.4 Dendrimers

These are one class of particles that are polymerized from a lot of individual monomers (Likos et al. 2001; Abbasi et al. 2014). There are more functional groups on their surface that can be used as chemical reaction sites. The binding of dendrimers usually occurs by administering the drug into the dendrimers or connecting the drugs to the functional groups on its surface. The utilization of dendrimers in the delivery of drugs can avoid the enzymatic hydrolysis of the drug in the body, which can lead to improvement of drug efficacy. It is seen that some proteins may expel massive nanoparticles once they enter into the blood. But nanoparticles with sizes less than 100 nm can be

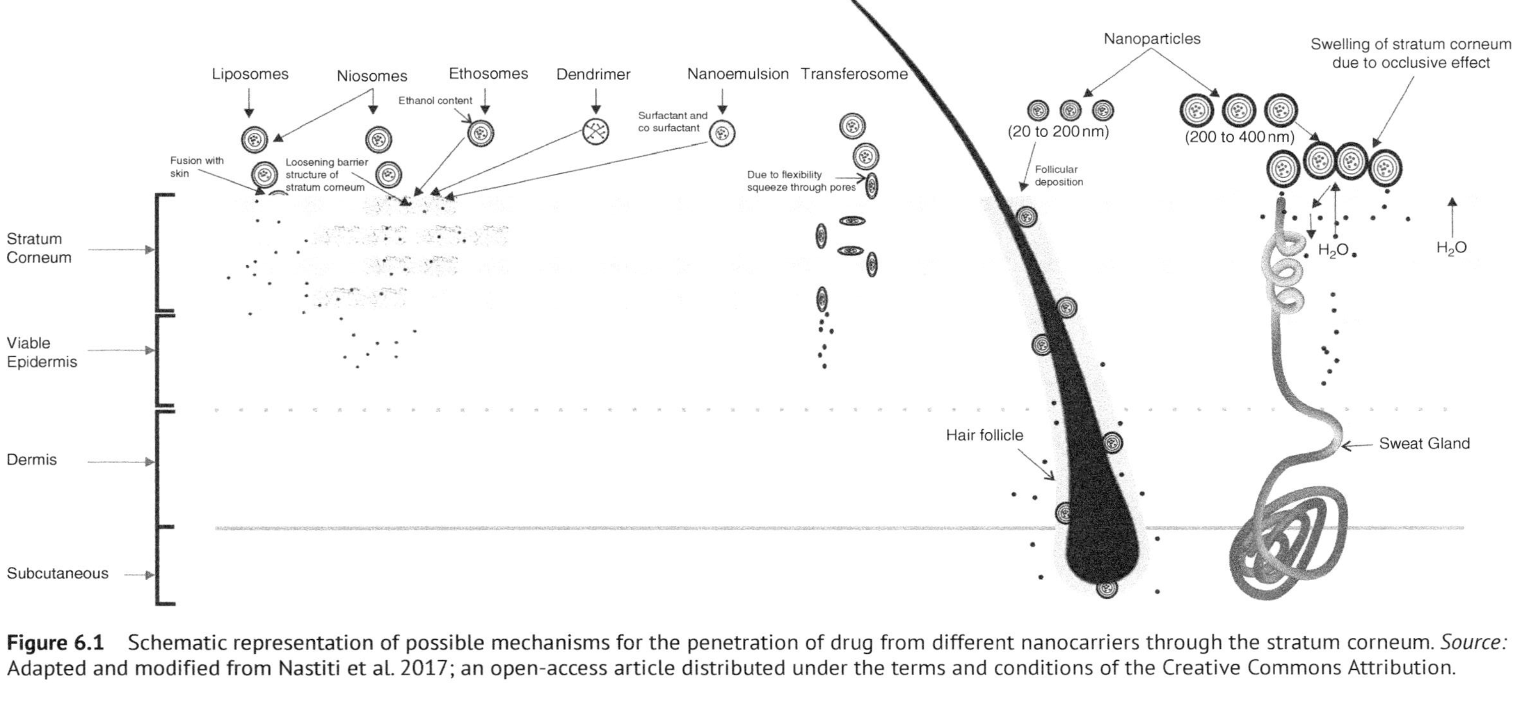

Figure 6.1 Schematic representation of possible mechanisms for the penetration of drug from different nanocarriers through the stratum corneum. *Source:* Adapted and modified from Nastiti et al. 2017; an open-access article distributed under the terms and conditions of the Creative Commons Attribution.

trapped inside solid tumors successfully, so they stay in the blood circulation for a longer period (Zhang et al. 2017). In this way, it significantly considered the size of the nanomolecule. Dendrimers can enhance the skin penetration of drug through its SC disruption property (Sun et al. 2012) (Figure 6.1).

6.5.1.5 Niosomes

Niosomal drug delivery is theoretically appropriate to various drugs which will act against many possible diseases. It can also be used as a pathway for poorly absorbable drugs to devise the new system of drug delivery. The niosomal shape is supposed to be sphere-shaped, and their mean diameter can be measured by using the technique of laser light scattering (Kazi et al. 2010). These carriers are able to transfer the drug through the skin by loosening the barrier structure of the skin and fusing with the skin surface (Tabbakhian et al. 2005) (Figure 6.1).

6.5.1.6 Nanoparticles

These are defined as particulate dispersed solid particles with a size in the range of 10–1000 nm (Kreuter 1994). The drug gets liquified, entrapped, condensed, or involved to a nanoparticle matrix. Therefore, the nanoparticles, spheres, or capsules can be obtained based on the preparation method. They can be composed of polymers, lipids, polysaccharides, and proteins (Pitt et al. 1981; Couvreur et al. 1995; Barratt 2000). These are supposed to breach the skin intracellularly through corneocytes, intercellularly around corneocytes, or by way of dermal structures like hair follicles (Palmer and DeLouise 2016). Nanoparticles having size between 200 and 400 nm can increase the permeation of drugs through occlusive effect (Jenning et al. 2000). On the other hand, nanoparticles ranging between 20 and 200 nm can penetrate in the hair follicle and be responsible for drug delivery (Alvarez-Roman et al. 2004b) (Figure 6.1).

Despite 60% oral absorption, high first-pass metabolisms and poor aqueous solubility resulted in 2% oral bioavailability of raloxifene. A study was designed for development of hydrogel containing raloxifene loaded with solid lipid nanoparticles (RL-SLNs) and penetration enhancer for transdermal delivery. Formation of SLN led to improvement in solubility and thermodynamic action because of drug amorphization and also by avoidance of the first-pass metabolism by the transdermal route. Cumulative effects will contribute to enhancement of permeation flux, controlling drug release, and bioavailability enhancement. Nanoparticles get synthesized by the solvent emulsification-evaporation approach. Then, they were evaluated further to find their various physicochemical properties, ex vivo permeation, and hydration studies *in vitro* toxicity assay, which in turn confirms the safety of formulation with negligible or no sign of toxicity. The data establishes that RL-SLN increases the solubility of RL. It simultaneously circumvents the first-pass metabolism, is delivered by the transdermal route, and also improves the transdermal permeation flux along with the penetration enhancer, and hence, RL-SLN by transdermal route can be an unconventional and safe delivery carrier of RL for patients with breast cancer and postmenopausal osteoporosis (Patel et al. 2019).

6.5.1.7 Nanoemulsions

A nanoemulsion is thermodynamically stable, and also it is said to be a transparent material. Nanoemulsions themselves serve as a permeation enhancer (Reza 2011). Nanoemulsion as transdermal formulation will differ a lot from other sub-micron carriers (Shaker et al. 2019). Surfactant and co-surfactant content in nanoemulsion can enhance skin permeation of drugs through alteration of the lipid bilayer structure of skin (Potts et al. 1991; Lane 2013) (Figure 6.1).

Nanoparticulate scaffolds of olmesartan medoxomil were successfully formulated by employing the melt-emulsification method along with the ultra-sonication using lipid blend as the lipid phases. The formulated nanoparticulate scaffolds with the smallest particle size and highest entrapment efficiency, loading efficiency, and drug release, as mentioned in the study, were obtained. The *in vivo* pharmacokinetic studies showed significant enhancement in bioavailability and may replace conventional oral dosage form, and the formulated optimized formulation was established to be stable and showed a shelf life of 566 days. The overall investigations of the present study showed that the developed formulation was an alternate for the drug delivery system in order to increase the olmesartan medoxomil bioavailability through the nanoparticulate scaffolds methodology (Hasnain et al. 2018).

6.6 Potentials of Nanotechnology

Potentials of nanotechnology can be described as the skin is a formidable barrier in the passage of substances in to and out of our body. It also acts as an interactive interface that can be configured to allow it to turn into a feasible pathway for drug transportation (Leite-Silva et al. 2012). The purpose of pathways penetration for topically applied elements into the skin can be described by many of the studies, for example the permeation of drugs deep into the skin, which in turn includes the allocation of drugs into the intact epidermis, including the skin appendages. These skin appendages have hair follicles and sweat glands which form a shunt pathway through the entire epidermis, occupying only 0.1% of the total human skin. Two pathways that pass over the intact barrier may be identified as the intercellular and transcellular route (Illel 1997).

The intercellular lipid route is in the corneocytes. Interlamellar regions in the SC with the linker regions contain less ordered lipids and more flexible hydrophobic chains. Fluid lipids in outer layer of the skin are mainly essential for the transepidermal distribution of the lipophilic/amphiphilic drug molecules (Xiang and Anderson 1998). The transcellular route considers crossing of the nanocarriers through the corneocytes and intervening lipids. The intracellular macromolecular matrix within the SC contains large numbers of keratin, which does not contribute directly as a skin barrier intact with the SC but provides strength to it. Transcellular diffusion is practically known to be unimportant for delivery of drugs through the transdermal route. The fine aqueous transepidermal channels are seen using a confocal laser scanning microscope. This lowest resistance passage presents between arrangements of corneocytes at the locations where such cellular groups are not laterally overlapped (Cevc and Vierl 2010).

6.7 Enhancement of TDD

6.7.1 Physical Approach

Technological advancement in recent decades helps to overcome the challenges in TDD. The physical method which permits the diffusion of drug molecules through the skin mainly involves energy of a mechanical, electrical, thermal, or magnetic nature. These techniques disrupt the skin membrane to allow drug transport. Some commonly used techniques are iontophoresis, electroporation, sonophoresis, laser radiation, and radiofrequency (Bharkatiya and Nema 2009; Alkilani et al. 2015).

6.7.2 Chemical Approach

Incorporation of diffusion enhancers will facilitate the absorption of drugs by altering the barrier property of the SC. A penetration enhancer should be pharmacologically nonreactive, nontoxic, non-irritating, non-allergic, odorless, tasteless, colorless, and also compatible with most drugs. They even have excellent solvent property. Permeation enhancers can enhance the permeability of drugs through the skin by various mechanisms. It includes interaction with intercellular lipids leading to disruption of their organisms and increasing fluidity (Gupta et al. 2019).

6.8 Contribution of Nanotechnology in TDD in the Future

TDD is an existing and challenging area. Further advancement in the delivery of transdermal drugs depends on the ability to overwhelm the challenges concerning the permeation and skin irritation of drug molecules. An increase in the use of novel methods of permeation enhancement with macromolecules and additional conventional molecules for a broader range of therapeutic indication is mostly needed for the pharmaceutical industry. One can also assume the first prodrug product to emerge on the market in the future. Development of newer prodrugs not only improve therapeutic levels of drugs but also avoid skin irritation (Paudel et al. 2010).

TDD systems have many methodologies to develop the bioavailability and raise a variety of drugs. The TDD system routes will overcome the challenges connected with a current popular system of drug delivery. It is the most preferred route of drug administration, and its upcoming market unquestionably has a hopeful future. The TDD system provides a promising substitute to the conventional drug delivery approach of oral administration and injection. But the commercial use of a TDD system is limited because only few drugs can be successfully delivered through the skin at the required rate (Escobar-Chávez et al. 2012a). Nanotechnology has the potential for small effects, but the maximum potentials are hypothetical at this point. Currently, the biggest pitfalls of the molecular manufacturing pattern have not yet been explored. Hence their benefits and drawbacks will remain the dominant focus of researchers. Various products of therapeutics based on self-regulation, ultra-adaptability, and nano-sized particles are being marketed. Some of them also have reached the clinical level. The first topical liposomal formulation of econazole (an antifungal drug) has been approved in Switzerland, while "VivaGel" is an example of a dendrimer-based formulation and was formulated by Starpharma (Gupta et al. 2012; Rupp et al. 2007).

An alternative one is the topically used NB-00X product (developed by NanoStat technology), which has proceeded for herpes labialis and was affected by herpes simplex virus I. IDEA AG, the biopharmaceutical company, announced the start of developing targeted therapeutics based on the novel transfer of some carriers (topically applied dosages of IDEA-033), and are currently in Phase III trials in Europe for the signs of treatment and symptoms related to osteoarthritis of the knee. Nevertheless, many questions arise, which remain unexplored and addressed. In the context of safety, the environmental effects and the potential effects on health are significant issues for the manufacturing of these particles. Finally, it is hoped that the application of nanotherapeutics is limitless, but the development of safety guidelines by the manufacturers should strongly be considered (Gupta et al. 2012, 2017).

Some mechanisms proposed for the improvement of nanocarriers' dermal penetration involve the evaporation of water at the topmost, comparatively dehydrated, skin layer that results in their reduction to appropriate sizes for enhanced para-cellular transport of drugs into the deeper hydrated layers of the skin. Ultimately, the nanocarriers get rehydrated to regain their original sizes with which they migrate again into the deeper skin layer to deliver their cargos, in an experiment conducted by Boakye et al. The analysis of physicochemical features of the nanocarriers via differential scanning calorimetry, Fourier transform infrared spectroscopy, and atomic force microscopy had confirmed the presence of the drug within the lipid bilayers of the nanocarriers. The presence of the drug in solubilized state may lead to the significantly ($p < 0.001$) potentiated skin deposition of diindolylmethane derivative ultraflexible nanocarriers in comparison to the polyethylene glycol solution. Additionally, the data of the current study revealed that the nanocarriers were extremely more deformable and malleable with respect to the plain nanocarriers (without the edge activator, tween 80) and the estimated deformability index of 13.03 was recorded which is in accordance to earlier studies (Boakye et al. 2016).

To raise the rate of drugs accessible for TDD, the use of nanocarriers became an excellent alternative source for delivering both lipophilic and hydrophilic drugs all over the SC with the opportunity of local or systemic effect for treating various diseases. TDD has practiced a healthy annual progression rate of 25%, which outpaces oral drug delivery (2%) and the inhalation market (20%), respectively (Grosh 2000). The development of such drug carriers will increase the usage of skin as a route of administration for the treatment of various conditions. However, subjective and objective analysis of these devices is essential to make sure that the scientific, regulatory, and consumer needs are met. The apparatus in development is costlier and complicated when compared with conventional transdermal patch therapies, since any permanent damage to the SC results in the loss of its barrier properties and hence its functions as a protective organ. Thus, for this new drug delivery expertise to prosper and compete with those already in the market, their safety, efficacy, transportability, user-friendliness, cost-effectiveness, and potential market have to be addressed.

6.9 Conclusion

The purpose of drug entrapment in the nanocarrier is the speedy delivery to target cells and reducing the toxicity from non-targeted organs. TDD is a systematic comfort method for delivering the drugs by drug formulation into the healthy skin. The greatest challenge that TDD faces is the obstructive nature of the skin, which restricts the entry of drugs. TDD has become a useful alternative for a drug delivery without fail on higher demand and requirements. As expressed by the classification of the biopharmaceutics system, the most significant parameters to consider for absorption are permeability and dissolution rate of the medication. Nanocarriers can significantly improve these parameters. Nanocarriers can penetrate biological membranes to deliver drugs for specific diseases. Considering the progress of TDD, the nanotechnology approach has gained importance in research. However, future investigations must make sure of the advantages and assess the risk factors for several drugs incorporated in nanoparticulate drug delivery systems for TDD.

References

Abbasi, E., Aval, S.F., Akbarzadeh, A. et al. (2014). Dendrimers: synthesis, applications, and properties. *Nanoscale Research Letters* 9 (1): 247.

Aillon, K.L., Xie, Y.M., El-Gendy, N. et al. (2009). Effects of nanomaterial physicochemical properties on *in vivo* toxicity. *Advanced Drug Delivery Reviews* 61 (6): 457–466.

Alkilani, A.Z., McCrudden, M.T., and Donnelly, R.F. (2015). Transdermal drug delivery: innovative pharmaceutical developments based on disruption of the barrier properties of the stratum corneum. *Pharmaceutics* 7 (4): 438–470.

Alvarez-Roman, R., Naik, A., Kalia, Y.N. et al. (2004a). Enhancement of topical delivery from biodegradable nanoparticles. *Pharmaceutical Research* 21: 1818–1825.

Alvarez-Roman, R., Naik, A., Kalia, Y.N. et al. (2004b). Skin penetration and distribution of polymeric nanoparticles. *Journal of Controlled Release* 99 (1): 53–62.

Ansel, H.C., Allen, L.V., and Popovich, N.G. (2002). Pharmaceutical Dosage Forms and Drug Delivery System, 7e. New York: Lipponcott Williams and Wilkins.

Barratt, G.M. (2000). Therapeutic applications of colloidal drug carriers. *Pharmaceutical Science and Technology Today* 3 (5): 163–171.

Behzadi, S., Serpooshan, V., Tao, W. et al. (2017). Cellular uptake of nanoparticles: journey inside the cell. *Chemical Society Reviews* 46 (14): 4218–4244.

Benson, H.A. (2006). Transfersomes for transdermal drug delivery. *Expert Opinion on Drug Delivery* 3 (6): 727–737.

Bharkatiya, M. and Nema, R.K. (2009). Skin penetration enhancement techniques. *Journal of Young Pharmacists* 1 (2): 110–115.

Boakye, C.H., Patel, K., Doddapaneni, R. et al. (2016). Ultra-flexible nanocarriers for enhanced topical delivery of a highly lipophilic antioxidative molecule for skin cancer chemoprevention. *Colloids and Surfaces B: Biointerfaces* 143: 156–167.

Bozzuto, G. and Molinari, A. (2015). Liposomes as nanomedical devices. *International Journal of Nanomedicine* 10: 975–999.

Brandner, J.M. (2009). Tight junctions and tight junction proteins in mammalian epidermis. *European Journal of Pharmaceutics and Biopharmaceutics* 72: 289–294.

Bulbake, U., Doppalapudi, S., Kommineni, N., and Khan, W. (2017). Liposomal formulations in clinical use: an updated review. *Pharmaceutics* 9 (2): 12.

Cevc, G. and Blume, G. (1992). Lipid vesicles penetrate into intact skin owing to the transdermal osmotic gradients and hydration force. *Biochimica et Biophysica Acta (BBA) - Biomembranes* 1104: 226–232.

Cevc, G. and Vierl, U. (2010). Nanotechnology and the transdermal route. A state of the art review and critical appraisal. *Journal of Controlled Release* 141 (3): 277–299.

Couvreur, P., Dubernet, C., and Puisieux, F. (1995). Controlled drug delivery with nanoparticles: current possibilities and future trends. *European Journal of Pharmaceutics and Biopharmaceutics* 41 (1): 2–13.

De Jong, W.H., Hagens, W.I., Krystek, P. et al. (2008). Particle size-dependent organ distribution of gold nanoparticles after intravenous administration. *Biomaterials* 29 (12): 1912–1919.

Del Rosso, J.Q. and Levin, J. (2011). The clinical relevance of maintaining the functional integrity of the stratum corneum in both healthy and disease-affected skin. *Journal of Clinical and Aesthetic Dermatology* 4 (9): 22.

des Rieux, A., Fievez, V., Garinot, M. et al. (2006). Nanoparticles as potential oral delivery systems of proteins and vaccines: a mechanistic approach. *Journal of Controlled Release* 116 (1): 1–27.

Desai, P., Patlolla, R.R., and Singh, M. (2010). Interaction of nanoparticles and cell-penetrating peptides with the skin for transdermal drug delivery. *Molecular Membrane Biology* 27 (7): 247–259.

Douglas, S.J., Illum, L., and Davis, S.S. (1985). Particle size and size distribution of poly (butyl 2-cyanoacrylate) nanoparticles. II. Influence of stabilizers. *Journal of Colloid and Interface Science* 103 (1): 154–163.

Dowling, A., Clift, R., Grobert, D. et al. (2004). Nanoscience and Nanotechnologies: Opportunities and Uncertainties. London: The Royal Society, The Royal Academy of Engineering.

Ehdaie, B. (2011). Enhanced delivery of transdermal drugs through human skin with novel carriers. *Journal of Pharmaceutical and Biomedical Sciences* 1 (8): 161–166.

El Maghraby, G.M., Williams, A.C., and Barry, B.W. (2000). Skin delivery of oestradiol from lipid vesicles: importance of liposome structure. *International Journal of Pharmaceutics* 204: 159–169.

Escobar-Chávez, J.J., Díaz-Torres, R., Rodríguez-Cruz, I.M. et al. (2012a). Nanocarriers for transdermal drug delivery. *Research and Reports in Transdermal Drug Delivery* 1: 3–17.

Escobar-Chávez, J.J., Rodríguez-Cruz, I.M. et al. (2012b). Nanocarrier systems for transdermal drug delivery. In: Recent Advances in Novel Drug Carrier Systems (ed. A.D. Sezer), 201–240. London: IntechOpen.

Friedman, A.D., Claypool, S.E., and Liu, R. (2013). The smart targeting of nanoparticles. *Current Pharmaceutical Design* 19 (35): 6315–6329.

Grosh, S. (2000). Transdermal drug delivery–opening doors for the future. *European Pharmaceutical Contractor* 4: 30–32.

Gupta, R. and Rai, B. (2018). In-silico design of nanoparticles for transdermal drug delivery application. *Nanoscale* 10 (10): 4940–4951.

Gupta, M., Agrawal, U., and Vyas, S.P. (2012). Nanocarrier-based topical drug delivery for the treatment of skin diseases. *Expert Opinion on Drug Delivery* 9 (7): 783–804.

Gupta, M., Sharma, V., and Chauhan, N.S. (2017). Promising novel nanopharmaceuticals for improving topical antifungal drug delivery. In: Nano-and Microscale Drug Delivery Systems (ed. A. Grumezescu), 197–228. Netherlands: Elsevier.

Gupta, R., Dwadasi, B.S., Rai, B., and Mitragotri, S. (2019). Effect of chemical permeation enhancers on skin permeability: in silico screening using molecular dynamics simulations. *Scientific Reports* 9 (1): 1456.

Haq, A. and Michniak-Kohn, B. (2018). Effects of solvents and penetration enhancers on transdermal delivery of thymoquinone: permeability and skin deposition study. *Drug Delivery* 25 (1): 1943–1949.

Haque, T. and Talukder, M.M. (2018). Chemical enhancer: a simplistic way to modulate barrier function of the stratum corneum. *Advanced Pharmaceutical Bulletin* 8 (2): 169.

Hasnain, M., Imam, S.S., Aqil, M. et al. (2018). Application of lipid blend-based nanoparticulate scaffold for oral delivery of antihypertensive drug: implication on process variables and in vivo absorption assessment. *Journal of Pharmaceutical Innovation* 13 (4): 341–352.

How, C.W., Rasedee, A., Manickam, S., and Rosli, R. (2013). Tamoxifen-loaded nanostructured lipid carrier as a drug delivery system: characterization, stability assessment and cytotoxicity. *Colloids and Surfaces B: Biointerfaces* 112: 393–399.

Hsu, C.Y., Wang, P.W., Alalaiwe, A. et al. (2019). Use of lipid nanocarriers to improve oral delivery of vitamins. *Nutrients* 11 (1): 68.

Illel, B. (1997). Formulation for transfollicular drug administration: some recent advances. *Critical Reviews in Therapeutic Drug Carrier Systems* 14 (3): 207–219.

Jenning, V., Schafer-Korting, M., and Gohla, S. (2000). Vitamin A-loaded solid lipid nanoparticles for topical use: drug release properties. *Journal of Controlled Release* 66 (2–3): 115–126.

Kazi, K.M., Mandal, A.S., Biswas, N. et al. (2010). Niosome: a future of targeted drug delivery systems. *Journal of Advanced Pharmaceutical Technology and Research* 1 (4): 374–380.

Kreuter, J. (1994). Nanoparticles. In: Encyclopedia of Pharmaceutical Technology (eds. J. Swarbrick and J.C. Boylan), 165–190. New York: Marcel Dekker Inc.

Kuchler, S., Radowski, M.R., Blaschke, T. et al. (2009). Nanoparticles for skin penetration enhancement – a comparison of a dendritic core-multishell-nanotransporter and solid lipid nanoparticles. *European Journal of Pharmaceutics and Biopharmaceutics* 71: 243–250.

Lane, M.E. (2013). Skin penetration enhancers. *International Journal of Pharmaceutics* 447 (1): 12–21.

Leite-Silva, V.R., De Almeida, M.M., Fradin, A. et al. (2012). Delivery of drugs applied topically to the skin. *Expert Review of Dermatology* 7 (4): 383–397.

Lemos, C.N., Pereira, F., Dalmolin, L.F. et al. (2018). Nanoparticles influence in skin penetration of drugs: *in vitro* and *in vivo* characterization. In: Nanostructures for the Engineering of Cells, Tissues and Organs (ed. A.M. Grumezescu), 187–248. Romania: William Andrew Publishing.

Likos, C.N., Schmidt, M., Löwen, H. et al. (2001). Soft interaction between dissolved flexible dendrimers: theory and experiment. *Macromolecules* 34 (9): 2914–2920.

Lissarrague, M.H., Garate, H., Lamanna, M.E. et al. (2013). Medicinal patches and drug nanoencapsulation: a noninvasive alternative. In: Nanomedicine for Drug Delivery and Therapeutics (ed. A.K. Mishra), 337–371. New York: Wiley-Scrivener.

Lopez, R.F., Seto, J.E., Blankschtein, D., and Langer, R. (2011). Enhancing the transdermal delivery of rigid nanoparticles using the simultaneous application of ultrasound and sodium lauryl sulfate. *Biomaterials* 32 (3): 933–941.

Lovrić, J., Bazzi, H.S., Cuie, Y. et al. (2005). Differences in subcellular distribution and toxicity of green and red emitting CdTe quantum dots. *Journal of Molecular Medicine* 83 (5): 377–385.

Matsui, T. and Amagai, M. (2015). Dissecting the formation, structure and barrier function of the stratum corneum. *International Immunology* 27: 269–280.

Mishra, B., Patel, B.B., and Tiwari, S. (2010). Colloidal nanocarriers: a review on formulation technology, types and applications toward targeted drug delivery. *Nanomedicine: Nanotechnology, Biology and Medicine* 6 (1): 9–24.

Morrow, D.I., McCarron, P.A., Woolfson, A.D., and Donnelly, R.F. (2007). Innovative strategies for enhancing topical and transdermal drug delivery. *The Open Drug Delivery Journal* 1: 36–59.

Mudhakir, D., Akita, H., Tan, E., and Harashima, H. (2008). A novel IRQ ligand-modified nanocarrier targeted to a unique pathway of caveolar endocytic pathway. *Journal of Controlled Release* 125 (2): 164–173.

Nag, O.K. and Awasthi, V. (2013). Surface engineering of liposomes for stealth behavior. *Pharmaceutics* 5 (4): 542–569.

Nastiti, C.M., Ponto, T., Abd, E. et al. (2017). Topical nano and microemulsions for skin delivery. *Pharmaceutics* 9 (4): 37.

Navya, P.N. and Daima, H.K. (2016). Rational engineering of physicochemical properties of nanomaterials for biomedical applications with nanotoxicological perspectives. *Nano Convergence* 3 (1): 1.

Nel, A., Xia, T., Mädler, L., and Li, N. (2006). Toxic potential of materials at the nanolevel. *Science* 311: 622–627.

Ng, K.W. and Lau, W.M. (2015). Skin deep: the basics of human skin structure and drug penetration. In: Percutaneous Penetration Enhancer's Chemical Methods in Penetration Enhancement (eds. N. Dragicevic and H.I. Maibach), 3–11. Berlin: Springer.

Nkanga, C.I., Bapolisi, A.M., Okafor, N.I., and Krause, R.W. (2019). General perception of liposomes: formation, manufacturing and applications. In: Liposomes-Advances and Perspectives (ed. A.D. Sez). London: IntechOpen.

Palmer, B.C. and DeLouise, L.A. (2016). Nanoparticle-enabled transdermal drug delivery systems for enhanced dose control and tissue targeting. *Molecules* 21 (12): 1719.

Pandey, V., Golhani, D., and Shukla, R. (2015). Ethosomes: versatile vesicular carriers for efficient transdermal delivery of therapeutic agents. *Drug Delivery* 22 (8): 988–1002.

Pastore, M.N., Kalia, Y.N., Horstmann, M., and Roberts, M.S. (2015). Transdermal patches: history, development and pharmacology. *British Journal of Pharmacology* 172 (9): 2179–2209.

Patel, R.P. and Baria, A.H. (2011). Formulation and evaluation consideration of transdermal drug delivery system. *International Journal of Pharmaceutical Research* 3: 1–9.

Patel, K.K., Gade, S., Anjum, M.M. et al. (2019). Effect of penetration enhancers and amorphization on transdermal permeation flux of raloxifene-encapsulated solid lipid nanoparticles: an ex vivo study on human skin. *Applied Nanoscience* 9 (6): 1383–1394.

Paudel, K.S., Milewski, M., Swadley, C.L. et al. (2010). Challenges and opportunities in dermal/transdermal delivery. *Therapeutic Delivery* 1 (1): 109–131.

Peralta, M.F., Guzmán, M.L., Pérez, A.P. et al. (2018). Liposomes can both enhance or reduce drugs penetration through the skin. *Scientific Reports* 8 (1): 13253.

Pfeiffer, C., Rehbock, C., Hühn, D. et al. (2014). Interaction of colloidal nanoparticles with their local environment: the (ionic) nanoenvironment around nanoparticles is different from bulk and determines the physico-chemical properties of the nanoparticles. *Journal of the Royal Society Interface* 11 (96): 20130931.

Pitt, C.G., Gratzl, M.M., Kimmel, G.L. et al. (1981). Aliphatic polyesters II. The degradation of poly (DL-lactide), poly (epsilon-caprolactone), and their copolymers in vivo. *Biomaterials* 2: 215–220.

Ponchel, G. and Irache, J.M. (1998). Specific and non-specific bioadhesive particulate systems for oral delivery to the gastrointestinal tract. *Advanced Drug Delivery Reviews* 34 (2–3): 191–219.

Potts, R.O., Mak, V.H., Francoeur, M.L., and Guy, R.H. (1991). Strategies to enhance permeability via stratum corneum lipid pathways. *Advances in Lipid Research* 24: 173–210.

Powers, K.W., Palazuelos, M., Moudgil, B.M., and Roberts, S.M. (2007). Characterization of the size, shape, and state of dispersion of nanoparticles for toxicological studies. *Nanotoxicology* 1 (1): 42–51.

Reza, K.H. (2011). Nanoemulsion as a novel transdermal drug delivery system. *International Journal of Pharmaceutical Sciences and Research* 2 (8): 1938–1946.

Roger, E., Lagarce, F., Garcion, E., and Benoit, J.P. (2010). Biopharmaceutical parameters to consider in order to alter the fate of nanocarriers after oral delivery. *Nanomedicine* 5 (2): 287–306.

Rupp, R., Rosenthal, S.L., and Stanberry, L.R. (2007). VivaGel™ (SPL7013 gel): a candidate dendrimer–microbicide for the prevention of HIV and HSV infection. *International Journal of Nanomedicine* 2 (4): 561–566.

Saini, R., Saini, S., and Sharma, S. (2010). Nanotechnology: the future medicine. *Journal of Cutaneous and Aesthetic Surgery* 3 (1): 32–33.

Salmaso, S. and Caliceti, P. (2013). Stealth properties to improve therapeutic efficacy of drug nanocarriers. *Journal of Drug Delivery* 2013: 374252.

Schmid-Wendtner, M.H. and Korting, H.C. (2006). The pH of the skin surface and its impact on the barrier function. *Skin Pharmacology and Physiology* 19 (6): 296–302.

Shaker, D.S., Ishak, R.A., Ghoneim, A., and Elhuoni, M.A. (2019). Nanoemulsion: a review on mechanisms for the transdermal delivery of hydrophobic and hydrophilic drugs. *Scientia Pharmaceutica* 87 (3): 17.

Sharma, V.K., Sarwa, K.K., and Mazumder, B. (2014). Fluidity enhancement: a critical factor for performance of liposomal transdermal drug delivery system. *Journal of Liposome Research* 24 (2): 83–89.

Singh, R. and Lillard, J.W. Jr. (2009). Nanoparticle-based targeted drug delivery. *Experimental and Molecular Pathology* 86 (3): 215–223.

Singh, A., Singh, M.P., Alam, G. et al. (2011). Expanding opportunities for transdermal delivery systems: an overview. *Journal of Pharmaceutical Research* 4: 1417–1420.

Soica, C., Coricovac, D., Dehelean, C. et al. (2016). Nanocarriers as tools in delivering active compounds for immune system related pathologies. *Recent Patents on Nanotechnology* 10 (2): 128–145.

Suk, J.S., Xu, Q., Kim, N. et al. (2016). PEGylation as a strategy for improving nanoparticle-based drug and gene delivery. *Advanced Drug Delivery Reviews* 99: 28–51.

Sun, M., Fan, A., Wang, Z., and Zhao, Y. (2012). Dendrimer-mediated drug delivery to the skin. *Soft Matter* 8 (16): 4301–4305.

Tabbakhian, M., Daneshamouz, S., Tavakoli, N., and Jaafari, M.R. (2005). Influence of liposomes and niosomes on the in vitro permeation and skin retention of finasteride. *Iranian Journal of Pharmaceutical Sciences* 1 (3): 119–130.

Touitou, E., Dayan, N., Bergelson, L. et al. (2000). Ethosomes-novel vesicular carriers for enhanced delivery: characterization and skin penetration properties. *Journal of Controlled Release* 65 (3): 403–418.

Üner, M. and Yener, G. (2007). Importance of solid lipid nanoparticles (SLN) in various administration routes and future perspectives. *International Journal of Nanomedicine* 2 (3): 289.

Verma, P. and Pathak, K. (2010). Therapeutic and cosmeceutical potential of ethosomes: an overview. *Journal of Advanced Pharmaceutical Technology and Research* 1 (3): 274–282.

Verma, S. and Utreja, P. (2019). Vesicular nanocarrier based treatment of skin fungal infections: potential and emerging trends in nanoscale pharmacotherapy. *Asian Journal of Pharmaceutical Sciences* 14 (2): 117–129.

Wu, X., Price, G.J., and Guy, R.H. (2009). Disposition of nanoparticles and an associated lipophilic permeant following topical application to the skin. *Molecular Pharmaceutics* 6: 1441–1448.

Xiang, T.X. and Anderson, B.D. (1998). Influence of chain ordering on the selectivity of dipalmitoylphosphatidylcholine bilayer membranes for permeant size and shape. *Biophysical Journal* 75 (6): 2658–2671.

Yadav, D., Sandeep, K., Pandey, D., and Dutta, R.K. (2017). Liposomes for drug delivery. *Journal of Biotechnology and Biomaterials* 7 (4): 276.

Zhang, J., Tang, H., Liu, Z., and Chen, B. (2017). Effects of major parameters of nanoparticles on their physical and chemical properties and recent application of nanodrug delivery system in targeted chemotherapy. *International Journal of Nanomedicine* 12: 8483.

7

Superparamagnetic Iron Oxide Nanoparticle-Based Drug Delivery in Cancer Therapeutics

Dipak Maity[1], *Atul Sudame*[2], *and Ganeshlenin Kandasamy*[3]

[1] *Department of Chemical Engineering, Institute of Chemical Technology Mumbai–Indian Oil Campus, IIT KGP Extension Centre, Bhubaneswar, Odisha, India*
[2] *Department of Mechanical Engineering, Shiv Nadar University, Greater Noida, UP, India*
[3] *Department of Biomedical Engineering, Vel Tech Rangarajan Dr. Sagunthala R&D Institute of Science and Technology, Chennai, TN, India*

7.1 Introduction

Over the past few decades, there has been tremendous progress in the development of nanomaterials to be used as effective anticancer drug delivery systems (DDs). Among different DDs, magnetic DDs (MDDs) have attained a great potential for clinical applications since they have the ability to improve drug loading and delivery effectively. Herein, the loading of the drugs inside MDDs can be done in two different ways: (i) by attaching the drugs on the surface of the magnetic nanoparticles (MNPs); or (ii) by co-encapsulating the drugs with MNPs inside the common drug delivery vehicles such as polymeric nanoparticles/micelles, niosomes, liposomes, and so on (Tavano et al. 2013; Kim et al. 2013; Pandey et al. 2016). Then, the drug delivery via MNPs can be performed by intravenous administration, followed by accumulation and release of drugs at targeted sites by applying a stimulus based on an external magnetic field. Several clinical trials of drug delivery via magnetic vehicles are widely carried out and also being conducted nowadays for efficient cancer treatment. Herein, the MNPs (which form the main core of MDDs) should be easily magnetized/demagnetized under an externally applied magnetic field (static/alternating) to effectively release the loaded drugs based on the generated mechanical vibration/induced heat (Knežević et al. 2019; Nardecchia et al. 2019). For this purpose, these MNPs should be "superparamagnetic" in nature, which can be attained by reducing their size below to a single-domain size.

Usually, such superparamagnetic MNPs are made of iron oxides and can otherwise be called superparamagnetic iron oxide nanoparticles (SPIONs). The surfaces of these SPIONs are commonly functionalized/coated with organic or inorganic surfactants during/after their synthesis to prevent the agglomeration due to magnetic dipole–dipole attractions for their effective use in the above-mentioned applications, including drug delivery (Mody et al. 2014; Theivasanthi and Alagar 2014; Li et al. 2015). Moreover, thermal vibration energy of SPIONs might overcome the energy barrier of the magnetic anisotropy to be rapidly magnetized or demagnetized under the application/removal of an external magnetic field. However, the physiological parameters of SPION-based

Nanobiotechnology in Diagnosis, Drug Delivery, and Treatment, First Edition. Edited by Mahendra Rai, Mehdi Razzaghi-Abyaneh, and Avinash P. Ingle.

MDDs such as size, surface characteristics, blood flow rate, and applied magnetic field strength are major influencing factors in effective drug delivery/release (Farah 2016).

In general, the role of the SPIONs present inside MDDs is to mainly act as a magnetic stimulus for drug release, after their navigation to the area of interest (i.e. cancer site) via magnetic targeting apart from passive/active targeting (Mahmoudi et al. 2011, García Casillas et al. 2014). Moreover, site-specific targeted agent-conjugated MDDs (i.e. active targeting in combination with magnetic targeting) could reduce the requirement of the higher dose of the chemotherapeutic drugs and the side effects associated with them due to their systemic distribution (Polyak and Friedman 2009). In addition to this, the SPIONs-based MDDs can also be simultaneously used in the treatment of cancer (in addition to chemotherapy via anticancer drugs) based on the induced heat (i.e. a therapeutic temperature of 45 °C) while applying an alternating magnetic field (AMF), which is called magnetic fluid hyperthermia (MFH) (Hervault and Thanh 2014; Zhang et al. 2016; Chang et al. 2018). Usually, the applied AMF below biological safe limit, so it is harmless and adaptable to any part of the human body (Lahiri et al. 2016). There are many current developments in the making of these SPIONs-based MDDs to reduce the disadvantages associated with conventional therapies in cancer treatments.

Therefore, the main aim of this chapter is to discuss the as-developed SPIONs based MDDs including surface-modified SPIONs, and SPIONs/anticancer drug co-encapsulating nanomaterials including polymeric nanoparticles/micelles, niosomes, liposomes, and other magnetic structures. In addition, detailed discussion about the delivery of different single/dual drugs (including curcumin [CUR], doxorubicin [DOX], daunorubicin [DRC], paclitaxel [PTX], and methotrexate [MTX]) via these SPIONs-based MDDs in the *in vitro/in vivo* conditions for cancer treatments through chemotherapy or/and MFH has been also given.

7.2 Magnetic Drug Delivery Systems

7.2.1 Surface-Modified SPIONs

Superparamagnetic iron oxide nanoparticles (SPIONs), especially magnetite (Fe_3O_4) or maghemite (γ-Fe_2O_3), are the most investigated MNPs for cancer treatments, simultaneously utilized as contrast agents in magnetic resonance imaging (MRI) and heat-inducing agents in MFH, since they have excellent chemical stability, biocompatibility, and biodegradability. Moreover, these SPIONs could also be used in the delivery of anticancer drugs by conjugating them on the surface of the SPIONs, (Wahajuddin and Arora 2012; Kandasamy and Maity 2015; Janko et al. 2019). Usually, SPIONs are coated with the different organic/inorganic capping agents (Figure 7.1) in order to maintain the biocompatibility and also to enhance their colloidal stability during drug delivery.

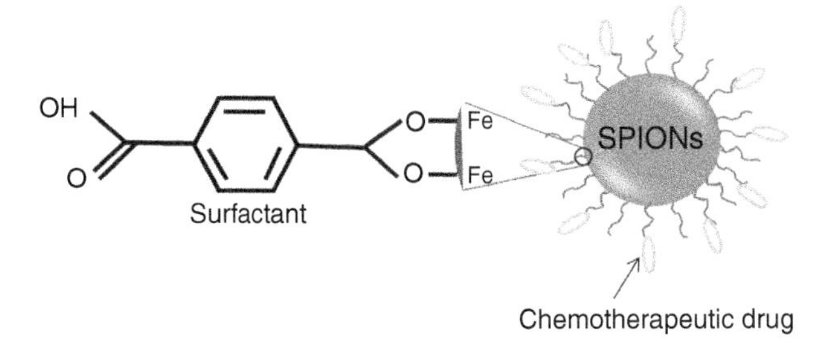

Figure 7.1 Schematic representation of surfactant-coated and CHD-conjugated SPIONs.

The drug-conjugated SPIONs could be easily delivered to the targeted cancer sites either via passive, active, or magnetic targeting strategies. In a passive targeting process, surface-modified SPIONs would be able to diffuse easily into leaky vasculatures of cancerous tissue by taking advantage of the enhanced permeability and retention (EPR) effect. Moreover, SPIONs are functionalized with targeting moieties to specifically target the cancerous cell sites for active targeting process. But in addition to this, the static magnetic fields could be used to accumulate the surface-modified SPIONs at the tumor site via a magnetic targeting process (Schleich et al. 2014). For example, Alexiou et al. (2007) have modified the surface of SPIONs with starch molecules and also conjugated them with a chemotherapeutic drug (i.e. mitoxantrone), which are further accumulated at the tumor sites, and consequently the drug is released by applying a magnetic field gradient (Alexiou et al. 2007). Similarly, the surface-modified porous MNPs (i.e. Fe_3O_4) and magnetic mesoporous silica nanoparticles are investigated for targeted drug delivery applications (Cheng et al. 2009; Tao and Zhu 2014). Additionally, SPIONs are conjugated with luminescent compounds and drugs are also investigated for magnetically targeted simultaneous fluorescence imaging and drug delivery (Shen et al. 2012).

7.2.2 SPIONs-Encapsulated Polymeric Nanoparticles/Micelles

Polymeric nanoparticles/micelles (PNPs/PMCs) are aggregates (via self-assembly) of amphiphilic molecules which are dispersed in an aqueous solution, where non-polar (i.e. hydrophobic) tails are located on the interior side and polar/ionic (i.e. hydrophilic) heads are on the exterior and in contact with water. The self-assembly process is mainly influenced by the concentration of the amphiphilic molecules (in the aqueous medium), and PNPs/PMCs will only be formed above their critical micelle concentration (CMC), wherein only a favorable state of entropy occurs based on the dehydration of the hydrophobic tails (Zasadzinski et al. 2001; Lombardo et al. 2015).

PNPs/PMCs (without encapsulating the SPIONs) are widely used in drug delivery applications. But, a few research studies have also reported the encapsulation of hydrophobic SPIONs along with anticancer drugs inside the PMCs to make MDDs (Palanisamy and Wang 2019). Herein, the drug release from the SPIONs/drugs encapsulated PMCs/PNPs (Figure 7.2) can be achieved by the action of magnetic/pH/thermal/mechanical stimulation. For example, Purushotham and Ramanujan (2010) have developed magnetite (Fe_3O_4) nanoparticles and doxorubicin (DOX, an anticancer drug) encapsulated thermoresponsive PNPs for multimodal cancer therapies by using a thermoresponsive polymer (poly-n-(isopropylacrylamide) [PNIPAM]) (Purushotham and Ramanujan 2010). This study has shown that the as-prepared thermoresponsive PNPs have released the drug inside the cancer cells via on/off switching mode of thermoresponsive polymer

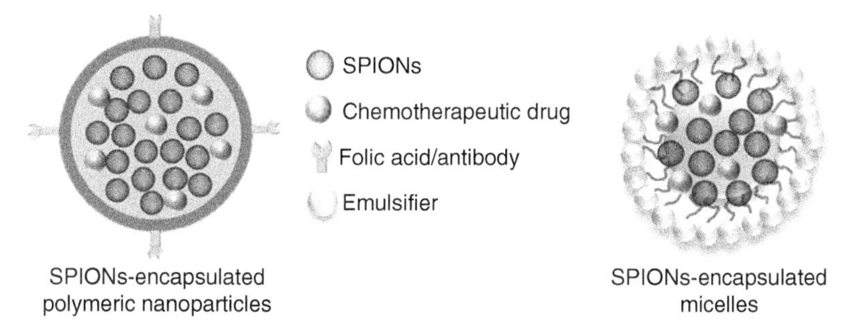

Figure 7.2 Schematic representation of SPIONs and CHD co-encapsulated polymeric nanoparticles/micelles.

at lower critical solution temperature (LCST) under the influence of an externally applied magnetic field. Moreover, the *in vitro* dual therapy (i.e. hyperthermia and chemotherapy) via thermoresponsive PNPs has resulted in effective cancer treatments.

Moreover, the SPION-encapsulated PMCs/PNPs could also be used as a heating agent in MFH, contrast agent in MRI, and actuating agent for thermoresponsive drug release at the tumor site (Li et al. 2013b; Asadi et al. 2016). Besides, the other type of imaging agents such quantum dots (QDs) or fluorophores could be co-encapsulated along with SPIONs and anticancer drugs in PMCs/PNPs to enhance the effectiveness of cancer diagnostics and therapeutics (Mandal et al. 2005; Asadi et al. 2016). Moreover, folic acid (FA) and/or antibodies (i.e. as targeting agents) can be conjugated on the outer surface of the PMCs/PNPs for active targeting of cancer cells. Moreover, the common procedures to synthesize SPIONs/drugs encapsulated PMCs/PNPs are nano-precipitation, dialysis, direct dissolution, and thin-film hydration (Talekar et al. 2011; Wakaskar 2017).

7.2.3 Magnetic Liposomes

Liposomes are made of a self-assembled lipid bilayer with vesicle-like structures – mainly composed of synthetic/natural phospholipids that are commercially used in the administration of nutrients. These liposomes could be used to encapsulate both hydrophobic (inside lipid bilayer) and hydrophilic (inside aqueous core) natured anticancer drugs and to release them consequently in a sustained manner. Liposomes have widely been used in drug delivery because of their low toxicity, good biocompatibility, and biodegradability (Bozzuto and Molinari 2015). Also, liposomes can fulfill the following basic requirements for anticancer therapy: (i) prolonged blood circulation and (ii) better bioavailability at the targeted sites (Hong et al. 2001).

Apart from drugs, the hydrophobic/hydrophilic SPIONs can also be encapsulated inside the liposomes to form magnetic liposomes. The initial studies of the encapsulation of the SPIONs inside liposomes are reported by De Cuyper and Joniau (1988, 1991). Later, the anticancer drug-loaded and targeting molecules (e.g. folic acid/antibody) conjugated magnetic liposomes are formed to be used in the concurrent targeted application in cancer treatment for chemotherapy, contrast enhancement via MRI, and heat induction via MFH (Figure 7.3) (Kumar et al. 2012; Akbarzadeh et al. 2013; Salim et al. 2014).

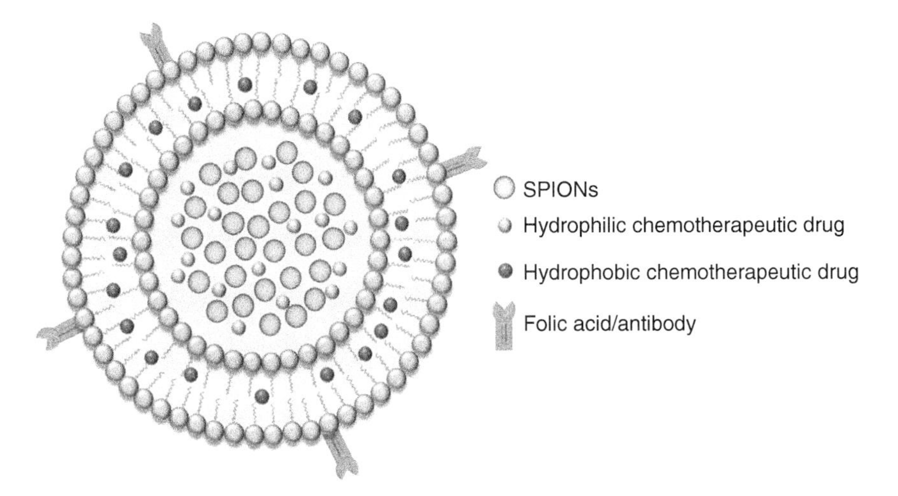

Figure 7.3 Schematic representation of hydrophobic/hydrophilic chemotherapeutic drug-encapsulated magnetic liposome for use in drug delivery.

Moreover, the temperature-sensitive lipid materials can also be used to encapsulate the SPIONs for developing stimuli-responsive magnetic liposomes for controlled drug release and MFH-based cancer treatment by the induced heat in a temperature range of 42–45 °C under the influence of an externally applied magnetic field (Ferreira et al. 2016; Shirmardi Shaghasemi et al. 2017). Furthermore, QDs/fluorophores can be encapsulated along with drugs and SPIONs to aid in fluorescence imaging along with MRI to enhance the sensitivity/accuracy in early detection/diagnosis of cancer cells (Hodenius et al. 2012). The general methods used for the preparation of magnetic liposomes are thin-film hydration and reverse-phase evaporation/double-emulsion method. For instance, DOX and magnetite nanoparticles encapsulated magnetic liposomes have been developed to suppress tumor growth and lung metastases via magnetic targeted drug delivery (Nobuto et al. 2004). The better antitumor effects of nanoparticle-based chemotherapy are observed as compared with standard methods of chemotherapy in the treatment of primary tumor/metastasis model. In another investigation, paclitaxel-encapsulated magnetic liposomes are prepared for the treatment of EMT-6 breast cancer (Zhang et al. 2005). Herein, the magnetic liposomes are easily accumulated under external applied magnetic fields which have resulted in better anticancer treatment in the xenograft model with fewer side effects compared to a conventional liposome formulation.

7.2.4 Magneto-Niosomes

Usually, liposomes suffer from a few critical drawbacks, such as high cost and lack of stability at different pH levels (Lombardo et al. 2019). To overcome this, nano-vesicles based on the non-ionic surfactants are developed and called as niosomes ("nios" – nonionic surfactant and "somes" – vesicles), where the anticancer drugs are encapsulated inside the closed bilayer structure. Moreover, these niosomes are inexpensive, relatively nontoxic, more stable, and biodegradable (Sankhyan and Pawar 2012; Rajni et al. 2019). In addition, the MNPs can be encapsulated inside the niosomes with or without chemotherapeutic drugs to form magneto-niosomes (Figure 7.4) (Tavano et al. 2013). Recently, magneto-niosomes are investigated for active/passive drug delivery in cancer

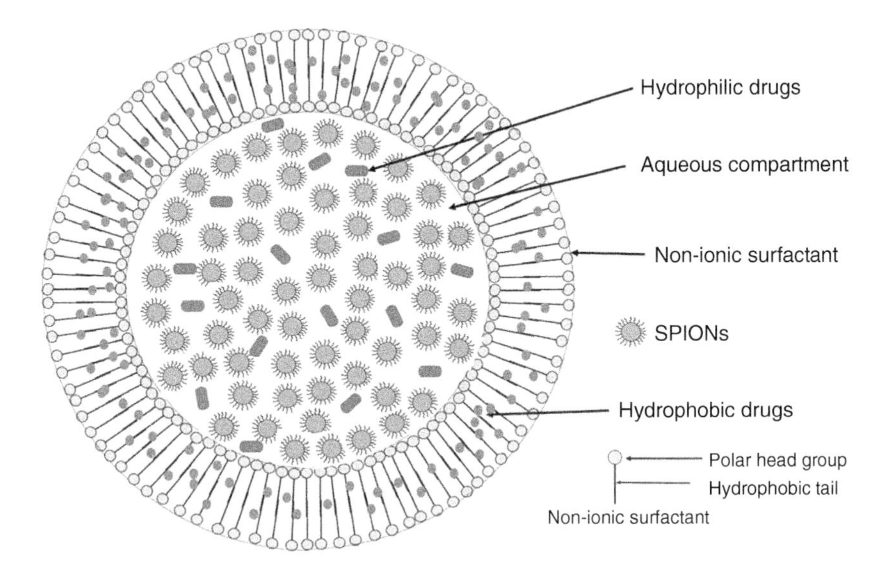

Figure 7.4 Schematic representation of the magneto-niosomes co-encapsulating SPIONs and chemotherapeutic drugs (hydrophilic and hydrophobic) for drug delivery.

treatment and subsequently they have displayed promising results in the stimuli-responsive drug release at the targeted tumor site under the influence of the externally applied magnetic field. Moreover, magneto-niosomes have shown effective results for combined therapies (i.e. chemotherapy and thermotherapy). Therefore, these magneto-niosomes have great potential to be developed as promising drug delivery systems for individual chemotherapy and combined chemo-thermo therapies in cancer treatment (Tavano et al. 2013).

In a recent investigation, TPGS-modified SPIONs and drug-loaded niosomes/pro-niosomes (i.e. the dry formulation of water-soluble carrier particles that are coated with surfactants) have been studied for targeted drug delivery in combination with hyperthermia to overcome multidrug resistance (MDR) in cancer cells (Liu et al. 2017). Besides, QDs/fluorophores could also be encapsulated along with the SPIONs and drugs inside magneto-niosomes to formulate magnetic nanoplatforms for bi-modal cancer diagnosis (i.e. combination of MRI and fluorescence imaging) and therapy (i.e. combination of localized MFH and chemotherapy).

7.2.5 Other Magnetic Nanostructures

The other magnetic nanostructures that are reported for magnetic-field-driven drug delivery include lipid–polymer hybrid nanoparticles, magnetic nanogel, MNP-entrapped multi-walled nanotubes (MWNTs), dendritic-magnetic nanocarriers, and MNP-encapsulated solid–lipid nanoparticles. In an investigation, Deok-Kong et al. (2013) have developed poly(D,L-lactide-co-glycolide) (PLGA)–lipid hybrid nanoparticles (PLHNPs) by incorporating them with superparamagnetic Fe_3O_4 nanoparticles and camptothecin (an anticancer drug), as shown in Figure 7.5, and further investigated for thermoresponsive drug release under the influence of the externally applied AMF. The induced heat (due to Neel and Brownian relaxations) through Fe_3O_4 nanoparticles (via AMF) breaks down the bonds in the polymer matrices for drug release (Deok-Kong et al. 2013). Moreover, the *in vitro* studies have showed better inhibition in MT2 mouse breast cancer cells by using chemotherapy (via camptothecin) under the influence of AMF compared to the cancer cell inhibition using only camptothecin.

In another study, a DOX- and SPION-encapsulated magnetic nanogel was developed for associated glutathione (GSH)/pH co-triggered drug delivery as shown in Figure 7.6, where the measured size and

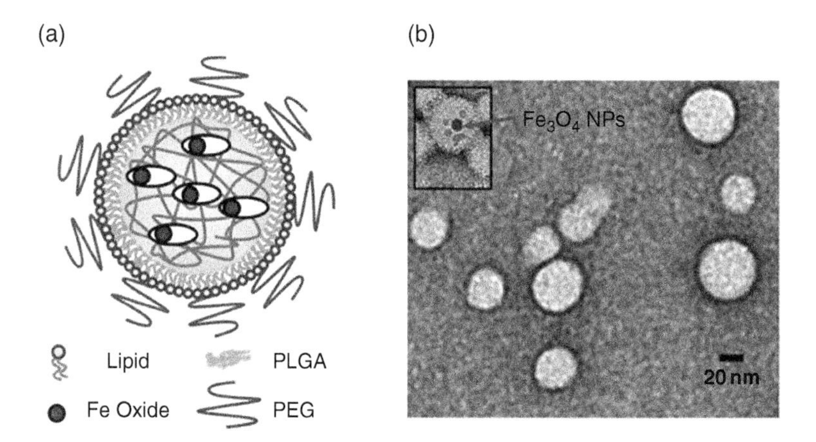

Figure 7.5 (a) Schematic representation of as-developed poly(D,L-lactide-co-glycolide) (PLGA)–lipid hybrid nanoparticles (PLHNPs) incorporated with superparamagnetic Fe_3O_4 nanoparticles and camptothecin, and (b) TEM image of PLHNPs (where insert image shows the entrapped Fe_3O_4 nanoparticles). *Source:* Adapted from Deok Kong et al. (2013) with permission from Elsevier.

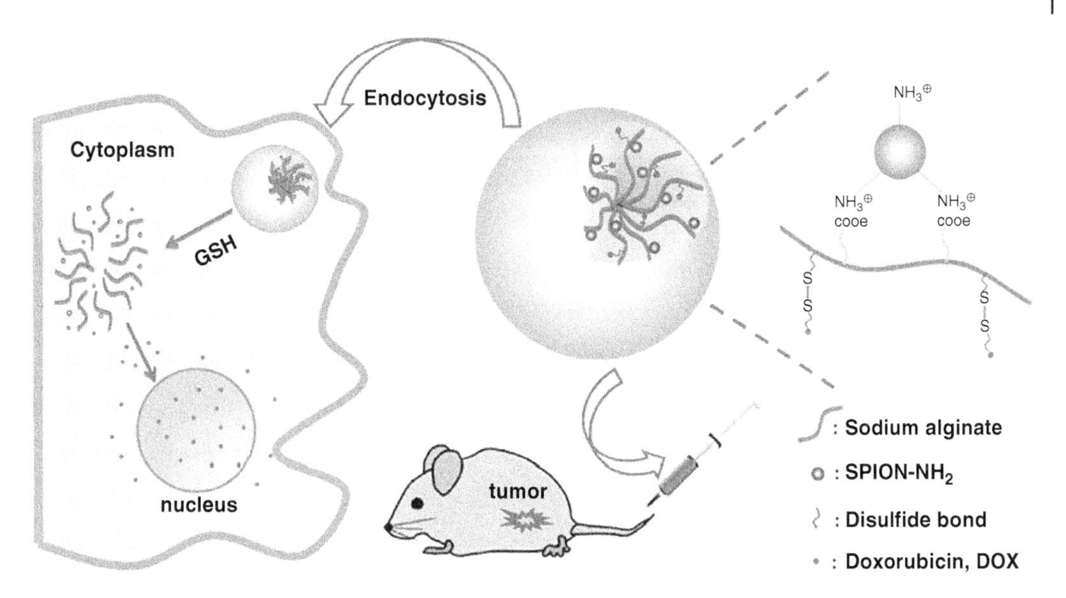

Figure 7.6 Schematic illustration of the synthesis of magnetic nanogel co-encapsulating DOX and SPIONs, and stimuli-responsive drug release characteristics after the cellular internalization. Source: Adapted from Chen et al. (2015) with permission from Royal Society of Chemistry.

surface charge of the nanogel are in the range of 120–320 nm with −40 mV, respectively (Huang et al. 2015). The nanogel has been found to be effective in inhibiting HeLa cells for at 2.3 μg ml^{-1} IC50 value.

Similarly, Yang et al. (2008) have incorporated Fe_3O_4 nanoparticles, and two chemotherapeutic drugs (i.e. 5-fluorouracil (5-FU) and cisplatin), inside folic acid (FA)-functionalized magnetic carbon nanotubes (MCNTs) for magnetic-targeted drug release in lymphatic system (Yang et al. 2008). The MCNTs navigated to the lymph nodes after the application of the external magnetic field and released the drugs in controlled manner. Moreover, the selected killing of the tumor cells in the lymph nodes was achieved via continuous release of chemotherapeutic drugs for several days. In another study, folate and iron di-functionalized multiwall carbon nanotubes (MWCNTs) have been found to be more effective (by sixfold) in *in vitro* studies for the magnetically targeted delivery of DOX inside the HeLa cells compared to the delivery of free doxorubicin (Li et al. 2011). Similarly, Seyfoori et al. have developed multifunctional magneto-/pH-responsive chitosan-coated nano-hybrid system (MpHNs) by combining multi-walled carbon nanotubes (CNT) and pH-sensitive nanogels for targeted DOX delivery at tumor sites, as shown in Figure 7.7 (Seyfoori et al. 2019). Herein, the MpHNs have displayed high drug-loading capacity, better targeting, and drug release characteristics in an acidic tumor environment.

7.3 Magnetic Delivery of Anticancer Drugs

Any of the magnetic drug delivery vehicles (i.e. surface-modified SPIONs, SPION-encapsulated polymeric nanoparticles/micelles, magnetic liposomes, magneto-niosomes, or other magnetic nanostructures) could be formulated with single/dual anticancer drugs. Moreover, different magnetic drug delivery approaches for single/dual anticancer drugs such as curcumin, doxorubicin paclitaxel, methotrexate, and daunorubicin, which are widely used in cancer treatment due to their better chemotherapeutic effects, are explained in the following sections.

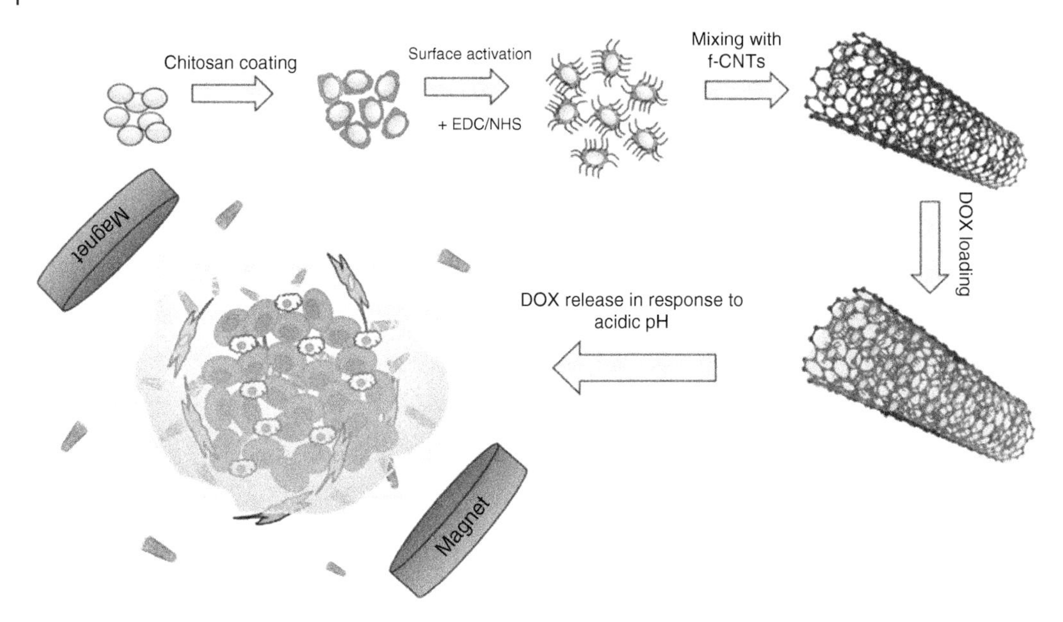

Figure 7.7 Schematic illustration of preparation of a magneto-/pH-responsive chitosan-coated nano-hybrid system (MpHNs) by combining multi-walled carbon nanotubes (CNT) and pH-sensitive nanogels for targeted DOX delivery. *Source:* Adapted from Seyfoori et al. (2019) with permission from Taylor and Francis.

7.3.1 Magnetic Delivery of Single Drugs

7.3.1.1 Delivery of Curcumin

Recently, curcumin has been encapsulated inside the magnetic alginate/chitosan nanoparticles to improve its biocompatibility and also uptake by breast adenocarcinoma cells (MDA-MB-231) (Song et al. 2018). Herein, the encapsulation efficiency of curcumin inside these nanoparticles was about 67.5% and MDA-MB-231 cancer cells have taken up the encapsulated curcumin threefold to sixfold better than its counterpart (i.e. free curcumin).

Similarly, Wang et al. (2014) have synthesized multifunctional magnetic/near-infrared (NIR)-based thermoresponsive core-shell hybrid nanogel for triggered drug release, where the SPIONs are clustered inside the core, and curcumin/fluorescent carbon dots are embedded on the porous carbon shell. That is coated with a thermosensitive poly(N-isopropylacrylamide) poly(NIPAM-AAm)-based gel layer (Figure 7.8a) to provide stability in aqueous media (Wang et al. 2014). The drug-loading content has been measured to be about $65.5\,\text{mg}\,\text{g}^{-1}$ (amount of drug in 1 g of dried nanocarrier). Also, the synthesized hybrid nanogels have shown controlled drug release through the action of stimulus with NIR light or AMF. In another study, curcumin/SPIONs are encapsulated inside the mixed micelles made up of d-α-Tocopheryl polyethylene glycol 1000 succinate (TPGS) and Pluronic F127 (Figure 7.8b), where 97% encapsulation efficiency is achieved for curcumin (Kandasamy et al. 2019).

Similarly, Hardiansyah et al. (2017) have loaded curcumin inside polyethylene glycol modified (PEGylated) magnetic liposomes and released it in a controlled manner during thermo-chemotherapy treatment. The encapsulation efficiency of curcumin inside the liposomes is about 76%, and these PEGylated magnetic liposomes have rapidly released curcumin at a thermoresponsive stimulus (i.e. at 45 °C), which consequently induced effective apoptosis in MCF-7 breast cancer cells (Hardiansyah et al. 2017). In a similar investigation, Montazerabadi et al. (2019) have encapsulated curcumin (CUR)/SPIONs inside the dendritic nanocarriers modified with folic acid (FA-mPEG-PAMAM

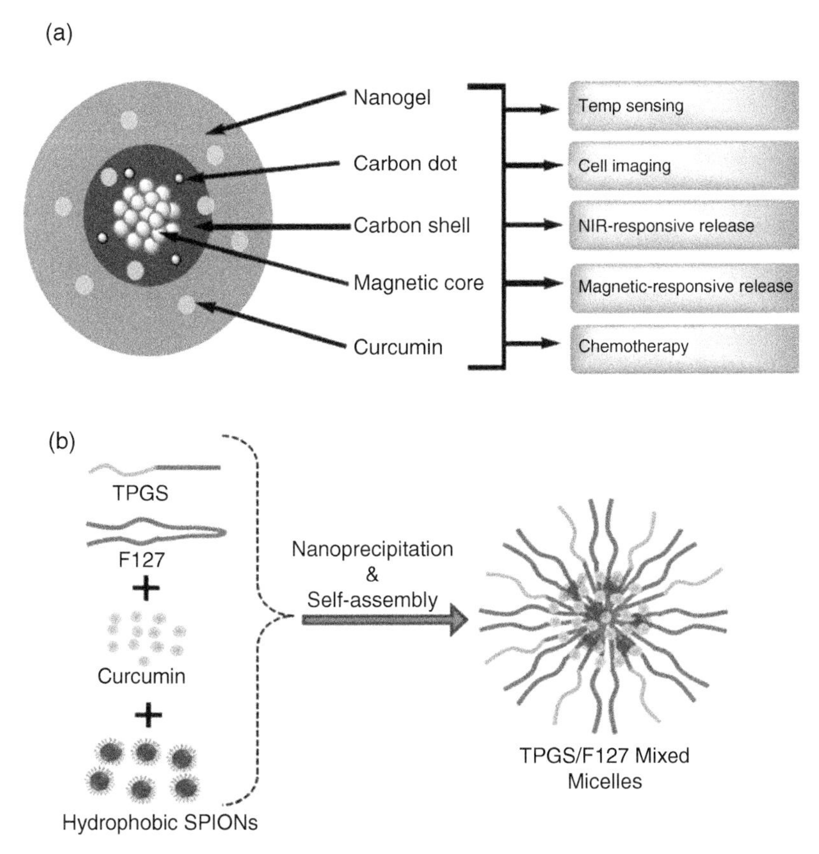

Figure 7.8 Schematic representation of (a) the encapsulation of curcumin into multifunctional core-shell hybrid nanogels (where the SPIONs are clustered inside the core and fluorescent carbon dots are embedded on the porous carbon shell and poly (N-isopropylacrylamide) poly (NIPAM-AAm)-based gel layer). (b) Synthesis of TPGS/F127 mixed micelles encapsulated with SPIONs and curcumin for drug delivery. *Source:* Adapted from Wang et al. (2014) and Kandasamy et al. (2019) with permission from Royal Society of Chemistry and Advanced Materials Letters.

G3-CUR@SPIONs) for concurrent targeted drug delivery and MFH-based cancer therapy. Herein, methoxy-PEGylated poly(amidoamine) (PAMAM) has been used as the polymer to make the dendritic nanocarriers, and the *in vitro* experimental results have displayed higher cellular uptake for these nanocarriers and better inhibition in MCF-7 human breast cancer cells and KB human nasopharyngeal carcinoma cell lines (due to effective drug release and combined therapy) (Montazerabadi et al. 2019).

Likewise, Zhao et al. (2014) have synthesized arginylglycylaspartic acid (RGD)-modified lipid–polymer hybrid MNPs for targeted delivery of 10-hydroxycamptothecin to human breast cancer (B-16 cell), where the calculated encapsulation efficiency for curcumin is measured to be 96%. Herein, the *in vitro* cytotoxicity studies have shown effective inhibition in the growth of the cancer cells via these hybrid nanoparticles (Zhao et al. 2014).

7.3.1.2 Delivery of Paclitaxel

Paclitaxel (PTX) and SPIONs are encapsulated inside PLGA nanoparticles to investigate the best effective way for targeted drug delivery in solid cancer tumor treatment (Schleich et al. 2014). Initially, different targeting methods are performed and compared individually and in combination, i.e.

passive targeting, active targeting of αvβ3 integrin, magnetic targeting through EPR effect, RGD grafting, and by placing the magnet on the tumor, respectively (Figure 7.9). The results obtained revealed that the as-prepared PLGA nanoparticles with combined active/magnetic targeting have shown to be eightfold more effective for the accumulation of PTX in tumor sites (in comparison to other individual targeting methods, i.e. passive, active or magnetic targeting), which resulted in the efficient inhibition of colon carcinoma (CT 26) tumor growth in xenografted mice models and prolonged their survival rate. Moreover, Ganipineni et al. (2018) have prepared PTX and SPION-loaded PEGylated PLGA nanoparticles (NPs; PTX/SPIO-NPs) for glioblastoma (GBM) therapy (Ganipineni et al. 2018). The calculated encapsulation efficiency for PTX inside these nanoparticles is ~ 30 ± 6% and the ex-vivo bio distribution studies have showed that these nanoparticles are found to be effective in increasing the accumulation of PTX inside GBM-bearing mice models via magnetic targeting, which prolonged their median survival time compared to the control treatments and passive targeting.

Analogously, Kulshrestha et al. (2012) have prepared PTX- and SPION-loaded thermosensitive magnetic liposomes (TSMLs) made up of 1,2-dipalmitoyl-sn-glycero-3-phosphocholine (DPPC) and 1-palmitoyl-2-oleoyl-sn-glycero-3-phospho-rac-glycerol (PG) by using the thin-film hydration method. The calculated encapsulation efficiency for PTX inside the TSMLs is 83 ± 3% and the *in vitro* studies have displayed that 100 nM PTX-loaded TSMLs are much effective in killing almost 89% of HeLa cancer cells under the externally applied AMF induction of thermotherapy in combination with chemotherapy (Kulshrestha et al. 2012). In similar fashion, Bano and colleagues have developed PTX- and MNP-loaded bovine serum albumin (BSA) nanocomposites (NCs) (with FA modified chitosan/carboxymethyl surface) for targeted drug delivery (Bano et al. 2016). These NCs have effectively enhanced the cancer inhibition rate on breast cancer cells. Moreover, Hussien et al. (2018) have developed pectin-conjugated magnetic-graphene oxide-based nano-hybrids for the delivery of PTX, where the calculated encapsulation efficiency of PTX is found to be

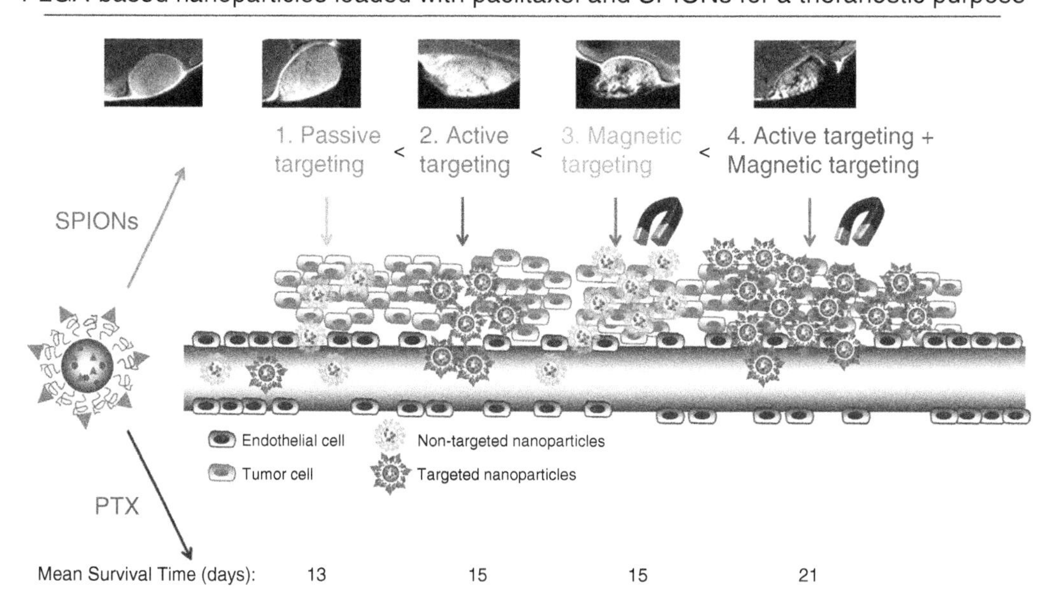

Figure 7.9 Schematic representation of SPIONs and paclitaxel loaded PLGA-based nanoparticles for cancer treatment application via different targeting methods. *Source:* Adapted from Schleich et al. (2014) with permission from Elsevier.

84.83 ± 5.33%. Moreover, the release studies have demonstrated the higher drug release from the nano-hybrids inside MCF-7 cancer cells at the endo-lysosomal pH than the normal physiological environment of the body. Also, the significant inhibition of the growth of MCF-7 cancer cells is confirmed from the cytotoxicity assays, realizing the higher suitability of these nano-hybrids in the treatment of cancer (Hussien et al. 2018).

7.3.1.3 Delivery of Doxorubicin (DOX)

Siminzar et al. (2019) have developed the DOX-encapsulated mucin-1 (MUC-1)-conjugated magnetic mesoporous silica nanocomposites (i.e. DOX-SPION@SiO2-MUC-1 NCs) for breast cancer treatment using targeted drug delivery, where the calculated encapsulation efficiency of DOX inside the nanocomposite have found to be ~ 83%. Also, the pH-responsive release studies have indicated a higher amount of DOX release in acidic tumor condition as compared to normal physiological environment. Moreover, the MUC-1-conjugated NCs are significantly internalized by the MUC-1 expressing MCF-7 cells as compared to their counterparts – i.e. MUC-1 negative MDA-MB-231 cells. Thus, the as-prepared DOX-SPION@SiO2-MUC-1 NCs are a potential nanoplatform for the targeted drug delivery for the cancer treatments (Siminzar et al. 2019). In another study, DOX-loaded PLGA polymer nanoparticles (DOX-PNPs) are prepared using the nano-precipitation method by using PLGA as polymer and BSA as the stabilizer, wherein the calculated encapsulation efficiency of DOX in DOX-PNPs is 80% (Sanson et al. 2011).

Moreover, the pH-dependent drug release studies have displayed a faster drug release at the endolysosomal pH of 4.0 and a slower drug release at body pH of 7.0. Also, the DOX-PNPs are found to be more effective in the treatment of MDA-MB-231 breast cancer cells in the *in vitro* studies than the free DOX-based treatment. Besides, the higher DOX loading and sustained release behavior through PEG-modified PLGA nanoparticles (PEG-PLGA) are obtained by the chemical conjugation of DOX rather than the physically incorporated DOX during their synthesis process (Lombardo et al. 2019). Moreover, Lin et al. (2016) have developed SPIONs/DOX-loaded and cell-penetrating peptide (CPP)-conjugated TSMLs using lipids of DPPC:MSPC:DSPE-PEG_{2000} for targeted drug delivery, where the encapsulation efficiency of DOX in TSMLs is calculated to be 87.23 ± 1.24% (Lin et al. 2016). Also, the temperature-sensitive drug release and *in vitro* biological studies have respectively showed a higher drug release at 42 °C temperature (induced via MFH by applying external AMF), and better inhibition of cancer cells at the same temperature compared to the cells treated at 37 °C or control formulation. Furthermore, the survival time of the MCF-7 xenograft mice that are treated with TSMLs at 42 °C temperature (induced via AMF) is enhanced by delaying the tumor progression.

Similarly, Ferjaoui et al. (2019) developed thermoresponsive core/shell magnetic nanoparticles (TMNPs) which are composed of Fe_3O_4 nanoparticles as core, and oligo (ethylene glycol) methacrylate (OEGMA) and 2-(2-methoxy) ethyl methacrylate (MEO2MA) as thermosensitive polymer shell for multimodal cancer therapy. TMNPs exhibited the drug release behaviors above low critical solution temperature (LCST) at about 41 °C (Figure 7.10a) while $P(MEO_2MA_{60}OEGMA_{40})$ coated Fe_3O_4 loaded with DOX have displayed very small amount of drug release below LCST but almost 100% of DOX are achieved at 42 °C in 52 hours (Figure 7.10b). TMNPs revealed effective temperature rise under externally applied AMF and thus could be used as a potential heating agent for magnetic fluid hyperthermia (MFH) application (Figure 7.10c). Moreover, TMNPs displayed reduced significant viability of SKOV-3 cells on exposure to free DOX and MFH at 37 and 41 °C after 24 hours (Figure 7.10d) (Ferjaoui et al. 2019).

In analogous fashion, Sanson and colleagues have prepared hydrophobic SPIONs and DOX-loaded magnetic polymersomes through the nanoprecipitation method – by using biodegradable poly(trimethylene carbonate)-b-poly(L-glutamic acid) (PTMC-b-PGA) block copolymers,

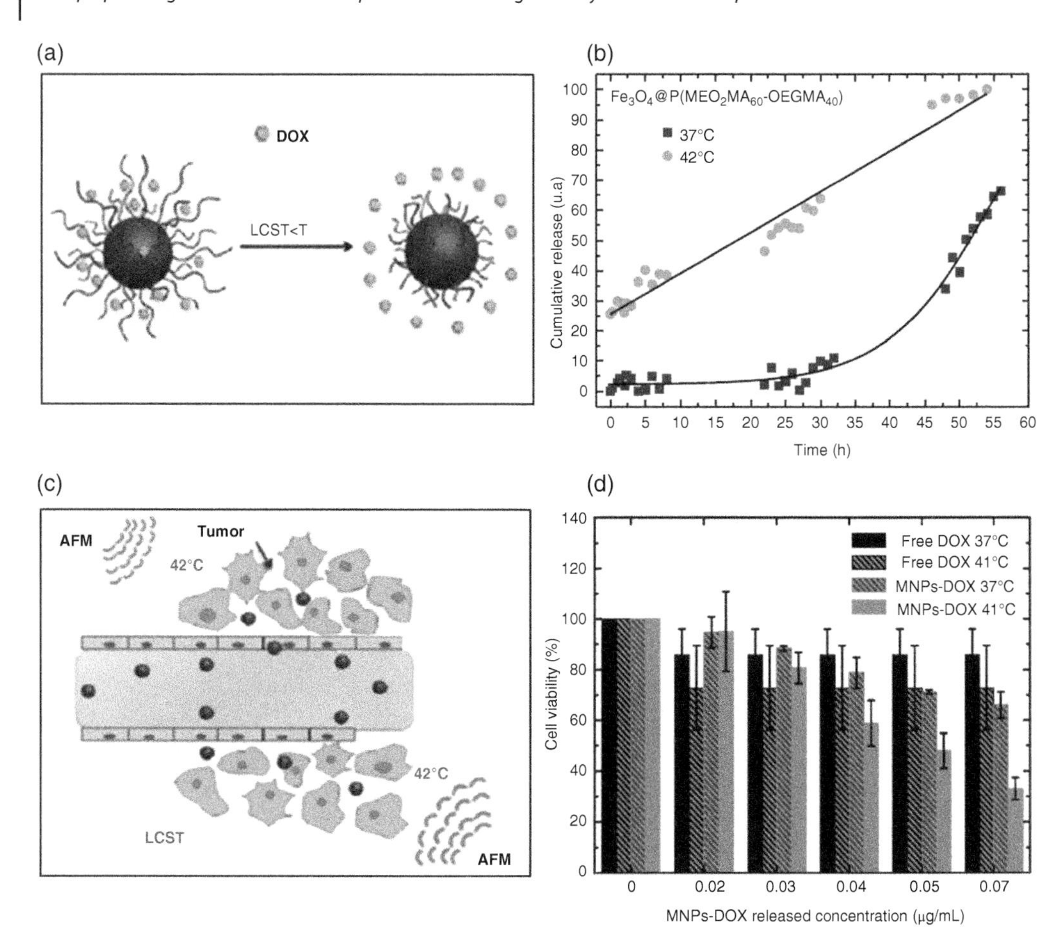

Figure 7.10 (a) Drug release behaviors of thermoresponsive core/shell magnetic nanoparticles (TMNPs) above low critical solution temperature (LCST). (b) Cumulative release curve of DOX from $Fe_3O_{4@}$ coated $P(MEO_2MA_{60}OEGMA_{40})$ TMNPs at 37 and 42 °C. (c) MFH treatment for solid tumor using TMNPs. (d) Cell Viability of SKOV-3 cells after exposure to free DOX, TMNPs at 37 and 41 °C after 24 hours. *Source:* Adapted from Ferjaoui et al. (2019) with permission from American Chemical Society.

where the calculated encapsulated efficiency of DOX in polymersomes is 22% (Sanson et al. 2011). Herein, the triggering via AMF has increased the release of the drugs in double amounts compared to the normal release conditions, which have made them suitable candidates for the application of magneto-chemotherapy in cancer treatment. In another similar study, Tavano et al. (2013) have prepared DOX and SPION-loaded magneto-niosomes – by using Tween 60 and Pluronic L64 as nonionic surfactants – for targeted drug delivery, where these magneto-niosomes have shown good drug encapsulation efficiency of 64%, and have exhibited excellent drug release behavior in controlled conditions (Tavano et al. 2013).

7.3.1.4 Delivery of Methotrexate

Basu et al. (2018) have prepared Fe_3O_4 modified PLGA-PEG nanocomposites (FPNCs) to improve the delivery of methotrexate (MTX) at targeted sites during cancer treatment, where these FPNCs are formed via double emulsion (w/o/w) method and 95% entrapment efficiency of MTX is observed at 1 : 1 polymer:drug ratio compared to the other ratios – i.e. 1 : 0.25 (81.03%) and 1 : 0.5

(89.5%) of ratio (Basu et al. 2018). Herein, the pH-dependent drug release over 72 hours have shown 86% of MTX release at pH of 4.6 (i.e. acidic tumor environment), while only 15% is observed at pH of 7.3 (i.e. physiological condition). Also, these FPNCs have showcased better treating efficacy in SK-BR-3 (breast adenocarcinoma) cancer compared to free MTX. In addition, Jiang et al. have prepared MTX-loaded Fe_3O_4/polypyrrole (PPy) nanospheres (MFPNs) for magnetic targeting-based drug delivery, where ~ 109 μg mg^{-1} of MTX is loaded encapsulated inside the MFPNs (Jiang et al. 2015). Moreover, the *in vitro* studies have displayed controlled drug release behavior, enhanced intracellular drug retention, and specific targeting ability to the HeLa cells. Thus, the MFPNs are a promising candidate for the targeting drug delivery in cancer treatment.

In another investigation, MTX-conjugated hyperbranched polyglycerol (HPG)-grafted SPIONs (SPIONs-g-HPG-MTX) are synthesized for drug delivery application (Li et al. 2013a). Initially, the SPIONs grafted with HPG are prepared by inorganic sol–gel reaction followed by thiol-ene click chemistry, and then, MTX is conjugated with the SPIONs through esterification reaction. SPIONs-g-HPG-MTX have displayed (i) higher cell uptake and (ii) 50% inhibition of human head and neck cancer (KB) cells at 0.5 mg ml^{-1} concentration, while attaining low cytotoxicity in normal cells – i.e. mouse macrophages (RAW 264.7) and 3T3 fibroblasts. Analogously, Souza et al. have surface modified the biodegradable SPIONs with poly(ethyleneglycol)-methyl-ether-poly(-caprolactone) (mPEG-PCL) copolymer, where MTX is entrapped inside a copolymer through hydrophobic interactions. The encapsulation efficiency of MTX in the surface-modified SPIONs is 98.6% and the their major drug release (of about 44%) is observed at 42 °C due to the thermosensitive nature of the mPEG-PCL copolymer (Souza et al. 2016).

Moreover, Zhu et al. (2009) have prepared MTX- and SPION-encapsulated TSMLs using DPPC and cholesterol through reverse-phase evaporation method for targeted drug delivery in skeletal muscular tissue (Zhu et al. 2009). The calculated encapsulation efficiency of MTX in TSML is 61.4% and temperature-dependent drug release studies up to 24 hours have shown a significant amount of MTX release at 41 °C under an externally applied magnetic field, while very slow MTX release is observed at 37 °C. Moreover, the tissue distribution and pharmacokinetics investigations have displayed the effective accumulation of the MTX in the skeletal muscular tissue. Thus, the TSML has been suggested as a potential candidate for the triggered drug release for the cancer treatment.

7.3.1.5 Delivery of Daunorubicin

Chen et al. (2011) have entrapped daunorubicin (DRC) onto the surface of oleic acid-coated/Pluronic F-127-modified SPIONs (DRC-SPIONs) for chemotherapy-based cancer treatment (Chen et al. 2011). The calculated encapsulation efficiency of DRC is found to be 76.2%. Moreover, the *in vitro* release studies have demonstrated an initial burst release behavior and then slow/sustained release of the DRC over several weeks from DRC-SPIONs. Additionally, the effective inhibition of the proliferation of K562 cancer cells have implied favorable antitumor properties of the drug-attached SPIONs in a dose-dependent manner. In another similar investigation, Hosseini et al. have fabricated DRC-loaded nanofibers formed by using poly (lactic acid) (PLA)/multi-walled carbon nanotubes (MWCNT)/Fe_3O_4 (PLA/MWCNT/Fe_3O_4) based NCs (as shown in Figure 7.11a) for the treatment of leukemia (Hosseini et al. 2016).

Herein, the calculated encapsulation efficiency of the DRC inside PLA/MWCNT/Fe_3O_4 nanofibers is 93–97%. Additionally, the studies have shown a higher release of DRC from the scaffold of nanofiber under externally applied magnetic field (as shown in Figure 7.11b) based on the non-Fickian transport mechanism. Moreover, the scaffold of the nanofibers is used for the investigation of the inhibition effect in leukemia K562 cell lines in the absence and presence of the externally applied

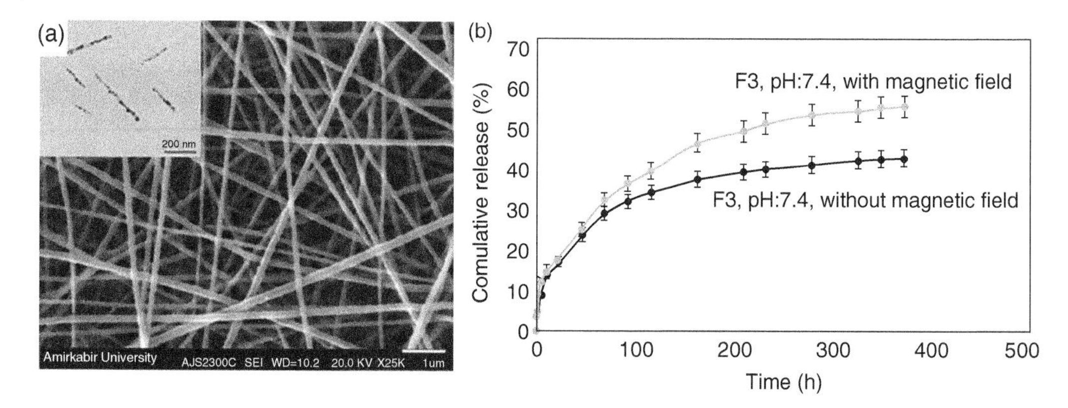

Figure 7.11 (a) SEM/TEM (inserted) image of using poly (lactic acid) (PLA)/multi-walled carbon nanotubes (PLA/MWCNT)/Fe_3O_4 nanofibers and (b) drug(daunorubicin/DRC) release behaviors from PLA/MWCNT/Fe_3O_4 nanofibers at pH of 7.4. *Source:* Adapted from Hosseini et al. (2016) with permission from Taylor and Francis.

AMF, where PLA/MWCNT/Fe_3O_4 nanofibers are capable of effectively killing the leukemia K562 cells under the presence of AMF based on the synergistic cytotoxic effects from thermotherapy and chemotherapy.

7.3.1.6 Delivery of Other Drugs

Mandriota et al. (2019) have developed cisplatin (platinum-based anticancer drug)-loaded superparamagnetic nanoclusters for cancer chemotherapy. Initially, hydrophobic SPIONs are synthesized by the solvothermal method, and then the nanoclusters of SPIONs (NCs-SPIONs) are formed through oil-phase evaporation-induced self-assembly. Later, NCs-SPIONs are surface-functionalized with a polydopamine layer (NCs-SPIONs@PDO) to improve their stability in an aqueous medium, and further loaded with cisplatin to form Cip-NCs-SPIONs (~$0.67 \pm 0.2\,\mu g$ of cisplatin is encapsulated per mg of NCs) for smart chemotherapy (Mandriota et al. 2019). In another investigation, the dual-targeted (i.e. combination of active and magnetic targeting) oleanolic acid-loaded octreotide-modified magnetic liposomes (O2MLs) are synthesized to increase the accumulation of anticancer drugs at the tumor site (Li et al. 2018). In O2MLs, (i) oleanolic acid is used as a model anti-cancer drug, (ii) octreotide has acted as a modified ligand for use in receptor-mediated targeting, and (iii) SPIONs are utilized as magnetic targeting agents. The encapsulation efficiency of oleanolic acid in O2MLs is calculated to be ~$94.3 \pm 0.31\%$. The *in vivo* investigation of O2MLs have shown higher cellular internalization, better antitumor effects, and lower systemic toxicity due to dual targeting compared to their counterparts such as free oleanolic acid, saline, and only octreotide-modified liposomes. Thus, the dual-targeting based O2MLs could be potential candidates for effective drug delivery in cancer treatments.

7.3.2 Magnetic Delivery of Dual Drugs

Pancreatic cancer is an aggressive disease known for forming highly chemo-resistant tumors, and it is complex and very tough to treat via chemotherapy. Gemcitabine (GEM) is used as the model drug for the treatment of pancreatic cancer. Usually, chemotherapy could be failing due to the development of chemo-resistance by pancreatic cells after multiple drug treatments including GEM, leading to poor survival of the patients. To overcome this challenge, Khan et al. (2019) developed curcumin-loaded SPIONs (SP-CUR) and SP-CUR+GEM formulations to effectively deliver dual anticancer agents to the pancreatic tumors and to simultaneously promote their uptake in cancer cells for reducing tumor growth and metastasis. The results have shown that the encapsulation efficiency of curcumin is ~80–90% and moreover, SP-CUR+GEM formulation is capable of effectively inhibiting the proliferation and migration of the pancreatic cancer cells compared to its counterpart (i.e. SP-CUR) (Khan et al. 2019).

In another investigation, Cui et al. (2016) have initially formed transferrin receptor-binding peptide(T7)-conjugated and PEGylated PLGA NPs and then loaded them with oleic acid (OA) coated MNPs and two chemotherapeutic drugs (i.e. PTX and CUR) to form PTX-CUR-MNPs-T7- PEGylated PLGA NPs for (i) magnetic targeting via externally applied alternating magnetic field (AMF) and (ii) active targeting with a transferrin receptor-binding peptide (as shown in Figure 7.12a) (Cui et al. 2016). The results have shown a 10-fold increased accumulation of drugs inside the glioma tumors (i.e. U87-Luc cells) and a 5-fold higher uptake by crossing the blood-brain barrier while utilizing PTX-CUR-MNPs-T7-PEGylated PLGA NPs compared to the non-targeted PLGA NPs peptide (Figure 7.12b). Moreover, the *in vivo* studies have displayed an increment in the survival rate of the male BALB/c nude mice (four to five weeks) due to the higher therapeutic efficacy via dual-targeted therapy using these PLGA NPs compared to the control group exposed to either active targeting or magnetic targeting alone (Figure 7.12c).

In another study, Balasubramanian et al. (2014) have developed 5-FU, curcumin, and SPION-encapsulated poly(D,L-lactic-co-glycolic acid) nanoparticles (5F-Cu-SPIONs-PLGA NPs) that are

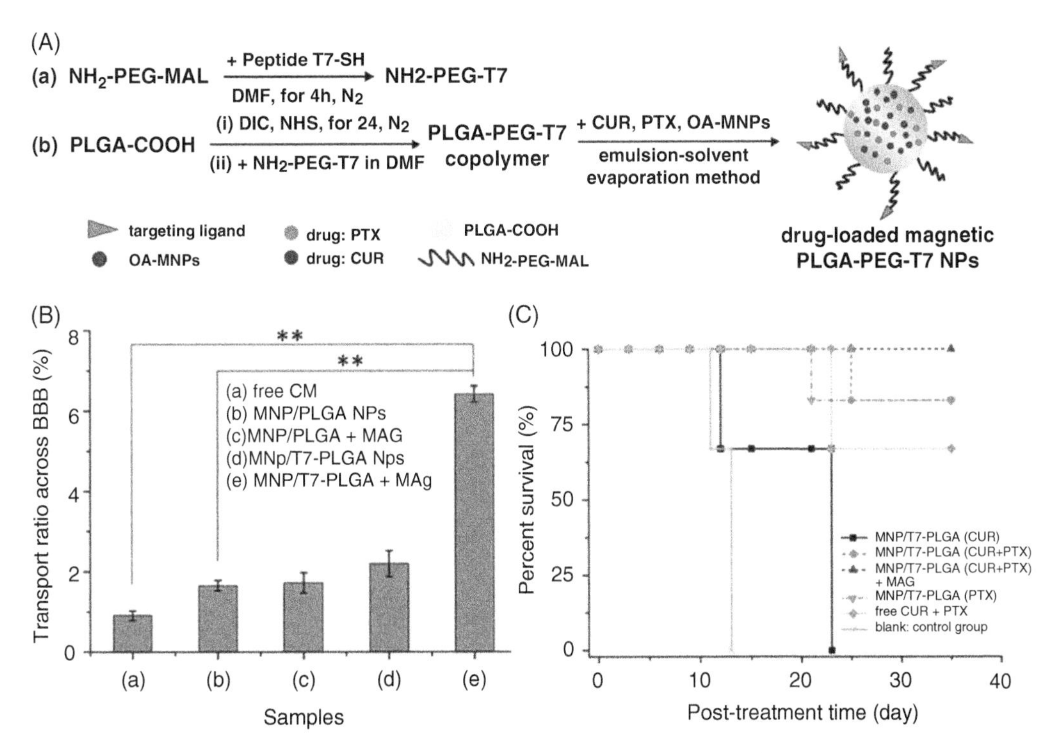

Figure 7.12 (A) Schematic representation of (a) synthesis of PLGA-PEG-T7 polymer and (b) drugs (PTX and CUR) encapsulated PEGylated PLGA NPs (i.e. PTX-CUR-SPIONs-T7- PEGylated PLGA NPs), (B) *in vitro* targeting efficiency (in terms of transport ration) of different samples (a) free CUR and PTX, (b) MNPs/PLGA NPs, (c) MNPs/PLGA NPs+magnetic field (MAG) (d) MNP/T7-PLGA NPs (e) MNP/T7-PLGA NPs+MAGinglioma cancer (U87-Luc cells) in four hours, and (C) *in vivo* study of overall survival of glioma-bearing mice for different samples. *Source:* Adapted from Cui et al. (2016) with permission from American Chemical Society.

Table 7.1 Superparamagnetic iron oxide nanoparticle-based AMF-induced in vitro drug release/MFH studies.

Sr. No.	Magnetic Nanoplatform	Drug loading	Magnetic Field and Frequency	Time	Mechanism	Cell viability Reduction	Ref.
(i)	Stealth and Folate-R targeted TMLs with hydrophilic IO NPs	DOX DLC=10%, DLE=85%	f=290 kKz B=15 mT H=12kA. m^{-1}	60 min	Membrane permeation at lipid T_m	KB and HeLa cell = 93% at 30 μM DOX + AMF (33.5°C) synergetic effect	(Pradhan et al. 2010)
(ii)	Polymersomes with surfactant-coated IONPs in hydrophobic membrane (30% wt.)	DOX DLE = 28% DLC=5.6 %	f=750 kHz B=14 mT H=11.2 $kA{\cdot}m^{-1}$	10 min, 3 times a day (2h interval between exposures)	Membrane permeation at polymer T_m (PTMC)	HeLa Cells = MH+DOX ca. 45% No MH ca. 30% MH ca. 5%	(Oliveira et al. 2013)
(iii)	Polymersomes with polyelectrolyte gel layers / citrated IONPs	IONPs + DOX DLE=73.3% DLC=9.3 %	f=37 kHz B=3.1 mT H=2.5 $kA.m^{-1}$	20 min	Membrane permeation T_m (distearin)	HeLa = 20% survival after MH vs. 45% at same [DOX]=10 μM	(Chiang et al. 2013)
(iv)	Fe_3O_4@PSMA grafted with polyA15 22 nm core	5-FU 5425 molecules/ NP	f=1.3 MHz B=41.3 mT H=33 $kA.m^{-1}$	15 min	H-bond breaking	MBT-2 50% reduction Anti-HER2 ligand	(Li et al. 2013c)
(v)	Magnetic PEG-PLA micelles covalently conjugated with MNP_S	DOX loading DLC 11.4% IO payload 4%	f=200 kHz B=1.9 mT H=1.5 $kA.m^{-1}$	–	T_c at 37°C T_m 40-56°C	A549 cells = Synergistic effect of DOX+AMF applied	(Kim et al. 2015)
(vi)	Magnetic alginate microbeads	DOX loading DLC= 0.34%	f=700 kHz B=10 mT H=8 $kA.m^{-1}$	2h	Gel permeation	MCF- 7 h breast Cell viabilities DOX + AMF (5.7%) Dox (63.3 %) AMF (59.4%)	(Brulé et al. 2011)

Source: Adapted from Mertz et al. (2017) with permission from Elsevier.

Table 7.2 Superparamagnetic iron oxide nanoparticle-based AMF-induced in vivo drug release/MFH studies.

Sr. No.	Magnetic Nanoplatform	Drug loading	Magnetic Field and Frequency	Time, Dose (mg/kg$^{-1)}$ administration route	Mechanism	Potency Results	Ref.
(i)	NC@MS@HPC USPIO (7 nm) nanoclusters	Gemcitabine Chemotherapy DLE 66% DLC 20 % wt	f=199 kHz B=17.5 mT H=14 kA.m^{-1} 30 min At t=2h inj.	2 h 5 mg·ml^{-1}; 100 μL i.t.	Pancreas xenografts tumor models in nude mice	Tumor apoptosis GEM+MHT 38% GEM 17.5% MHT 14.7%	(Kim et al. 2014)
(ii)	NC@Polymer Fe3O4@Ppyr-PEG-FA 70 nm	DOX Chemotherapy	f=230 kHz B=10 mT H=8 kA·m^{-1} 20 min	45 days NPs 5 mg·kg^{-1} DOX 0.13 mg·kg^{-1} i.t	Subcutaneous Xenograft myeloma cell line in mice	Tumor suppression after 12 days Survival 200 days vs controls no MHT/DOX, 100days	(Hayashi et al. 2014)
(iii)	NANOGEL Supramolecular magnetic nanogel 70-160 nm	DOX Chemotherapy H-bonds and pi pi stacking	f=500kHz B=46.3 mT H=37 kA·m^{-1} 10 min at 36h inj.	15 days 150 μg·kg^{-1} NPs 2.8 μg·kg-1 DOX i.v. injection	Colon DLD1 cancer xeno-grafts model on mices	Tumor suppression over 15 days With double injection	(Lee et al. 2013)
(iv)	DEC capsules PVA shell +IO NPs 5 nm 150 nm DEC	DOX/PTX Dual Chemotherapy IVO24 peptide targeting	f=50 kHz B=10 mT H=8 kA·m^{-1} 10 min at t= 24 h post inj.	30 days 100 μl saline solution at 2 wt % DEC i.t.	MCF-7 breast carcinoma tumor model	In vivo tumor model = suppression when PTX+DOX+AMF+IVO24	(Hu et al. 2012)
(v)	TMLs loaded with magnetite 9 nm	Docetaxel (DOC) chemotherapy	f=478 kHz B=8 mT H=6.4kA·m^{-1} P=1 kW 35 min.	28 days 100 μL of DML 187mg·ml^{-1} -magnetite 56.8- 568.5 μg·mL^{-1} (DOC)	Xenograft MKN45 gastric cancer cells in mice	Tumor suppression at day 10	(Yoshida et al. 2010)

Source: Adapted from Mertz et al. (2017) with permission from Elsevier.

functionalized with two cancer-specific ligands (i.e. folate- and transferrin) for drug delivery application (Balasubramanian et al. 2014). The encapsulated efficiencies of 5FU and curcumin inside these nanoparticles are determined as 63% and 71%, respectively. Moreover, the folate-/transferrin-functionalized PLGA NPs have shown higher uptake by MCF-7 cancer and glial (G1) cells compared to their non-functionalized counterparts. Also, the cytotoxicity studies have shown a sharp decline in the viability of the cancer cells treated with PLGA NPs (that are exposed to magnetic field) as compared to their controls. In a recent study, Rahimi et al. (2017) have encapsulated DOX/MTX-loaded and glutamic acid-coated iron oxide nanoparticles inside dendritic chitosan-g-mPEG based magnetic nanocarriers (i.e. DOX-MTX-Fe_3O_4-DCP NPs) for dual drug-based chemotherapeutics (Rahimi et al. 2017). Herein, the encapsulation efficiencies of doxorubicin and methotrexate inside the DCP NPs are found to be 95.96% and 67.91%. Also, the drug release studies have demonstrated a higher percentage (80.36% of DOX and 40.30% of MTX) of drug release at acidic pH of 5.0 (i.e. tumor environment) for ~ 400 hours. However, only 19.47% of DOX and 33.74% of MTX are released at the physiological pH of 7.4. Moreover, the effectiveness of dual drug-based chemotherapy via DOX-MEX-Fe_3O_4-DCP NPs is validated through cytotoxicity assay, cellular uptake, image staining, cell cycle, and apoptosis analysis.

In this chapter, we have discussed different types of SPION-encapsulated drug delivery vehicles for chemo-thermo therapeutics. The key role of SPIONs is to carry and release chemotherapeutic drugs (CHDs) at targeted tumor sites under externally stimuli/applied AMF for multimodal cancer treatment via CHDs and MFH. SPIONs-based AMF induced *in vitro* and *in vivo* drug release/MFH studies are summarized in Tables 7.1 and 7.2, respectively (Mertz et al. 2017).

7.4 Conclusion

To summarize, we have initially discussed the as-developed SPIONs-based MDDs, including surface-modified SPIONs, polymeric nanoparticles/micelles with SPIONs, magnetic liposomes, magneto-niosomes, and other magnetic structures. Then, we have reviewed in detail the magnetic delivery of the different single/dual drugs (including curcumin, doxorubicin, daunorubicin, paclitaxel, and methotrexate) via these SPIONs-based MDDs in the *in vitro/in vivo* conditions for cancer treatments through chemotherapy or/and MFH. Herein, we have particularly highlighted the effectiveness of the encapsulation of the anticancer drugs inside the SPION-based MDDs and their consequent usage in the treatment of different cancers. In the near future, these SPION-based MDDs can be loaded with multiple imaging/therapeutic agents to formulate multifunctional magnetic nanoplatforms for multi-modal diagnostics (i.e. combination of MRI, fluorescence, computed tomography [CT], and radionuclide-based imaging) and therapy (i.e. combination of chemotherapy, localized MFH, photo dynamic/thermal therapy, and sonodynamic therapy) for more efficient cancer treatments.

References

Akbarzadeh, A., Rezaei-Sadabady, R., Davaran, S. et al. (2013). Liposome: classification, preparation, and applications. *Nanoscale Research Letters* 8 (1): 102.

Alexiou, C., Jurgons, R., Seliger, C. et al. (2007). Delivery of superparamagnetic nanoparticles for local chemotherapy after intraarterial infusion and magnetic drug targeting. *Anticancer Research* 27 (4 A): 2019–2022.

Asadi, H., Khoee, S., and Deckers, R. (2016). Polymer-grafted superparamagnetic iron oxide nanoparticles as a potential stable system for magnetic resonance imaging and doxorubicin delivery. *RSC Advances* 6 (87): 83963–83972.

Balasubramanian, S., Girija, A.R., Nagaoka, Y. et al. (2014). Curcumin and 5-fluorouracil-loaded, folate- and transferrin-decorated polymeric magnetic nanoformulation: a synergistic cancer therapeutic approach, accelerated by magnetic hyperthermia. *International Journal of Nanomedicine* 9 (1): 437.

Bano, S., Afzal, M., Waraich, M.M. et al. (2016). Paclitaxel loaded magnetic nanocomposites with folate modified chitosan/carboxymethyl surface; a vehicle for imaging and targeted drug delivery. *International Journal of Pharmaceutics* 513 (1–2): 554–563.

Basu, T., Singh, S., and Pal, B. (2018). Fe_3O_4 @ PLGA-PEG nanocomposite for improved delivery of methotrexate in cancer treatment. *ChemistrySelect* 3 (29): 8522–8528.

Bozzuto, G. and Molinari, A. (2015). Liposomes as nanomedical devices. *International Journal of Nanomedicine* 10: 975.

Brulé, S., Levy, M., Wilhelm, C. et al. (2011). Doxorubicin Release Triggered by Alginate Embedded Magnetic Nanoheaters: A Combined Therapy. *Advanced Materials* 23 (6): 787–790.

Chang, D., Lim, M., Goos, J.A.C.M. et al. (2018). Biologically targeted magnetic hyperthermia: potential and limitations. *Frontiers in Pharmacology* 9: 831.

Chen, B., Wang, J., Chen, B. et al. (2011). Synthesis and antitumor efficacy of daunorubicin-loaded magnetic nanoparticles. *International Journal of Nanomedicine* 6: 203–211.

Chen, L., Xue, Y., Xia, X. et al. (2015). A redox stimuli-responsive superparamagnetic nanogel with chemically anchored DOX for enhanced anticancer efficacy and low systemic adverse effects. *Journal of Materials Chemistry B* 3 (46): 8949–8962.

Cheng, K., Peng, S., Xu, C., and Sun, S. (2009). Porous hollow Fe_3O_4 nanoparticles for targeted delivery and controlled release of cisplatin. *Journal of the American Chemical Society* 131 (30): 10637–10644.

Chiang, W.H., Huang, W.C., Chang, C.W. et al. (2013). Functionalized polymersomes with outlayered polyelectrolyte gels for potential tumor-targeted delivery of multimodal therapies and MR imaging. *Journal of Controlled Release* 168 (3): 280–288.

Cui, Y., Zhang, M., Zeng, F. et al. (2016). Dual-targeting magnetic PLGA nanoparticles for codelivery of paclitaxel and curcumin for brain tumor therapy. *ACS Applied Materials and Interfaces* 8 (47): 32159–32169.

De Cuyper, M. and Joniau, M. (1988). Mechanistic aspects of the adsorption of phospholipids onto lauric acid stabilized magnetite nanocolloids. *European Biophysics Journal* 15 (5): 311–319.

De Cuyper, M. and Joniau, M. (1991). Mechanistic aspects of the adsorption of phospholipids onto lauric acid stabilized Fe_3O_4 nanocolloids. *Langmuir* 7 (4): 647–652.

Deok Kong, S., Sartor, M., Jack Hu, C.-M. et al. (2013). Magnetic field activated lipid–polymer hybrid nanoparticles for stimuli-responsive drug release. *Acta Biomaterialia* 9 (3): 5447–5452.

Farah, F.H. (2016). Magnetic microspheres: a novel drug delivery system. *Journal of Analytical & Pharmaceutical Research* 3 (5): 00067.

Ferjaoui, Z., Jamal Al Dine, E., Kulmukhamedova, A. et al. (2019). Doxorubicin-loaded thermoresponsive superparamagnetic nanocarriers for controlled drug delivery and magnetic hyperthermia applications. *ACS Applied Materials and Interfaces* 11 (34): 30610–30620.

Ferreira, R.V., Martins, T.M.D.M., Goes, A.M. et al. (2016). Thermosensitive gemcitabine-magnetoliposomes for combined hyperthermia and chemotherapy. *Nanotechnology* 27 (8): 085105.

Ganipineni, L.P., Ucakar, B., Joudiou, N. et al. (2018). Magnetic targeting of paclitaxel-loaded poly(lactic-co-glycolic acid)-based nanoparticles for the treatment of glioblastoma. *International Journal of Nanomedicine* 13: 4509–4521.

García Casillas, P.E., Armendariz, I.O., Gonzalez, C.C. et al. (2014). Magnetic nanostructures for biomedical applications. In: *Microspheres: Technologies, Applications and Role in Drug Delivery Systems* (eds. L.F. Fraceto and D.R. de Araújo), 137. Nova Science Publishers.

Hardiansyah, A., Yang, M.-C., Liu, T.-Y. et al. (2017). Hydrophobic drug-loaded PEGylated magnetic liposomes for drug-controlled release. *Nanoscale Research Letters* 12 (1): 355.

Hayashi, K., Nakamura, M., Miki, H. et al. (2014). Magnetically Responsive Smart Nanoparticles for Cancer Treatment with a Combination of Magnetic Hyperthermia and Remote-Control Drug Release. *Theranostics* 4 (8): 834–844.

Hervault, A. and Thanh, N.T.K. (2014). Magnetic nanoparticle-based therapeutic agents for thermo-chemotherapy treatment of cancer. *Nanoscale* 6 (20): 11553–11573.

Hodenius, M., Würth, C., Jayapaul, J. et al. (2012). Fluorescent magnetoliposomes as a platform technology for functional and molecular MR and optical imaging. *Contrast Media and Molecular Imaging* 7 (1): 59–67.

Hong, M.-S., Lim, S.-J., Lee, M.-K. et al. (2001). Prolonged blood circulation of methotrexate by modulation of liposomal composition. *Drug Delivery* 8 (4): 231–237.

Hosseini, L., Mahboobnia, K., and Irani, M. (2016). Fabrication of PLA/MWCNT/Fe_3O_4 composite nanofibers for leukemia cancer cells. *International Journal of Polymeric Materials and Polymeric Biomaterials* 65 (4): 176–182.

Hu, S.H., Liao, B.J., Chiang, C.S. et al. (2012). Core-Shell Nanocapsules Stabilized by Single-Component Polymer and Nanoparticles for Magneto-Chemotherapy/Hyperthermia with Multiple Drugs. *Advanced Materials* 24 (27): 3627–3632.

Huang, J., Xue, Y., Cai, N. et al. (2015). Efficient reduction and pH co-triggered DOX-loaded magnetic nanogel carrier using disulfide crosslinking. *Materials Science and Engineering C* 46: 41–51.

Hussien, N.A., Işıklan, N., and Türk, M. (2018). Pectin-conjugated magnetic graphene oxide nanohybrid as a novel drug carrier for paclitaxel delivery. *Artificial Cells, Nanomedicine, and Biotechnology* 46 (sup1): 264–273.

Janko, C., Ratschker, T., Nguyen, K. et al. (2019). Functionalized superparamagnetic iron oxide nanoparticles (SPIONs) as platform for the targeted multimodal tumor therapy. *Frontiers in Oncology* 9 (February): 1–9.

Jiang, H., Zhao, L., Gai, L. et al. (2015). Conjugation of methotrexate onto dedoped Fe_3O_4/PPy nanospheres to produce magnetic targeting drug with controlled drug release and targeting specificity for HeLa cells. *Synthetic Metals* 207: 18–25.

Kandasamy, G. and Maity, D. (2015). Recent advances in superparamagnetic iron oxide nanoparticles (SPIONs) for in vitro and in vivo cancer nanotheranostics. *International Journal of Pharmaceutics* 496 (2): 191–218.

Kandasamy, G., Sudame, A., and Maity, D. (2019). SPIONs and curcumin co-encapsulated mixed micelles based nanoformulation for biomedical applications. *Advanced Materials Letters* 10 (3): 185–188.

Khan, S., Setua, S., Kumari, S. et al. (2019). Superparamagnetic iron oxide nanoparticles of curcumin enhance gemcitabine therapeutic response in pancreatic cancer. *Biomaterials* 208: 83–97.

Kim, D.H., Guo, Y., Zhang, Z. et al. (2014). Temperature-Sensitive Magnetic Drug Carriers for Concurrent Gemcitabine Chemohyperthermia. *Advanced Healthcare Materials* 3 (5): 714–724.

Kim, D.H., Vitol, E.A., Liu, J. et al. (2013). Stimuli-responsive magnetic nanomicelles as multifunctional heat and cargo delivery vehicles. *Langmuir* 29 (24): 7425–7432.

Kim, H.C., Kim, E., Jeong, S.W. et al. (2015). Magnetic nanoparticle-conjugated polymeric micelles for combined hyperthermia and chemotherapy. *Nanoscale* 7 (39): 16470–16480.

Knežević, N., Gadjanski, I., and Durand, J.O. (2019). Magnetic nanoarchitectures for cancer sensing, imaging and therapy. *Journal of Materials Chemistry B* 7 (1): 9–23.

Kulshrestha, P., Gogoi, M., Bahadur, D., and Banerjee, R. (2012). In vitro application of paclitaxel loaded magnetoliposomes for combined chemotherapy and hyperthermia. *Colloids and Surfaces B: Biointerfaces* 96: 1–7.

Kumar, D., Sharma, D., Singh, G. et al. (2012). Lipoidal soft hybrid biocarriers of supramolecular construction for drug delivery. *International Scholarly Research Notices Pharmaceutics* 2012: 1–14.

Lahiri, B.B., Muthukumaran, T., and Philip, J. (2016). Magnetic hyperthermia in phosphate coated iron oxide nanofluids. *Journal of Magnetism and Magnetic Materials* 407: 101–113.

Lee, J.H., Chen, K.J., Noh, S.H. et al. (2013). On-Demand Drug Release System for In Vivo Cancer Treatment through Self-Assembled Magnetic Nanoparticles. *Angewandte Chemie International Edition* 52 (16): 4384–4388.

Li, R., Wu, R., Zhao, L. et al. (2011). Folate and iron difunctionalized multiwall carbon nanotubes as dual-targeted drug nanocarrier to cancer cells. *Carbon* 49 (5): 1797–1805.

Li, M., Neoh, K.-G., Wang, R. et al. (2013a). Methotrexate-conjugated and hyperbranched polyglycerol-grafted Fe_3O_4 magnetic nanoparticles for targeted anticancer effects. *European Journal of Pharmaceutical Sciences* 48 (1–2): 111–120.

Li, X., Li, H., Yi, W. et al. (2013b). Acid-triggered core cross-linked nanomicelles for targeted drug delivery and magnetic resonance imaging in liver cancer cells. *International Journal of Nanomedicine* 8: 3019–3031.

Li, T.J., Huang, C.C., Ruan, P.W. et al. (2013c). In vivo anti-cancer efficacy of magnetite nanocrystal-based system using locoregional hyperthermia combined with 5-fluorouracil chemotherapy. *Biomaterials* 34 (32): 7873–7883.

Li, W., Hinton, C.H., Lee, S.S. et al. (2015). Surface engineering superparamagnetic nanoparticles for aqueous applications: design and characterization of tailored organic bilayers. *Environmental Science: Nano* 3: 1–20.

Li, L., Wang, Q., Zhang, X. et al. (2018). Dual-targeting liposomes for enhanced anticancer effect in somatostatin receptor II-positive tumor model. *Nanomedicine* 13 (17): 2155–2169.

Lin, W., Xie, X., Yang, Y. et al. (2016). Thermosensitive magnetic liposomes with doxorubicin cell-penetrating peptides conjugate for enhanced and targeted cancer therapy. *Drug Delivery* 23 (9): 3436–3443.

Liu, H., Tu, L., Zhou, Y. et al. (2017). Improved bioavailability and antitumor effect of docetaxel by TPGS modified proniosomes: in vitro and in vivo evaluations. *Scientific Reports* 7 (March): 43372.

Lombardo, D., Kiselev, M.A., Magazù, S., and Calandra, P. (2015). Amphiphiles self-assembly: basic concepts and future perspectives of supramolecular approaches. *Advances in Condensed Matter Physics* 2015: 1–22.

Lombardo, D., Kiselev, M.A., and Caccamo, M.T. (2019). Smart nanoparticles for drug delivery application: development of versatile nanocarrier platforms in biotechnology and nanomedicine. *Journal of Nanomaterials* 2019: 1–26.

Mahmoudi, M., Sant, S., Wang, B. et al. (2011). Superparamagnetic iron oxide nanoparticles (SPIONs): development, surface modification and applications in chemotherapy. *Advanced Drug Delivery Reviews* 63 (1–2): 24–46.

Mandal, S.K., Lequeux, N., Rotenberg, B. et al. (2005). Encapsulation of magnetic and fluorescent nanoparticles in emulsion droplets. *Langmuir* 21 (9): 4175–4179.

Mandriota, G., Di Corato, R., Benedetti, M. et al. (2019). Design and application of cisplatin-loaded magnetic nanoparticle clusters for smart chemotherapy. *ACS Applied Materials and Interfaces* 11 (2): 1864–1875.

Mertz, D., Sandre, O., and Bégin-Colin, S. (2017). Drug releasing nanoplatforms activated by alternating magnetic fields. *Biochimica et Biophysica Acta (BBA) - General Subjects* 1861 (6): 1617–1641.

Mody, V.V., Cox, A., Shah, S. et al. (2014). Magnetic nanoparticle drug delivery systems for targeting tumor. *Applied Nanoscience* 4 (4): 385–392.

Montazerabadi, A., Beik, J., Irajirad, R. et al. (2019). Folate-modified and curcumin-loaded dendritic magnetite nanocarriers for the targeted thermo-chemotherapy of cancer cells. *Artificial Cells, Nanomedicine, and Biotechnology* 47 (1): 330–340.

Nardecchia, S., Sánchez-Moreno, P., de Vicente, J. et al. (2019). Clinical trials of thermosensitive nanomaterials: an overview. *Nanomaterials* 9 (2): 191.

Nobuto, H., Sugita, T., Kubo, T. et al. (2004). Evaluation of systemic chemotherapy with magnetic liposomal doxorubicin and a dipole external electromagnet. *International Journal of Cancer* 109 (4): 627–635.

Oliveira, H., Pérez-Andrés, E., Thevenot, J. et al. (2013). Magnetic field triggered drug release from polymersomes for cancer therapeutics. *Journal of Controlled Release* 169 (3): 165–170.

Palanisamy, S. and Wang, Y.-M. (2019). Superparamagnetic iron oxide nanoparticulate system: synthesis, targeting, drug delivery and therapy in cancer. *Dalton Transactions* 48 (26): 9490–9515.

Pandey, H., Rani, R., and Agarwal, V. (2016). Liposome and their applications in cancer therapy human and animal health. *Archives of Biology and Technology* 5959 (59): 16150477–16150477.

Polyak, B. and Friedman, G. (2009). Magnetic targeting for site-specific drug delivery: applications and clinical potential. *Expert Opinion on Drug Delivery* 6 (1): 53–70.

Pradhan, P., Giri, J., Rieken, F. et al. (2010). Targeted temperature sensitive magnetic liposomes for thermo-chemotherapy. *Journal of Controlled Release* 142 (1): 108–121.

Purushotham, S. and Ramanujan, R.V. (2010). Thermoresponsive magnetic composite nanomaterials for multimodal cancer therapy. *Acta Biomaterialia* 6 (2): 502–510.

Rahimi, M., Safa, K.D., and Salehi, R. (2017). Co-delivery of doxorubicin and methotrexate by dendritic chitosan-g-mPEG as a magnetic nanocarrier for multi-drug delivery in combination chemotherapy. *Polymer Chemistry* 8 (47): 7333–7350.

Rajni, S., Singh, D.J., Prasad, D.N., and Shabnam, H. (2019). Advancement in novel drug delivery system: niosomes. *Journal of Drug Delivery and Therapeutics* 9: 995–1001.

Salim, M., Minamikawa, H., Sugimura, A., and Hashim, R. (2014). Amphiphilic designer nano-carriers for controlled release: from drug delivery to diagnostics. *Medicinal Chemistry Communications* 5 (11): 1602–1618.

Sankhyan, A. and Pawar, P. (2012). Recent trends in niosome as vesicular drug delivery system. *Journal of Applied Pharmaceutical Science* 2 (6): 20–32.

Sanson, C., Diou, O., Thévenot, J. et al. (2011). Doxorubicin loaded magnetic polymersomes: theranostic nanocarriers for MR imaging and magneto-chemotherapy. *ACS Nano* 5 (2): 1122–1140.

Schleich, N., Po, C., Jacobs, D. et al. (2014). Comparison of active, passive and magnetic targeting to tumors of multifunctional paclitaxel/SPIO-loaded nanoparticles for tumor imaging and therapy. *Journal of Controlled Release* 194: 82–91.

Seyfoori, A., Sarfarazijami, S., and Seyyed Ebrahimi, S.A. (2019). pH-responsive carbon nanotube-based hybrid nanogels as the smart anticancer drug carrier. *Artificial Cells, Nanomedicine, and Biotechnology* 47 (1): 1437–1443.

Shen, J.M., Guan, X.M., Liu, X.Y. et al. (2012). Luminescent/magnetic hybrid nanoparticles with folate-conjugated peptide composites for tumor-targeted drug delivery. *Bioconjugate Chemistry* 23 (5): 1010–1021.

Shirmardi Shaghasemi, B., Virk, M.M., and Reimhult, E. (2017). Optimization of magneto-thermally controlled release kinetics by tuning of magnetoliposome composition and structure. *Scientific Reports* 7 (1): 1–10.

Siminzar, P., Omidi, Y., Golchin, A. et al. (2019). Targeted delivery of doxorubicin by magnetic mesoporous silica nanoparticles armed with mucin-1 aptamer. *Journal of Drug Targeting* 28: 1–29.

Song, W., Su, X., Gregory, D. et al. (2018). Magnetic alginate/chitosan nanoparticles for targeted delivery of curcumin into human breast cancer cells. *Nanomaterials* 8 (11): 907.

Souza, M.A., Santos, H.T., Pretti, T.S. et al. (2016). Magnetic nanoparticles surface modified with biodegradable polymers for controled methotrexate delivery in cancer therapy. *Journal of Nanopharmaceutics and Drug Delivery* 3 (1): 77–84.

Talekar, M., Kendall, J., Denny, W., and Garg, S. (2011). Targeting of nanoparticles in cancer. *Anti-Cancer Drugs* 22 (10): 949–962.

Tao, C. and Zhu, Y. (2014). Magnetic mesoporous silica nanoparticles for potential delivery of chemotherapeutic drugs and hyperthermia. *Dalton Transactions* 43 (41): 15482–15490.

Tavano, L., Vivacqua, M., Carito, V. et al. (2013). Doxorubicin loaded magneto-niosomes for targeted drug delivery. *Colloids and Surfaces B: Biointerfaces* 102: 803–807.

Theivasanthi, T. and Alagar, M. (2014). Innovation of superparamagnetism in lead nanoparticles. *Physics and Technical Sciences* 1 (3): 39–45.

Wahajuddin and Arora, S. (2012). Superparamagnetic iron oxide nanoparticles: magnetic nanoplatforms as drug carriers. *International Journal of Nanomedicine* 7: 3445–3471.

Wakaskar, R.R. (2017). Passive and active targeting in tumor microenvironment. *International Journal of Drug Development and Research* 9 (2): 37–41.

Wang, H., Yi, J., Mukherjee, S. et al. (2014). Magnetic/NIR-thermally responsive hybrid nanogels for optical temperature sensing, tumor cell imaging and triggered drug release. *Nanoscale* 6 (21): 13001–13011.

Yang, F., Fu, D.L., Long, J., and Ni, Q.X. (2008). Magnetic lymphatic targeting drug delivery system using carbon nanotubes. *Medical Hypotheses* 70 (4): 765–767.

Yoshida, M., Watanabe, Y., Sato, M. et al. (2010). Feasibility of chemohyperthermia with docetaxel-embedded magnetoliposomes as minimally invasive local treatment for cancer. *International Journal of Cancer* 126 (8): 1955–1965.

Zasadzinski, J.A., Kisak, E., and Evans, C. (2001). Complex vesicle-based structures. *Current Opinion in Colloid and Interface Science* 6 (1): 85–90.

Zhang, J.Q., Zhang, Z.R., Yang, H. et al. (2005). Lyophilized paclitaxel magnetoliposomes as a potential drug delivery system for breast carcinoma via parenteral administration: in vitro and in vivo studies. *Pharmaceutical Research* 22 (4): 573–583.

Zhang, Y., Yu, J., Bomba, H.N. et al. (2016). Mechanical force-triggered drug delivery. *Chemical Reviews* 116 (19): 12536–12563.

Zhao, Y., Lin, D., Wu, F. et al. (2014). Discovery and in vivo evaluation of novel RGD-modified lipid-polymer hybrid nanoparticles for targeted drug delivery. *International Journal of Molecular Sciences* 15 (10): 17565–17576.

Zhu, L., Huo, Z., Wang, L. et al. (2009). Targeted delivery of methotrexate to skeletal muscular tissue by thermosensitive magnetoliposomes. *International Journal of Pharmaceutics* 370 (1–2): 136–143.

8

Virus-Like Nanoparticle-Mediated Delivery of Cancer Therapeutics

Yasser Shahzad[1], Abid Mehmood Yousaf[1], Talib Hussain[1], and Syed A.A. Rizvi[2]

[1] *Department of Pharmacy, COMSATS University Islamabad, Lahore Campus, Lahore, 54000, Pakistan*
[2] *School of Pharmacy, Hampton University, Virginia, 23608, USA*

8.1 Introduction

Cancer remains a serious global health problem and a leading cause of death worldwide (Khan et al. 2017; Neubi et al. 2018). In 2018, the National Cancer Institute (NCI) of the United States estimated a 35% mortality due to cancer in the reported cases (NCI 2018). Typical cancer treatment includes chemotherapy, radiation, and surgery (Khan et al. 2017; Rizvi and Qureshi 2018; Rockson et al. 2019; Farran et al. 2020). Surgical procedures have limited success, whilst radiation and chemotherapy are effective yet have severe side effects, which hampers quality of life (Rohovie et al. 2017). Chemotherapy's success is related to overcoming various challenges such as selectivity, unfavorable biodistribution, limited access to the tumor site, and multidrug resistance (Danhier et al. 2010; Shahzad et al. 2014; Parayath and Amiji 2017; Khan et al. 2017). With the increasing understanding of cancer, the past decades have seen tremendous growth in drug delivery and targeting of the cancer cells, fueled by advancement in nanotechnology and nanobiotechnology (Yoo et al. 2011; Qamar et al. 2019).

Nanotechnology refers to the science of small particles, generally in the size range of 1–100 nm, which is rapidly expanding with a plethora of applications in various fields (Park 2007). Most notably, the emergence of nanotechnology in medicine, dubbed nanomedicine, has revolutionized the course of therapy by improving the targeting of drugs to their pathological sites, thereby ensuring that healthy cells remain unaffected (Mirza and Siddiqui 2014; Parodi et al. 2017; Rizvi and Saleh 2018). Nanomedicine utilizes various nanosized materials that help in various applications, including drug delivery, diagnosis, imaging, and biosensing (De Jong and Borm 2008; Oliveira Jr et al. 2014; Holzinger et al. 2014). These nanosized carriers include polymeric nanoparticles, liposomes, and micelles among many others, with the ability to cargo the active therapeutic agents to the tumorous tissue (Langer 1990; Neubi et al. 2018; Grodzinski et al. 2019). All of these synthetic nanocarriers possess their own benefits and limitations.

Nanobiotechnological advancement enlightens the use of bioinspired and biomimetic nanocarriers, ranging from pathogens (bacteria and viruses) to mammalian cells (erythrocytes, immune cells, stem cells, platelets), for their applications in drug delivery and targeting (Narayanan and Han 2017a). These natural particulates are biodegradable and have the ability to carry therapeutic

Nanobiotechnology in Diagnosis, Drug Delivery, and Treatment, First Edition. Edited by Mahendra Rai, Mehdi Razzaghi-Abyaneh, and Avinash P. Ingle.

payloads to the target site with fewer toxic effects. They also possess their own delivery mechanisms that are different from conventional nanocarriers (Yoo et al. 2011). The bioinspired nanocarriers are highly optimized for specific *in vivo* performance, and their structural features are compelling for use as drug delivery vehicles in tumor targeting (Singh et al. 2019). Pathogens such as bacteria and viruses have the capacity to avoid the host immune system and enter their desired target cell (Hornef et al. 2002; Mudhakir and Harashima 2009).

The present chapter focuses on a succinct overview of the current status of research and development in the area of virus-like nanoparticles (VLNPs) in drug delivery. Particularly, the utility of virus and its components in delivering specific payload (drugs, diagnostic agents, etc.) to a target site in the body has been discussed.

8.2 Viruses as Bioinspired Delivery Vehicles

Viruses are small infectious microscopic organisms consisting of nucleic acid molecules encapsulated in capsid proteins by self-assembling to form a stable cage or shell (Green et al. 2014). Viruses exist as non-free-living organisms having a simple structural and genetic organization (Steinmetz and Manchester 2011). By default, viruses have no cellular metabolism; instead, they act as intracellular parasites. That is why viruses require self-replication for their survival, and they achieve this with a stealthy entrance into host cells for transferring genetic information. This lifestyle has endowed viruses with the unique ability to coordinate with the cellular functions (Sainsbury 2017) and overriding the cellular processes without being detected by the host immune system. This is how viruses drop their genetic payload into the host cell's cytoplasm or nucleus, and this property can be exploited for therapeutic delivery. Thus, virus–host interactions have potential biotechnological and biomedical applications (Narayanan and Han 2017a).

Viruses have remained the focus of researchers for more than 100 years, and we now know the structure and functions of various viruses that exist to date. For many years, virus infections stayed lethal, and the ultimate goal of the field of medicine was to save humankind from fatalities from viruses. As early as 1950, by the virtue of further discoveries and knowledge about viruses, scientists had begun to think of using viruses against bacteria; a classic example is phage viruses that attack bacteria. However, this concept faded away quickly because of antibiotics, which were and still are highly effective in addressing various bacterial infections. The use of viruses resurfaced in the 1970s with the introduction of virus-mimicking or -like particles (Ludwig and Wagner 2007; Steinmetz and Manchester 2011). For example, empty viruses were initially used for vaccination purposes, as an alternative to attenuated live viruses because their antigenicity is comparable to the parent virus (Yoo et al. 2011). Since then, viruses have been used as a surrogate to nanocarriers for nanotechnological and nanobiotechnological applications (Madamsetty et al. 2019). Virus-like particles (VLPs) or, more accurately stated, VLNPs, being nanosized entities, are originated from animal, plant, and bacteriophage viruses (Ding et al. 2018). Various viruses that have been modified into VLNPs are briefly discussed in this section. For the clarity of readers, we have focused on VLNPs that are specifically derived from various viruses, and are not just structurally similar to viruses.

8.2.1 Plant-Based Virus-Like Nanoparticles

Viruses that attack plants are usually parasites on hosts, thus known as phytophages. Phytophages consume hosts' proteins to survive and are generally non-infectious to animals and humans (Narayanan and Han 2017a). Plant viruses have undergone extensive research and development for their potential use in pharmaceutical nanotechnology (Hefferon 2018). Plant viruses exist in spherical, icosahedral, helical, or tubular shapes and are generally in the size range of 10 nm to several microns. They have variable stabilities to pH, temperature, salt, solvent, and protease degradation and they also differ from each other in terms of plasticity, assembly–disassembly parameter, and electrostatic interactions (Strable and Finn 2009). Plant viruses as nanoparticles have shown great potential not only in the medical field, but they also possess features that can be exploited as materials for electronics and optics (Narayanan and Han 2017a; Narayanan and Han 2017b; Czapar and Steinmetz 2017). Plant viruses of both icosahedral and helical morphologies offer unmet attributes which make them fantastic biomaterials for drug delivery and imaging purposes. Empty icosahedral virus cages can encapsulate a variety of drugs and deliver their payload under appropriate physiological conditions (Shukla et al. 2017). On the other hand, helical-shaped plant viruses have a high aspect ratio, and their rod-like or filamentous shape gives them potential to be used as bio-template for novel nanostructured material production (Narayanan and Han 2018).

Among various plant viruses, cowpea mosaic virus (CPMV) is the most extensively studied (Hefferon 2018). Recently, CPMV nanoparticles have shown their ability to accumulate in the solid tumor and elicit local immune response within the surrounding microenvironment (Lizotte et al. 2015). This study revealed that the CPMV nanoparticle can activate both innate and acquired immune system against metastatic cancers including melanoma, and breast and ovarian carcinoma. Various other plant virus nanoparticles have shown their potential in targeting the tumor tissue and delivering cancer chemotherapeutics at the targeting site, as exemplified in Table 8.1.

Table 8.1 Various plant virus nanoparticles and their potential applications in targeting the tumor tissue in different cancers.

Plant virus nanoparticle	Activity	References
A. Icosahedral viruses		
Cowpea mosaic virus-like nanoparticle	Solid tumors (melanoma, breast and ovarian carcinoma)	Lizotte et al. (2015)
Doxorubicin-loaded red clover necrotic mosaic virus-like nanoparticle	Ovarian cancer, melanoma	Madden et al. (2017)
Hibiscus chlorotic ringspot virus-like nanoparticle loaded with doxorubicin	Ovarian cancer	Ren et al. (2007)
Physalis mottle virus-like nanoparticle	Targeted cancer imaging	Hu et al. (2019)
Cowpea chlorotic mottle virus-like nanoparticle loaded with fluorescent dye IR780 iodide	Targeted cancer imaging and photothermal therapy	Wu et al. (2017)
Cucumber mosaic virus-like nanoparticle loaded with doxorubicin	Ovarian cancer	Zeng et al. (2013)
B. Helical/filamentous Viruses		
Tobacco mosaic virus-like nanoparticle loaded with doxorubicin	Breast cancer	Bruckman et al. (2016)
Potato virus X nanoparticle conjugated with A647 and PEG	Ovarian cancer	Shukla et al. (2012)
Papaya mosaic virus-like nanoparticle showing intrinsic immune-stimulatory effect	Melanoma	Lebel et al. (2016)

8.2.2 Animal-Based Virus-Like Nanoparticles

Viruses have evolved to interact with various cellular proteins, control various intracellular machinery, and transfer their genetic material to replicate. These properties have led to the development of mammalian virus-based nanoparticles for gene therapy. It is noteworthy that mammalian viruses are infectious to humans due to virus–host interactions, thereby eliciting pathogenic effects (Guenther et al. 2014), thus only a limited number of viruses have been studied for their potential use as carriers. Notable viral species that have been extensively studied include Hepatitis B virus (HBV), human papillomaviruses (HPVs) (Koudelka et al. 2015), and Sindbis virus (Lundstrom 2015).

8.2.3 Phage Virus-Like Nanoparticles

Bacteriophages are widely accepted to be the most abundant microorganisms on the face of the earth (Hyman and Denyes 2018). Approximately 96% of all known bacteriophages are of the order *Caudovirales* and are tailed viruses containing dsDNA genomes (Ackermann 1998). Recently, several bacteriophages have been developed into VLNPs for drug delivery, targeting, and imaging purposes (Manchester and Singh 2006). Among various bacteriophages, icosahedral shaped Qβ, MS2 and *Salmonella typhimurium* P22 phage viruses, and filamentous M13 phage virus are the most notable (Rohovie et al. 2017). Since all bacteriophages attack bacteria, they are generally considered safe to be used in humans (Czapar and Steinmetz 2017).

8.3 Virus-Like Nanoparticle (VLNP) Production

VLNPs are designed to mimic the native conformation that is morphologically similar to the parent virus but devoid of parent characteristics, such as self-replication and infectious genome (Ding et al. 2018). Various methods have been proposed to fabricate VLNPs which include genetic engineering, bioconjugate chemistry, infusion, mineralization, and self-assembly (Jeevanandam et al. 2018). Here, we briefly discuss various methods of VLNP fabrication.

Genetic engineering is a viable technique to alter viral genomes and subsequently viral protein patterns in the capsid. CRISPR (Clustered Regularly Interspaced Short Palindromic Repeats) genetic sequences have enabled insertion of whole protein or protein domains in viral capsid and more recently production of chimeric VLNPs with two distinct protein expressions in a single viral capsid (Dickmeis et al. 2015; Neubi et al. 2018). Bioconjugation, which is considered as an alternative to genetic engineering, is another fantastic technique to target and modify both natural and synthetic amino acids on virus capsids (Biabanikhankahdani et al. 2018). Biochemical conjugation has been classically achieved through altering the viral genome via functional groups such as cysteine, lysine, glutamic acid, and aspartic acid (Chen et al. 2018). The most recent intervention in conjugation is the use of biomolecules including antigens and enzymes for converting viruses into nanocarriers for drug targeting and delivery purposes (Walper et al. 2015).

Infusion is a highly effective method of converting an infectious virus into a non-infectious VLNP. The method involves the removal of the viral genome and insertion of a therapeutic entity into the viral capsid for targeted delivery purposes (Lizotte et al. 2015). Another useful approach of VLNP production with desired size and shape is biomineralization. Biomineralization utilizes mineralization-directing peptides and fine-tuned electrostatic techniques for the fabrication of interior and exterior surfaces of the VLNP and generally results in rod or icosahedral morphologies (Altintoprak et al. 2015). Viral proteins have the ability to assemble and disassemble, a property that has been exploited to fabricate VLNPs. Self-assembly of viral proteins has been used to encapsulate various entities such as gold nanoparticles, drugs, dyes, and quantum dots for therapeutic and imaging purposes (Luque et al. 2014).

8.4 VLNP-Mediated Cancer Drug Delivery

Viruses are evolved to evade the host immune system in a variety of ways for their own survival. A specific virus can target a specific host and it is plausible that they can target a specific cell. For example, human viruses have the capability of entering the body and targeting a specific cell type. On the contrary, plant viruses lack the inherent ability to target human cells, however they can act as carriers for chemotherapeutic agents to target cancer cells. Thus, VLNPs derived from plant viruses, phage viruses, or even human viruses, can be used to carry drugs and target the cells (Franzen and Lommel 2009). Here, we have discussed viruses of plant, phage, and human origin that have been studied as carriers for chemotherapeutic agents and their ability to target a specific tumor.

8.4.1 Plant Virus-Derived VLNP-Mediated Delivery of Cancer Therapeutics

For many years, plants have been investigated to generate vaccines and biopharmaceuticals. In recent times, this ongoing research on plants developed into a novel and diversified approach by incorporating plant viruses into nanoparticles to deliver drug molecules in cancer cells (Hefferon 2014; Bruckman et al. 2016; Franke et al. 2017; Hefferon 2018). Being a cost-effective, safe, and efficient medium, this sophisticated technique is readily employed for a variety of purposes and considered as a prospective vehicle for the delivery of medicines. The plant viruses are biocompatible and biodegradable, having the potential for scalable manufacturing, making them an ideal candidate to deliver cargos of therapeutic agents to cancer cells and tissues (Hefferon 2014; Bruckman et al. 2016; Franke et al. 2017; Hefferon 2018).

Among extensively studied icosahedral plant viruses for the development of nanoparticles, CPMV demonstrated significant potential, ascribed to its structural and chemical stability, ease of production, and lack of toxicity (Gonzalez et al. 2009). The following study reported, for the first time, the direct use of VLNP as a cancer immunotherapy agent. The suitability of CPMW is being analyzed which is confirmed by results. The findings of this study indicated that CPMV-based nanoparticles are immunogenic in nature, hence it can be effectively used as a monotherapy as well as it can serve as a nanocarrier (Lizotte et al. 2016).

Brunel et al. (2010) described the fabrication of multivalent CPMV nanoparticles by hydrazine ligation strategy and the resultant VLNP showed ability to specifically target vascular endothelial growth factor receptors (VEGFRs). The optimized VLNP (FP3), comprised of 133 copies of VEGFR-1 ligand, 55 copies of a PEGylated peptide, and a total of 188 fluorescent dyes, showed excellent binding with the human colon carcinoma (HT-29) tumor *ex vivo* as compared with the P3 VLNP (without VEGFR targeting peptide). They further tested FP3 for its *in vivo* performance in nude/WEHI mice by injecting them, and the VLNP was allowed to circulate for two hours. It was observed that there was a significant accumulation of FP3 VLNP in the tumor, however, no P3 VLNP was detected in the tumor tissue. These findings suggested that multivalent VLNP can target tumors *in vivo* and could possibly be used as a chemotherapeutic carrier.

Cowpea chlorotic mottle virus (CCMV), an icosahedral shaped plant virus, has also been utilized as nanocarriers for chemotherapeutic agents. For example, Wu et al. (2017) demonstrated tumor targeting by genetically modified CCMV derived VLNP. The viral capsid was decorated with homing peptide F3 by fusion expression and the VLNP was encapsulated with near-infrared (NIR) fluorescent dye IR780 iodide. The resultant VLNP showed excellent targeting to the nucleolin receptor overexpressed on the surface of Michigan Foundation Cancer-7 (MCF-7) tumor cells, thus endorsing the suitability of CCMV-derived VLNP for potential photothermal therapy of tumorous tissues. Barwal and co-workers (2016) also demonstrated the suitability of CCMV-derived VLNPs as tumor-targeting carriers loaded with a chemotherapeutic agent. CCMV-derived VLNPs were fabricated by ligand bioconjugation and the nanocarrier was loaded with doxorubicin conjugated gold nanoparticles, folic acid, and a fluorescent dye. The chimeric VLNP showed excellent internalization in the folate positive receptor MFC-7 cell lines, and high cytotoxicity was achieved in the cancer cells.

Cucumber mosaic virus (CMV) is also an icosahedral virus with an average size of 29 nm. Zeng and co-workers (2013) demonstrated the first report on CMV-derived VLNPs encapsulated with a chemotherapeutic agent, namely doxorubicin. The CMV-derived VLNP was decorated with folic acid as targeting moiety while the inner core was filled with doxorubicin through a RNA-doxorubicin conjugate for cancer targeting (FA-CMV-DOX). Results displayed a sustained doxorubicin release over five days at physiological pH. The efficacy of CMV-derived VLNP against ovarian cancer in OVCAR-3 BALB/c nude mouse xenograft model was investigated through histological alterations and TUNEL assessment, which revealed a significant uptake in ovarian cancer, thus leading to improved antitumor activity compared with free doxorubicin (free dox) and doxorubicin-loaded VLNP without conjugated folic acid (CMV-DOX), as depicted in Figure 8.1.

Filamentous nanomaterials have recently shown superior pharmacokinetics and tumor-targeting capabilities (Geng et al. 2007; Christian et al. 2009; Truong et al. 2015). In this context, filamentous plant viruses including tobacco mosaic virus (TMV), potato virus X (PVX), and papaya mosaic virus (PapMV) may provide better opportunities for tumor targeting and therapy. Bruckman and

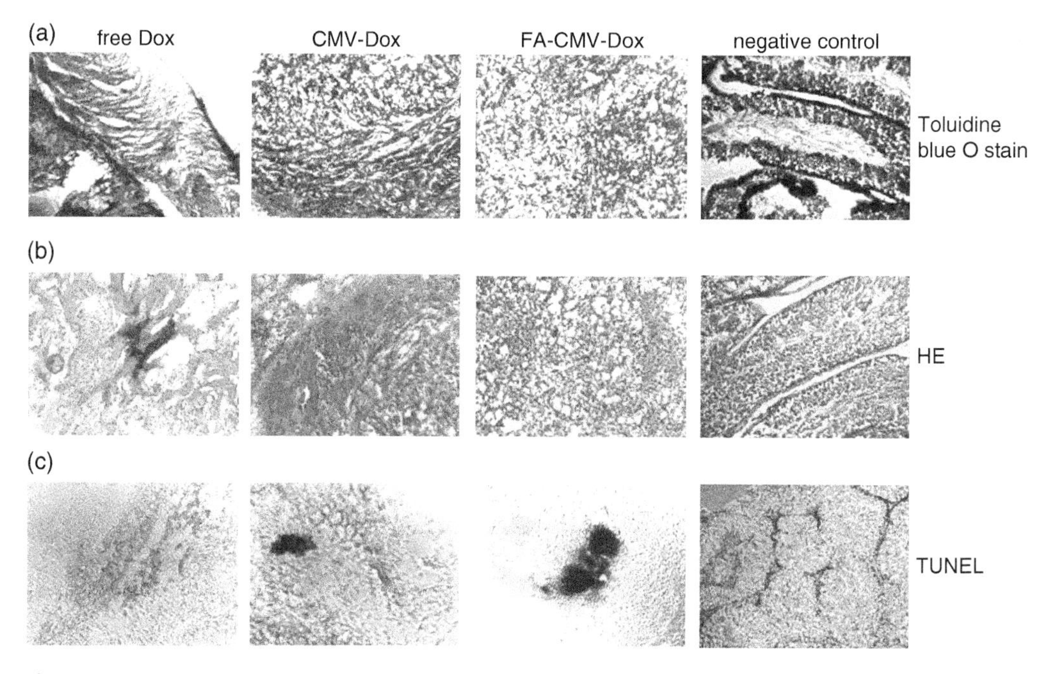

Figure 8.1 (a) Toluidine blue O stained tumor samples; (b) Hematoxylin and Eosin (HE) stained tumor samples; (c) TUNEL assay results 21 days post treatment, comparing all negative control with all three treatment groups. *Source:* Adapted from Zeng et al. (2013); with copyright permission from Elsevier.

co-workers (2016) developed a delivery approach for a chemotherapeutic agent, i.e. doxorubicin using a plant-based virus, namely TMV, to target breast cancer. The results demonstrated overall safety profile along with effective drug cargo delivery and cancer cell killing owing to longer incubation times and higher nanoparticle uptake by cells. In another study, TMV has been used as a therapeutic option to deliver cisplatin in platinum-resistant ovarian cancer cells (Franke et al. 2017). Ovarian cancer is the most fatal malignancy and leading cause of death in women (Jayson et al. 2014; Franke et al. 2017). The poor prognosis and failure of alternate treatments led the researchers to explore the potential of TMV for the specific delivery of cisplatin to cancer cells. The use of TMV nanoparticles provides an effective alternate platform. Hence, cisplatin delivery via TMV nanoparticles provides a therapeutic level of the drug at the target site by circumventing the mechanism of resistance. TMV is gaining popularity owing to its advantageous features such as scalable manufacturing, cost-effectiveness, and environmentally safe production (Franke et al. 2017).

Czapar et al. (2016) also reported the use of TMV as a drug delivery carrier for phenanthriplatin, a potent anticancer drug. Platinum-based anticancer drugs face obstacles of dose-limiting toxicity and drug resistance. To overcome such impediment, there is a need to devise a promising drug delivery approach which is safe and effectively kills cancer cells, along with inhibition of tumor growth. In this work, investigation has been made to explore the suitability of TMV. The biodegradable nature, along with excellent blood and tissue compatibility (Bruckman et al. 2014) of TMV, make it a promising and attractive alternative for drug delivery (Czapar et al. 2016). The enhanced efficacy along with successful delivery to cancer cells confirmed the suitability of TMV (Czapar et al. 2016).

Non-Hodgkin's lymphoma is primarily treated with chemotherapy (Kernan et al. 2017), which is facing certain challenges. To address such limitations of systemic therapy, there is a dire need to develop an alternate system. Therefore, Kernan et al. (2017) formulated a nanoparticle drug delivery system using the nucleoprotein component of TMV. The successful *in vitro* results demonstrated effective killing of non-Hodgkin's lymphoma. Improved efficacy and mitigation of off-target toxicities through TMV provide an avenue for targeted therapies (Kernan et al. 2017).

Shukla and co-workers (2012) demonstrated tumor targeting using PVX-derived VLNP and compared it with CPMV-derived VLNP. Both VLNPs were tested in NCR nu/nu mice that were xenografted with HT-29 tumors on each flank. Results showed distinct biodistribution of both VLNPs, and their tumor targeting and penetration were significantly different from each other, with PVX-derived VLNP showing superior targeting and proving effective in reducing the tumor. The difference in performance of CPMV- and PVX-derived VLNPs was due to the fact that both have different geometries and surface charge, and as mentioned earlier, filamentous nanocarriers are somehow found to be more effective compared to their spherical counterparts.

The intrinsic immune-stimulatory features of plant VLPs confer therapeutic activity toward various tumors (Lebel et al. 2016). PapMV displayed tremendous immune-stimulatory potential against cancer cells. The performance of PapMV correlates with a marked decrease in melanoma progression and prolonged survival (Lebel et al. 2016). Overall, plant virus-derived VLNPs have great potential to target various tumors along with the ability to deliver chemotherapeutic agents at the targeting sites. Since plant viruses only target the plants, they are deemed safer to be used in humans as nanocarriers.

8.4.2 Phage Virus-Derived VLNP-Mediated Delivery of the Cancer Therapeutics

VLNPs derived from phage viruses are promising tools to deliver chemotherapeutic drugs in the treatment of various types of cancers. For instance, a genome-free capsid or outer coat of bacteriophage MS2 virus possesses numerous physiognomies that makes it adequate for exploitation in delivery applications, such as simplistic production and adaptation, the ability to furnish multiple copies of targeting ligands, and the capability to transport a variety of payloads (Aanei et al. 2016). Also, the bacteriophage MS2 capsid acts as a propitious monomer. Noninfectious genome-free capsids with the amendable amino acid composition are obtained in *Escherichia coli* by replication after recombinant technology (Wen 2016). About 180 genome-free capsids of bacteriophage MS2 virus polymerize to constitute a spherical hollow structure which is 27 nm in diameter and bears 32 pores, each having a pore size of 2 nm (Wu et al. 2009). These pores provide a platform for the installation of small drug molecules on the inner surface of the structure without disturbing the assembly of monomers. In this way, this assembly can be utilized as an effective drug-delivery vehicle (Valegård et al. 1990). Both the inner and outer surfaces of this spherical assembly can be utilized for the attachment of small molecules through covalent alterations of particular amino acid residues. For example, investigators have accomplished linking of positron emission tomography radioisotopes (Hooker et al. 2008), magnetic resonance imaging contrast agents (Carrico et al. 2008), and fluorescent dyes (Kovacs et al. 2007; Tong et al. 2009) to the interior surface. On the other hand, the exterior surface has been modified with peptides (Carrico et al. 2008) and DNA aptamers (Tong et al. 2009), which can avidly bind to particular moieties or receptors on cancer cells, as well as polyethylene glycol (PEG) chains that prevent this protein assembly from the antibody attack (Kovacs et al. 2007). The drug molecules are attached on the inner surface with the hope that the drug would be sheltered from degradation and from unnecessary interactions with the healthy cells. Wesley and coworkers (2009) employed these genome-free capsids for delivery of Taxol. Taxol is a potent chemotherapeutic agent used in the treatment of a variety of cancers (Khayat et al. 2000). The drug was attached to the interior surface of wild-type bacteriophage MS2 protein coat. The interior of the capsid was modified to a cysteine residue (N87C) to expose a sulfhydryl group (-SH) for the binding of the drug. Taxol has low water solubility; however, it is essential to solubilize it in water to enable protein bio-conjugation. Accordingly, the investigators designed a linker carrying three functionalities: a moiety for binding to protein, a charged moiety for enhancing water solubility, and a cleavable link to Taxol. In a nutshell, a water-soluble derivative of Taxol was synthesized and attached to the interior surface of MS2 viral protein coat. The drug was released when this drug-loaded capsid was incubated with MCF-7 cells.

Bacteriophage MS2 capsids have also been conjugated with anti-EGFR antibodies (Ab) to construct nanoparticles for studying biodistribution of conjugates in breast cancer models in mice (Aanei et al. 2016). First, the researchers attached nitrophenol (NP) moieties to anti-EGFR antibody (Ab) via lysine modification. Then, these nitrophenol moieties were reduced to produce aminophenol-antibody conjugate (AP-Ab). The Ap-Ab were attached to *p*-aminophenylalanine (*p*aF) MS2 to constitute MS2-Ab conjugate (Figure 8.2). MS2-ab conjugates were radiolabeled with ^{64}Cu isotopes and administered parenterally to mice carrying breast cancer models for investigating biodistribution. This study concluded that the MS2 agent exhibited excellent stability in physiological environment up to 48 hours and targeted attaching to EGFR receptors on cancer cell surface in *in vitro* test. The capsids demonstrated long circulation times (10–15% ID/g in serum at one day) and moderate tumor uptake (2–5% ID/g). Nevertheless, the targeting antibodies did not result in ameliorated uptake *in vivo* in spite of *in vitro* improvements.

In a study carried out by Joel and co-authors, it is reported that direct intratumoral injection through convection-enhanced delivery (CED) of anticancer drugs such as doxorubicin is a promising way of drug delivery; it has embraced some degree of success (Finbloom et al. 2018). Numerous studies have revealed that attaching of a drug to a nano-particulated polymeric matrix enhances drug delivery efficacy via CED. Accordingly, the researchers of this study utilized three different

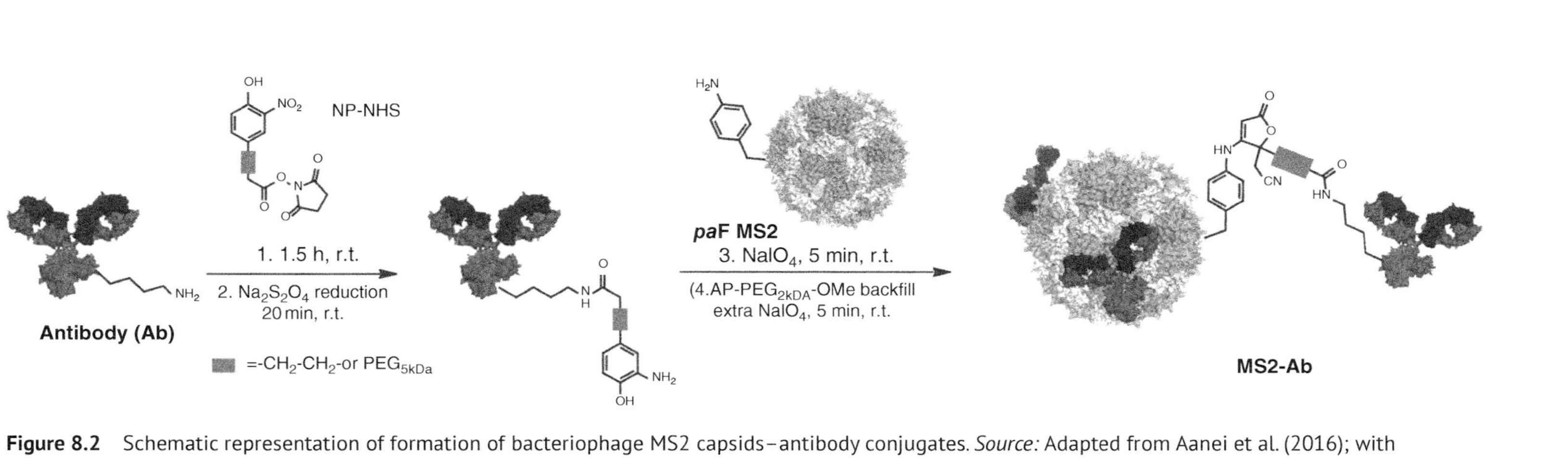

Figure 8.2 Schematic representation of formation of bacteriophage MS2 capsids–antibody conjugates. *Source:* Adapted from Aanei et al. (2016); with copyright permission from American Chemical Society.

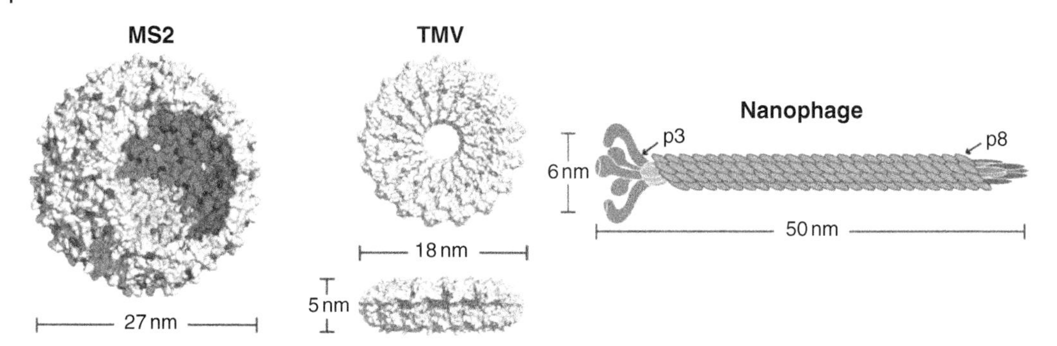

Figure 8.3 Three different structures of nanocarriers composed of bacteriophage: MS2 spheres, TMV disks, and nanophage filamentous rods. *Source:* Adapted from Finbloom et al. (2018); an open access article, permission allowed under Creative Commons Attribution License from.

VLNPs for their capability as CED drug delivery vehicles for the treatment of glioblastoma, which is considered as one of the deadliest and difficult to treat cancer (Rizvi et al. 2018). The panel of three different VLNPs was consisting of MS2 assembly, tobacco mosaic virus (TMV) disks, and nanophage filamentous rods (NFR) (Figure 8.3). All these doxorubicin-loaded VLNPs furnished appropriate drug delivery and cell uptake *in vitro*.

Amongst these three VLNPs, doxorubicin-loaded TMV disks showed the best survival rate response in glioblastoma-affected mice treated via CED. MS2 assembly, conjugated to doxorubicin, was second to show survival rate response. However, this research underscores the capability of these VLNPs and emphasizes more detailed investigations into how the morphological features of VLNPs monitor their drug delivery characteristics.

8.4.3 Animal Virus-Derived VLNP-Mediated Delivery of Cancer Therapeutics

A hepatitis E virus-like structure was developed by Chen et al. (2016) for conjugating breast cancer-targeting group LXY30. The protein coat was expressed in an insect cell to generate VLNPs having the site-directed mutation in at least one of the five loops on the protrusion moiety. This resulted in surface alteration via thiol-based annexing to the modified and exposed cysteine residue. Cysteine mutation was proposed on the following five residues: Y485C, T489C, S533C, N573C, and T586C. Among these five mutated moieties, N573C moiety exhibited the highest affinity for surface conjugation; however, all the mutants were capable of forming VLNPs. The LXY30 group was conjugated to N573C moiety, and a NIR stain marker to monitor the track of VLNP. This study showed that LXY30 conjugated VLNPs interacted breast cancer cells in culture more avidly as compared to unconjugated VLNPs. Moreover, targeting of breast cancer cells *in vivo* was also more efficient with LXY30 conjugated VLNPs (Chen et al. 2016).

Biabanikhankahdani et al. (2016) developed and characterized doxorubicin-loaded pH-responsive VLNPs for enhanced targeted drug delivery to the tumor cells (Biabanikhankahdani et al. 2016). The VLNPs were based on shortened hepatitis B virus core antigen (tHBcAg), presenting folic acid for sustained drug delivery. Folic acid was linked to a pentadecapeptide chain having nanoglue attached on tHBcAg nanoparticles to improve the performance such as specificity and efficacy of the drug-loaded VLNPs (Figures 8.4 and 8.5). This study showed that doxorubicin and polyacrylic

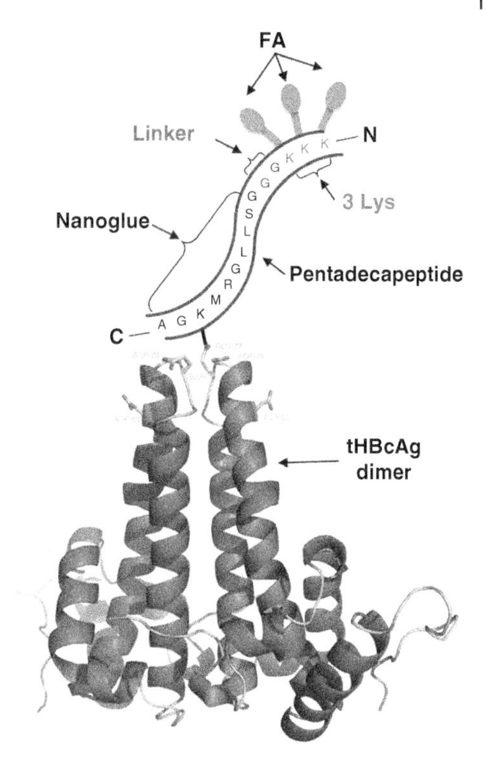

Figure 8.4 Schematic illustration of self-assembling tHBcAg dimer consisting of pentadecapeptide containing the nanoglue, linker, and 3 Lys with a tHBcAg dimer with folic acid (FA) molecules conjugated to the free Lys residues at the N-terminus of the pentadecapeptide. *Source:* Adapted from Biabanikhankahdani et al. (2016); an open access article, permission granted under Creative Commons Attribution License from.

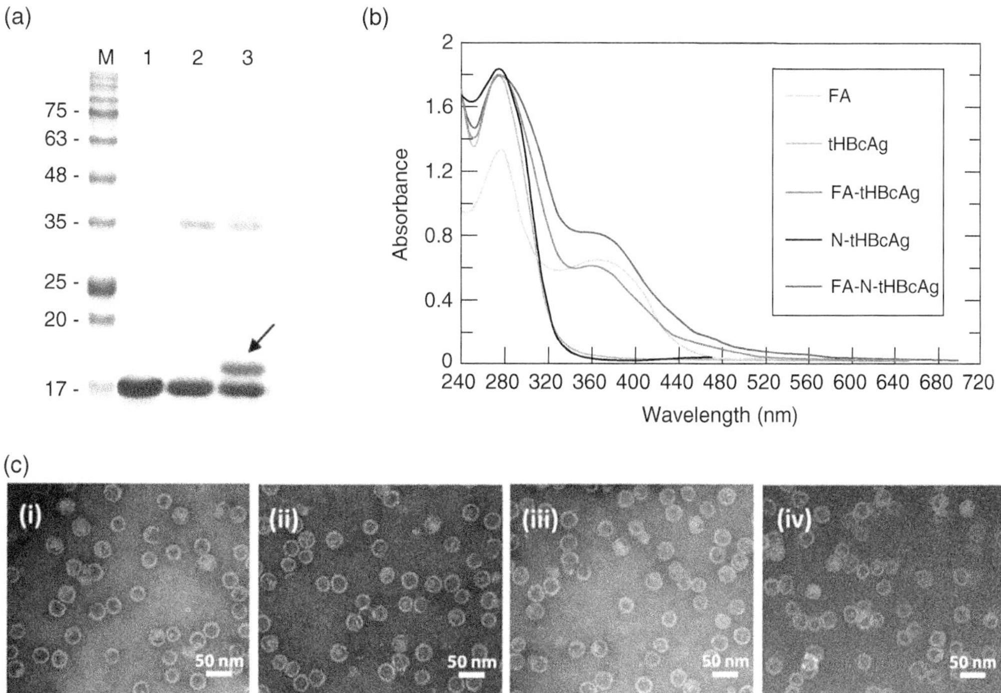

Figure 8.5 (a) An SDS-polyacrylamide gel of the tHBcAg nanoparticles cross-linked with the pentadecapeptide containing the nanoglue showing a shifted band (arrow) of approximately 1 kDa above the 17 kDatHBcAg. (b) UV spectra of the FA and its conjugated nanoparticles of various types. (c) Electron micrographs (TEM) of various tHBcAg nanoparticles, white bars indicate 50 nm. *Source:* Adapted from Biabanikhankahdani et al. (2016); an open access article, permission granted under Creative Commons Attribution License from.

acid-loaded tHBcAg VLNPs furnished a controlled drug release *in vitro* owing to the presence of polyacrylic acid, and ameliorated the uptake of the drug in colorectal cancerous tissue; accordingly, resulting in enhanced antitumor activity. Doxorubicin and polyacrylic acid can be loaded together in VLNPs without any alteration of the drug molecule; thus, the safety and specific efficacy of the drug is preserved. The nanoglue helps in exposing the tumor-targeting moiety on the exterior surface of VLNPs (Biabanikhankahdani et al. 2016).

In another study, Biabanikhankahdani et al. displayed doxorubicin on the HBcAg VLNPs by employing the hexahistidine-tag (His-tag) exposed on the surface of VLNPs (Biabanikhankahdani et al. 2017). His-tags were exploited to display the drug via nitrilotriacetic acid (NTA) conjugation. His-tags acted as pH-responsive nanolinks which liberated the drug from VLNPs in a sustained manner. The VLNPs, tagged with His-tags, loaded with doxorubicin and NTA, and linked with FA, were capable to specifically attach and liberate doxorubicin into ovarian cancer cells through FA-receptor mediated endocytosis. This resulted in increased deposition of the drug in the cells which ameliorated antitumor effects. This research showed that NTA and doxorubicin can be conveniently displayed on His-tagged VLNPs by a simple Add-and-Display step with enhanced coupling efficiency (Figure 8.6). Also, doxorubicin was only liberated at acidic pH in a sustained manner (Biabanikhankahdani et al. 2017).

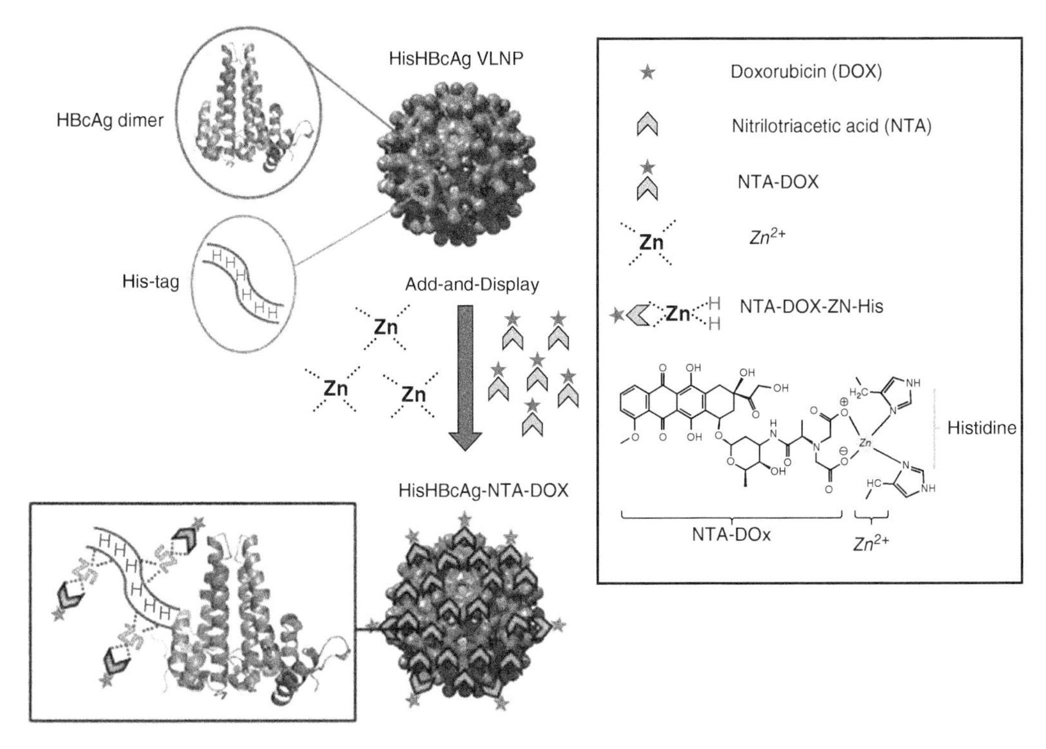

Figure 8.6 Schematic representation of the Add-and-Display method for immobilization of doxorubicin non-covalently on His-tagged VLNPs. *Source:* Adapted from Biabanikhankahdani et al. (2017); an open access article, permission granted under Creative Commons Attribution License from.

Guenther, C.M., Kuypers, B.E., Lam, M.T. et al. (2014). Synthetic virology: engineering viruses for gene delivery. *Wiley Interdisciplinary Reviews: Nanomedicine and Nanobiotechnology* 6: 548–558.

Hefferon, K. (2014). Plant virus expression vector development: new perspectives. *BioMed Research International* 2014: 785382.

Hefferon, K.L. (2018). Repurposing plant virus nanoparticles. *Vaccines* 6 (1): 11.

Holzinger, M., Le Goff, A., and Cosnier, S. (2014). Nanomaterials for biosensing applications: a review. *Frontiers in Chemistry* 2: 63.

Hooker, J.M., O'Neil, J.P., Romanini, D.W. et al. (2008). Genome-free viral capsids as carriers for positron emission tomography radiolabels. *Molecular Imaging and Biology* 10: 182–191.

Hornef, M.W., Wick, M.J., Rhen, M., and Normark, S. (2002). Bacterial strategies for overcoming host innate and adaptive immune responses. *Nature Immunology* 3: 1033–1040.

Hu, H., Masarapu, H., Gu, Y. et al. (2019). Physalis mottle virus-like nanoparticles for targeted cancer imaging. *ACS Applied Materials and Interfaces* 11: 18213–18223.

Hyman, P. and Denyes, J. (2018). Bacteriophages in nanotechnology: history and future. In: *Bacteriophages* (eds. D.R. Harper, S.T. Abedon, B.H. Burrowes and M.L. Mcconville), 1–31. Springer International Publishing.

Jayson, G.C., Kohn, E.C., Kitchener, H.C., and Ledermann, J.A. (2014). Ovarian cancer. *The Lancet* 384: 1376–1388.

Jeevanandam, J., Pal, K., and Danquah, M.K. (2018). Virus-like nanoparticles as a novel delivery tool in gene therapy. *Biochimie* 157: 38–47.

Kernan, D.L., Wen, A.M., Pitek, A.S., and Steinmetz, N.F. (2017). Featured article: delivery of chemotherapeutic vcMMAE using tobacco mosaic virus nanoparticles. *Experimental Biology and Medicine* 242: 1405–1411.

Khan, I.U., Khan, R.U., Asif, H. et al. (2017). Co-delivery strategies to overcome multidrug resistance in ovarian cancer. *International Journal of Pharmaceutics* 533 (1): 111–124.

Khayat, D., Antoine, E.-C., and Coeffic, D. (2000). Taxol in the management of cancers of the breast and the ovary. *Cancer Investigation* 18: 242–260.

Koudelka, K.J., Pitek, A.S., Manchester, M., and Steinmetz, N.F. (2015). Virus-based nanoparticles as versatile nanomachines. *Annual Review of Virology* 2: 379–401.

Kovacs, E.W., Hooker, J.M., Romanini, D.W. et al. (2007). Dual-surface-modified bacteriophage MS2 as an ideal scaffold for a viral capsid-based drug delivery system. *Bioconjugate Chemistry* 18: 1140–1147.

Langer, R. (1990). New methods of drug delivery. *Science* 249: 1527–1533.

Lebel, M.-È., Chartrand, K., Tarrab, E. et al. (2016). Potentiating cancer immunotherapy using papaya mosaic virus-derived nanoparticles. *Nano Letters* 16: 1826–1832.

Lizotte, P.H., Wen, A.M., Sheen, M.R. et al. (2015). In situ vaccination with cowpea mosaic virus nanoparticles suppresses metastatic cancer. *Nature Nanotechnology* 11 (3): 295–303.

Lizotte, P.H., Wen, A.M., Sheen, M.R. et al. (2016). In situ vaccination with cowpea mosaic virus nanoparticles suppresses metastatic cancer. *Nature Nanotechnology* 11: 295–303.

Ludwig, C. and Wagner, R. (2007). Virus-like particles-universal molecular toolboxes. *Current Opinion in Biotechnology* 18: 537–545.

Lundstrom, K. (2015). Alphaviruses in gene therapy. *Viruses* 7: 2321–2333.

Luque, D., De La Escosura, A., Snijder, J. et al. (2014). Self-assembly and characterization of small and monodisperse dye nanospheres in a protein cage. *Chemical Science* 5: 575–581.

Madamsetty, V.S., Mukherjee, A., and Mukherjee, S. (2019). Recent trends of the bio-inspired nanoparticles in cancer theranostics. *Frontiers in Pharmacology* 10: 1264.

Madden, A.J., Oberhardt, B., Lockney, D. et al. (2017). Pharmacokinetics and efficacy of doxorubicin-loaded plant virus nanoparticles in preclinical models of cancer. *Nanomedicine* 12: 2519–2532.

Manchester, M. and Singh, P. (2006). Virus-based nanoparticles (VNPs): platform technologies for diagnostic imaging. *Advanced Drug Delivery Reviews* 58: 1505–1522.

Mirza, A.Z. and Siddiqui, F.A. (2014). Nanomedicine and drug delivery: a mini review. *International Nano Letters* 4: 94.

Mudhakir, D. and Harashima, H. (2009). Learning from the viral journey: how to enter cells and how to overcome intracellular barriers to reach the nucleus. *The AAPS Journal* 11 (1): 65–77.

Narayanan, K.B. and Han, S.S. (2017a). Helical plant viral nanoparticles-bioinspired synthesis of nanomaterials and nanostructures. *Bioinspiration and Biomimetics* 12: 031001.

Narayanan, K.B. and Han, S.S. (2017b). Icosahedral plant viral nanoparticles-bioinspired synthesis of nanomaterials/nanostructures. *Advances in Colloid and Interface Science* 248: 1–19.

Narayanan, K.B. and Han, S.S. (2018). Recombinant helical plant virus-based nanoparticles for vaccination and immunotherapy. *Virus Genes* 54: 623–637.

NCI 2018, 'National Cancer Institute', Cancer Statistics [Online]. Available: https://www.cancer.gov/about-cancer/understanding/statistics [Accessed 9 May 2019].

Neubi, G.M.N., Opoku-damoah, Y., Gu, X. et al. (2018). Bio-inspired drug delivery systems: an emerging platform for targeted cancer therapy. *Biomaterials Science* 6: 958–973.

Oliveira, O.N. Jr., Iost, R.M., Siqueira, J.R. Jr. et al. (2014). Nanomaterials for diagnosis: challenges and applications in smart devices based on molecular recognition. *ACS Applied Materials and Interfaces* 6: 14745–14766.

Parayath, N.N. and Amiji, M.M. (2017). Therapeutic targeting strategies using endogenous cells and proteins. *Journal of Controlled Release* 258: 81–94.

Park, K. (2007). Nanotechnology: what it can do for drug delivery. *Journal of Controlled Release* 120: 1.

Parodi, A., Molinaro, R., Sushnitha, M. et al. (2017). Bio-inspired engineering of cell-and virus-like nanoparticles for drug delivery. *Biomaterials* 147: 155–168.

Qamar, S.A., Asgher, M., Khalid, N., and Sadaf, M. (2019). Nanobiotechnology in health sciences: Current applications and future perspectives. *Biocatalysis and Agricultural Biotechnology* 22: 101388.

Ren, Y., Wong, S.M., and Lim, L.Y. (2007). Folic acid-conjugated protein cages of a plant virus: a novel delivery platform for doxorubicin. *Bioconjugate Chemistry* 18: 836–843.

Rizvi, S.A.A. and Saleh, A.M. (2018). Applications of nanoparticle system in drug delivery technology. *Saudi Pharamceutical Journal* 26 (1): 64–70.

Rizvi, S.A.A. and Qureshi, Z. (2018). Pilocytic astrocytoma. *Consultant* 58 (10): 283–284.

Rizvi, S.A.A., Zafar, S., Ahmed, S.S. et al. (2018). Glioblastoma Multiforme. *Consultant* 58 (11): 318–320.

Rockson, S.G., Keeley, V., Kilbreath, S. et al. (2019). Cancer-associated secondary lymphoedema. *Nature Reviews Disease Primers* 5: 1–16.

Rohovie, M.J., Nagasawa, M., and Swartz, J.R. (2017). Virus-like particles: next-generation nanoparticles for targeted therapeutic delivery. *Bioengineering and Translational Medicine* 2: 43–57.

Sainsbury, F. (2017). Virus-like nanoparticles: emerging tools for targeted cancer diagnostics and therapeutics. *Therapeutic Delivery* 8 (12): 1019–1021.

Shahzad, Y., Khan, I.U., Hussain, T. et al. (2014). Bioactive albumin-based carriers for tumour chemotherapy. *Current Cancer Drug Targets* 14: 752–763.

Shukla, S., Ablack, A.L., Wen, A.M. et al. (2012). Increased tumor homing and tissue penetration of the filamentous plant viral nanoparticle Potato virus X. *Molecular Pharmaceutics* 10: 33–42.

Shukla, S., Myers, J.T., Woods, S.E. et al. (2017). Plant viral nanoparticles-based HER2 vaccine: immune response influenced by differential transport, localization and cellular interactions of particulate carriers. *Biomaterials* 121: 15–27.

Singh, S.P., Sirbaiya, A.K., and Mishra, A. (2019). Bioinspired smart nanosystems in advanced therapeutic applications. *Pharmaceutical Nanotechnology* 7: 246–256.
Steinmetz, N.F. and Manchester, M. (2011). *Viral Nanoparticles: Tools for Material Science and Biomedicine*. New York: Jenny Stanford Publishing https://doi.org/10.1201/9780429067457.
Strable, E. and Finn, M. (2009). Chemical modification of viruses and virus-like particles. In: *Viruses and Nanotechnology. Current Topics in Microbiology and Immunology*, vol. 327 (eds. M. Manchester and N.F. Steinmetz). Springer, Berlin, Heidelberg Viruses and Nanotechnology.
Tong, G.J., Hsiao, S.C., Carrico, Z.M., and Francis, M.B. (2009). Viral capsid DNA aptamer conjugates as multivalent cell-targeting vehicles. *Journal of the American Chemical Society* 131: 11174–11178.
Truong, N.P., Whittaker, M.R., Mak, C.W., and Davis, T.P. (2015). The importance of nanoparticle shape in cancer drug delivery. *Expert Opinion on Drug Delivery* 12: 129–142.
Valegård, K., Liljas, L., Fridborg, K., and Unge, T. (1990). The three-dimensional structure of the bacterial virus MS2. *Nature* 345: 36–41.
Walper, S.A., Turner, K.B., and Medintz, I.L. (2015). Enzymatic bioconjugation of nanoparticles: developing specificity and control. *Current Opinion in Biotechnology* 34: 232–241.
Wen, AM 2016,'Engineering Virus-Based Nanoparticles for Applications in Drug Delivery, Imaging, and Biotechnology', Electronic Thesis or Dissertation. Case Western Reserve University, OhioLINK Electronic Theses and Dissertations Center. [Accessed 12 Nov 2019].
Wu, W., Hsiao, S.C., Carrico, Z.M., and Francis, M.B. (2009). Genome-free viral capsids as multivalent carriers for taxol delivery. *Angewandte Chemie International Edition* 48: 9493–9497.
Wu, Y., Li, J., Yang, H. et al. (2017). Targeted cowpea chlorotic mottle virus-based nanoparticles with tumor-homing peptide F3 for photothermal therapy. *Biotechnology and Bioprocess Engineering* 22: 700–708.
Yoo, J.-W., Irvine, D.J., Discher, D.E., and Mitragotri, S. (2011). Bio-inspired, bioengineered and biomimetic drug delivery carriers. *Nature Reviews Drug Discovery* 10: 521–535.
Zeng, Q., Wen, H., Wen, Q. et al. (2013). Cucumber mosaic virus as drug delivery vehicle for doxorubicin. *Biomaterials* 34: 4632–4642.

9

Magnetic Nanoparticles: An Emergent Platform for Future Cancer Theranostics

Parinaz Nezhad-Mokhtari[1], Fatemeh Salahpour-Anarjan[1], Armin Rezanezhad[2], and Abolfazl Akbarzadeh[3]

[1] *Department of Medical Nanotechnology, Faculty of Advanced Medical Sciences, Tabriz University of Medical Sciences, Tabriz, Iran*
[2] *Faculty of Mechanical Engineering, Department of Materials Science and Engineering, University of Tabriz, Tabriz, Iran*
[3] *Stem Cell Research Center, Tabriz University of Medical Sciences, Tabriz, Iran*

9.1 Introduction

All chemical compounds or elements of our earth display some magnetic properties under certain situations. In this context, magnetic nanoparticles (MNPs) have gained a huge interest all over the world due to their promising applications in the biomedical field. Considering this, the main focus has been given on materials having ferri- or ferromagnetic, superparamagnetic (SPM), and superferrimagnetic properties at ambient temperature. In this regard, three types of materials exist. First are metals; the metallic elements which possess ferromagnetism at room temperature include iron, nickel, and cobalt. The nanoparticle (NP) preparation of these metals and their use in biomedical applications hereof is probable because of their favorable magnetic behaviors (Hadjipanayis et al. 2008; Li et al. 2013). Since such NPs present a strong tendency to oxidation in non-magnetic oxides (antiferromagnetic FeO, NiO, CoO), an oxidation-protective layer is needed. Also, due to the toxicity of metallic NPs, they play a limited role in medicine (Tran and Webster 2010). Ferromagnetic alloys such as FePt, FeNi, CoPt, or FeCo are the second type of ferromagnetic materials. The MNP preparation consisting of ferromagnetic alloys is illustrated in the literature by numerous research groups (Wang et al. 2014). To date, none of those nanostructures has found access in biomedical usage, mainly owing to two reasons: (i) some of the ferromagnetic alloys (such as FeCoCr, AlNiCo, CoPt) display a hard-magnetic behavior (a remnant magnetization and coercivity), leading to potential particle agglomeration and causing vessel embolism risk for the patient; and (ii) most of the alloys with favorable magnetic behavior have toxic components (such as Co or Ni) which prevent the usage of these materials in the human body.

The third kind of materials are oxides; magnetic oxide materials could be divided into mixed oxides by various crystal structures (such as the magnetic garnets and the ferrites) as well as the pure metallic oxides. Because the saturation magnetization of all garnets is very small, these materials are not appropriate for biomedical uses. The ferrites indicate hard- or soft-magnetic behavior, depending on their composition. In spite of some groups having found soft-magnetic ferrites with favorable magnetic properties, they have only been used for certain biomedical applications, and

Nanobiotechnology in Diagnosis, Drug Delivery, and Treatment, First Edition. Edited by Mahendra Rai, Mehdi Razzaghi-Abyaneh, and Avinash P. Ingle.

hence very few reports could be found in the literature (Rodrigues et al. 2015; Ruthradevi et al. 2017). Representative hard-magnetic ferrites having favorable magnetic behavior are strontium-, barium-, or cobalt-ferrite for medicinal usages. Meanwhile, cobalt-ferrite ($CoFe_2O_4$) was reported to have fewer toxic effects than Sr- or Ba-ferrite. NPs of this material were found to have enhancing applications for biomedical aims, for example as minimal invasive tumor treatments for magnetic hyperthermia and as diagnostic agents in lab-on-a-chip (Salunkhe et al. 2013).

Generally, four various oxides have to be mentioned: Fe_2O_3 (iron[III] oxide) and Fe_3O_4 (iron[II,III] oxide), as well as the rather unsteady FeO (iron[II] oxide) and Fe_2O (iron[I] oxide). Numerous Fe_2O_3 phases exist, including α-, β-, γ-, or γ-Fe_2O_3, which all demonstrate various magnetic behavior. Of the iron oxides, only Fe_2O_3 (maghemite) and Fe_3O_4 (magnetite) illustrate ferromagnetic behavior or, more specifically, ferrimagnetism owing to the spinell structure (a subcategory of the cubic lattice). The complete work on the iron oxide nature and its features is reported by Cornell and Schwertmann (Cornell and Schwertmann 2003). For the first time, the preparation of iron oxide MNPs was studied in 1980 by (Khalafalla and Reimers (1980) and Massart (1980). Then, various preparation routes were developed, and thus developed MNPs with favorable magnetic properties have been used for numerous biomedical applications (Krishnan 2010).

Considering these facts, the present chapter aims to discuss various related aspects like magnetic properties of MNPs, advantages of MNPs in biomedicine, various methods for the preparation of MNPs, and modification of MNPs. Moreover, special focus has been given to biomedical applications of MNPs.

9.2 Magnetic Properties of MNPs

Besides other factors like the shape or magnetic anisotropy, the magnetic behavior of MNPs is also determined by the size of particles. Numerous areas of homogeneous magnetization are structured for macroscopic particles in the size range of micrometer (μm) and above. These so-called magnetic domains are separated via Bloch walls (Bloch 1932). Owing to the formation of this domain, the magnetic stray field of the particle is decreased and the formation of the domain in the lack of an external magnetic field is energetically suitable compared to a homogeneously magnetized particle. All domains magnetization directions in the particle are oriented statistically, which leads to compensation of all magnetic moments within the particle, resulting in no external magnetization of the particle without an outer magnetic field.

By reducing the dimensions of the magnetic particle, the comparative proportion of wall energy to that of the entire particle energy enhances. Owing to energetic reasons, below a critical particle size no magnetic domains can be made and the whole particle in one direction displays a spontaneous magnetization. The magnetization direction of these so-called single domain particles is illustrated with the particle's crystal lattice and is named "the easy axis." For the single domain particle formation, the critical size is given via the material-specific magnetic anisotropy K and the form factor of the particle (ratio of particle length in various directions related to the magnetic field) (Kittel 1946). For spherical and cubic magnetite particles, the theoretical upper limit for the single domain particle formation is around 80 nm (Fabian et al. 1996), which was experimentally confirmed by Dutz (2008).

An additional diminution of the particle size leads to the decrease of the particle's magnetic anisotropy energy. In this case, a certain possibility is that for finite temperatures, the thermal energy exceeds the anisotropy energy owing to thermic changes, and the spontaneous variations in the magnetization orientation of particle can be observed (Néel 1949). This leads to a thermally prompted temporal attenuation (relaxation) of the remnant magnetization (M_R) following Eq. (9.1):

$$M_R(t) = M_R(t=0) \times e^{-t/\tau N} \tag{9.1}$$

The so-called Néel relaxation time τ_N, after which M_R reaches a value near to zero, could be estimated from the ratio of the anisotropy energy ($K \times V$) to the thermal energy ($k \times T$) with the k (Boltzmann constant) and T (temperature) following Eq. (9.2), where τ_0 is the minimum natural relaxation time of 10^{-9} s:

$$\tau_N = \tau_0 \times e^{(K \times V)/(k \times T)} \tag{9.2}$$

Hence, the magnetic behavior of very small particles strongly depends on the relation of τ_N (Néel relaxation time) and t_M (measurement time). If $t_M \ll \tau_N$, for relaxation processes there is no sufficient time and the particles reveal a steady hysteretic behavior. If $t_M > \tau_N$, the Neel relaxation occurs, leading to M_R attenuation and so no coercivity could be detected. This phenomenon is named superparamagnetism (SPM). SPM particles display no remnant magnetization and coercivity in quasi-static measurements, but a pronounced hysteresis can be observed when exposed to a high-frequency alternating magnetic field (AMF).

A particular case of magnetism could happen if small SPM particles form a bigger cluster. These clusters in the absence of an outer magnetic field reveal SPM behavior with no remainder magnetization or coercivity. If the particles are exposed to an external field, depending on the particle interactions' strength, a collective magnetism may result, and the clusters illustrate ferrimagnetic behavior by an observable hysteresis. The curves in Figure 9.1 present the differences between ferri/ferromagnetic and superparamagnetic properties and compare their magnetization properties.

The superferrimagnetism is characteristic of magnetic multicore particles (Dutz et al. 2007; Koplovitz et al. 2019) and such particles reveal very hopeful properties for biomedical uses (Khatami et al. 2019; Nasrin et al. 2019; Satheeshkumar et al. 2019). From the above observation it becomes clear, that the particle size plays a vital role in the magnetic behavior of MNPs. Moreover, the size distribution is an essential factor for the resulting magnetic features. The detailed considerations on the size distribution influence theory were assessed by Hergt et al. (2008) and then the influence in experiments was developed in various studies (Müller et al. 2013; Xiao-Li et al. 2015).

9.3 Advantages of MNPs in Biomedicine

It is currently well-known that NPs (~10 nm) have stronger physical properties like mechanical strength, more corrosion resistance, and more chemical activity. Additionally, their size is much less than biomolecules such as gene, virus, cell, etc. For example, the typical cell size, virus size, and protein molecule size are within the range of 10–100 μm, 20–400 nm, and 20–50 nm, respectively, while gene dimension lies between 10–100 nm long and 2 nm wide. This size range eases to bound or coat biomolecules on NPs. Such coated biomolecules (so-called protein) are mentioned as "labeled" or "tagged." In addition, if these used NPs are magnetic then they will have further benefits. The domain size of most of the magnetic materials is between 4 and 15 nm. A single-domain MNP will act like a tiny magnet. With the help of an outer magnetic field, the tagged molecules can be directed to the desired site. The only precaution one has to take is to select a magnetic material having low toxicity and an ability for binding biomolecules. Currently, many such MNPs are involved in various

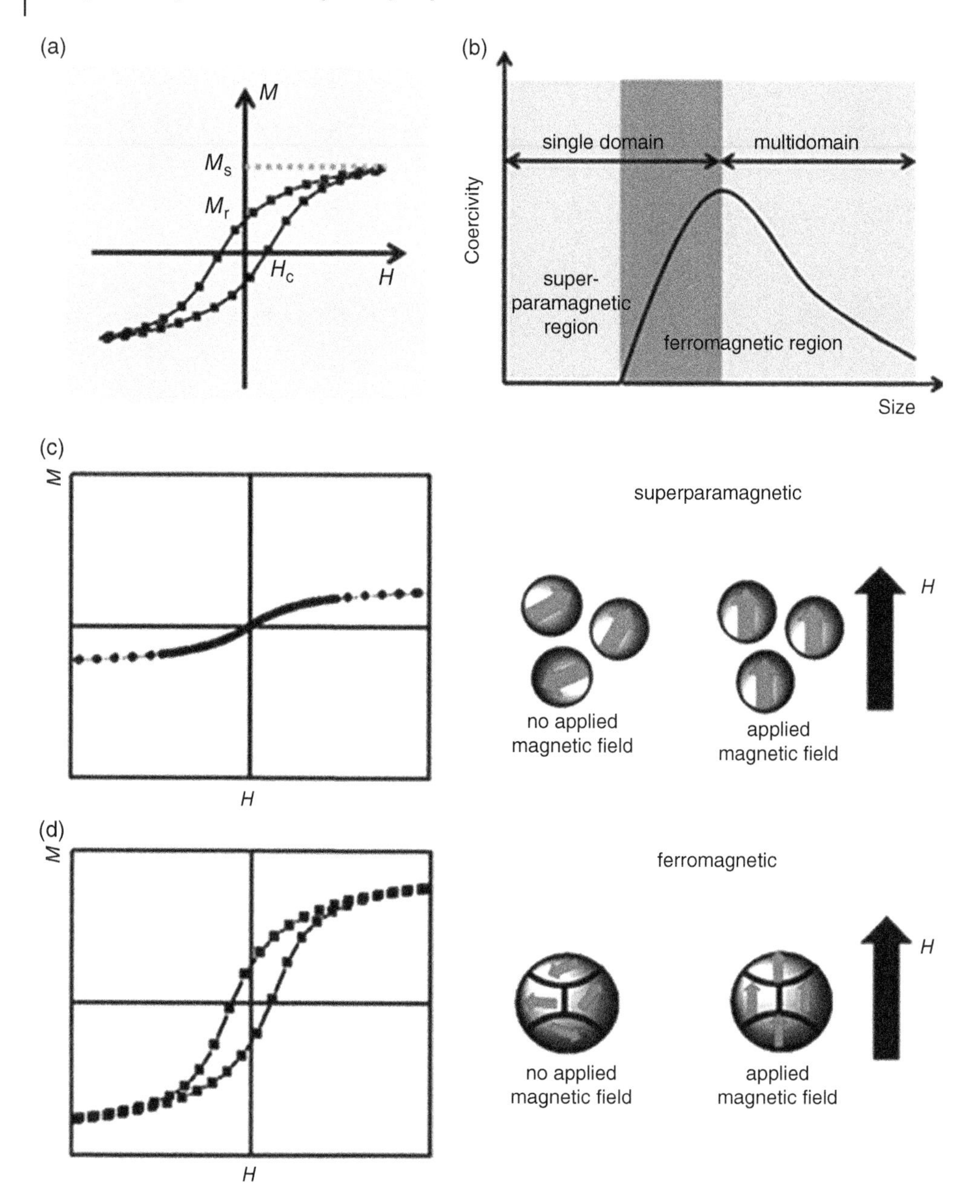

Figure 9.1 (a) Typical hysteresis loop of a ferri/ferromagnetic material. (b) Schematic illustration of the dependence of coercivity on particle size. (c and d) Hysteresis loop of a superparamagnetic nanomaterial and ferri/ferromagnetic material under a magnetic field, the magnetic moments of the domains of ferri/ferromagnetic particles and single-domain SPM particles are aligned. After removal of the magnetic field, ferri/ferromagnetic particles maintain a net magnetization unlike superparamagnetic particles (H is magnetic field and M is the magnetic moment per unit volume).

biomedical applications. Interaction between the biomolecules and MNPs usually takes place in liquid surroundings. Hence, MNP dispersion in water or other biocompatible liquid is applied. Such dispersion is the so-called magnetic fluid, ferrofluid, or SPM fluid (Mehta 2017).

Since a very large number of MNPs in the SPM fluid are dispersed, they form aggregates owing to magnetic and van der Waals attractions. Subsequently, the durability of the fluid will be impaired.

Table 9.1 Classification of iron oxide particles according to their size.

Iron oxide particles (IOPs)	Diameter
Nano-sized very small superparamagnetic particles of iron oxide (VSPIO)	20 nm
Ultra-small superparamagnetic particles of iron oxide (USPIO)	20–50 nm
Small superparamagnetic particles of iron oxide (SPIO)	60–250 nm
Micro-sized particles of iron oxide (MPIO)	1–8 μm

To avoid this, the MNP surface is modified either via surfactant coating or assigned an electric charge (Singh M et al. 2011; Singh V et al. 2011). MNPs or ferrofluid synthesis should be performed, keeping in mind the above requirements as well as basic physics behind a specific application. Since size is the main factor in engineered iron oxide particles (IOPs), IOPs are commonly categorized into four ranges, referred to in Table 9.1.

Due to similar size, very-small SPM particles of iron oxide (VSPIO) and ultra-small SPM particles of iron oxide (USPIO) are often classified together as USPIO. Ultra-small and small size ranges of IOPs are used mostly in medical applications (Stirrat et al. 2014). If the size of MNPs is in between 50 and 250 nm, it can be rapidly uptaken by resident macrophages in the lymphoreticular organs including the spleen, lymph nodes, and liver. Size of MNPs as contrast agent is a key property; USPIO particles are small enough to escape from phagocytosis and result in a much longer half-life up to 36 hours (McLachlan et al. 1994), whereas the half-life of large particles of superparamagnetic iron oxide (SPIO) is as low as two hours in humans. Even the non-SPM NPs may lead to undesired agglomeration. Magnetic moments of cobalt and manganese may increase toxicity, therefore prevention of their clinical and biomedical usage is sometimes required. Accordingly, replacing with a new class of magnetic theranostic agents for clinical application is necessary. The various range of IOP particles listed in Table 9.1 can be an efficient alternative.

9.4 Preparation of MNPs

In general, there are several physical and chemical approaches that have been proposed for the synthesis of MNPs. All these approaches can be categorized into two broad groups: top-down and bottom-up approach. One of the simple approaches to produce dispersion of MNPs is a grinding manner. Here, bulk ferro/ferrimagnetic materials or their salts are grounded down to the nanoscale by using strong mechanical shear forces, applying ball mills or attrition mills. The main impediment of this technique is that the resulting nanomaterials can be contaminated by the container and balls (Rezaie et al. 2020). Some of the most common chemical MNPs synthesis (bottom-up method) approaches for biomedical uses include co-precipitation, sol-gel, thermal decomposition, hydrothermal/solvothermal, microemulsion, sonochemical, electrochemical, aerosol/vapor phase rout, polyol, flow injection synthesis (FIS), microwave technique, etc.

These approaches produce narrowly dispersed and spherical MNPs. Ferromagnetic metals like iron, nickel, and cobalt have larger saturation magnetization but they are toxic, hence generally iron oxides in the form of magnetite (Fe_3O_4) or maghemite (γFe_2O_3) are applied in biomedical uses. Researchers have found that in place of the iron oxides – which are also known as ferrites – another mixed ferrite like $MnFe_2O_4$ are more appropriate for certain medicinal uses. In almost all applications of MNPs, large numbers of particles dispersed in suitable liquid medium are applied. Meanwhile, when these particles are single or sub-domain sized they behave like tiny magnets. When they are close together they form aggregates, owing to van der Waal and dipole–dipole attractions. To inhibit aggregation each of these small magnets can be coated by a surfactant or electrically charged to prepare steric and Coulomb repulsion. Considering all these aspects scientists have produced MNPs through co-precipitation or thermal diffusion approaches. These MNPs are functionalized using various methods, for instance direct binding (Rezaie et al. 2020), Hong's technique (Zhang et al. 2006), bioremediation (Darwesh et al. 2019), and so on. In the following sections salient features of these particles and their potential applications in biomedicine and biotechnology are discussed.

9.4.1 Physical Methods

The synthesis of MNPs over physical approaches follows a top-down method where bulk materials, or large particles, are broken into nanosized particles. Though appropriate for large-scale productions, these ways, described in detail elsewhere (Hajalilou et al. 2014), are recognized to produce MNPs of non-uniform size distributions and include time-consuming and expensive technologies. Electron beam lithography (EBL) and milling are instances of these methods.

9.4.1.1 Milling

Mechanical alloying is a primary method to form NPs of various materials (Baláž et al. 2013). This appears to have been performed first for iron oxide NPs in 1965 with Papell (Stephen 1965), who milled Fe_3O_4 by heptane and oleic acid for 19 days. The average size of thus-prepared NPs was quite large (about 135 nm), and a very wide particle size distribution of 60–240 nm was formed. The high-energy ball milling route, which encompasses a chemical reaction through milling, is often considered a mechanosynthesis or mechanochemical synthesis. It is possible to synthesize advanced biomaterials and achieve a very good microstructure via this technique. This is owing to the frequent fracturing and cold re-welding of powder particles, transferring of biomaterials via diffusion of components, generating of severe plastic deformation, creating a wide range of lattice defects through the milling process (Vázquez et al. 2012). Furthermore, the mass transformed through the milling process enables commercial compounds to be made at ambient temperature, improving their properties, particularly soft-magnetic materials, via subsequent heat treatment. It is flexible for large-scale production, low-cost, and simple methodology. This method is an ideal manner and needs optimization of a few factors to achieve the essential phase: absence of remainders, and control of particle size (Abdollah et al. 2015). The different milling methods have been briefly discussed below.

9.4.1.2 Wet Milling

There are a few reports about the wet milling technique. The various proper liquids might be applied depending on the material. The solvents addition (i.e. water) and/or excesses of salt species are identified to promote effective milling and to work as anti-agglomeration agents (Chen et al. 2007). Water is essential for oxidation purposes because it generates high vapor pressure at ambient temperature. There are few reports involving MNPs with Fe mechanical milling in water (Hajalilou et al. 2016). Milling time, in wet-milling procedures, has also been recognized as a determinant factor for the single-phase pure MNPs production (Can et al. 2010; Janot and Guérard 2002), with non-reduced Fe free MNPs being produced for milling times longer than 48 hours.

9.4.1.3 Dry Milling

De Carvalho et al. (2013) demonstrated that a magnetite sample can be formed by applying a planetary mill (Fritsch Pulveristte 7 premium line) with an angular velocity of 300 rpm. In the beginning, the magnetite particles were obtained via ball milling in a hardened steel vial by the powder with 20:1 mass ratio and the 200 rpm rotation speed. The powder was milled for 10, 40, 60, and 96 hours (De Carvalho et al. 2013).

9.4.1.4 Electron Beam Lithography (EBL)

EBL could be defined as the conversion of bulk high purity iron materials or films, deposited on the surface of a substrate, to iron oxide (Fe_3O_4) NPs over the electron beam emission across the material (Rishton et al. 1997). The electron beam causes the initial iron precursors' evaporation and the one-dimensional arrays' formation of shape-controlled superparamagnetic iron oxide nanoparticles (SPIONs). Two-dimensional arrangements have, however, been reached by Krása et al. (2009) over the EBL copulation via radio-frequency sputtering.

9.4.2 Chemical Methods

MNPs produced with chemical approaches are considered bottom-up methods where molecular entities containing iron condensate, in a series of conditions, produce full nanometric-sized SPM NPs. Condensation reactions have, however, been described in the literature to be carried out in either solution/aqueous or gas phases, the former using solvents to mediate chemical reactions and the latter promoting precursor's nucleation in the gas phase, typically needing higher reaction energies (Marcelo et al. 2019).

Chemical synthesis approaches have been broadly applied to form MNPs owing to their capacity to provide reasonable amounts of MNPs cost-effectively in a rather straightforward method (Rezaie et al. 2018). The MNPs synthesis via chemical approaches is commonly based on the so-called "LaMer model" (LaMer and Dinegar 1950): the stable nuclei formation of the desired materials in a solution owing to chemical reactions, their following growth into small nanocrystals, and the final size enhance by Ostwald ripening, which basically consists of small crystals dissolving and redepositing onto larger crystals (Willard et al. 2004). Chemical synthesis methods could provide control over the size, composition, morphology, shape, crystallinity, magnetic properties, and colloidal stability of the MNPs with tuning various parameters, including nature, concentration of reacting agents and stabilizing surfactants, the mixing and pH of the solution, the reaction time, temperature, etc. Though, chemical synthesis techniques could also present some disadvantages in comparison to further methods, for example, the possible incorporation of intermediate reaction products in the MNPs through the process, the complexity of the various steps that must be followed in order to achieve MNPs with the desired properties, and sometimes challenging reproducibility of the chemical reactions (Frey et al. 2009; Alonso et al. 2018). In recent years, there have been numerous reviews on MNP synthesis by chemical approaches (Figure 9.2).

9.4.2.1 Coprecipitation

One of the oldest and most broadly applied methods for the MNPs chemical synthesis is coprecipitation. This is a convenient method to achieve large amounts of MNPs directly dispersed in aqueous media (Alonso et al. 2018). This method is probably the simplest and most efficient chemical pathway to achieve magnetic particles. Iron oxides (either Fe_3O_4 or γFe_2O_3) are typically prepared with an aging stoichiometric mixture of ferric and ferrous salts in an aqueous medium. Coprecipitation synthesis includes two major steps: first, the nuclei formation and second, the nuclei growth. In order to form monodisperse MNPs, these two phases should be separated. In coprecipitation synthesis, this separation is not straightforward, and as a result, MNPs with a wide size distribution are usually achieved. The Fe_3O_4 formation chemical reaction may be written as reaction 9.R1.

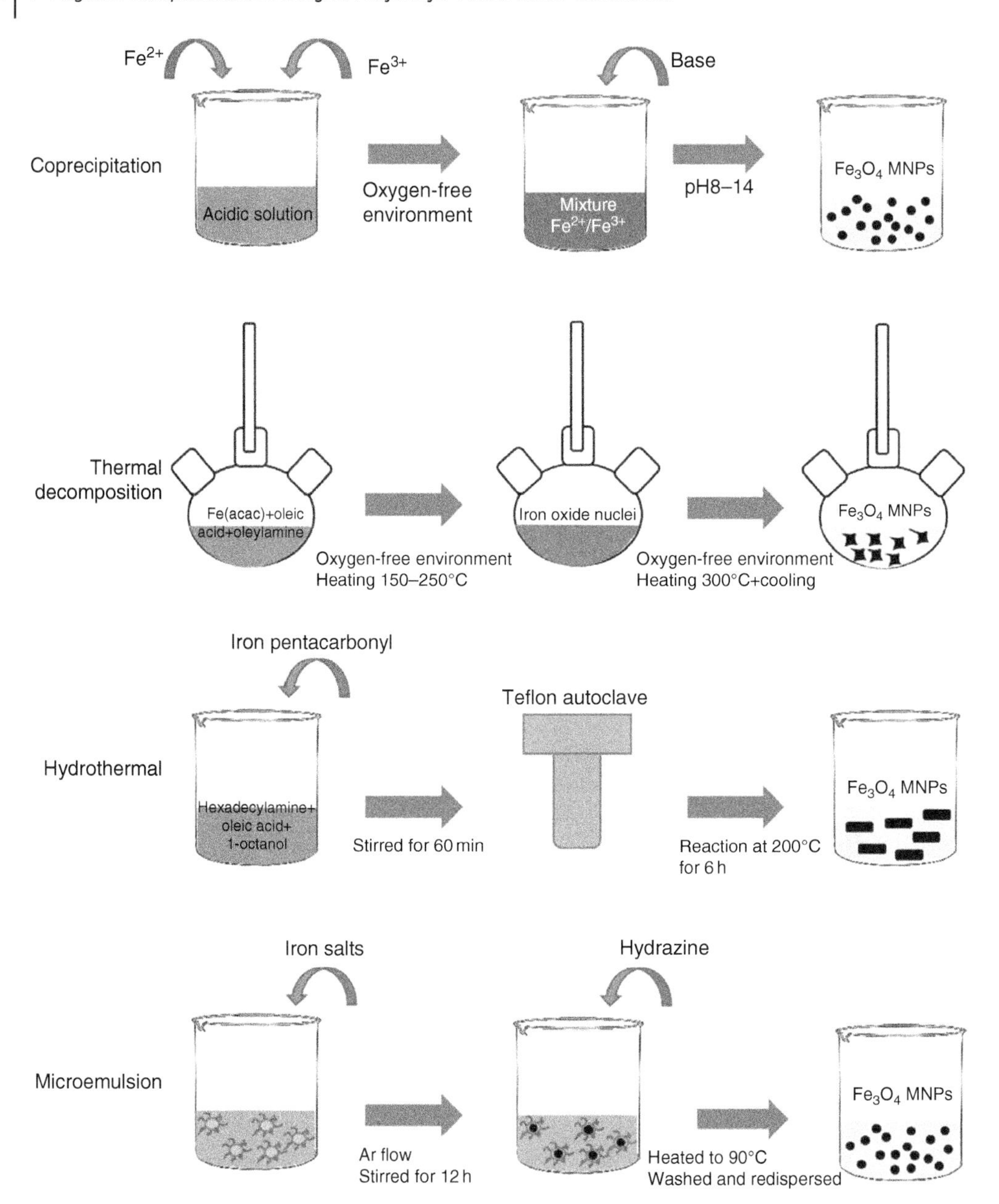

Figure 9.2 Schematic illustration of the chemical synthesis approaches described in this section.

$$M\ (=Fe^{2+}, Mn^{2+}, Co^{2+}, Ni^{2+}, Mg^{2+}, Ca^{2+}, \ldots) + 2Fe^{3+} + 8OH- \leftrightarrow MFe_2O_4 + 4H_2O \quad (9.R1)$$

According to the thermodynamics of this reaction, complete precipitation of Fe_3O_4 should be expected at a pH between 8 and 14, with a stoichiometric ratio of 2 : 1 (Fe^{3+}/Fe^{2+}) in a non-oxidizing oxygen environment. Though, Fe_3O_4 is not very steady and is sensitive to oxidation. Magnetite is transformed into maghemite (γFe_2O_3) in the oxygen presence (reaction 9.R2).

$$Fe_3O_4 + 2H^+ \leftrightarrow \gamma Fe_2O_3 + Fe^{2+} + H_2O \quad (9.R2)$$

Moreover, oxidation in the air is not the only way to transform magnetite (Fe_3O_4) into maghemite (Fe_2O_3). Different electron or ion transfers depending upon the pH of the suspension are involved, according to reaction 9.R2. Under acidic and anaerobic conditions, surface Fe^{2+} ions are desorbed as hexa-aqua complexes in solution, whereas under basic conditions, theoxidation of magnetite involves the oxidation-reduction of the surface of magnetite. The oxidation of ferrous ions is always correlated with migration of cations over the lattice framework, creating cationic vacancies to maintain the charge balance (Shi et al. 1999; Sirivat and Paradee 2019).

The main benefit of the coprecipitation process is that a large amount of NPs could be synthesized. Though the particle size distribution control is limited, because only kinetic parameters are controlling the growth of the crystal. In the coprecipitation process, two steps are involved: (i) a short burst of nucleation occurs when the species concentration reaches critical supersaturation; and (ii) there is a slow nuclei growth with diffusion of the solutes to the crystal surface. To prepare monodisperse iron oxide NPs, these two steps should be separated, i.e. nucleation should be avoided through the period of growth. The coprecipitation technique has been applied in various syntheses, usually at ambient or slightly elevated temperatures, and efforts have been prepared to reduce the polydispersity with size fractionation after the coprecipitation step (Alonso et al. 2018; Arteaga-Díaz et al. 2019). The shape, size, and composition of the resulting NPs very much depend upon:

- the type of salts used, e.g. chloride, sulfate, perchlorate, or nitride
- the Fe^{2+}/Fe^{3+} ratio
- the mixing order
- the mixing rate
- the reaction temperature
- the pH value

Though the coprecipitation technique is one of the successful and classical techniques for the synthesis of MNPs with high saturation magnetization, more attention should be paid to overcoming the shortcomings of this technique, such as the broad particle size distribution of products, and the strong base utilization in the reaction process.

9.4.2.2 Sol-Gel

The sol-gel process is an appropriate wet route for the synthesis of nanostructured metal oxides. This process is based on the hydroxylation and condensation of molecular precursors in solution, originating a "sol" of nanometric particles. Further condensation and inorganic polymerization lead to a three-dimensional metal oxide network denominated wet "gel." Because these reactions are done at ambient temperature, further heat treatments are required to obtain the final crystalline state. From the literature, it is obvious that the gel properties are very dependent upon the structure created through the sol stage of the sol-gel process. Similarly to coprecipitation methods, typical metal precursors include alkoxides and metals salts such as chlorides, nitrates, sulfates, and acetates that undergo hydrolysis followed by polycondensation. Although these reactions could be run at room temperature, additional heat treatment is important to render MNPs to a crystalline phase (Hench and West 1990).

The main factors that influence the growth reactions, kinetics, condensation reactions, hydrolysis, and subsequently, the structure and properties of the gel are solvent, temperature, nature, the concentration of the salt precursors employed, pH, and agitation (Livage et al. 1988). This technique proposal has some benefits, including (i) the possibility to gain materials with a predetermined structure according to conditions of the experiment; (ii) the possibility to achieve pure amorphous phases, monodispersity, and good control of the particle size; (iii) the microstructure control and the reaction products' homogeneity; and (iv) the possibility to embed molecules, which maintain their durability and properties within the sol-gel matrix (Chanéac et al. 1995; Kakihana 1996).

9.4.2.3 Hydrothermal/Solvothermal

The hydrothermal method is a high-temperature and/or high-pressure process, where metal precursors are dissolved in aqueous media and whose conditions are promising for the high crystalline hydrophilic MNP formation. This process is carried inside a Teflon-sealed stainless-steel autoclave, where water is in a supercritical/sub-supercritical state, and Fe^{2+} ions undergo hydrolysis, forming hydroxide intermediates ($FeOH_n$), followed by dehydration to form the desired ferrite materials (Liang et al. 2010). There are two main routes for the formation of ferrites via hydrothermal conditions: hydrolysis and oxidation or neutralization of mixed-metal hydroxides. These two reactions are very similar, except that ferrous salts are applied in the first technique. In this method, the reaction conditions, including solvent, time, and temperature typically have essential effects on the products (Rezaie et al. 2019). Furthermore, the MNP growth could be also controlled using ligands and surfactants. In these processes, the reaction conditions including the solvent nature, time, and temperature have a strong influence on the characteristics of the achieved MNPs. So, this method combines some of the benefits of both coprecipitation and thermal decomposition, and in the last years, hydrothermal approaches have been applied to synthesize a wide variation of MNPs with good water solubility and high crystallinity.

Though this method in general permits the achievement of water-dispersible MNPs with high crystallinity and very good magnetic response (close to the bulk material), the size of the achieved MNPs could sometimes be too large for certain uses (e.g. biomedical) and give problems with their colloidal durability. Furthermore, the reaction period through hydrothermal synthesis tends to be quite long (several hours) and controlling the shape and aspect ratio of the achieved MNPs could be not as straightforward as in the case of thermal decomposition, particularly for MNPs smaller than 30 nm. Despite this, hydrothermal synthesis presents itself as a very reliable, environmentally friendly route, and cost-effective to synthesize MNPs for several applications (Feng and Xu 2001; Xu and Lin 2013).

9.4.2.4 Microemulsion

The synthesis of MNPs by microemulsion is an area of increasing interest. Microemulsions are thermodynamically stable colloidal suspensions in which two initially immiscible liquids coexist in one phase thanks to the surfactants. In the water-in-oil microemulsion method, a stable dispersion of two immiscible solvents (water/oil) stabilized with a surfactant is prepared. In this case, the aqueous phase is dispersed as small nanodroplets (typically 1–100 nm in diameter), which contain the MNPs' precursors inside, and are surrounded by the surfactant molecules, forming the so-called "micelles," which act as nanoreactors. The synthesis of the MNPs takes place inside these micelles. The stabilized micelles provide confinement that limits particle nucleation, growth, and agglomeration. Then, a second emulsion or a base is added to the solution to generate a precipitate. This addition will make the micelles collide, break, and coalesce, favoring the formation of nanocrystals inside them. Finally, by adding a solvent such as acetone, the nanoparticles can be extracted. In this way, MNPs with sizes of between 1 and 100 nm can be obtained (Li and Park 1999).

Microemulsions can be of two types: direct (oil dispersed in water, O/W) and reversed (water dispersed in oil, W/O); with both being used for the synthesis of MNPs and other types of nanomaterials. Commonly used surfactants include cetyltrimethylammonium bromide (CTAB), sodium dodecyl sulphate (SDS), and polyvinylpyrrolidone (PVP), which are known to play significant roles in reaction dynamics, particle stabilization, and size control. Control over sizes and shapes is achieved by varying not only the iron precursor concentration but also the surfactants and solvents (Okoli et al. 2012).

9.4.2.5 Electrochemical

The electrochemical pathway to MNP synthesis is based on the reduction and/or oxidation of an iron-based sacrificial electrode, immersed in an electrolyte (Gopi et al. 2016). The revealed approaches of MNP production take two forms. First, a sacrificial iron cathode and an iron anode are immersed in an electrolyte including a 60 °C aqueous solution of Me_4NCl (tetramethylammonium chloride), and the 1–15 V potential (current density of 10–200 mA cm^{-2}) is used for 30 minutes. In the second method, rather than the iron coming from a sacrificial anode, it is formed with electrochemical decomposition of the electrolyte itself. In this case, carbon electrodes were applied and the electrolyte was a $Fe(NO_3)_3{\cdot}9H_2O$ in ethanol solution. The 62 V potential was applied and led to Fe_3O_4 NPs being collected on the cathode. The average NPs diameter, which ranged from 4.5 to 9 nm, was determined via both the $Fe(NO_3)_3{\cdot}9H_2O$ concentration and the current density. The particles formed this way appeared to be relatively monodisperse.

9.5 Coating of Magnetic Nanoparticles

Besides the need to generate MNPs with required shapes, size, and monodispersity, it is of prime importance that MNPs should be biocompatible for medical uses. The coating of MNPs is an approach to prevent or decrease toxicity and undesired toxicant side effects (Jiang et al. 2019). However, the main purpose of the coating is inhibiting the aggregation, to retain particles within a colloidal suspension and to provide a surface for targeting ligands and drug molecules conjugation. The most studied surface coatings are polymer-based, such as dextran and its derivatives (Chiu et al. 2019; Yalcin 2019), proteins (Bychkova et al. 2019), polyethylene glycol (PEG) (Dabbagh et al. 2019) and PEG copolymers (Licciardi et al. 2019), and antifouling polymers (Low et al. 2019).

The bio-functional layers that shell MNP cores have three purposes: (i) during the circulation in the body protect the therapeutic and targeting agents; (ii) improve their delivery efficiency; and (iii) in vital organs avoid undesired accumulation. Generally, coating procedures for MNPs can be divided into covalent and adsorptive methods. Covalent approaches could be more subdivided into grafting-to, grafting-from, and grafting-during approaches (Figure 9.3) (Biehl et al. 2018).

Prior functionalization of the MNP surface is needed for covalent attachment of a polymeric shell. The most premier example is the thin SiO_2 shell synthesis on the surface which could be prepared by using the Stöber process (Stöber et al. 1968). If functional silane precursors are applied, the resulting SiO_2 surface shows further functional groups such as thiols or amines, which could, later on, be performed for grafting procedures of polyelectrolytes (Ureña-Benavides et al. 2016). For grafting-to, the relevant polyelectrolyte is bio-functionalized by a suitable end group capable of reacting by the modified MNPs surface, while in grafting-from methods, the MNPs surface is bio-functionalized by an initiator, followed by the next surface-initiated polymerization.

Covalent grafting-to could be obtained through polyelectrolytes end capped by triethoxysilanes, which could be bound to the modified MNP surface (silica precoating). Grafting-from could can be recognized via bio-functionalization of the MNP surface through initiators for polymerization, for instance, N-(2-aminoethyl)-2-bromo-2-methylpropanamide, which has been applied for the ATRP (Atom Transfer Radical Polymerization) of carboxybetaine methacrylate from iron oxide NPs (Zhang et al. 2011). For grafting-through, polymerizable groups could be presented for instance with MPS (γ-methacryloxy-propyl-trimethoxysilane) condensation (Chen et al. 2015).

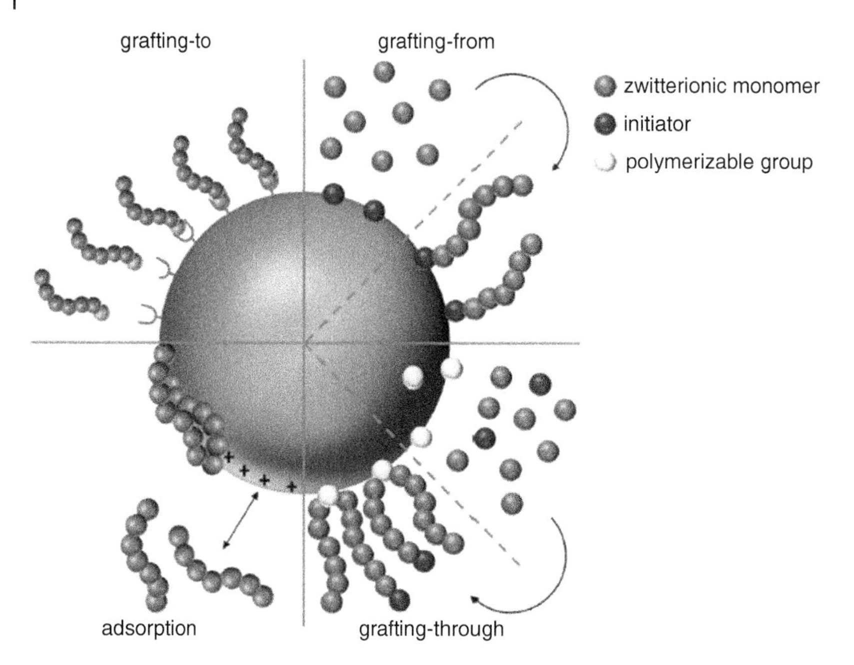

Figure 9.3 Schematic demonstration of various grafting approaches of polymers to nanoparticle surfaces.

The most public manner to attach polyelectrolytes to MNP surfaces is physisorption or chemisorption with either iron ions complexation at the surface, electrostatic interactions between MNPs and polymer, or with developing hydrophobic interactions (van der Waals forces) (Figure 9.4) (Biehl et al. 2018). Particular examples are the chemisorption of polymers featuring carboxylic acid moieties, as for model immobilized polydehydroalanine on pristine MNPs revealed by von der Lühe et al. (2015), or applied MNPs which were stabilized by oleic acid and immobilized amphiphilic zwitterionic polymers with hydrophobic interactions developed by Pombo-García et al. (2016) at the hydrophobic surface of the MNPs. Other approaches were successfully synthesized a dual responsive and intelligent nanocarriers via reversible addition-fragmentation chain transfer (RAFT) polymerization of N-isopropyl acrylamide and methacrylic acid monomers and Fe_3O_4 NPs (Ahmadkhani et al. 2018).

9.6 Biomedical Applications of MNPs

9.6.1 Targeted Drug Delivery

Through the cancer chemotherapeutic treatment, therapeutic agents having high cytotoxic activities need to be delivered into individual cancer cells to kill or damage cancer cells. In conventional ways, the drug's accumulation in both healthy and tumor tissue is often equivalent owing to the non-specific property of anticancer drugs injected into the blood circulation (Chanéac et al. 1995). This phenomenon provides a promotion to the well-known side effects, for instance that normal healthy cells are attacked in the treatment procedures. Targeted delivery of drug molecules with MNPs can increase the drug specificity and decrease this side effect (Sharma et al. 2019). Therapeutic compounds are encapsulated, or attached to biocompatible MNPs, and the applied magnetic fields are focused on particular targets in vivo. The fields capture the particle complex

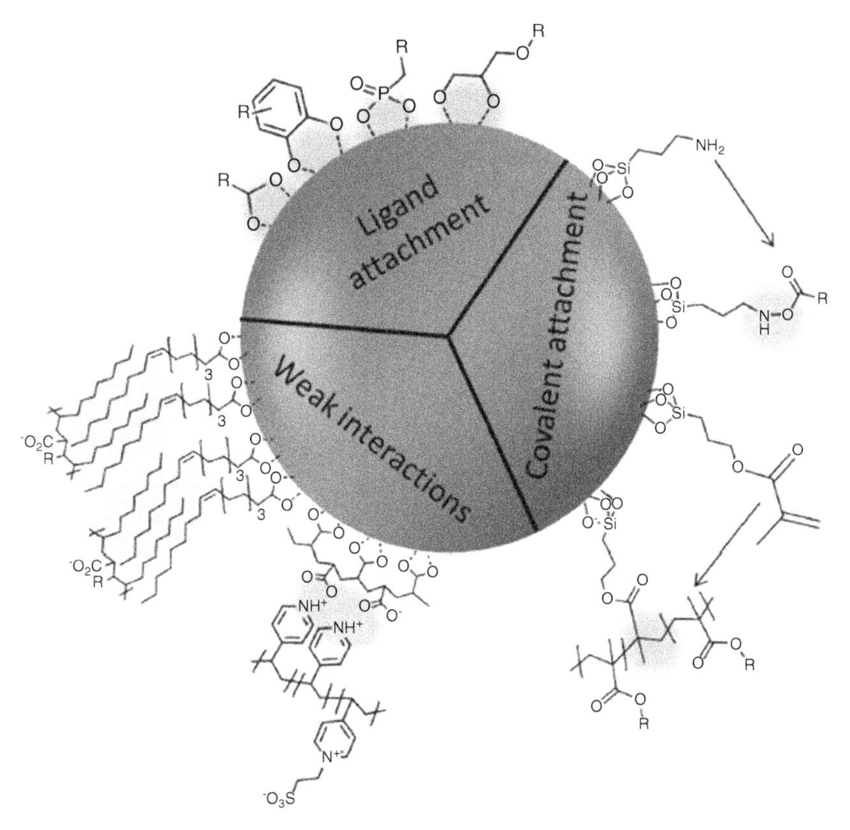

Figure 9.4 Schematic demonstration of used immobilization methods for polyzwitterions at the surface of MNPs.

and result in improved delivery to the target position (Figure 9.5) (Singh et al. 2014). Targeted drug delivery with MNPs could protect the drug molecules from the environment and improve the distribution, showing higher internalization with tumor cells than healthy cells and allowing the usage of the therapeutic agents at low adequate doses to decrease the chemotherapy toxicity (Pankhurst et al. 2003).

In this regard, Tiwari et al. (2019) described a new multifunctional magneto-fluorescent (MFCS) nanocarrier (MFCSNPs-FA-CHI-5FU) that contains MFCSNPs targeted with folic acid (FA), modified with chitosan (CHI), and with the loading of 5-flouoruracil (5-FU) in dual-mode targeted drug delivery and imaging. MFCS nanocarriers display SPM behavior and multicolor emission which are profitable for bioimaging and magnetic resonance imaging (MRI), respectively. In vitro drug release assays illustrated a pH-activated drug release and MRI revealing excellent dose-dependent signal enhancement in T2-weighted images. This proposes that MFCSNPs-FA-CHI-5FU nanocarriers could be applied as T2-weighted negative contrast agents in cancer diagnosis. Gawali et al. (2019) reported the pH-labile ascorbic acid-coated MNP (AMNPs) development for doxorubicin hydrochloride (DOX) delivery to cancer cells. The uniqueness of this drug delivery system (DDS) lies in the covalent DOX conjugation through hydrazone and carbamate bonds, resulting in a sustained and slow drug release profile at various environmental acidities. AMNPs revealed high colloidal durability in aqueous and cell culture media and have good magnetic field responsivity and protein resistance characteristics.

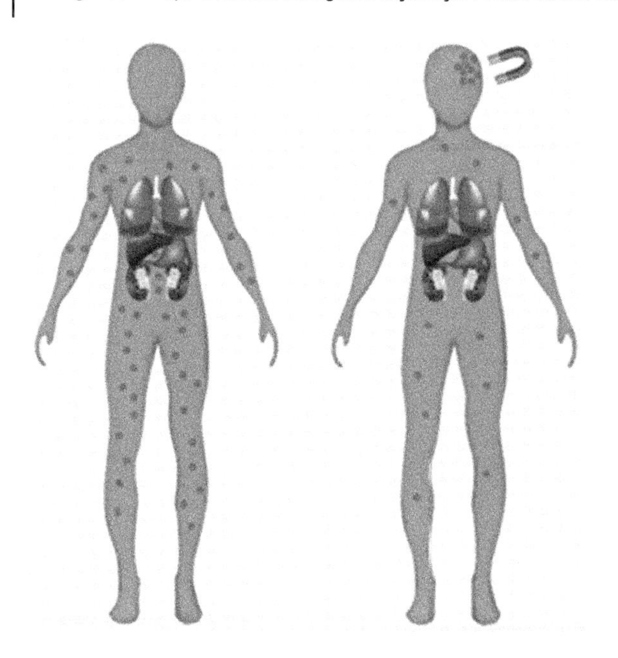

Figure 9.5 Magnetic targeting. No accumulation of MNPs occurs in the magnetic field absence, wbhereas under the influence of this field, MNPs alone or in combination with therapeutic cargo accumulate at a destined site.

In another work, the novel Au-coated Fe_3O_4 NPs capped with pH/redox dual-responsive nanomicelles was designed for intracellular co-delivery of doxorubicin and 6-mercaptopurine (Ghorbani et al. 2018). The developed nanocarrier showed many promising capabilities including a narrow size distribution range, high drug loading capacities, and stimuli-responsive drug release.

9.6.2 Passive Targeting

Passive targeting may be considered very important since the free MNP accumulation in the cancer area can enhance MRI detection and simplify the cancer diagnosis. Actually, the tumor tissues increase their own blood supply from the existing vascular system. Cancer-induced angiogenesis creates abnormal vessels with current fissures and during this leaky vasculature state, MNPs may accumulate in the cancer tissues (Chanéac et al. 1995). But, passive targeting depends also on various physicochemical MNP features including hydrophobicity, the surface charge, and the particle size. Small MNPs (<20 nm) are excreted with kidneys while medium-sized MNPs from 30 to 150 nm have been found in the bone marrow, heart, kidney, and stomach.

In contrast, large MNPs tend to accumulate in the liver and spleen. Additionally, solid tumor tissues display the enhancement permeability and retention (EPR) effect for MNPs of the suitable size. The EPR effect is owing to the defective vasculature and deficient lymphatic drainage system in solid tumor cells causing enhanced pore size in the endothelial wall. In consequence, the large size of MNPs does not readily penetrate the tumor cells while small MNPs can simply enter and leave cancer tissues. It is proposed that MNPs ranging from around 10 to 100 nm favorably accumulate in the cancer tissues as opposed to normal cells (Singh et al. 2014). Relating to the other MNP properties, it seems that positively charged and hydrophobic molecules due to opsonization have short circulation periods, while negatively charged and hydrophilic MNPs have long circulation periods (Sivakumar et al. 2018). In summary, though the leaky vasculature is a vast benefit for targeting cancer tissues, passive targeting may not allow for appropriate amounts of MNPs to accumulate in cancer tissues for successful molecular MRI and treatment.

9.6.3 Active Targeting

A hopeful approach to reach tumors is the active targeting development of cancer tissues via magnetic drug treatment or during interaction mediated with ligands including antibodies, folate, peptides, and lectins as presented with MNPs. In magnetic drug treatments, the drug molecules are conjugated on the MNP surface or can be encapsulated with a magnetic molecule into the NP platform including micelles, liposomes, or dendrimers. An outer magnetic field used on surface cancer cells should attract and maintain MNPs loaded with active molecules into the region of interest. An outer magnetic force could also increase magnetic targeted aerosol delivery to particular lung regions other than the airways or lung periphery. The magnetic aerosol inhalation could be applied not only for lung cancer treatment but also in other non-malignant pathologies including chronic obstructive pulmonary disease, cystic fibrosis, and respiratory infections (Caravan et al. 1999).

Another way to enhance anticancer drugs' accumulation into cancer is to apply NP platforms by bio-targeting agents that bind specially to cancer cells. The MNPs' platform use will not only enhance drug concentration in the cancer cells but also expose the tumor itself in MRI. Huang et al. (2011) developed a macromolecular platform carrying chlorotoxin and a MRI contrast agent, Gd, to target particular cancers. In nude mice bearing C6 glioma or liver tumor models, they presented that their MRI platform could be applied to diagnose matrix metalloproteinase-2 (MMP-2)-expressing tumors. One can, thus, envisage an anticancer NP platform by a MRI contrast agent and a double action peptide ligand. Such ligand can selectively bind to the proteins of the cancer cell surface and trigger off cell death. The best instance is a synthetic ligand of cell surface, nucleolin, identified as N6L. Certainly, surface nucleolin expression is increased in several cancer cells and shows a role in the apoptotic process (antitumor activity).

9.6.4 Cancer Diagnostics

The presence of both Fe^{2+} and Fe^{3+} combinations cause Fe_3O_4 NPs to differ from other MNPs. Having unique properties such as low toxicity, biocompatibility, special magnetic behavior, and capable function in multifunctional modalities, SPIONs are being applied in the wide areas of biomedical applications such as imaging, magnetic separation, targeted drug and gene delivery, tissue repair, cell proliferation, biosensors, and therapeutics (Sivakumar et al. 2018). MNPs are a versatile class of nano-scale material in recent decades that are under extensive development for improved diagnosis of a wide range of diseases, especially cancer, to predict disease progression and to monitor response to treatment and guide therapeutic intervention. MNPs applied as tracking probes and contrast agents act as diagnosing in MRI and as heat-producing particles in magnetic hyperthermia therapy, therefore MNPs are commonly referred to as "theranostics." Early detection and efficient treatment for cancer are the most important issues to improve the magnetic properties as theranostic agents. Trying to use low concentrations of MNP contrast agents can significantly improve methods of cancer diagnosis and treatment by decreasing probably toxicity (Vallabani and Singh 2018).

9.6.5 Magnetic Resonance Imaging (MRI)

Imaging is widely used in investigation and medical applications because scientists need to visualize the progress of new therapeutic agents. MRI is one of the most used and powerful tools for noninvasive clinical diagnosis owing to its high degree of soft-tissue contrast, spatial resolution, depth of penetration, and diagnostic or prognostic of diseases. One of the important advantages of

MRI imaging that made it a favorite imaging tool at a clinical level is that it doesn't use ionizing radiation that would cause minimal side effects. The mechanism of MRI is based on the presence of the proton and the difference in relaxation times (T_1 and T_2) of protons after being exposed to an external magnetic field. Most living organisms are made of water and fat and the hydrogen nuclei consist only of a proton. Hydrogen nuclei are not only positively charged, but also have magnetic spin. MRI employs this magnetic spin property of protons of hydrogen to create images. Protons become excited by an external magnetic field and their return to their equilibrium state happens at different rates: T_1 (spin-lattice or longitudinal relaxation) and T_2 (spin-spin or transverse relaxation) can create a signal that is processed to form an image of the organs. The intensity of the signal depends on the density of those nuclei in that specific part. MRI instruments by using this information create construct images to help diagnosis and treatment of patients.

However, MRI suffers from a low signal-to-noise ratio that results in low temporal resolution. Therefore, it is essential to find good contrast agents to enhance MRI signals by decreasing T_1 and T_2 relaxation times and to increases longitudinal relaxivity (r1) and transverse relaxivity (r2) rates through Eq. (9.3) and enhance resolution and produce sharper images (Hingorani et al. 2015; Shabestari-Khiabani et al. 2017).

$$\frac{1}{T_i} = r_i\left[C\right] \quad i = 1,2 \tag{9.3}$$

In this equation, T_i, is seconds, [C] is the concentration of the paramagnetic compound measured in millimoles per liter, and r_i is the relaxivity in $LmM^{-1}\ S^{-1}$ (Caravan et al. 1999; Lauffer 1987). T_1-weighted imaging is useful in obtaining morphological information imaging and solid organ pathologies such as identifying fatty tissue, assessing the cerebral cortex, and characterizing focal spleen and liver lesions and pancreatic neuroendocrine tumors (Owen et al. 2001; Chenevert et al. 2002), whereas T_2-weighted image is useful in imaging soft tissues and detecting edema and inflammation through uptake by inflammatory cells (Garcia-Diez et al. 2000; Einarsdóttir et al. 2004). Coating of MNPs exploits to prevent aggregate particles within a colloidal suspension and also to provide a surface for conjugation of drug molecules and targeting ligands. Carbohydrate or polymer construction has been used to coat iron oxide core of USPIOs, total particle diameter of less than 50 nm. Using antibody or ligand labeling, MNPs are able to be targeted to specific cell-surface markers. Coating of MNPs impacts biodistribution, magnetic behavior, and thus the potential applications of the particle which causes these materials to be non-toxic, biocompatible, and targetable (Anarjan 2019).

Based on magnetic properties, contrast agents divide into two categories: paramagnetic such as gadolinium (Gd^{3+}) and manganese (Mn^{2+}) ions; and superparamagnetic compounds such as iron oxide nanoparticles (IONPs). Paramagnetic compounds have a relatively larger T1-shortening effect compared to their T2-shortening effect ($T1 > T2$) and provide positive or bright contrast and T1-weighted images, while superparamagnetic nanoparticles have a relatively larger T2-shortening effect compared to their T1-shortening effect ($T1 < T2$) and cause negative or dark contrast and T2-weighted images. Paramagnetic compounds are usually conjugated to the carrier by chelation. Their clearance is usually through the renal excretion pathway. Free ions of Gd^{3+} are cytotoxicity and incomplete exertion increases transmetalation. Also, there exists homeostasis between injected IONPs and iron-associated proteins of the body such as hemoglobin and apoferritin transferrin which influences the degradation and clearance of IONPs from the body.

Application of Hybrid T1/T2 dual MRI contrast agent is a more effective strategy to highlight areas of tissue inflammation on MRI. A study has reported a hybrid T1/T2 dual MRI contrast agent by the encapsulation of SPIONs (T2 contrast agent) into an iron-based coordination polymer with a T1-weighted signal as SPION@coordination polymer NPs (Borges et al. 2015). This hybrid presented improved relaxometry and low cytotoxicity, so could be a compatible contrast agent for MRI. The coating is an impressive issue to reduce toxicity and to increase uptake degree in special tissue by attaching associated ligands. Coating of contrast agents can have various impacts, for example in a study, researchers could reduce the cytotoxicity of amphiphilic low molecular weight polyethylenimine (C12-PEI2K) in PEI-coated IONP as a contrast agent by using lactose without reducing MR imaging capability (Du et al. 2016).

One of the most major applications of MRI is the long-term fate of stem cell tracking and their differentiation studying in the clinical setting. For example, studying the long-term fate and potential mechanisms of mesenchymal stem cells (MSCs) after transplantation is essential for the improving functional benefits of stem cell-based stroke treatment. In a study, researchers have used SPION-loaded cationic polymersomes to label green fluorescent protein (GFP)-expressing MSCs to characterize MRI ability to reflect survival, long-term fate, and potential mechanisms of MSCs in ischemic stroke therapy (Duan et al. 2017). It is proven that MRI can verify the biodistribution and migration of grafted cells when pre-labeled with SPION-loaded polymersomes and the dynamic change of low signal volume on MRI can reflect the tendency of cell survival and apoptosis. They also noted that the overestimated long-term survival may be due to the presence of iron-laden macrophages around the cell graft.

9.6.6 Magnetic Hyperthermia Therapy

The normal human body temperature ranges between 36.5 and 37.5 °C (97.7–99.5 °F) (Hutchison et al. 2008). Hyperthermia is an approach in cancer therapy whereby the temperature increases throughout the target tissue up to 42–45 °C. By raising the temperature for about 30 minutes, the viability of cancerous cells is reduced and the produced heat can provoke cell apoptosis or make cancer cells more sensitive to the effects of radiation and/or certain anticancer drugs (Franckena et al. 2009). Even the temperature of the surrounding environment can be increased up to 55–60 °C, which normal healthy cells can tolerate very well, but cancerous cells cannot (Vallabani and Singh 2018). The first use of IONPs for magnetic hyperthermia treatment of cancers has reported by Professor Gilchrist in 1957 (Franckena et al. 2009).

Non-uniform heat distribution in the tumor area and eventual damage to healthy adjacent tumor cells are the weaknesses of treatment with hyperthermia. It has been proven that hyperthermia could be more effective by using MNP-based drugs in cancer treatment. In 2004, the first clinical magnetic fluid hyperthermia treatment system was developed at Charité – Medical University of Berlin (Gneveckow et al. 2004). Due to tunable properties, biocompatibility, and low cost, MNPs have been known as suitable candidates in therapeutic hyperthermia treatment, moreover having much higher colloidal stability causes superparamagnetic particles to have applied in hyperthermia studies (Abenojar et al. 2016). MNPs generate heat by hysteresis loss when placed in a high-frequency ~1 MHz magnetic field, so the heating should be restricted to the accumulation place of MNPs (Zare and Sattarahmady 2016). MNPs containing drugs not only can accumulate in tumor targets under an external magnetic field (targeted drug delivery) but also can oscillate using an AMF and produce heat (hyperthermia treatment). In magnetic hyperthermia, the heating can occur by any of the three mechanisms:

1) Eddy current heating (induction heating): heating due to the usage of an alternating pulsed magnetic field;

2) Frictional heating: heating induced by the interaction between the MNPs and the surrounding; and
3) Relaxation and hysteretic losses of the magnetic MNPs (Abenojar et al. 2016).

The rate of absorption energy when ferrofluid of MNPs is exposed to an external magnetic field is called a specific absorption rate (SAR). Achieving the maximum SAR by using MNPs allows us to use less injectable ferrofluid. On the other hand, the produced heat by MNPs due to AFM mainly depends on their SAR values. SAR defines the transformation of magnetic energy into heat and is usually based on power losses due to the Brownian (τ_B) and Néel (τ_N) relaxation times which is defined as effective relaxation time (τ_{eff}). When the direction of magnetic moments under the applied magnetic field direction is changed, the heat produces through Néel relaxation (τ_N) by reorientation to the equilibrium position, while Brownian relaxation (τ_B) happens when the physical rotation of particles within a ferrofluid, therefore the ferrofluid viscosity influences on the movement of the particles within a ferrofluid (Torres et al. 2019). By using linear response models through Néel and Brownian relaxation times can predict optimum SAR values for MNPs and reduce the ferrofluid dose which is an important challenge for *in vivo* application. Experimentally, the SAR (watts per gram [$\mathrm{W\,g^{-1}}$]) is calculated according to the following Eq. (9.4):

$$SAR = \frac{C}{M}\frac{\Delta T}{\Delta t} \tag{9.4}$$

In this Eq. 9.4, C is the specific heat capacity of the medium (assumed equal to that of water, C water = $4185\,\mathrm{J\,l^{-1}\,K^{-1}}$), m is the concentration of the transition metal of magnetic material in solution ($\mathrm{g\,l^{-1}}$), and $\Delta T/\Delta t$ is the slope of the initial linear section of the temperature versus time curve (Cardoso et al. 2018).

Also, the amount and intensity of produced heat depend on the number and magnetization properties of MNPs, magnetic field strength, the frequency of the AMF, and the physical properties of the ferrofluid such as magnetic and hydrodynamic radius of the MNPs, particle size distribution, ferrofluid viscosity, shape, and crystal structure. On the other hand, MNPs are capable of conjugating with various ligands. For example, with MNPs conjugated with antibodies to cancer-specific antigens, MNPs uptake by tumors cells improved, therefore hyperthermia together with MNPs can act more effectively (Figure 9.6).

There are many kinds of research about what IONPs have accomplished in magnetic hyperthermia to improve its application and efficacy. In a study, hydrophilic and surface-functionalized SPM iron oxide nanoparticles (SPIOs) have been used as magnetic fluid hyperthermia (MFH) for treating liver cancer (Kandasamy et al. 2018). In this study, the heating efficacies of the SPIOs have been studied in calorimetric MFH (C-MFH) with respect to their concentrations, dispersion medium, surface coatings, and applied AMFs. Researchers demonstrated the SPIOs coated by 3,4 diaminobenzoic acid (34DABA) have good cytocompatibility for 24-/48- hour incubation periods and higher killing efficiency of 61–88% (via MFH) in HepG2 liver cancer cells compared to their treatment with only AMF/water-bath-based thermotherapy. Also, the 34DABA-coated SPIOs have shown very promising heat-inducing agents for MFH-based thermotherapy and could be used as helpful nanomedicine in cancer treatments. This research shows that material is used to coat SPIOs is an important parameter in SPIOs efficacy.

In a study, core/shell structure MNPs composed of manganese ferrite/gold as a core and covered with a lipid bilayer as a shell and forming magnetoliposomes with sizes of around 120 nm were observed to transport drugs in combined chemo/photo/hyperthermia therapy in cancer therapy (Rio et al. 2019). This multifunctional nanosystem provides magnetic guidance and produces local heat, and in the next step to respond to rising temperature triggers encapsulated drug (antitumor

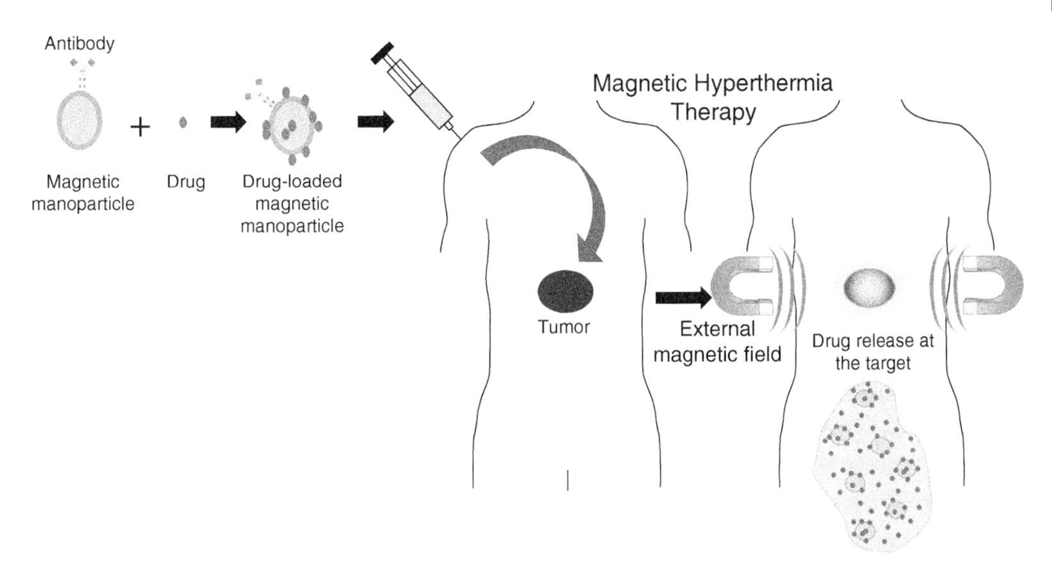

Figure 9.6 The mechanism of hyperthermia treatment by using MNPs.

tricyclic lactones) release, and resulted in synergistic cytotoxic effects against the human tumor cell lines HCT15 (colorectal adenocarcinoma) and NCI-H460 (non-small cell lung cancer).

9.7 Conclusions

MNPs, owing to their unique physicochemical advantages, are considered as one of the most favorable nanoparticulate systems for biomedical uses because of magnetic driving through the external field. For this reason, engineered MNP development by using controllable synthesis approaches and careful tuning of their features is in constant improvement. This chapter provides an up-to-date overview of the possible approaches to successfully construct favorable MNPs for future clinical applications. The first required step to design MNPs with the correct physicochemical properties such as shape, size, and structure is adjusting the parameters of the synthesis process. A deeper study on the magnetic characterization of MNPs is essential, in order to better perceive the MNP interaction with the external magnetic field applied as a driving/delivery biosystem. The promising property of superparamagnetism trends MNPs interesting nanomaterials, as verified by a large number of present studies focused on the detailed examination of magnetic properties, including coercive field, saturation magnetization, and remnant magnetization. These factors are critical for the interplay with the used magnetic field, permitting the magnetic-driven nanocarriers' modulation in the biological situation. Afterward some successful exams in animals by favorable results, drug delivery via MNP coating is presently undergoing preliminary human trials, however it will take some time before they are clinically accessible. Also, essential steps must be assumed in hyperthermia treatment.

The future challenges and applications in MNP-based biomedicine permit the definitive clarification of the toxicological aspects in nanoscale of nanosystems both in *in vitro* and *in vivo* trials, as well as the long-term durability determination of the functionalized MNPs before they can be extensively presented for clinical usage. Moreover, interdisciplinary methodologies are required to guarantee that the trials applied in the laboratory can more obviously match the predictable conditions that would be handled *in vivo*, permitting suitable steps toward uses. Considerable contributions of mathematical modeling of smart and more complex platforms, by the objective of considerate, or more exactly, the compound effects and interactions, will define whether, in the final examination, a given biomedical approach could be effectively applied.

References

Abdollah, H., Mansor, H., Reza, E.-K., and Taghi, M.M. (2015). Effect of milling atmosphere on structural and magnetic properties of Ni–Zn ferrite nanocrystalline. *Chinese Physics B* 24 (4): 048102.

Abenojar, E.C., Wickramasinghe, S., Bas-Concepcion, J., and Samia, A.C.S. (2016). Structural effects on the magnetic hyperthermia properties of iron oxide nanoparticles. *Progress in Natural Science: Materials International* 26 (5): 440–448.

Ahmadkhani, L., Akbarzadeh, A., and Abbasian, M. (2018). Development and characterization dual responsive magnetic nanocomposites for targeted drug delivery systems. *Artificial Cells, Nanomedicine, and Biotechnology* 46 (5): 1052–1063.

Alonso, J., Barandiarán, J.M., Barquín, L.F., and García-Arribas, A. (2018). Magnetic nanoparticles, synthesis, properties, and applications. In: *Magnetic Nanostructured Materials* (eds. A.A. El-Gendy, J.M. Barandiarán and R.L. Hadimani), 1–40. UK: Elsevier.

Anarjan, F.S. (2019). Active targeting drug delivery nanocarriers: ligands. *Nano-Structures and Nano-Objects* 19: 100370.

Arteaga-Díaz, S.J., Meramo-Hurtado, S.I., León-Pulido, J. et al. (2019). Environmental assessment of large scale production of magnetite (Fe3O4) nanoparticles via coprecipitation. *Applied Sciences* 9 (8): 1682.

Baláž, P., Achimovičová, M., Baláž, M. et al. (2013). Hallmarks of mechanochemistry: from nanoparticles to technology. *Chemical Society Reviews* 42 (18): 7571–7637.

Biehl, P., Von der Lühe, M., Dutz, S., and Schacher, F.H. (2018). Synthesis, characterization, and applications of magnetic nanoparticles featuring polyzwitterionic coatings. *Polymers* 10 (1): 91.

Bloch, F. (1932). *Zur Theorie des Austauschproblems und der Remanenzerscheinung der Ferromagnetika*, 295–335. Springer.

Borges, M., Yu, S., Laromaine, A. et al. (2015). Dual T 1/T 2 MRI contrast agent based on hybrid SPION@ coordination polymer nanoparticles. *RSC Advances* 5 (105): 86779–86783.

Bychkova, A.V., Lopukhova, M.V., Wasserman, L.A. et al. (2019). Interaction between immunoglobulin G and peroxidase-like iron oxide nanoparticles: physicochemical and structural features of the protein. *Biochimica et Biophysica Acta (BBA) - Proteins and Proteomics* 1868: 140300.

Can, M.M., Ozcan, S., Ceylan, A., and Firat, T. (2010). Effect of milling time on the synthesis of magnetite nanoparticles by wet milling. *Materials Science and Engineering: B* 172 (1): 72–75.

Caravan, P., Ellison, J.J., McMurry, T.J., and Lauffer, R.B. (1999). Gadolinium (III) chelates as MRI contrast agents: structure, dynamics, and applications. *Chemical Reviews* 99 (9): 2293–2352.

Cardoso, V.F., Francesko, A., Ribeiro, C. et al. (2018). Advances in magnetic nanoparticles for biomedical applications. *Advanced Healthcare Materials* 7 (5): 1700845.

Chanéac, C., Tronc, E., and Jolivet, J. (1995). Thermal behavior of spinel iron oxide-silica composites. *Nanostructured Materials* 6 (5–8): 715–718.

Chen, D., Ni, S., and Chen, Z. (2007). Synthesis of Fe_3O_4 nanoparticles by wet milling iron powder in a planetary ball mill. *China Particuology* 5 (5): 357–358.

Chen, Y., Xiong, Z., Zhang, L. et al. (2015). Facile synthesis of zwitterionic polymer-coated core–shell magnetic nanoparticles for highly specific capture of N-linked glycopeptides. *Nanoscale* 7 (7): 3100–3108.

Chenevert, T.L., Meyer, C.R., Moffat, B.A. et al. (2002). Diffusion MRI: a new strategy for assessment of cancer therapeutic efficacy. *Molecular Imaging* 1 (4): 336–343.

Chiu, C.-Y., Chung, T.-W., Chen, S.-Y., and Ma, Y.-H. (2019). Effects of PEGylation on capture of dextran-coated magnetic nanoparticles in microcirculation. *International Journal of Nanomedicine* 14: 4767.

Cornell, R.M. and Schwertmann, U. (2003). *The Iron Oxides: Structure, Properties, Reactions, Occurrences and Uses*. Wiley.

Dabbagh, A., Hedayatnasab, Z., Karimian, H. et al. (2019). Polyethylene glycol-coated porous magnetic nanoparticles for targeted delivery of chemotherapeutics under magnetic hyperthermia condition. *International Journal of Hyperthermia* 36 (1): 104–114.

Darwesh, O.M., Matter, I.A., and Eida, M.F. (2019). Development of peroxidase enzyme immobilized magnetic nanoparticles for bioremediation of textile wastewater dye. *Journal of Environmental Chemical Engineering* 7 (1): 102805.

De Carvalho, J., De Medeiros, S., Morales, M. et al. (2013). Synthesis of magnetite nanoparticles by high energy ball milling. *Applied Surface Science* 275: 84–87.

Du, J., Zhu, W., Yang, L. et al. (2016). Reduction of polyethylenimine-coated iron oxide nanoparticles induced autophagy and cytotoxicity by lactosylation. *Regenerative Biomaterials* 3 (4): 223–229.

Duan, X., Lu, L., Wang, Y. et al. (2017). The long-term fate of mesenchymal stem cells labeled with magnetic resonance imaging-visible polymersomes in cerebral ischemia. *International Journal of Nanomedicine* 12: 6705.

Dutz, S. (2008). *Nanopartikel in der Medizin*. Hamburg: Verlag Dr. Kovač.

Dutz, S., Andrä, W., Hergt, R. et al. (2007). Influence of dextran coating on the magnetic behaviour of iron oxide nanoparticles. *Journal of Magnetism and Magnetic Materials* 311 (1): 51–54.

Einarsdóttir, H., Karlsson, M., Wejde, J., and Bauer, H.C. (2004). Diffusion-weighted MRI of soft tissue tumours. *European Radiology* 14 (6): 959–963.

Fabian, K., Kirchner, A., Williams, W. et al. (1996). Three-dimensional micromagnetic calculations for magnetite using FFT. *Geophysical Journal International* 124 (1): 89–104.

Feng, S. and Xu, R. (2001). New materials in hydrothermal synthesis. *Accounts of Chemical Research* 34 (3): 239–247.

Franckena, M., Fatehi, D., de Bruijne, M. et al. (2009). Hyperthermia dose-effect relationship in 420 patients with cervical cancer treated with combined radiotherapy and hyperthermia. *European Journal of Cancer* 45 (11): 1969–1978.

Frey, N.A., Peng, S., Cheng, K., and Sun, S. (2009). Magnetic nanoparticles: synthesis, functionalization, and applications in bioimaging and magnetic energy storage. *Chemical Society Reviews* 38 (9): 2532–2542.

Garcia-Diez, A., Mendoza, L.R., Villacampa, V. et al. (2000). MRI evaluation of soft tissue hydatid disease. *European Radiology* 10 (3): 462–466.

Gawali, S.L., Barick, K.C., Shetake, N.G. et al. (2019). pH-labile magnetic nanocarriers for intracellular drug delivery to tumor cells. *ACS Omega* 4 (7): 11728–11736.

Ghorbani, M., Mahmoodzadeh, F., Nezhad-Mokhtari, P., and Hamishehkar, H. (2018). A novel polymeric micelle-decorated Fe 3 O 4/Au core–shell nanoparticle for pH and reduction-responsive intracellular co-delivery of doxorubicin and 6-mercaptopurine. *New Journal of Chemistry* 42 (22): 18038–18049.

Gneveckow, U., Jordan, A., Scholz, R. et al. (2004). Description and characterization of the novel hyperthermia-and thermoablation-system for clinical magnetic fluid hyperthermia. *Medical Physics* 31 (6): 1444–1451.

Gopi, D., Ansari, M.T., and Kavitha, L. (2016). Electrochemical synthesis and characterization of cubic magnetite nanoparticle in aqueous ferrous perchlorate medium. *Arabian Journal of Chemistry* 9: S829–S834.

Hadjipanayis, C.G., Bonder, M.J., Balakrishnan, S. et al. (2008). Metallic iron nanoparticles for MRI contrast enhancement and local hyperthermia. *Small* 4 (11): 1925–1929.

Hajalilou, A., Hashim, M., Ebrahimi-Kahrizsangi, R., and Sarami, N. (2014). Synthesis and structural characterization of nano-sized nickel ferrite obtained by mechanochemical process. *Ceramics International* 40 (4): 5881–5887.

Hajalilou, A., Mazlan, S.A., Lavvafi, H., and Shameli, K. (2016). Preparation of magnetic nanoparticle. In: *Field Responsive Fluids as Smart Materials* (eds. A. Hajalilou, S. Amri Mazlan, H. Lavvafi and K. Shameli), 121–126. Cham: Springer.

Hench, L.L. and West, J.K. (1990). The sol-gel process. *Chemical Reviews* 90 (1): 33–72.

Hergt, R., Dutz, S., and Röder, M. (2008). Effects of size distribution on hysteresis losses of magnetic nanoparticles for hyperthermia. *Journal of Physics: Condensed Matter* 20 (38): 385214.

Hingorani, D.V., Bernstein, A.S., and Pagel, M.D. (2015). A review of responsive MRI contrast agents: 2005–2014. *Contrast Media and Molecular Imaging* 10 (4): 245–265.

Huang, R., Han, L., Li, J. et al. (2011). Chlorotoxin-modified macromolecular contrast agent for MRI tumor diagnosis. *Biomaterials* 32 (22): 5177–5186.

Hutchison, J.S., Ward, R.E., Lacroix, J. et al. (2008). Hypothermia therapy after traumatic brain injury in children. *New England Journal of Medicine* 358 (23): 2447–2456.

Janot, R. and Guérard, D. (2002). One-step synthesis of maghemite nanometric powders by ball-milling. *Journal of Alloys and Compounds* 333 (1–2): 302–307.

Jiang, Z., Shan, K., Song, J. et al. (2019). Toxic effects of magnetic nanoparticles on normal cells and organs. *Life Sciences* 220: 156–161.

Kakihana, M. (1996). Invited review "sol-gel" preparation of high temperature superconducting oxides. *Journal of Sol-Gel Science and Technology* 6 (1): 7–55.

Kandasamy, G., Sudame, A., Luthra, T. et al. (2018). Functionalized hydrophilic superparamagnetic iron oxide nanoparticles for magnetic fluid hyperthermia application in liver cancer treatment. *ACS Omega* 3 (4): 3991–4005.

Khalafalla, S. and Reimers, G. (1980). Preparation of dilution-stable aqueous magnetic fluids. *IEEE Transactions on Magnetics* 16 (2): 178–183.

Khatami, M., Alijani, H.Q., Fakheri, B. et al. (2019). Super-paramagnetic iron oxide nanoparticles (SPIONs): Greener synthesis using Stevia plant and evaluation of its antioxidant properties. *Journal of Cleaner Production* 208: 1171–1177.

Kittel, C. (1946). Theory of the structure of ferromagnetic domains in films and small particles. *Physical Review* 70 (11–12): 965.

Koplovitz, G., Leitus, G., Ghosh, S. et al. (2019). Nano ferromagnetism: single domain 10 nm ferromagnetism imprinted on superparamagnetic nanoparticles using chiral molecules (small 1/2019). *Small* 15 (1): 1970004.

Krása, D., Wilkinson, C.D., Gadegaard, N. et al. (2009). Nanofabrication of two-dimensional arrays of magnetite particles for fundamental rock magnetic studies. *Journal of Geophysical Research: Solid Earth* 114 (B2): 1–11.

Krishnan, K.M. (2010). Biomedical nanomagnetics: a spin through possibilities in imaging, diagnostics, and therapy. *IEEE Transactions on Magnetics* 46 (7): 2523–2558.

LaMer, V.K. and Dinegar, R.H. (1950). Theory, production and mechanism of formation of monodispersed hydrosols. *Journal of the American Chemical Society* 72 (11): 4847–4854.

Lauffer, R.B. (1987). Paramagnetic metal complexes as water proton relaxation agents for NMR imaging: theory and design. *Chemical Reviews* 87 (5): 901–927.

Li, Y. and Park, C.-W. (1999). Particle size distribution in the synthesis of nanoparticles using microemulsions. *Langmuir* 15 (4): 952–956.

Li, Y., Hu, Y., Huang, G., and Li, C. (2013). Metallic iron nanoparticles: flame synthesis, characterization and magnetic properties. *Particuology* 11 (4): 460–467.
Liang, M.-T., Wang, S.-H., Chang, Y.-L. et al. (2010). Iron oxide synthesis using a continuous hydrothermal and solvothermal system. *Ceramics International* 36 (3): 1131–1135.
Licciardi, M., Scialabba, C., Puleio, R. et al. (2019). Smart copolymer coated SPIONs for colon cancer chemotherapy. *International Journal of Pharmaceutics* 556: 57–67.
Livage, J., Henry, M., and Sanchez, C. (1988). Sol-gel chemistry of transition metal oxides. *Progress in Solid State Chemistry* 18 (4): 259–341.
Low, S.C., Ng, Q.H., and Tan, L.S. (2019). Study of magnetic-responsive nanoparticle on the membrane surface as a membrane antifouling surface coating. *Journal of Polymer Research* 26 (3): 70.
von der Lühe, M., Günther, U., Weidner, A. et al. (2015). SPION@ polydehydroalanine hybrid particles. *RSC Advances* 5 (40): 31920–31929.
Marcelo, G.A., Lodeiro, C., Capelo, J.L. et al. (2019). Magnetic, fluorescent and hybrid nanoparticles: from synthesis to application in biosystems. *Materials Science and Engineering C* 106: 110104.
Massart, R. (1980). Preparation of aqueous ferrofluids without using surfactant-behavior as a function of the pH and the counterions. *Comptes Rendus Hebdomadaires des Seances de l Academie des Sciences Serie C* 291 (1): 1–3.
McLachlan, S.J., Morris, M.R., Lucas, M.A. et al. (1994). Phase I clinical evaluation of a new iron oxide MR contrast agent. *Journal of Magnetic Resonance Imaging* 4 (3): 301–307.
Mehta, R. (2017). Synthesis of magnetic nanoparticles and their dispersions with special reference to applications in biomedicine and biotechnology. *Materials Science and Engineering: C* 79: 901–916.
Müller, R., Dutz, S., Neeb, A. et al. (2013). Magnetic heating effect of nanoparticles with different sizes and size distributions. *Journal of Magnetism and Magnetic Materials* 328: 80–85.
Nasrin, S., Chowdhury, F.-U.-Z., and Hoque, S. (2019). Study of hyperthermia temperature of manganese-substituted cobalt nano ferrites prepared by chemical co-precipitation method for biomedical application. *Journal of Magnetism and Magnetic Materials* 479: 126–134.
Néel, L. (1949). Influence des fluctuations thermiques sur laimantation de grains ferromagnetiques tres fins. *Comptes Rendus Hebdomadaires Des Seances De L Academie Des Sciences* 228 (8): 664–666.
Okoli, C., Sanchez-Dominguez, M., Boutonnet, M. et al. (2012). Comparison and functionalization study of microemulsion-prepared magnetic iron oxide nanoparticles. *Langmuir* 28 (22): 8479–8485.
Owen, N., Sohaib, S., Peppercorn, P. et al. (2001). MRI of pancreatic neuroendocrine tumours. *The British Journal of Radiology* 74 (886): 968–973.
Pankhurst, Q.A., Connolly, J., Jones, S.K., and Dobson, J. (2003). Applications of magnetic nanoparticles in biomedicine. *Journal of Physics D: Applied Physics* 36 (13): R167.
Pombo-García, K., Weiss, S., Zarschler, K. et al. (2016). Zwitterionic polymer-coated ultrasmall superparamagnetic iron oxide nanoparticles with low protein interaction and high biocompatibility. *ChemNanoMat* 2 (10): 959–971.
Rezaie, E., Rezanezhad, A., Ghadimi, L.S. et al. (2018). Effect of calcination on structural and supercapacitance properties of hydrothermally synthesized plate-like $SrFe_{12}O_{19}$ hexaferrite nanoparticles. *Ceramics International* 44 (16): 20285–20290.
Rezaie, E., Rezanezhad, A., Hajalilou, A. et al. (2019). Electrochemical behavior of $SrFe_{12}O_{19}/CoFe_2O_4$ composite nanoparticles synthesized via one-pot hydrothermal method. *Journal of Alloys and Compounds* 789: 40–47.
Rezaie, E., Hajalilou, A., Rezanezhad, A. et al. (2020). Magnetorheological studies of polymer nanocomposites. In: *Rheology of Polymer Blends and Nanocomposites* (eds. S. Thomas, C. Sarathchandran and N. Chandran), 263–294. UK: Elsevier.

Rio, I.S., Rodrigues, J.M., Rodrigues, A.R.O., Coutinho, P. J., Castanheira, E., and Queiroz, M. J. R. (2019). Novel magnetoliposomes containing magnetic nanoparticles covered with gold as nanocarriers for new antitumor tricyclic lactones RICT 2019. International Conference on Medicinal Chemistry, Nantes, France, Jul 3–5 2019.

Rishton, S., Lu, Y., Altman, R. et al. (1997). Magnetic tunnel junctions fabricated at tenth-micron dimensions by electron beam lithography. *Microelectronic Engineering* 35 (1–4): 249–252.

Rodrigues, A.R.O., Gomes, I.T., Almeida, B.G. et al. (2015). Magnetic liposomes based on nickel ferrite nanoparticles for biomedical applications. *Physical Chemistry Chemical Physics* 17 (27): 18011–18021.

Ruthradevi, T., Akbar, J., Kumar, G.S. et al. (2017). Investigations on nickel ferrite embedded calcium phosphate nanoparticles for biomedical applications. *Journal of Alloys and Compounds* 695: 3211–3219.

Salunkhe, A., Khot, V., Thorat, N. et al. (2013). Polyvinyl alcohol functionalized cobalt ferrite nanoparticles for biomedical applications. *Applied Surface Science* 264: 598–604.

Satheeshkumar, M., Kumar, E.R., Indhumathi, P. et al. (2019). Structural, morphological and magnetic properties of algae/$CoFe_2O_4$ and algae/Ag-Fe-O nanocomposites and their biomedical applications. *Inorganic Chemistry Communications* 111: 107578.

Shabestari-Khiabani, S., Farshbaf, M., Akbarzadeh, A., and Davaran, S. (2017). Magnetic nanoparticles: preparation methods, applications in cancer diagnosis and cancer therapy. *Artificial Cells, Nanomedicine, and Biotechnology* 45 (1): 6–17.

Sharma, G., Parchur, A.K., Jagtap, J.M. et al. (2019). Hybrid nanostructures in targeted drug delivery. In: *Hybrid Nanostructures for Cancer Theranostics* (eds. R.A. Bohara and N. Thorat), 139–158. UK: Elsevier.

Shi, Y., Ding, J., Liu, X., and Wang, J. (1999). $NiFe_2O_4$ ultrafine particles prepared by co-precipitation/ mechanical alloying. *Journal of Magnetism and Magnetic Materials* 205 (2–3): 249–254.

Singh, M., Kumar, M., Štěpánek, F. et al. (2011). Liquid-phase synthesis of nickel nanoparticles stabilized by PVP and study of their structural and magnetic properties. *Advanced Materials Letters* 2: 409–414.

Singh, V., Seehra, M., Bali, S. et al. (2011). Magnetic properties of (Fe, Fe–B)/γ-Fe_2O_3 core shell nanostructure. *Journal of Physics and Chemistry of Solids* 72 (11): 1373–1376.

Singh, D., McMillan, J.M., Kabanov, A.V. et al. (2014). Bench-to-bedside translation of magnetic nanoparticles. *Nanomedicine* 9 (4): 501–516.

Sirivat, A. and Paradee, N. (2019). Facile synthesis of gelatin-coated Fe_3O_4 nanoparticle: effect of pH in single-step co-precipitation for cancer drug loading. *Materials and Design* 181: 107942.

Sivakumar, D., Naidu, K.C.B., Nazeer, K.P. et al. (2018). Structural characterization and dielectric studies of superparamagnetic iron oxide nanoparticles. *Journal of the Korean Ceramic Society* 55 (3): 230–238.

Stephen, P.S. (1965). Low viscosity magnetic fluid obtained by the colloidal suspension of magnetic particles. Google Patents.

Stirrat, C.G., Vesey, A.T., and McBride, O. (2014). Ultra-small superparamagnetic particles of iron oxide in magnetic resonance imaging of cardiovascular disease. *Journal of Vascular Diagnostics and Interventions* 2: 99–112.

Stöber, W., Fink, A., and Bohn, E. (1968). Controlled growth of monodisperse silica spheres in the micron size range. *Journal of Colloid and Interface Science* 26 (1): 62–69.

Tiwari, A., Singh, A., Debnath, A. et al. (2019). Multifunctional magneto-fluorescent nanocarriers for dual mode imaging and targeted drug delivery. *ACS Applied Nano Materials* 2 (5): 3060–3072.

Torres, T.E., Lima, E., Calatayud, M.P. et al. (2019). The relevance of Brownian relaxation as power absorption mechanism in magnetic hyperthermia. *Scientific Reports* 9 (1): 3992.

Tran, N. and Webster, T.J. (2010). Magnetic nanoparticles: biomedical applications and challenges. *Journal of Materials Chemistry* 20 (40): 8760–8767.

Ureña-Benavides, E.E., Lin, E.L., Foster, E.L. et al. (2016). Low adsorption of magnetite nanoparticles with uniform polyelectrolyte coatings in concentrated brine on model silica and sandstone. *Industrial and Engineering Chemistry Research* 55 (6): 1522–1532.

Vallabani, N.S. and Singh, S. (2018). Recent advances and future prospects of iron oxide nanoparticles in biomedicine and diagnostics. *3 Biotech* 8 (6): 279.

Vázquez, E., Gómez, X., and Ciurana, J. (2012). An experimental analysis of process parameters for the milling of micro-channels in biomaterials. *International Journal of Mechatronics and Manufacturing Systems* 5 (1): 46–65.

Wang, C., Meyer, J., Teichert, N. et al. (2014). Heusler nanoparticles for spintronics and ferromagnetic shape memory alloys. *Journal of Vacuum Science and Technology B, Nanotechnology and Microelectronics: Materials, Processing, Measurement, and Phenomena* 32 (2): 020802.

Willard, M., Kurihara, L., Carpenter, E. et al. (2004). Chemically prepared magnetic nanoparticles. *International Materials Reviews* 49 (3–4): 125–170.

Xiao-Li, L., Yong, Y., Jian-Peng, W. et al. (2015). Novel magnetic vortex nanorings/nanodiscs: synthesis and theranostic applications. *Chinese Physics B* 24 (12): 127505.

Xu, Z. and Lin, J. (2013). Hydrothermal synthesis, morphology control and luminescent properties of nano/microstructured rare earth oxide species. *Reviews in Nanoscience and Nanotechnology* 2 (4): 225–246.

Yalcin, S. (2019). Dextran coated iron oxide nanoparticle for delivery of miR-29a to breast cancer cell line. *Pharmaceutical Development and Technology* 24 (8): 1032–1037.

Zare, T. and Sattarahmady, N. (2016). A mini-review of magnetic nanoparticles: applications in biomedicine. *Basic and Clinical Cancer Research* 7 (4): 29–39.

Zhang, L., He, R., and Gu, H.-C. (2006). Oleic acid coating on the monodisperse magnetite nanoparticles. *Applied Surface Science* 253 (5): 2611–2617.

Zhang, X., Lin, W., Chen, S. et al. (2011). Development of a stable dual functional coating with low non-specific protein adsorption and high sensitivity for new superparamagnetic nanospheres. *Langmuir* 27 (22): 13669–13674.

10

Chitosan Nanoparticles: A Novel Antimicrobial Agent

Divya Koilparambil, Sherin Varghese, and Jisha Manakulam Shaikmoideen

School of Biosciences, Mahatma Gandhi University, Kottayam, Kerala, India

10.1 Introduction

Chitin was discovered for the first time in 1811 by Henri Braconnot while conducting research in mushrooms. Later in 1859, Professor C. Rouget found that alkali treatment of chitin yielded a substance that, unlike chitin, can be dissolved in acids. Further, Hoppe Seiler called this deacetylated chitin as "chitosan" (Divya et al. 2014). Chitin is the most widely spread biopolymer in nature after cellulose. It is the major component of cuticles of insects, fungal cell walls, yeast, or green algae (Einbu and Vårum 2008). It is also present in crab and shrimp shells (Wang and Xing 2007). On the other hand, chitosan does not occur abundantly in nature. It has been found only in cell walls of certain fungi (Muzzarelli et al. 1986). Chitosan can be obtained at industrial scale from shrimp and other crustaceans like crab, krill, and lobsters (Paulino et al. 2006; Yen et al. 2009; Limam et al. 2011; Tarafdar and Biswas 2013). Reports of microbial synthesis of chitosan from fungus, bacteria, and also from silkworms, beetle have also been documented (Paulino et al. 2006; Liu et al. 2012; Moussa et al. 2013). Figure 10.1 shows a schematic representation of the extraction of chitosan from crustacean shell waste.

Chitosan, the deacetylated product of chitin, is a linear amino polysaccharide of (1→4) linked 2-acetamido-2-deoxy-β-D-glucopyranose (GlcNAc; A units) and 2-amino-2-deoxy-β-D-glucopyranose (Glc N; D units) (Aranaz et al. 2009; Badawy and Rabea 2011). Chitosan contains at least 60% **D** units (Kumirska et al. 2011). The molar fraction of **D** units is expressed as the degree of deacetylation (DD) (Aranaz et al. 2009). It is an important characteristic that influences the performance of chitosan in many applications (Kumirska et al. 2010). DD can be determined by potentiometric titration (Zhang et al. 2011), IR (Baxter et al. 1992), UV (Kasaai 2009), gel permeation chromatography (Kumar 1999), ^{1}H-liquid-state NMR (nuclear magnetic resonance), and solid-state ^{13}C NMR (Ottey et al. 1996). The presence of the free amine groups along the chitosan chain makes it, unlike chitin, soluble in diluted acidic solvents (Divya et al. 2018). The molecular weight and viscosity development in aqueous solutions also play a significant role in the biochemical and pharmacological application of chitosan. Other major parameters include crystallinity, ash content, moisture content, heavy metal content, and so on (Rinaudo 2006).

Chitosan is used as an antimicrobial agent on account of its biodegradability, non-toxicity, ability to form metal complexes, and antimicrobial properties. Chitosan exhibits widespread activity

Nanobiotechnology in Diagnosis, Drug Delivery, and Treatment, First Edition. Edited by Mahendra Rai, Mehdi Razzaghi-Abyaneh, and Avinash P. Ingle.

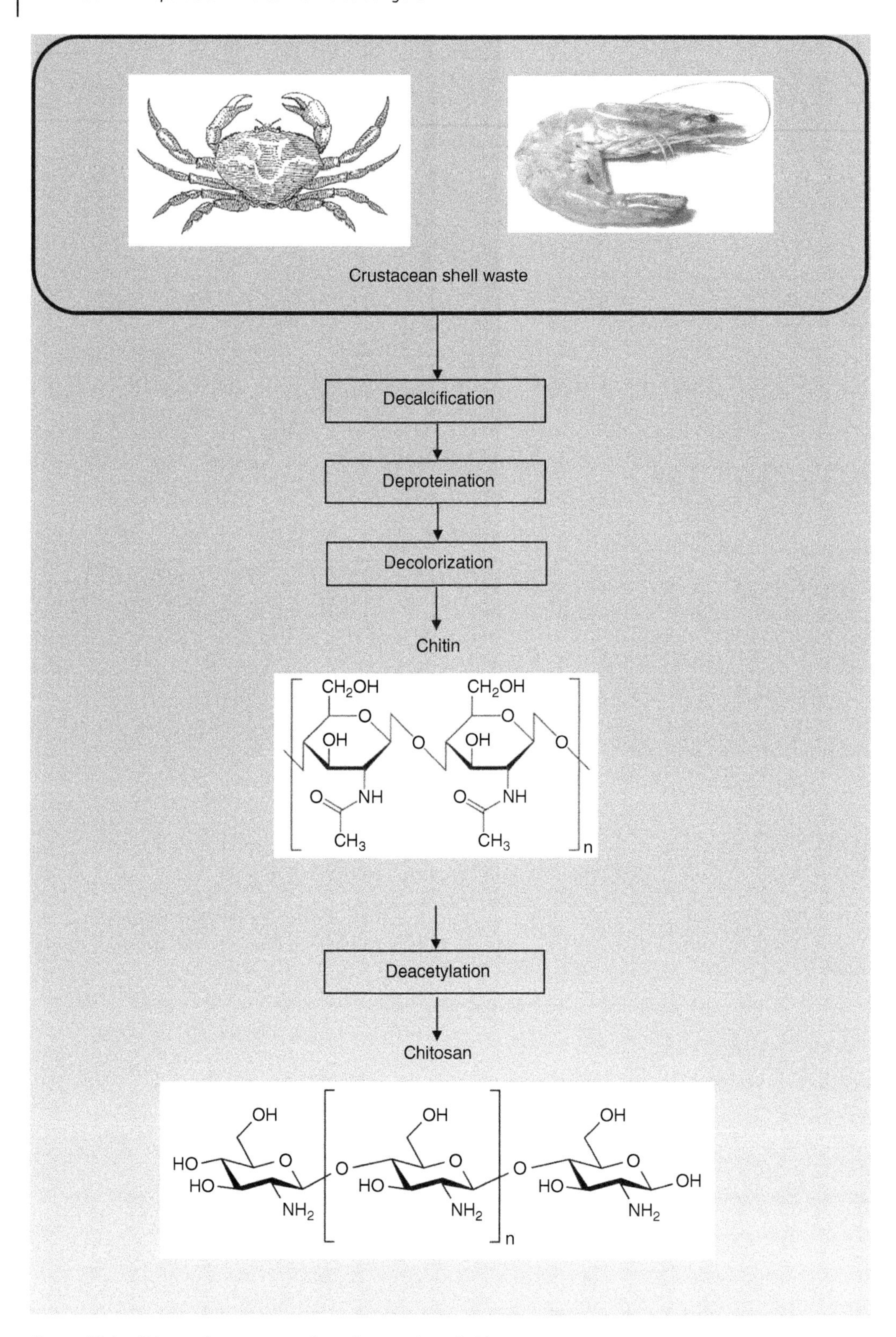

Figure 10.1 Schematic representation of extraction of chitosan from crustacean shell waste.

against bacteria, filamentous fungi, and yeasts but has lower toxicity for mammalian cells (Kong et al. 2010). Several factors like species of microorganism, concentration, pH, solvent, and molecular mass influence the antimicrobial activity of chitosan (Kaur et al. 2013). In spite of these characteristics, the use of chitosan as an antimicrobial agent is limited. This is because chitosan shows antimicrobial activity at acidic conditions and not at neutral pH because of the loss of positive charge on amino groups at neutral pH (Jeon et al. 2014).

Therefore, to overcome these problems, researchers have focused on the synthesis of chitosan nanoparticles (ChNPs). ChNPs were first described in 1994 by Ohya and co-workers (1994). They used ChNPs prepared by emulsifying and cross-linking for intravenous delivery of the anticancer drug 5- fluorouracil (Grenha 2010). Since then many methods have been employed for the synthesis of ChNPs. Different methods like ionic gelation (Calvo et al. 1997; Koukaras et al. 2012), microemulsion (De et al. 1999), emulsification solvent diffusion (Niwa et al. 1993; El-Shabouri 2002), polyelectrolyte complex (PEC) formation (Erbacher et al. 1998), synthesis with carboxymethylcellulose, coacervation, and reverse micellar (Brunel et al. 2008) have been developed for synthesis of ChNPs. Among these, the ionic gelation method is the preferred approach as the nanoparticles can be obtained spontaneously without using high temperature, organic solvents, or sonication (Sailaja et al. 2011; Tiyaboonchai 2013). The small size and quantum size effect of ChNPs help them to exhibit superior antimicrobial activity compared to chitosan and chitin (Huang et al. 2009).

This chapter deals with the latest advances in antimicrobial activities of ChNPs and their various nanoformulations. The various factors affecting the antimicrobial activity and mode of action of ChNPs have been also discussed in parts regarding active compound chitosan, the effect of the size of the NP, and the target microorganism. Finally, the various applications of ChNPs in medical treatment and water treatment are discussed in brief.

10.2 Bioactivities of ChNPs

Due to their biodegradability and broad-spectrum antimicrobial activity, ChNPs have found applications in many fields. Some of the prominent applications include (i) as antibacterial agents, gene delivery vectors, and carriers for protein release and drugs; (ii) as a wound-healing material for the prevention of opportunistic infection and for enabling wound healing; (iii) as effective antimicrobial activity against many bacteria, fungi, and viruses; and (iv) as an additive in antimicrobial textiles for producing clothes for healthcare and other professionals. ChNPs finds its major applicability as an antimicrobial agent. The emergence of antibiotic-resistant microorganisms has produced the need to find safe and effective alternatives. ChNPs are not only non-toxic and cost-effective but also have a wide spectrum of activity (Qi et al. 2004; Wei et al. 2009). It is also used in hospital-related textile manufacturing, as drug delivery systems, in orthopedic surgery as wound dressing material, etc. (Surendiran et al. 2009; Ali et al. 2011; Rodrigues et al. 2012; Tan et al. 2013). Some of the important bioactivities of ChNPs are discussed here.

10.2.1 Antimicrobial Activity of ChNPs

The recent years have seen an upsurge in ChNP research owing to its non-toxicity, biodegradability, and its multifarious applicability. Antimicrobial activity is one of the dominant properties of ChNPs. It has its noteworthiness in the context of emerging antibiotic-resistant microorganisms worldwide. Chitosan possesses all the characteristics of an ideal antimicrobial polymer: (i) easily and inexpensively synthesized; (ii) stable in long-term usage and can be stored at the temperature for its intended application; (iii) not soluble in water for a water-disinfection application; (iv) does not decompose to and/or emit toxic products; (v) should not be toxic or irritating to those who are handling it; (vi) can be regenerated upon loss of activity; and (vii) biocidal to a broad spectrum of pathogenic microorganisms in short times of contact (Kenawy et al. 2007). The antimicrobial properties of ChNPs have been reported to be higher than chitosan due to their small and compact particles and high surface charge (Qi et al. 2004; Kong et al. 2010).

The antimicrobial properties of ChNPs can be employed in a wide array of fields. But the commercialization of ChNPs requires a detailed understanding regarding the mechanism of antimicrobial activity.

10.2.1.1 Antibacterial Activity of ChNPs

ChNPs exhibit broad-spectrum inhibitory activity against both Gram-negative and Gram-positive bacteria (Sarwar et al. 2014). ChNPs were reported to have high antibacterial activity against medically important pathogens like *Klebsiella pneumoniae*, *Escherichia coli*, *Staphylococcus aureus*, and *Pseudomonas aeruginosa* (Divya et al. 2017). The antibacterial activity of ChNP-impregnated antibacterial paper was studied by Ma et al. (2010). The author demonstrated that ChNPs had antibacterial activity against *E. coli* and *S. aureus* and no colony was found at 10 $mg\,ml^{-1}$ concentration for *E. coli*.

Fish and its products like fish fingers and fish burgers have limited storage time due to undesired changes during frozen storage. ChNP-coated fish fingers were shown to have improved shell life and counts of psychrophilic bacteria, coliform bacteria, and proteolytic bacteria were found to be much less than in normal fish fingers. Also, they were found to be free of *Salmonella* sp., *S. aureus*, yeast, and mold (Abdou et al. 2012). Chitosan derived from shrimp has been generally recognized as safe (GRAS) for use in food according to the US Food and Drug Administration (US-FDA 2013).

One of the major advantages of ChNPs is their ability to form nanocomposites with metal ions. Metal chitosan complexes have shown to improve the antimicrobial activity of chitosan. Metal nanoparticles are very effective as antimicrobial agents (Divya et al. 2016), but their limiting factor is increased toxicity at nanoscale. This disadvantage could be overcome by integrating ChNPs into the metals and preparing metal nano-chitosan complexes (Katas et al. 2018). Metal chitosan complexes like chitosan–Ag^{+} complex (Yi et al. 2003; Chen et al. 2005), chitosan–Zn^{2+}complex (Wang et al. 2004), and chitosan–Cu^{2+} (Domek et al. 1984; Gu et al. 2007; Pinto et al. 2013) exhibit high antibacterial activity *in vitro*.

Du et al. (2009) performed a comparative analysis of the antibacterial activity of chitosan, ChNPs, and ChNPs loaded with different metal compounds like silver nitrate, copper sulfate, zinc sulfate, manganese sulfate, iron sulfate, and chlortetracycline. All the metal-loaded ChNPs except Fe^{2+}-loaded ChNPs showed higher activity than ChNPs and chitosan. Silver-loaded ChNPs showed the highest activity with minimum inhibitory concentration (MIC) of 3 and 6 $\mu g\,ml^{-1}$ against *E. coli* and *S. aureus*, respectively. The zeta potential of the nanoparticle-loaded with Ag^{+} was the highest, followed by Cu^{2+} loaded, Zn^{2+} loaded, and Mn^{2+} loaded. The nanoparticle-loaded Fe^{2+} had the lowest zeta potential, however this was still higher than the zeta potential of the naked ChNPs. The antibacterial activity is thus directly proportional to the zeta potential of nanoparticles, showing a clear association between the two. Copper-loaded carboxymethyl chitosan (Cu-CMC) nanoparticles prepared by chelation under aqueous conditions showed 99% efficiency against *S. aureus*, 68.9% more than copper ions alone and 6.1% more than CMCs (Gu et al. 2007). A report on the comparative analysis of the antibacterial activity of zinc nanoparticles, ChNPs, zinc neomycin nanoparticles, and zinc chitosan neomycin nanoparticles showed that zinc chitosan neomycin nanoparticles were the most effective antimicrobial agent against *Bacillus subtilis*, *S. aureus*, *E. coli*, and *P. aeruginosa* (Swathi et al. 2013). Nevertheless, ChNPs showed higher antibacterial activity than zinc nanoparticles, indicating a greater capacity of chitosan to act against different strains of bacteria than zinc (Abdelhady 2012).

In another study, chitosan magnetic nanoparticles were synthesized by incorporation method and the antibiotic streptomycin was added to them. The streptomycin-coated chitosan magnetic nanoparticles (strep-CS-MNPs) showed enhanced antibacterial activity against methicillin-resistant *S. aureus* (MRSA). Strep-CS-MNPs showed a zone of inhibition of 32 mm diameter which was almost double than 17 mm diameter of free streptomycin (Hussein-Al-Ali et al. 2014). Similarly, Chitosan -α -iron oxide nanoparticles synthesized by self-assembly method were reported to show enhanced antibacterial activity against Gram-positive (*S. aureus*) and Gram-negative (*E. coli*) bacteria when compared to individual chitosan and α- Fe_2O_3 nanoparticles (Kavitha et al. 2013).

Prema and Thangapandiyan (2013) synthesized gold nanoparticles with and without non-toxic, biocompatible, natural stabilizing agents like chitosan, CMC, and starch, and the bactericidal activity was comparatively evaluated. The maximum growth inhibition zone (30 mm) was recorded against *E. coli*, *K. pneumoniae*, and *Vibro vulnificus* for chitosan-capped gold nanoparticles followed by CMC and starch (Prema and Thangapandiyan 2013). Even though many different metal composites are tried, the most frequently used and most effective are chitosan-based silver nanoparticles (Ch-AgNPs). This is because the binding interaction between chitosan and AgNPs will lead to the stabilization of Ch-AgNPs composite thus enabling the AgNPs attached to the polymer chain to disperse in solution when the composite is dissolved completely (Chowdappa et al. 2014).

Wei et al. (2009) reported the enhanced antibacterial activity of Ch-AgNPs compared to ChNPs and chitosan-based gold nanoparticles (Ch-AuNPs) against *E. coli*, *B. subtilis*, and *S. aureus*. Here, the MIC of Ch-AgNPs was found to be 10 $\mu g\,ml^{-1}$ for all three bacteria, however, it was 48 $\mu g\,ml^{-1}$, 24 $\mu g\,ml^{-1}$, and 12 $\mu g\,ml^{-1}$ respectively against *E. coli*, *B. subtilis*, and *S. aureus* for ChNPs and 96 $\mu g\,ml^{-1}$, 48 $\mu g\,ml^{-1}$, and 48 $\mu g\,ml^{-1}$ against *E. coli*, *B. subtilis*, and *S. aureus* for Ch-AuNP. Owing to its non-toxicity, ChNPs could be effectively used in hospital-related textiles like bandages, hospital linen, etc. A study reported by Ali et al. (2011) investigated the effect of ChNPs and silver (Ag) loaded ChNPs. According to this, the ChNP-treated fabric showed 90% antibacterial activity. Ag-loaded ChNPs, on the other hand, showed almost 100% activity. This is due to the synergetic effect of silver ions and ChNPs. Another study demonstrated that the antibacterial activity of ChNPs and copper-loaded ChNPs was significantly higher than chitosan and doxycycline. Moreover, the antibacterial activity of Cu-loaded ChNPs was higher than ChNPs alone. This is because the zeta potential of Cu-loaded ChNPs is higher than ChNPs due to absorption of Cu ions into ChNPs. The higher surface charge density of Cu-loaded ChNPs actually enhances the affinity for negatively charged bacterial membranes (Qi et al. 2004). Anitha et al. (2009) reported the antibacterial activity of ChNPs against *S. aureus*. They evaluated the antibacterial activity of ChNP and its water-soluble derivatives such as O-carboxymethyl ChNPs (O-CMC) and N-O-carboxymethyl ChNPs (N-O-CMC). It was seen that N-O-CMC had higher antibacterial activity with no colony formation at 1 $mg\,ml^{-1}$ concentration. This was attributed to the high degree of substitution of carboxymethyl groups in N-O-CMC.

10.2.1.2 Antifungal Activity of ChNPs

The activity of ChNPs against fungi is mainly by the disruption of the cell wall and cell membrane. Palma-Guerrero et al. (2009) reported that chitosan inhibits spore germination and mycelia growth in fungus by penetrating the fungal cell and causing lysis and cell death. They also found that different cell types (conidia, germ tubes, and vegetative hyphae) exhibited differential sensitivity to chitosan, with ungerminated conidia being the most sensitive.

Low molecular weight (LMW) ChNP and high molecular weight (HMW) ChNP have shown activity against *Candida albicans*, *Aspergillus niger*, and *Fusarium solani* (Yien et al., 2012). The antifungal efficacy of oleoyl-ChNP against *Nigrospora sphaerica*, *Botryosphaerica dothidea*,

Nigrospora oryzae, *Alternaria tenussima*, *Gibberella zeae*, and *Fusarium culmorum* was tested by Xing et al. (2016) and four phytopathogens *N. sphaerica*, *B. dothidea*, *N. oryzae*, *A. tenussima* were found to be chitosan sensitive, whereas, *G. zeae* and *F. culmorum* were chitosan resistant. Chitosan-silver nanoparticle composites tested positive for their activity against mango anthracnose pathogen *Collectotrichum gleosporides* (Chowdappa et al. 2014).

10.2.1.3 Antiviral Activity of ChNPs

As mentioned above, the antibacterial activity of ChNPs has been explored fully, however, the antiviral efficacy of ChNPs is not studied at that extent. Therefore, there is scant literature available in this field. Mori et al. (2013) reported the antiviral activity of AgNPs/CS composites against H1N1 influenza A virus. Silver binds directly to the viral glycoprotein envelope and inhibits viral penetration into the host cell. Small-sized AgNPs also causes spatial restriction of binding. The AgNPs/CS composites could directly inhibit the viral contact with the host cell (Mori et al. 2013). Similarly, ChNPs prepared by ionotropic gelation of chitosan induced by Foscarnet were successfully used to detect the antiviral activity against HCMV (human cytomegalovirus) strain AD-169. The nanoparticles also showed no toxicity against non-infected HELF (human embryonic lung fibroblast) cells (Russo et al. 2014).

10.2.2 Anticancer Activity of ChNPs

Doxorubicin is normally utilized for cancer treatments. However, it also showed some undesirable reactions, for example, cardiotoxicity. To diminish these side effects, the drug has been exemplified in ChNPs. These nanoparticles can improve the ingestion of doxorubicin in the entire small intestine (Feng et al. 2013). The nanoparticle builds the endurance time of drug conjugates or the free drug and decreases unfavorable responses of the drug (Mitra et al. 2001). Moreover, lung cancer is one of the most common reasons for disease demise in developed nations (Jemal et al. 2008). Practically 80% of all these lung diseases are non-small cell lung cancer. Treatment of lung disease with paclitaxel (chemotherapy sedate) uncovers clear movement for non-small cell lung malignant growth at later stages. The transient incitement of the bloodstream by intracellular nanoparticle aggregates, prompting improved trapping capacity in pulmonary capillaries, has been built up as the mechanism by which nanoparticles containing paclitaxel obliterate lung tumors (Yang et al. 2009). These reports have also demonstrated that, under acidic tumor conditions, nanoparticles containing paclitaxel become progressively forceful and firmly associate with negatively charged tumor cells (Yang et al. 2009).

Chitosan core-shell nanocomposites with the least toxicity to typical cell lines were found to be suitable for antitumor treatment, drug targeting, and cell imaging (Eid et al. 2018). Chitosan/polyvinyl alcohol doped AgNPs arranged by gamma radiation are potential antitumor agents against prostatic malignancy cells (Abaza et al. 2018). Biogenic AgNPs conjugated with chitosan indicated considerable cytotoxicity toward MCF-7 (Michigan Cancer Foundation) (bosom malignancy cell lines) (Bilal et al. 2019). Vijayan et al. (2019) revealed that chitosan/biogenic AgNP (Ch/Bio-AgNP) conjugate demonstrated magnificent anticancer action on MDA MB (MD Anderson Metastasis Breast) cancer cells and Si Ha cells. Ch/Bio-AgNPs indicated improved movement contrasted with Bio-AgNPs and this can be clarified based on upgraded bioavailability and dependability. Upregulation of p53 and p38 genes was explained by gene expression studies and their role in the regulation of cancer cell proliferation and apoptosis was very much delineated in the investigation.

10.2.3 Other Biomedical Applications

Protein-based drugs are relentlessly hydrolyzed by enzymes in the gastrointestinal tract. When these medications are encased in ChNPs, they are not easily harmed by gastric enzymes. ChNPs can altogether upgrade the stability of the drug. The ChNPs can control drug discharge, improve the biodegradation of proteins, and upgrade the absorption of hydrophilic substances through the epithelial layer. They are being examined for the conveyance of drugs that exert their activity in the stomach (Sinha et al. 2004).

Chitosan-encapsulated gentamicin with both antimicrobial and cancer prevention agent exercises has been set up for lung conveyance (Huang et al. 2016). Since fucoidan (sulfated polysaccharide in seaweed) has antioxidants to expel the dynamic oxygen created by gentamicin (Yuan and Macquarrie 2015) an examination has been performed to look at the discharge properties of nanoparticles delivered by embodying gentamicin and fucoidan in chitosan. The delivered nanoparticles were found to have expanded antimicrobial activity and decreased systemic toxicity, demonstrating guarantee for the treatment of pneumonia infections.

Although dietary polyphenols show assorted pharmacological potential, for example antioxidant properties, anti-inflammatory effects, and anticipation of cardiovascular illness, cancer, and Alzheimer's disease, their moderate paces of assimilation and poor bioavailability avert the utilization of these synthetic substances as orally administered drugs (Rastogi and Jana 2016). In this context, to tackle this issue Liang et al. (2017) encapsulated tea-inferred polyphenols into ChNPs for oral delivery. They summarized that ChNPs can upgrade the stability of tea polyphenols and guard against oxidation or deterioration in the gastrointestinal tract. These encapsulations likewise prompted direct take-up of polyphenolic compounds at the tight epithelial intersections by epithelial cells through endocytosis. N-2-HACC and N,O-carboxymethyl chitosan (CMC) nanoparticles have been orchestrated and assessed as a vaccine adjuvant for Newcastle Disease Vaccine (NDV) and Irresistible Bronchitis Vaccine (IBV). The immune responses in chicken uncovered that these nanoparticles containing NDV/IBV can incite better intranasal immunization of IgG and IgA antibodies, and improve the multiplication of lymphocytes (Zhao et al. 2017). Chitosan is a biodegradable biopolymer that can invigorate an immune response. Additionally, ChNPs in blend with plasmid DNA improve antigen-specific immunity (Tao et al. 2013).

10.3 Factors Affecting the Antimicrobial Activity of ChNPs

There are numerous reports on the antimicrobial activity of ChNPs and chitosan nanoformulations, from various sources tested under diverse conditions. The antimicrobial activity of chitosan is influenced by various intrinsic and extrinsic factors. Among the intrinsic factors, type of chitosan, its molecular weight, degree of deacetylation, viscosity, solvent and concentration, size of nanoparticles, and zeta potential are considered most important, wheras extrinsic factors are related to environmental conditions, such as pH, temperature, ionic strength, metal ions, etc. The antimicrobial activity is also dependent on microbial factors like the species of microorganisms and the age of culture (Odds 2003; Zivanovic et al. 2004; Tsai et al. 2006; Fouad 2008). All these factors are discussed in the following sections.

10.3.1 Intrinsic Factors

10.3.1.1 Molecular Weight

The effect of the molecular weight of chitosan on antibacterial activity is the most explored factor. Even though it has been established that LMW chitosan has the high activity, the exact molecular weight is not standardized. It has been demonstrated that LMW chitosan of less than 10 kDa showed greater antimicrobial activity (Liu et al. 2006; Badawy and Rabea 2009; Honary et al. 2011; de Paz et al. 2011). LMW chitosan of 4.6 kDa exhibited better antibacterial activity against *E. coli*,

Pseudomonas aureofaciens, *Enterobacter agglomerans*, *B. subtilis*, *Candida krusei*, and *Fusarium oxysporum* f. sp. *radicis lycopersici* (Tikhonov et al. 2006). Fei Liu et al. (2001) demonstrated that the antimicrobial activity of chitosan against *E. coli* increased with increasing MW (molecular weight), up to a MW of 91.6 kDa; above that value, there was an inverse relationship between both.

10.3.1.2 Degree of Deacetylation

A clear relation between DD and antimicrobial activity has been established by researchers (Oh et al. 2001). The increase in the DD suggests that the number of amino groups on chitosan also increases (Morimoto and Shigemasa 1997; Fei Liu et al. 2001). The increase in the number of amino groups means the number of protonated groups of chitosan has increased. This leads to an increased chance of interaction between chitosan and negatively charged cell membranes of microorganisms (Sekiguchi et al. 1993).

The OD (optical density) versus the culture time for the chitosan with different DD against *E. coli* was studied by Fei Liu et al. (2001). The OD of the medium with five types of different DD chitosan was observed. The curves demonstrate that the OD decreased gradually with the percent of the DD of chitosan. So, the antibacterial activity was increased respectively. The results show that the heightening of DD causes increase of the -NH_2 concentration and then increases the -$NH_3{}^+$. Thus, more -$NH_3{}^+$ positive charges may have bound to the negatively charged bacterial surface to cause agglutination (Morimoto and Shigemasa 1997).

10.3.1.3 Concentration of ChNPs

The concentration of ChNPs used has a direct relationship with its antimicrobial activity. With the increase in the concentration, the antimicrobial activations of the chitosan increases. Chitosan concentrations higher than 200 ppm could almost kill all of the bacteria. It was reported that the chitosan concentration of 20 ppm itself had antimicrobial activity (Liu et al. 2006).

10.3.1.4 Particle Size and Zeta Potential of Nanoparticles

Particle size and zeta potential are the important properties that influence the antimicrobial activity of nanoparticles. A study on the influence of particle size and zeta potential on antimicrobial effect was done on *C. albicans*, *F. solani*, and *A. niger* by using nanoparticles with different particle sizes and zeta potential. Several factors affect the particle size of nanoparticles. These include the concentration and molecular weight of chitosan. The results of the study showed that the size of nanoparticles, especially HMW ChNPs, was greatly influenced by the concentration of chitosan which was added into a constant amount of TPP (tripolyphosphate).

A linear relationship was also observed where an increase in concentration would increase particle size. A similar relationship was also observed with the molecular weight of chitosan in which the effect on particle size was also very prominent. These linear relationships enable easy manipulation of nanoparticle size for application in different fields (Gan et al. 2005; Yien et al. 2012). The smaller the size of the nanoparticles the larger the surface area in contact with the bacterial cells and hence, the greater the interaction with the cells (Panáček et al. 2006).

10.3.2 Extrinsic Factors

10.3.2.1 pH

Lower pH increases the antimicrobial activity of ChNPs (Sudarshan et al. 1992). Inhibition of five foodborne pathogens – *S. aureus*, *E. coli*, *Yersinia enterocolitica*, *Listeria monocytogenes*, and *Salmonella typhimurium* – by chitosan at different pH were evaluated. It was found that pH 5.5 was the most effective. This suggests that the antimicrobial activity comes from the cationic nature of chitosan (Wang 1992).

10.3.2.2 Temperature

Higher temperature has been shown to enhance antimicrobial activity. Reports on the effect of temperature on the antibacterial activity of chitosan against *E. coli* has been seen (Tsai and Su 1999). The cell suspension containing 150 ppm chitosan incubated at 4, 15, 25, and 37 °C for various time intervals and the surviving cells were counted. The antibacterial activity was found to be directly proportional to the temperature. At the temperatures of 25 and 37 °C, the *E. coli* cells were completely killed within 0.5 and 1 hours, respectively. However, at lower temperatures (4 and 15 °C) the number of *E. coli* declined within the first five hours and then stabilized. Hence, it could be concluded that the reduced antimicrobial activity resulted from the decreased rate of interaction between chitosan and cells at a lower temperature.

10.3.2.3 Time

The antibacterial activity of fungal chitosan against two foodborne pathogens, *S. typhi* and *S. aureus,* was analyzed using tetrazolium/formazan test and the relationship with time was noted. The absorbance of formed formazan color at 480 nm revealed that, after the first hour of incubation, the absorbance was 1.82 for *S. aureus* and 1.79 for *S. typhi*. The absorbance continued to decrease down to zero after four hours for *S. typhi*, and five hours for *S. aureus*, from the incubation time with fungal chitosan (Moussa et al. 2013).

10.3.3 Microbial Factors

10.3.3.1 Species of Microorganism

Even though chitosan has a broad spectrum of antimicrobial activity, the inhibitory efficiency of chitosan toward various microorganisms is different. While some studies have reported that chitosan has stronger antibacterial activity against Gram-negative bacteria (No et al. 2002; Chung et al. 2004), some reports showed that Gram-positive bacteria were more susceptible (Zhong et al. 2008). Generally, bacteria appear to be less sensitive to the antimicrobial activity of chitosan than fungi (Roller and Covill 1999).

10.3.3.2 Cell Age

The age of the culture has a great influence on the antimicrobial activity of chitosan for any given microbial species. The age of a bacterial culture affected its susceptibility to chitosan, with cells in the late exponential phase being most sensitive to chitosan (Tsai and Su 1999). A related result was reported by Chen and Chou (2005) where *S. aureus* in its late exponential phase showed more susceptibility in comparison with cells in mid-exponential phase. The difference in cell surface electron negativity varies with the growth phase of microorganism causing the difference in susceptibility of cells toward chitosan (Yang et al. 2007).

The above-mentioned factors are the most important factors affecting the antimicrobial activity of ChNPs. In addition, there are other factors like chelating capacity of chitosan (Kurita 1998; Rabea et al. 2003), polycationic structure of chitosan (Young and Kauss 1983; Wang et al. 2004; Kong et al. 2008; Takahashi et al. 2008), hydrophilic/hydrophobic characteristics of chitosan (Dutta et al. 2004; Ignatova et al. 2006), the physical state of chitosan whether soluble or in solid-state (Qi et al. 2004; Chung et al. 2005; Xie et al. 2007; Sadeghi et al. 2008), ionic strength (Chung et al. 2003; Xing et al. 2009), etc. Table 10.1 shows a brief summary of different factors affecting the antimicrobial activity of ChNPs.

Table 10.1 Factors affecting antimicrobial activity of chitosan.

Factors	Condition	Effect on antimicrobial property	References
Molecular weight	Below 10 kDa above 7 kDa	Permeability increases at low MW causing leakage of intracellular substances and subsequent cell death	Uchida et al. (1989)
Degree of deacetylation	High DD	Increase in number of protonated groups due to increase in amino groups at high DD	Sudarshan et al. (1992) and Sekiguchi et al. (1993)
Concentration	High concentration	Interaction between chitosan and cells are more due to presence of more amount of chitosan	Tsai and Su (1999)
Particle size	Small size	As size decreases, surface area of chitosan in contact with microorganism increases	No et al. (2002)
Zeta potential	High zeta potential	Increase in zeta potential of nanoparticles has a positive effect on activity	Qi et al. (2004)
Type	Nanoformulations of chitosan	Synergistic effect of both chitosan and participating component	Zhang et al. (2007)
pH	Acidic pH	Hurdle effect of inflicting acid stress on target organism	Kong et al. (2008)
Temperature	High temperature (37 °C)	Decreased rate of interaction between chitosan and microorganism at low temperature	Ali et al. (2011)
Reaction time	More time	Rate of interaction increases with reaction time	Moussa et al. (2013)
Species of microorganism	Fungi > Gram-negative bacteria > Gram-positive bacteria	Chitosan could act directly on cell wall of fungus while action on Gram-positive and Gram-negative bacteria are more complicated due to cell surface characteristics	Roller and Covill (1999), No et al. (2002), and Zhong et al. (2008)
Cell age	Late exponential phase	Electronegativity of cell surface vary with growth phase	Tsai and Su (1999)

10.4 Mode of Action of ChNPs

Regardless of its high effectiveness as an antimicrobial agent, the precise mechanism of action of ChNPs against microorganisms remains vaguely defined. In this context, it would be appropriate to make a brief review of the established modes of antimicrobial action of ChNPs. This aspect can be divided into two: part of the active component, chitosan, and part of the microorganism.

10.4.1 Part of Active Component: Chitosan

Though the exact mechanism of action of chitosan is still not clear, six main mechanisms of actions that act independently have been proposed. They are:

i) interactions between the positively charged chitosan molecules and negatively charged moieties on the microbial cell outer membranes leading to changes in the cell membrane structure and permeability. It results in the leakage of proteinaceous and other intracellular constituents, thereby challenging the biochemical and physiological competency of the bacteria, leading to loss of replicative ability and eventual death;

ii) chitosan acts as a chelating agent that selectively binds to trace metals and subsequently inhibits the production of toxins and microbial growth;
iii) chitosan activates several defense processes in the host tissue;
iv) LMW chitosan penetrates the cytosol of the microorganisms and binds with DNA, resulting in the interference with the synthesis of mRNA and proteins;
v) chitosan on the surface of the cell can form an impermeable polymeric layer which alters the cell permeability and prevents nutrients from entering the cell; and
vi) as the ability of chitosan to adsorb the electronegative substances in the cell and flocculate them, it disturbs the physiological activities of the microorganism leading to their death (Badawy and Rabea 2011).

Electron microscopical examination of ChNP-treated microorganisms proposes that it could cause cell membrane damage in bacteria and fungi. Transmission electron microscopy (TEM) image of *E. coli* and *S. aureus* treated with LMW and HMW for two hours showed evidence of bacterial membrane damage (Sarwar et al. 2014). In spite of the difference in cell membrane composition and constitution, both Gram-negative and Gram-positive bacterial cell membrane has been disrupted and compromised by ChNPs. Liu et al. (2012) reported that chitosan was able to diffuse into fungal cells and alter cell membrane integrity, thus inducing leakage of intracellular material. The scanning electron microscopy (SEM) images revealed that *Rhizactonia solani* with loose and coarse mycelium became dense and smooth in the presence of chitosan. The TEM image showed abnormalities and cellular disorganization in the presence of chitosan. Membrane permeability assays have divulged that chitosan has the potential to damage and permeate through cell membranes (Chung et al. 2004).

Chitosan can bind to metals by chelation. The amine groups in chitosan are responsible for the uptake of metal ions. This mechanism is more efficient at high pH as unprotonated amine groups are present and donate electrons to metal ions. The chitosan forms a complex with metal ions surrounding bacteria and blocks nutrient flow leading to cell death (Cuero et al. 1991; Roller and Covill 1999; Lim and Hudson 2003). Chitosan was found to stimulate the activity of chitin deacetylase which is involved in chitosan biosynthesis in *Rhizopus stolonifer*. This process disturbs the biosynthesis of chitin and makes cell walls more viscoelastic (El-Ghaouth et al. 1992). Treatment with chitosan induces reactive oxygen species (ROS) production and cell death in young protonemal tissues and gametophores of moss *Physcomitrella patens* infected with fungi *Botrytis cinera* (Ponce De León and Montesano 2013).

LMW chitosan could penetrate cells of microorganism and bind to DNA and interfere with transcription and translation. The validity of this mechanism is still controversial, as experiments with both supporting and rejecting this hypothesis could be found. In research using radio-labeled chitosan applied to fungal cells, the ability of chitosan to interact with DNA was suggested. The glucosamine residues of radio-labeled chitosan as recovered from nuclei of the cell (Hadwiger et al. 1986). But when HMW and LMW chitosan were treated with *E. coli*, it was found that HMW chitosan that stacked the cell wall inhibited cell growth, while, LMW chitosan that penetrated through without stacking the cell wall accelerated cell growth (Tokura et al. 1996). This is supporting evidence for the ability of chitosan to inhibit cell growth by blocking nutrient flow-through cell walls.

Raafat et al. (2008) carried out a transcriptional response pattern to detect changes in *S. aureus* treated with chitosan. Chitosan treatment was found to reduce bacterial growth rate as there was downregulation in macromolecular biosynthesis as well as the metabolism of carbohydrates, amino acids, nucleotides, and nucleic acids.

10.4.2 Part of Microorganisms

The antibacterial activity of chitosan is in close coordination with the cell surface characteristics of microorganisms. The antimicrobial activity of chitosan follows a sequence of elementary events in the lethal action: (i) adsorption of chitosan onto the cell surface; (ii) diffusion through the cell wall; (iii) adsorption onto the cytoplasmic membrane; (iv) disruption of the cytoplasmic membrane; (v) leakage of the cytoplasmic constituents; and (vi) death of the cell (Kong et al. 2010).

A work on antimicrobial activity of ChNPs on different bacteria showed that chitosan was more active against the Gram-negative bacterium *P. aeruginosa*. This is due to the influx of smaller-sized nanoparticles into the cell wall of Gram-negative bacteria which consists of an outer membrane and single peptidoglycan layer, while the Gram-positive cell wall has many layers of peptidoglycan (Usman et al. 2013). The bacterial membrane serves as a structural component, which may become compromised by treatment with biocides like chitosan. Therefore, the measure of the release of intracellular components is a good indicator of membrane integrity. Small ions such as potassium and phosphate tend to leach out first, followed by large molecules such as DNA, RNA, and other materials. Since these nucleotides have strong UV absorption at 260 nm, they are described as "260 nm absorbing materials." The detection of 260 nm absorbing materials can be easily done using a UV-Vis spectrophotometer (Chen and Cooper 2002). Sarwar et al. (2014) reported that CS–TPP NPs were capable of damaging the cell membranes of *E. coli* and *S. aureus*. The absorption at 260 nm increased for up to 80 minutes and then decreased gradually.

The most probable target of chitosan interaction with cell membranes is phospholipids and proteins. There are mainly three modes of interaction between polycationic chitosan and negatively charged liposomes: (i) simple surface binding due to opposite charges; (ii) charge interaction in addition to penetration of polymer causing alteration in membrane permeability; and (iii) formation of polymer-lipid micelles thereby completely disrupting the membrane (Kong et al. 2010). The activity of ChNPs against fungi is mainly by disruption of the cell wall and cell membrane. Guerrero et al. (2009) reported that chitosan inhibits spore germination and mycelia growth in fungus by penetrating the fungal cell and causing lysis and cell death. They also found that different cell types (conidia, germ tubes, and vegetative hyphae) exhibited differential sensitivity to chitosan, with ungerminated conidia being the most sensitive.

10.5 Conclusion

ChNPs research is a growing field of nanotechnology. The multifaceted applicability of ChNPs has made it a popular choice in many fields. The applicability of ChNPs extends over a wide range, including biomedical applications. Another reason for the widespread use of ChNPs is the resource of natural products. Owing to its natural origin, ChNPs has an inherent instability. The characteristics of ChNPs also change drastically depending on a variety of factors like the process of synthesis, the concentration of chitosan used, size of the nanoparticle formed, etc. These problems make it almost impossible to compare the results of different experiments and establish a relationship between the physiological behavior of ChNPs and its properties. Nevertheless, it is possible to deduce a general recommendation in connection with the properties of ChNPs for a certain application from the available literature.

The antimicrobial property of chitosan has received greater recognition in recent years. The worldwide emergence of antibiotic-resistant microbes has elicited the need for a safe alternative. According to a recent statistic, there is not a single identified microorganism that is not resistant to at least one antibiotic. ChNPs have proved to be very effective and safe antimicrobial agents with a wide spectrum of activity against bacteria, fungi, and even viruses. The antimicrobial activity of ChNPs depends on various factors like molecular weight, DD, concentration, size, and zeta potential of NP, pH, temperature, reaction time, species of microorganism, and cell age. These factors are not fully explained due to a lack of knowledge on the exact mechanism of the antimicrobial activity of ChNPs. There is also a lack of standardized protocol for elucidating the mechanism of action of ChNPs. This makes it difficult to conclude the exact relation between the antimicrobial activity and its proposed mechanism of action. In this regard, a standard set of protocols is required to be fixed or an existing method has to be fine-tuned, like the Clinical and Laboratory Standards Institute (CLSI) methods for dilution antimicrobial susceptibility test from bacteria that grow aerobically.

ChNPs, to be accepted as an antimicrobial agent for commercial application, must have a clear correlation with its mode of activity. The future works have to be channeled toward clarifying the molecular details and their relevance to the antimicrobial activity of chitosan. Furthermore, the collaboration and participation of various research organizations, industry, and government regulatory agencies need to be established, as this association is the key to the success of antimicrobial mechanisms.

References

Abaza, A., Hegazy, E.-A., Mahmoud, G.-A., and Elsheikh, B. (2018). Characterization and antitumor activity of chitosan/poly (vinyl alcohol) blend doped with gold and silver nanoparticles in treatment of prostatic cancer model. *Journal of Pharmacy and Pharmacology* 6: 659–673.

Abdelhady, M. (2012). Preparation and characterization of chitosan/zinc oxide nanoparticles for imparting antimicrobial and UV protection to cotton fabric. *International Journal of Carbohydrate Chemistry* 2012: 1–6.

Abdou, E.S., Osheba, A., and Sorour, M. (2012). Effect of chitosan and chitosan-nanoparticles as active coating on microbiological characteristics of fish fingers. *International Journal of Applied Science and Technology* 2 (7): 158–169.

Ali, S.W., Rajendran, S., and Joshi, M. (2011). Synthesis and characterization of chitosan and silver loaded chitosan nanoparticles for bioactive polyester. *Carbohydrate Polymers* 83: 438–446.

Anitha, A., Rani, V.D., Krishna, R. et al. (2009). Synthesis, characterization, cytotoxicity and antibacterial studies of chitosan, O-carboxymethyl and N, O-carboxymethyl chitosan nanoparticles. *Carbohydrate Polymers* 78: 672–677.

Aranaz, I., Mengíbar, M., Harris, R. et al. (2009). Functional characterization of chitin and chitosan. *Current Chemical Biology* 3: 203–230.

Badawy, M.E. and Rabea, E.I. (2009). Potential of the biopolymer chitosan with different molecular weights to control postharvest gray mold of tomato fruit. *Postharvest Biology and Technology* 51: 110–117.

Badawy, M.E. and Rabea, E.I. (2011). A biopolymer chitosan and its derivatives as promising antimicrobial agents against plant pathogens and their applications in crop protection. *International Journal of Carbohydrate Chemistry* 2011: 1–30.

Baxter, A., Dillon, M., Anthony Taylor, K.D., and Roberts, G.A.F. (1992). Improved method for i.r. determination of the degree of *N*-acetylation of chitosan. *International Journal of Biological Macromolecules* 14: 166–169.

Bilal, M., Zhao, Y., Rasheed, T. et al. (2019). Biogenic nanoparticle-chitosan conjugates with antimicrobial, antibiofilm, and anticancer potentialities: development and characterization. *International Journal of Environmental Research and Public Health* 16: 598.

Brunel, F., Véron, L., David, L. et al. (2008). A novel synthesis of chitosan nanoparticles in reverse emulsion. *Langmuir* 24: 11370–11377.

Calvo, P., Remuñán-López, C., Vila-Jato, J.L., and Alonso, M.J. (1997). Novel hydrophilic chitosan-polyethylene oxide nanoparticles as protein carriers. *Journal of Applied Polymer Science* 63: 125–132.

Chen, Y.L. and Chou, C.C. (2005). Factors affecting the susceptibility of *Staphylococcus aureus* CCRC 12657 to water soluble lactose chitosan derivative. *Food Microbiology* 22: 29–35.

Chen, C.Z. and Cooper, S.L. (2002). Interactions between dendrimer biocides and bacterial membranes. *Biomaterials* 23: 3359–3368.

Chen, S., Wu, G., and Zeng, H. (2005). Preparation of high antimicrobial activity thiourea chitosan–Ag^+ complex. *Carbohydrate Polymers* 60: 33–38.

Chowdappa, P., Gowda, S., Chethana, C., and Madhura, S. (2014). Antifungal activity of chitosan-silver nanoparticle composite against *Colletotrichum gloeosporioides* associated with mango anthracnose. *African Journal of Microbiology Research* 8: 1803–1812.

Chung, Y.C., Wang, H.L., Chen, Y.M., and Li, S.L. (2003). Effect of abiotic factors on the antibacterial activity of chitosan against waterborne pathogens. *Bioresource Technology* 88: 179–184.

Chung, Y.-C., Su, Y.-P., Chen, C.-C. et al. (2004). Relationship between antibacterial activity of chitosan and surface characteristics of cell wall. *Acta Pharmacologica Sinica* 25: 932–936.

Chung, Y.-C., Kuo, C.-L., and Chen, C.-C. (2005). Preparation and important functional properties of water-soluble chitosan produced through Maillard reaction. *Bioresource Technology* 96: 1473–1482.

Cuero, R., Duffus, E., Osuji, G., and Pettit, R. (1991). Aflatoxin control in preharvest maize: effects of chitosan and two microbial agents. *The Journal of Agricultural Science* 117: 165–169.

De Paz, L.E.C., Resin, A., Howard, K.A. et al. (2011). Antimicrobial effect of chitosan nanoparticles on *Streptococcus mutans* biofilms. *Applied and Environmental Microbiology* 77: 3892–3895.

De, T.K., Ghosh, P.K., Maitra, A. and Sahoo, S.K. 1999, Process for the preparation of highly monodispersed polymeric hydrophilic nanoparticles. Google Patents.

Divya, K., Rebello, S., and Jisha, M.S. (2014). A simple and effective method for extraction of high purity chitosan from shrimp shellwaste. *International Journal of Environmental Engineering* 1 (4): 86–90.

Divya, K., Kurian, L.C., Vijayan, S., and Jisha, M.S. (2016). Green synthesis of silver nanoparticles by *Escherichia coli*: analysis of antibacterial activity. *Journal of Water and Environmental Nanotechnology* 1 (1): 63–74. https://doi.org/10.7508/jwent.2016.01.008.

Divya, K., Vijayan, S., and Jisha, M.S. (2017). Antimicrobial properties of chitosan nanoparticles: mode of action and factors affecting activity. *Fibers and Polymers* 18 (2): 221–230.

Divya, K., Vijayan, S., Nair, S.J., and Jisha, M.S. (2018). Optimization of chitosan nanoparticle synthesis and its potential application as germination elicitor of *Oryza sativa* L. *International Journal of Biological Macromolecules* 124: 1053–1059.

Domek, M.J., Lechevallier, M.W., Cameron, S.C., and Mcfeters, G.A. (1984). Evidence for the role of copper in the injury process of coliform bacteria in drinking water. *Applied and Environmental Microbiology* 48: 289–293.

Du, W.-L., Niu, S.-S., Xu, Y.-L. et al. (2009). Antibacterial activity of chitosan tripolyphosphate nanoparticles loaded with various metal ions. *Carbohydrate Polymers* 75: 385–389.

Dutta, P.K., Dutta, J., and Tripathi, V. (2004). Chitin and chitosan: chemistry, properties and applications. *Journal of Scientific and Industrial Research* 63: 20–31.

Eid, M.-M., El-Hallouty, S.-M., El-Manawaty, M. et al. (2018). Preparation conditions effect on the physico-chemical properties of magnetic–plasmonic core–shell nanoparticles functionalized with chitosan: green route. *Nano-Structures and Nano-Objects* 16: 215–223.

Einbu, A. and Vårum, K.M. (2008). Characterization of chitin and its hydrolysis to GlcNAc and GlcN. *Biomacromolecules* 9: 1870–1875.

El Ghaouth, A., Arul, J., Grenier, J. et al. (1992). Effect of chitosan and other polyions on chitin deacetylase in Rhizopus stolonifer. *Experimental Mycology* 16 (3): 173–177.

El-Shabouri, M.H. (2002). Positively charged nanoparticles for improving the oral bioavailability of cyclosporin-A. *International Journal of Pharmaceutics* 249: 101–108.

Erbacher, P., Zou, S., Bettinger, T. et al. (1998). Chitosan-based vector/DNA complexes for gene delivery: biophysical characteristics and transfection ability. *Pharmaceutical Research* 15: 1332–1339.

Fei Liu, X., Lin Guan, Y., Zhi Yang, D. et al. (2001). Antibacterial action of chitosan and carboxymethylated chitosan. *Journal of Applied Polymer Science* 79: 1324–1335.

Feng, C., Wang, Z., Jiang, C. et al. (2013). Chitosan/O-carboxymethyl chitosan nanoparticles for efficient and safe oral anticancer drug delivery: in vitro and in vivo evaluation. *International Journal of Pharmaceutics* 457: 158–167.

Fouad, D.R.G. (2008). *Chitosan as an Antimicrobial Compound: Modes of Action and Resistance Mechanisms*. Mathematisch-Naturwissenschaftliche Fakultät, Universität Bonn.

Gan, Q., Wang, T., Cochrane, C., and Mccarron, P. (2005). Modulation of surface charge, particle size and morphological properties of chitosan–TPP nanoparticles intended for gene delivery. *Colloids and Surfaces B: Biointerfaces* 44: 65–73.

Grenha, A. (2010). Chitosan nanoparticles: a survey of preparation methods. *Journal of Drug Targeting* 20: 291–300.

Gu, C., Sun, B., Wu, W. et al. (2007). Synthesis, characterization of copper-loaded carboxymethyl-chitosan nanoparticles with effective antibacterial activity. *Macromolecular Symposia*, Wiley Online Library 254: 160–166.

Hadwiger, L., Kendra, D., Fristensky, B., and Wagoner, W. (1986). Chitosan both activates genes in plants and inhibits RNA synthesis in fungi. In: *Chitin in Nature and Technology* (eds. R. Muzzarelli, C. Jeuniaux and G.W. Gooday), 209–214. Springer.

Honary, S., Ghajar, K., Khazaeli, P., and Shalchian, P. (2011). Preparation, characterization and antibacterial properties of silver-chitosan nanocomposites using different molecular weight grades of chitosan. *Tropical Journal of Pharmaceutical Research* 10: 69–74.

Huang, L., Cheng, X., Liu, C. et al. (2009). Preparation, characterization, and antibacterial activity of oleic acid-grafted chitosan oligosaccharide nanoparticles. *Frontiers of Biology in China* 4: 321–327.

Huang, Y., Li, R., Chen, J., and Chen, J. (2016). Recent advances in chitosan-based nano particulate pulmonary drug delivery. *Carbohydrate Polymers* 138: 114–122.

Hussein-Al-Ali, S.H., El Zowalaty, M.E., Hussein, M.Z. et al. (2014). Synthesis, characterization, controlled release, and antibacterial studies of a novel streptomycin chitosan magnetic nanoantibiotic. *International Journal of Nanomedicine* 9: 549.

Ignatova, M., Starbova, K., Markova, N. et al. (2006). Electrospun nano-fibre mats with antibacterial properties from quaternised chitosan and poly (vinyl alcohol). *Carbohydrate Research* 341: 2098–2107.

Jemal, A., Siegel, R., Ward, E. et al. (2008). Cancer statistics 2008. *CA: a Cancer Journal for Clinicians* 58: 71–96.

Jeon, S.J., Oh, M., Yeo, W.S. et al. (2014). Underlying mechanism of antimicrobial activity of chitosan microparticles and implications for the treatment of infectious diseases. *PLoS One* 9: e92723.

Kasaai, M.R. (2009). Various methods for determination of the degree of N-acetylation of chitin and chitosan: a review. *Journal of Agricultural and Food Chemistry* 57: 1667–1676.

Katas, H., Moden, N.Z., Lim, C.S. et al. (2018). Biosynthesis and potential applications of silver and gold nanoparticles and their chitosan-based nanocomposites in nanomedicine. *Journal of Nanotechnology* 2018: 1–13.

Kaur, P., Choudhary, A., and Thakur, R. (2013). Synthesis of chitosan-silver nanocomposites and their antibacterial activity. *International Journal of Scientific & Engineering Research* 4 (4): 869–872.

Kavitha, A., Prabu, H.G., and Babu, S.A. (2013). Synthesis of low-cost iron oxide: chitosan nanocomposite for antibacterial activity. *International Journal of Polymeric Materials* 62: 45–49.
Kenawy, E.R., Worley, S., and Broughton, R. (2007). The chemistry and applications of antimicrobial polymers: a state-of-the-art review. *Biomacromolecules* 8: 1359–1384.
Kong, M., Chen, X.G., Xue, Y.P. et al. (2008). Preparation and antibacterial activity of chitosan microshperes in a solid dispersing system. *Frontiers of Materials Science in China* 2: 214–220.
Kong, M., Chen, X.G., Xing, K., and Park, H.J. (2010). Antimicrobial properties of chitosan and mode of action: a state of the art review. *International Journal of Food Microbiology* 144: 51–63.
Koukaras, E.N., Papadimitriou, S.A., Bikiaris, D.N., and Froudakis, G.E. (2012). Insight on the formation of chitosan nanoparticles through ionotropic gelation with tripolyphosphate. *Molecular Pharmaceutics* 9: 2856–2862.
Kumar, M.N.V.R. (1999). Chitin and chitosan fibres: a review. *Bulletin of Materials Science* 22: 905–915.
Kumirska, J., Czerwicka, M.G., Kaczyński, Z. et al. (2010). Application of spectroscopic methods for structural analysis of chitin and chitosan. *Marine Drugs* 8: 1567–1636.
Kumirska, J., Weinhold, M.X., Thaming, J., and Stepnowski, P. (2011). Biomedical activity of chitin/chitosan based materials—influence of physicochemical properties apart from molecular weight and degree of *N*-acetylation. *Polymers* 3: 1875–1901.
Kurita, K. (1998). Chemistry and application of chitin and chitosan. *Polymer Degradation and Stability* 59: 117–120.
Liang, J., Yan, H., Puligundla, P. et al. (2017). Applications of chitosan nanoparticles to enhance absorption and bioavailability of tea polyphenols: a review. *Food Hydrocolloids* 69: 286–292.
Lim, S.-H. and Hudson, S.M. (2003). Review of chitosan and its derivatives as antimicrobial agents and their uses as textile chemicals. *Journal of Macromolecular Science, Part C: Polymer Reviews* 43: 223–269.
Limam, Z., Selmi, S., Sadok, S. et al. (2011). Extraction and characterization of chitin and chitosan from crustacean by-products: Biological and physicochemical properties. *African Journal of Biotechnology* 10 (4): 640–647.
Liu, N., Chen, X.-G., Park, H.-J. et al. (2006). Effect of MW and concentration of chitosan on antibacterial activity of *Escherichia coli*. *Carbohydrate Polymers* 64: 60–65.
Liu, S., Sun, J., Yu, L. et al. (2012). Extraction and characterization of chitin from the beetle *Holotrichia parallela* motschulsky. *Molecules* 17: 4604–4611.
Ma, Y., Liu, P., Si, C., and Liu, Z. (2010). Chitosan nanoparticles: preparation and application in antibacterial paper. *Journal of Macromolecular Science, Part B Physics* 49: 994–1001.
Mitra, S., Gaur, U., Ghosh, P.C., and Maitra, A.N. (2001). Tumour targeted delivery of encapsulated dextran-doxorubicin conjugate using chitosan nanoparticles as carrier. *Journal of Controlled Release* 74: 317–323.
Mori, Y., Ono, T., Miyahira, Y. et al. (2013). Antiviral activity of silver nanoparticle/chitosan composites against H1N1 influenza A virus. *Nanoscale Research Letters* 8: 1–6.
Morimoto, M. and Shigemasa, Y. (1997). Characterization and bioactivities of chitin and chitosan regulated their degree of deacetylation. *Kobunshi Ronbunshu* 54: 621–631.
Moussa, S.H., Tayel, A.A., Al-Hassan, A.A., and Farouk, A. (2013). Tetrazolium/formazan test as an efficient method to determine fungal chitosan antimicrobial activity. *Journal of Mycology* 2013: 1–8.
Muzzarelli, R.A.A., Jeuniaux, C., and Gooday, G.W. (1986). *Chitin in Nature and Technology*. New York: Plenum Publishing Corporation.
Niwa, T., Takeuchi, H., Hino, T. et al. (1993). Preparations of biodegradable nanospheres of water-soluble and insoluble drugs with D, L-lactide/glycolide copolymer by a novel spontaneous emulsification solvent diffusion method, and the drug release behavior. *Journal of Controlled Release* 25: 89–98.

No, H.K., Park, N.Y., Lee, S.H., and Meyers, S.P. (2002). Antibacterial activity of chitosans and chitosan oligomers with different molecular weights. *International Journal of Food Microbiology* 74: 65–72.

Odds, F. (2003). Synergy, antagonism, and what the chequerboard puts between them. *Journal of Antimicrobial Chemotherapy* 52: 1–1.

Oh, H.I., Kim, Y.J., Chang, E.J., and Kim, J.Y. (2001). Antimicrobial characteristics of chitosans against food spoilage microorganisms in liquid media and mayonnaise. *Bioscience, Biotechnology, and Biochemistry* 65: 2378–2383.

Ohya, Y., Shiratani, M., Kobayashi, H., and Ouchi, T. (1994). Release behavior of 5-fluorouracil from chitosan-gel nanospheres immobilizing 5-fluorouracil coated with polysaccharides and their cell specific cytotoxicity. *Journal of Macromolecular Science, Part A: Pure and Applied Chemistry* 31: 629–642.

Ottey, M.H., Vårum, K.M., and Smidsrød, O. (1996). Compositional heterogeneity of heterogeneously deacetylated chitosans. *Carbohydrate Polymers* 29: 17–24.

Palma-Guerrero, J., Huang, I.-C., Jansson, H.-B. et al. (2009). Chitosan permeabilizes the plasma membrane and kills cells of *Neurospora crassa* in an energy dependent manner. *Fungal Genetics and Biology* 46: 585–594.

Panáček, A., Kvitek, L., Prucek, R. et al. (2006). Silver colloid nanoparticles: synthesis, characterization, and their antibacterial activity. *The Journal of Physical Chemistry B* 110: 16248–16253.

Paulino, A.T., Simionato, J.I., Garcia, J.C., and Nozaki, J. (2006). Characterization of chitosan and chitin produced from silkworm crysalides. *Carbohydrate Polymers* 64: 98–103.

Pinto, R.J., Daina, S., Sadocco, P. et al. (2013). Antibacterial activity of nanocomposites of copper and cellulose. *BioMed Research International* 2013: 1–7.

Ponce De León, I. and Montesano, M. (2013). Activation of defense mechanisms against pathogens in mosses and flowering plants. *International Journal of Molecular Sciences* 14: 3178–3200.

Prema, P. and Thangapandiyan, S. (2013). In-vitro antibacterial activity of gold nanoparticles capped with polysaccharide stabilizing agents. *International Journal of Pharmacy and Pharmaceutical Sciences* 5: 310–314.

Qi, L., Xu, Z., Jiang, X. et al. (2004). Preparation and antibacterial activity of chitosan nanoparticles. *Carbohydrate Research* 339: 2693–2700.

Raafat, D., Von Bargen, K., Haas, A., and Sahl, H.-G. (2008). Insights into the mode of action of chitosan as an antibacterial compound. *Applied and Environmental Microbiology* 74: 3764–3773.

Rabea, E.I., Badawy, M.E.-T., Stevens, C.V. et al. (2003). Chitosan as antimicrobial agent: applications and mode of action. *Biomacromolecules* 4: 1457–1465.

Rastogi, H. and Jana, S. (2016). Evaluation of physicochemical properties and intestinal permeability of six dietary polyphenols in human intestinal colon adenocarcinoma Caco-2 cells. *European Journal of Drug Metabolism and Pharmacokinetics* 41: 33–43.

Rinaudo, M. (2006). Chitin and chitosan: properties and applications. *Progress in Polymer Science* 31: 603–632.

Rodrigues, S., Dionísio, M., López, C.R., and Grenha, A. (2012). Biocompatibility of chitosan carriers with application in drug delivery. *Journal of Functional Biomaterials* 3: 615–641.

Roller, S. and Covill, N. (1999). The antifungal properties of chitosan in laboratory media and apple juice. *International Journal of Food Microbiology* 47: 67–77.

Russo, E., Gaglianone, N., Baldassari, S. et al. (2014). Preparation, characterization and in vitro antiviral activity evaluation of foscarnet-chitosan nanoparticles. *Colloids and Surfaces B: Biointerfaces* 118: 117–125.

Sadeghi, A., Dorkoosh, F., Avadi, M. et al. (2008). Preparation, characterization and antibacterial activities of chitosan, N-trimethyl chitosan (TMC) and N-diethylmethyl chitosan (DEMC) nanoparticles loaded with insulin using both the ionotropic gelation and polyelectrolyte complexation methods. *International Journal of Pharmaceutics* 355: 299–306.

Sailaja, A., Amareshwar, P., and Chakravarty, P. (2011). Different techniques used for the preparation of nanoparticles using natural polymers and their application. *International Journal of Pharmacy and Pharmaceutical Sciences* 3: 45–50.

Sarwar, A., Katas, H., and Zin, N.M. (2014). Antibacterial effects of chitosan–tripolyphosphate nanoparticles: impact of particle size molecular weight. *Journal of Nanoparticle Research* 16: 1–14.

Sekiguchi, S., Miura, Y., Kaneko, H. et al. (1993). Molecular weight dependency of antimicrobial activity by chitosan oligomers. In: *Food Hydrocolloids* (eds. K. Nishinari and E. Doi), 71–76. Springer.

Sinha, V.R., Singla, A.K., Wadhawan, S. et al. (2004). Chitosan microspheres as a potential carrier for drugs. *International Journal of Pharmaceutics* 274: 1–33.

Sudarshan, N., Hoover, D., and Knorr, D. (1992). Antibacterial action of chitosan. *Food Biotechnology* 6: 257–272.

Surendiran, A., Sandhiya, S., Pradhan, S., and Adithan, C. (2009). Novel applications of nanotechnology in medicine. *The Indian Journal of Medical Research* 130: 689–701.

Swathi, V., Vidyavathi, M., Prasad, T., and Kumar, R.S. (2013). Design, characterization and evaluation of metallic nano biocomposites of neomycin. *Journal of Applied Solution Chemistry and Modeling* 2: 136–144.

Takahashi, T., Imai, M., Suzuki, I., and Sawai, J. (2008). Growth inhibitory effect on bacteria of chitosan membranes regulated with deacetylation degree. *Biochemical Engineering Journal* 40: 485–491.

Tan, H., Ma, R., Lin, C. et al. (2013). Quaternized chitosan as an antimicrobial agent: antimicrobial activity, mechanism of action and biomedical applications in orthopedics. *International Journal of Molecular Sciences* 14: 1854–1869.

Tao, W., Ziemer, K.S., and Gill, H.S. (2013). Gold nanoparticle-M2e conjugate coformulated with CpG induces protective immunity against influenza a virus. *Nanomedicine* 9: 237–251.

Tarafdar, A. and Biswas, G. (2013). Extraction of Chitosan from Prawn Shell Wastes and Examination of its Viable Commercial Applications. *International Journal on Theoretical and Applied Research in Mechanical Engineering* 2: 2319–3182.

Tikhonov, V.E., Stepnova, E.A., Babak, V.G. et al. (2006). Bactericidal and antifungal activities of a low molecular weight chitosan and its N-/2 (3)-(dodec-2-enyl) succinoyl/-derivatives. *Carbohydrate Polymers* 64: 66–72.

Tiyaboonchai, W. (2013). Chitosan nanoparticles: a promising system for drug delivery. *Naresuan University Journal: Science and Technology* 11: 51–66.

Tokura, S., Ueno, K., Miyazaki, S., and Nishi, N. (1996). Molecular weight dependent antimicrobial activity by chitosan. In: *New Macromolecular Architecture and Functions* (eds. M. Kamachi and A. Nakamura), 199–207. Springer.

Tsai, G.J. and Su, W.H. (1999). Antibacterial activity of shrimp chitosan against *Escherichia coli*. *Journal of Food Protection* 62: 239–243.

Tsai, G.J., Tsai, M.T., Lee, J.M., and Zhong, M.Z. (2006). Effects of chitosan and a low-molecular-weight chitosan on *Bacillus cereus* and application in the preservation of cooked rice. *Journal of Food Protection* 69: 2168–2175.

Uchida, Y. Izume, M., Ohtakara A. (1989). Preparation of chitosan oligomers with purified chitosanase and its application. In: *Proceedings of 4th Internationl Conference on Chitin/Chitosan*. Essex, UK: Elsevier Applied Science, pp. 372–382.

US Food and Drug Administration (US-FDA) 2013,Shrimp derived chitosan GRAS notification [Online]. http://www.fda.gov/food/ingredients_packaging_labelling/GRAS/notice/inventory/ucm34.

Usman, M.S., El Zowalaty, M.E., Shameli, K. et al. (2013). Synthesis, characterization, and antimicrobial properties of copper nanoparticles. *International Journal of Nanomedicine* 8: 4467.

Vijayan, S., Divya, K., and Jisha, M.S. (2019). In vitro anticancer evaluation of chitosan/biogenic silver nanoparticle conjugate on Si Ha and MDA MB cell lines. *Applied Nanoscience*: 1–14.

Wang, G.H. (1992). Inhibition and inactivation of five species of foodborne pathogens by chitosan. *Journal of Food Protection* 55: 916–919.

Wang, X. and Xing, B. (2007). Importance of structural makeup of biopolymers for organic contaminant sorption. *Environmental Science and Technology* 41: 3559–3565.

Wang, X., Du, Y., and Liu, H. (2004). Preparation, characterization and antimicrobial activity of chitosan–Zn complex. *Carbohydrate Polymers* 56: 21–26.

Wei, D., Sun, W., Qian, W. et al. (2009). The synthesis of chitosan-based silver nanoparticles and their antibacterial activity. *Carbohydrate Research* 344: 2375–2382.

Xie, Y., Liu, X., and Chen, Q. (2007). Synthesis and characterization of water-soluble chitosan derivate and its antibacterial activity. *Carbohydrate Polymers* 69: 142–147.

Xing, K., Chen, X.G., Liu, C.S. et al. (2009). Oleoyl-chitosan nanoparticles inhibits *Escherichia coli* and *Staphylococcus aureus* by damaging the cell membrane and putative binding to extracellular or intracellular targets. *International Journal of Food Microbiology* 132: 127–133.

Xing, K., Shen, X., Zhu, X., et al. (2016). Synthesis and in vitro antifungal efficacy of oleoyl-chitosan nanoparticles against plant pathogenic fungi. International *Journal of Biological Macromolecules* 82: 830–836.

Yang, T.C., Li, C.F., and Chou, C.C. (2007). Cell age, suspending medium and metal ion influence the susceptibility of *Escherichia coli* O157: H7 to water-soluble maltose chitosan derivative. *International Journal of Food Microbiology* 113: 258–262.

Yang, R., Shim, W.S., Cui, F.D. et al. (2009). Enhanced electrostatic interaction between chitosan-modified PLGA nanoparticle and tumor. *International Journal of Pharmaceutics* 371: 142–147.

Yen, M.T., Yang, J.H., and Mau, J.L. (2009). Physicochemical characterization of chitin and chitosan from crab shells. *Carbohydrate Polymers* 75: 15–21.

Yi, Y., Wang, Y., and Liu, H. (2003). Preparation of new crosslinked chitosan with crown ether and their adsorption for silver ion for antibacterial activities. *Carbohydrate Polymers* 53: 425–430.

Yien, I.L., Mohamad, Z.N., Sarwar, A. et al. (2012). Antifungal activity of chitosan nanoparticles and correlation with their physical properties. *International Journal of Biomaterials* 2012: 1–9.

Young, D.H. and Kauss, H. (1983). Release of calcium from suspension-cultured Glycine max cells by chitosan, other polycations, and polyamines in relation to effects on membrane permeability. *Plant Physiology* 73: 698–702.

Yuan, Y. and Macquarrie, D. (2015). Microwave assisted extraction of sulfated polysaccharides (fucoidan) from *Ascophyllum nodosum* and its antioxidant activity. *Carbohydrate Polymers* 129: 101–107.

Zhang, L., Jiang, Y., Ding, Y. et al. (2007). Investigation into the antibacterial behaviour of suspensions of ZnO nanoparticles (ZnO nanofluids). *Journal of Nanoparticle Research* 9: 479–489.

Zhang, Y., Zhang, X., Ding, R. et al. (2011). Determination of the degree of deacetylation of chitosan by potentiometric titration preceded by enzymatic pretreatment. *Carbohydrate Polymers* 83: 813–817.

Zhao, K., Li, S., Li, W. et al. (2017). Quaternized chitosan nanoparticles loaded with the combined attenuated live vaccine against Newcastle disease and infectious bronchitis elicit immune response in chicken after intranasal administration. *Drug Delivery* 24: 1574–1586.

Zhong, Z., Xing, R., Liu, S. et al. (2008). Synthesis of acyl thiourea derivatives of chitosan and their antimicrobial activities *in vitro*. *Carbohydrate Research* 343: 566–570.

Zivanovic, S., Basurto, C., Chi, S. et al. (2004). Molecular weight of chitosan influences antimicrobial activity in oil-in-water emulsions. *Journal of Food Protection* 67: 952–959.

11

Sulfur Nanoparticles: Biosynthesis, Antibacterial Applications, and Their Mechanism of Action

Priti Paralikar and Mahendra Rai

Nanobiotechnology Laboratory, Department of Biotechnology, SGB Amravati University, Amravati, Maharashtra, India

11.1 Introduction

Despite increased knowledge of microbial pathogenesis, the application of modern therapeutics, mortality, and morbidity associated with microbial infection remains high. The reason is the evolution of resistant strains capable of inactivating the antibiotics, which can prevent them. It is predicted that the problem of multidrug-resistant microbes will cause a higher mortality rate than cancer by 2050 (www.theweek.co.uk/antibiotics/51908/antibiotics-resistance-could-cause-10m-deaths-a-year-by-2050). Multidrug resistance is a condition that occurs when microbes are resistant to more than one antibiotic. The reason behind this is the sequential use of the same antibiotics for years against particular diseases. The antibiotic resistance occurs because bacteria, fungi, viruses, or protozoa cannot be fully inhibited or killed by the antibiotic. They become resistant by adapting their structure and function in such a way that prevents them from being killed by the antibiotics (Fair and Tor 2014; Landecker and Blackman 2016; Bader et al. 2016; Ma and Hughes 2018). These resistant microbes are now able to survive, multiply, spread, and cause infectious diseases in the community even in the presence of antibiotics. The patients with drug-resistant bacterial infections are at high risk of dying from infection and may lead to longer and costly hospital treatment. The Centers for Disease Control and Prevention (CDC) estimated that 23 000 deaths are caused by drug-resistant bacteria in the United States alone in a year (Colomb-Cotinat et al. 2016; Lim et al. 2016; https://www.cdc.gov/drugresistance/threat-report-2013/index.html). If we start estimating a worldwide number of illnesses and deaths in patients then the numbers are likely higher.

Antibiotics play a crucial role in treating infectious diseases but unfortunately, available antibiotics lead to the development of resistance in microbes. The prime reason for developing resistance is an inappropriate use of antibiotics, even before primary diagnosis (Lee et al. 2013; Llor and Bjerrum 2014; MacGowan and Macnaughton 2017). The increasing prevalence of antimicrobial resistance is one of the greatest problems for healthcare systems worldwide (Davies and Davies 2010; Manjunath et al. 2011; Buzayan and El-Garbulli 2014; Ventola 2015; Frieri et al. 2017).

In this context, applications of nanotechnology are gaining importance to prevent the consequences of antibiotic resistance. Nanomaterials are one showing promising applications in the field of antimicrobial treatment therapies (Seil and Webster 2012; Zhu et al. 2014; Rai et al. 2018).

Nanobiotechnology in Diagnosis, Drug Delivery, and Treatment, First Edition. Edited by Mahendra Rai, Mehdi Razzaghi-Abyaneh, and Avinash P. Ingle.

By using nanoparticles, high antimicrobial activities can be achieved with low concentrations. In the present scenario, nanoparticles are called "a marvel of modern medical specialty." The nanoparticles were found to be ideal for the treatment of bacterial infections (Gao et al. 2014; Wang et al. 2017; Gugala and Turner 2018; Sajid et al. 2018). Huge research has been carried out on different nanoparticles, which possess antimicrobial ability.

Various types of organic and inorganic nanoparticles have been utilized as antibacterial agents. Among the different types of nanoparticles, silver is the most intensively investigated for its antibacterial activity against Gram-positive and Gram-negative bacteria. Moreover, silver nanoparticles found to be very effective against multidrug-resistant bacterial species (Rai et al. 2012; Kumari et al. 2017; Kale and Jagtap 2018). Antimicrobial nanoparticles recently being used have numerous modes of action. Nanoparticles target bacteria by affecting their genetic material and consequently their cell wall structure, metabolic pathways, and many other components that when disrupted could cause death (Nisar et al. 2019; Varier et al. 2019). The mechanism of antibacterial activity varies from nanoparticle to nanoparticle and is not yet fully understood (Wang et al. 2017). However, some proposed mechanisms relate to the physical structure of the nanoparticles (i.e. membrane-damaging abrasiveness of the nanoparticles), while others relate to the enhanced release of antibacterial metal ions from nanoparticle surfaces. The specific surface area of nanoparticles increases as the particle size decreases, allowing for greater material interaction with the surrounding environment which leads to efficient results (Rai et al. 2018).

Sulfur is the most versatile element used in various biomedical fields. Its versatility can attribute to its unique properties. Sulfur has gained a reputation since it is among the most useful medicines known to humans. It has a long history of use for a variety of dermatological disorders, as an ingredient in acne ointments, in antidandruff shampoos, and as an antidote for acute exposure to radioactive material (Keri and Shiman 2009). Sulfur aids in wound healing via keratin and it also has a history of folk usage as a remedy for skin rashes. The keratolytic activity of sulfur is due to H_2S (hydrogen sulfide) which forms on direct interaction between sulfur and keratinocytes (Lin et al. 1988). Sulfur can induce various histological changes, including hyperkeratosis, acanthosis and dilation of dermal vessels (Parcell 2002). It is also used alone or in combination with agents such as sodium sulfacetamide or salicylic acid and demonstrated efficacy for the treatment of many dermatological conditions such as acne, rosacea, seborrheic dermatitis, etc. (Gupta and Nicol 2004). Apart from medicinal applications, it has a wide range of applications in various other fields such as the production of sulfuric acid, pulp and paper industry, petrol refining, etc. It is a vital component in the production of dyes, fungicides, and different agrochemicals (Orton 1974; Griffith et al. 2015).

Sulfur has many potent applications when it is used in nanoform. Sulfur nanoparticles (SNPs) are used in the preparation of nanocomposite for lithium batteries (Yong et al. 2007; Chen et al. 2013), used for modification of carbon nanotubes (Barkauskas et al. 2007), sulfur nanowires are used for production of hybrid materials with useful properties, i.e. for gas sensor and catalytic properties (Santiago et al. 2006), and in pharmaceutical industries (Ilardi et al. 2014). The recent emergence of nanotechnology has provided a new therapeutic modality in SNPs for use in medicine. Current research in nanomaterials having good antimicrobial activity has opened a new approach in the pharmaceutical field. Sulfur is the oldest known antimicrobial agent, but little research has been carried out on SNPs as antimicrobial agents.

A potential new weapon to fight multidrug-resistant bacteria comes in a very small package, commonly referred to as nanoparticles. The use of nanoparticles in biomedical science has revealed new avenues that could bring lifesaving solutions and improve the quality of life. The nanoparticles were reported to have potent antimicrobial potential, and hence nanoparticles have been established as a promising approach to solve this problem. This chapter focuses on the biosynthesis of SNPs, their proposed mechanism of action, and the antibacterial application of SNPs.

11.2 Mechanisms of Antibiotic Resistance and Combination Therapy

As discussed earlier, it was noticed that bacteria can adapt and protect themselves against various antibiotics through the development of antibiotic-resistant mechanisms. So far scientists have identified four main types of mechanisms through which bacteria can acquire drug resistance. First, the interaction of the antibiotic with its specific biomolecular target can be prevented either by mutational alteration of the target protein (Lambert 2005), or by binding some other proteins to the target, thereby changing target conformation and impeding the association of the antibiotic (Connell et al. 2003; Nikaido 2009). Second, the antibiotic itself can be enzymatically modified or degraded (Wright 2005). Third, overexpression of bacterial drug efflux pumps combined with reduced porin expression causing decreased drug permeability was also found to be associated with antibiotic resistance (Kumar and Schweizer 2005; Fernández and Hancock 2012; Sun et al. 2014). Fourth, acquired resistant genes may cause bacteria to develop alternative pathways for their metabolic/growth requirements that bypass the reaction inhibited by the antibiotic (Tenover 2006).

These issues have become an obstacle to treat infections caused by resistant pathogens. Without impediment, a new development in present treatments and novel strategies are required to crack this crisis. To avoid this problem, the use of combination therapy (a combination of antibiotics and nanoparticles) for the treatment of such antibiotic-resistant infections is preferred. Combination therapy is utilized to broaden the antimicrobial effect of antibiotics, to prevent antibiotic resistance, and to minimize toxicity. The use of such synergistic combinations in antimicrobial therapy is frequently used for the treatment of diverse infections. Combining nanoparticles with antibiotics opens up the fascinating option to treat multidrug-resistant pathogens. Materials at the nanoscale range have unique physical properties that correspond to size range with biomolecule and cellular systems (Daniel and Astruc 2004; Khan et al. 2017). This feature makes them appealing materials for therapeutic applications.

There are three approaches for combination of antibiotics and nanoparticles: (i) to have synergistic effect, simultaneous administration of antibiotics and nanoparticles; (ii) the functionalization of biocompatible nanoparticles with antibiotics; and (iii) the functionalization of nanoparticles which have natural antimicrobial properties with antibiotics, in order to get potential effect (Morones-Ramirez et al. 2013).

11.3 Biosynthesis of Sulfur Nanoparticles (SNPs)

Nowadays, renewed interest has been generated in developing sustainable, eco-friendly, and green methods for the synthesis of nanoparticles using a biological route. The main advantages of synthesis of nanoparticles via biological approaches include easy, rapid synthesis, controlled toxicity, and high biocompatibility (Shukla and Iravani 2017). The bio-directed synthesis of nanoparticles has garnered profound interest in the secrets that were inspired by nature, which has led to the development of biomimetic approaches and advanced nanomaterials. As far as biological synthesis of nanoparticles is concerned, to date a variety of biological systems have been used for the synthesis of different nanoparticles, which mainly include actinomycetes (Anasane et al. 2016; Golińska et al. 2016; Manimaran and Kannabiran 2017; Wypij et al. 2017, 2018), fungi (Yadav et al. 2015; Moghaddam et al. 2015; Siddiqi and Husen 2016; Elamawi et al. 2018; Khan et al. 2018), bacteria (Husseiney et al. 2007; Deepak et al. 2011; Priyadarshini et al. 2011), yeasts (Moghaddam et al. 2015), and plants (Devanesan et al. 2017; Paralikar and Rai 2017; Suryavanshi et al. 2017; Kale and Jagtap 2018; Shayegan-Mehr et al. 2018; Vishveshvar et al. 2018). Among these, plants are found to be an excellent natural source for the synthesis of nanoparticles due to their easy availability and the wide range of secondary metabolites present in them. Plant parts such as leaves, roots, stem, fruits, and seeds have been used for nanoparticle synthesis, as their extracts are rich sources of phytoconstituents such as proteins, vitamins, flavonoids, polyphenols, amino acids, organic acids, tannic acid, etc., which act as reducing, capping, and stabilizing agents enabling the desired size and shape of nanoparticles (Akhtar et al. 2013; Gnanajobith et al. 2013; Nagajyothi et al. 2013; Sadeghi et al. 2015).

As far as the synthesis of SNPs is concerned, little work has been done in this area and therefore scant literature is available. However, most chemical methods have been exploited for the synthesis of SNPs and only a few reports are available concerning the biological synthesis of SNPs. Hence, there is a wide scope in the field of biogenic synthesis of SNPs using different plants and microbes.

Recently, Paralikar et al. (2019) demonstrated the biosynthesis of SNPs by the coprecipitation method using sodium thiosulfate in the presence of *Catharanthus roseus* leaf extract. In another study, Paralikar and Rai (2017) also reported the biosynthesis of SNPs from sodium polysulfide in the presence of *C. roseus* leaf extract. Moreover, among the biological synthesis of SNPs, Suryavanshi et al. (2017) reported the antibacterial activity of SNPs synthesized from the fungal extract of *Colletotrichum* sp. against food-borne pathogenic bacteria *Listeria monocytogenes* and *Salmonella typhi*. Araj et al. (2015) developed a biogenic approach for the synthesis of SNPs. Here, the aqueous leaf extract of citrus was acidified with dilute HCl which was later added to sodium thiosulfate with mild stirring leads to precipitation of SNPs. Likewise, Awwad et al. (2015) also demonstrated the synthesis of SNPs using sodium thiosulfate in the presence of *Albizia julibrissin* fruit extract. The study resulted in the formation of monoclinic SNPs with an average size of 20 nm and spherical shape. Facts like the involvement of hazardous chemicals, such as H_2S gas and many others, in the chemical approaches of SNPs synthesis, and unavailability of reports on the biogenic synthesis of SNPs, have attracted the attention of researchers toward the development of a novel and eco-friendly biological approach for rapid and mass level synthesis of SNPs. The mechanism of biosynthesis of SNPs has been given in Figure 11.1. Moreover, a variety of biomaterials used for the synthesis of SNPs are shown in Table 11.1.

11.4 Antibacterial Application of SNPs

The vast popularity of nanoparticles lies in their medical applications. It makes them prime candidates for the fight against various unwelcome microbial invaders, particularly bacteria of the human body because of their pathogen-sized proportions and they become highly toxic to microbes. Considering these facts, antibacterial properties of SNPs have been elaborated here.

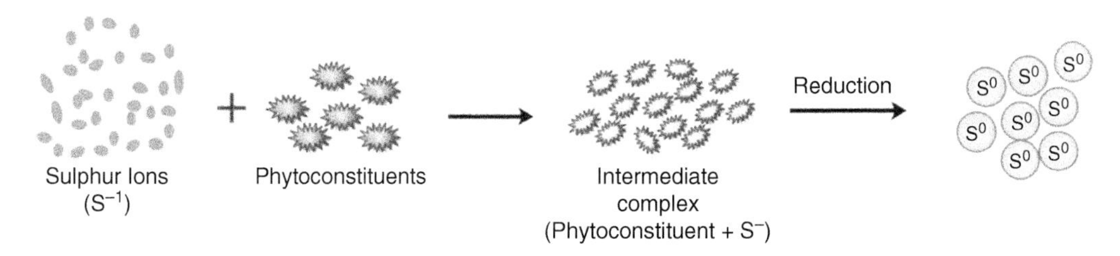

Figure 11.1 Schematic representation of the mechanism for the biosynthesis of SNPs.

Table 11.1 Various biomaterials used for synthesis of SNPs.

Size of SNPs (nm)	Shape of SNPs	Synthesis biomaterial	References
20	Spherical	*Albizia julibrissin* fruit extract	Awwad et al. (2015)
8	Spherical	Leaf extract of citrus	Araj et al. (2015)
50	Spherical	Fungal extract of *Colletotrichum* sp.	Suryavanshi et al. (2017)
70–80	Spherical	*Catharanthus roseus* leaf extract	Paralikar and Rai (2017)
20–86	Spherical	*Catharanthus roseus* leaf extract	Paralikar et al. (2019)
2–15	Spherical	*Ficus bengalensis* leaf extract	Tripathi et al. (2018)
10–40	Spherical	*Punica granatum* peels	Salem et al. (2016)
40	Spherical	*Acanthophyllum bracteatum*	Kouzegaran and Farhadi (2017)

Like other nanoparticles, SNPs also showed potential antibacterial activity. Only a few reports are available demonstrating the evaluation of the antibacterial potential of SNPs. Recently, Paralikar et al. (2019) evaluated the efficacy of SNPs alone and in combination with commercially available antibiotics against clinical isolates of urinary tract infection-causing bacteria, namely *Escherichia coli, Proteus mirabilis, Pseudomonas aeruginosa, Klebsiella pneumoniae, Enterococcus faecalis,* and *Staphylococcus aureus*. The results revealed that the efficacy of SNPs increases in combination with antibiotics against the tested uropathogens. Jaiswal et al. (2019) reported the wound healing application of SNPs. In this study, carrageenan-based functional wound-healing hydrogel films were prepared by incorporating chitosan-capped SNPs. The results revealed that hydrogel-based antibacterial wound dressing plays an important role in providing additional protection against infection which can be life-threatening. The prepared dressing was found to be worth it because of their unique physicochemical and functional properties such as antibacterial, non-cytotoxic, biodegradability, biocompatibility and high absorption, and moisture retention.

Shankar and Rhim (2018) tested the efficacy of chitosan SNP composite film on food-borne pathogenic Gram-positive (*L. monocytogenes*) bacteria and Gram-negative (*E. coli*) bacteria using a viable colony count method. It was observed that the chitosan/SNP nanocomposite films showed enhanced antibacterial activity against food-borne pathogenic bacteria, *E. coli*, and *L. monocytogenes*. Especially, chitosan/SNP composite film showed the complete removal of *E. coli* and *L. monocytogenes* within 6 and 12 hours of treatment, respectively. Similarly, Suryavanshi et al. (2017) reported the antibacterial activity of SNPs synthesized from the fungal extract of *Colletotrichum* sp. against food-borne pathogenic bacteria *L. monocytogenes* and *S. typhi*. The study revealed that SNPs were found to be highly effective against *S. typhi* compared to *L. monocytogenes*. Another study involving the use of SNPs as an antibacterial agent was performed by Paralikar and Rai (2017). The authors reported antibacterial activity of SNPs synthesized from sodium polysulfide against common pathogenic bacteria, namely *E. coli* and *S. aureus*. Suleiman et al. (2015) demonstrated the antibacterial activity of SNPs against Gram-positive (*S. aureus*) and Gram-negative (*E. coli* and *P. aeruginosa*) bacteria. They found that SNPs showed significant antibacterial activity against the Gram-positive bacterium *S. aureus*, whereas these nanoparticles did not show any activity against *E. coli* and *P. aeruginosa* even at high concentrations, i.e. up to $800\,\mu g\,ml^{-1}$.

Roy Choudhury et al. (2012) reported the antibacterial effect of two different sizes and surface-modified chemically synthesized SNPs against multi-drug resistant Gram-negative bacilli including *E. coli, K. pneumoniae, Acinetobacter baumannii, Stenotrophomonas maltophilia,* and *Enterobacter aerogenes* using the agar dilution method and the broth micro-dilution method. The study revealed that PEGylated SNPs showed antibacterial effect against all tested Gram-negative bacilli at a concentration between 9.41 and 18.82 $mg l^{-1}$ in the case of the broth micro-dilution method, whereas in the case of the agar dilution method, SNPs showed uniform minimum inhibitory value (18.82 $mg l^{-1}$) against all tested bacteria. Similarly, in another study, Roy Choudhury et al. (2013a) claimed antimicrobial activity of orthorhombic and monoclinic nanoallotropes of sulfur against a series of Gram-negative (*E. coli, K. pneumoniae, A. baumannii, S. maltophilia,* and *P. aeruginosa*) and Gram-positive (*S. aureus* and *Staphylococcus kloosii*) bacteria.

In another study, Roy Choudhury et al. (2013b) reported that SNPs were capable of showing antibacterial activity, while elemental sulfur failed to show any bacterial growth inhibition against Gram-positive and Gram-negative bacteria. Thakur et al. (2015) demonstrated the antibacterial activity of SNPs against *S. aureus* and *E. coli*. Further, they tried to study the modes of action of SNPs against these bacteria using a scanning electron microscope. They reported membrane rupture leading to the lysis of cell membranes, which is one of the most important factors responsible for the decline in bacterial growth.

11.5 Possible Mechanisms for Antibacterial Action

Several metal nanoparticles were extensively studied for their antimicrobial mechanism of action. Various possible modes of action have been proposed for different nanoparticles. Nanoparticles target bacteria by affecting their genetic material and consequently their cell wall structure, metabolic pathways, and many other components that when disrupted could cause death. The mechanism of this antibacterial activity varies from nanoparticle to nanoparticle. For all varieties of nanoparticles, the antibacterial mechanism is not fully understood. While some proposed mechanisms relate to the physical structure of the nanoparticles (i.e. membrane-damaging abrasiveness of the nanoparticles), others relate to the enhanced release of antibacterial metal ions from nanoparticle surfaces. The specific surface area of a dose of nanoparticles increases as the particle size decreases, allowing for greater material interaction with the surrounding environment (Rai et al. 2015; Gugala and Turner 2018).

In this context, the mechanism of action for SNPs against microbes was not much studied. Rai et al. (2016) reviewed the possible mechanism for the antimicrobial action of SNPs. Elemental sulfur was most widely used in the ancient era for medicinal purposes (Gupta and Nicol 2004). The high concentration of sulfur is considered to be toxic for microbes due to the generation of H_2S gas, which is a volatile compound, after adsorption by microbial cells (Roy Choudhury et al. 2011, 2012). There are more theories available on the mode of action of elemental sulfur. According to them, elemental sulfur is taken up by pathogens and then gets oxidized to form pentathionic acid or reduced to form H_2S. Further, H_2S promotes obstruction or oxidation of free –SH groups of many essential enzymes within the cell and disrupts an important intermediary metabolism in mitochondria. Also, elemental sulfur is responsible for the cross-linking of proteins or lipids with free radical of sulfur (McCallan 1949; Libenson et al. 1953).

Recently, Paralikar et al. (2019) performed a study emphasizing the antimicrobial activity of SNPs against urinary tract infection-causing bacteria. In this study, the zeta potential of uropathogenic bacterial isolates was measured to monitor the effect of synthesized SNPs on the membrane potential of bacteria. To maintain bacterial cell growth and metabolic activities, bacterial cell surface potential plays an important role. The change in potential alters membrane permeability, which subsequently leads to leakage of cell components followed by cell death (Halder et al. 2015). Intact bacterial cells were found to maintain stable membrane potential. Treatment of synthesized

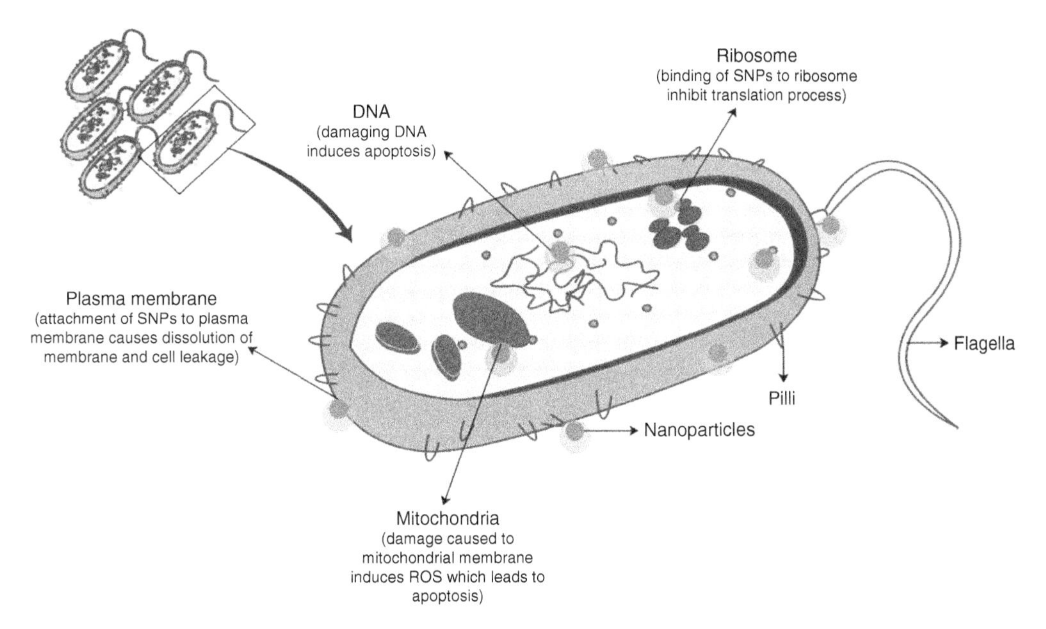

Figure 11.2 Diagrammatic illustration of the possible mechanism involved in the antibacterial action of SNPs.

SNPs alters the surface potential of bacterial cells, which could be attributed to bacterial membrane damage. The observed results are in support of the antibacterial mechanism of SNPs, that nanoparticles caused membrane damage which ultimately leads to cell death (Paralikar et al. 2019). The possible antibacterial mechanism of SNPs is illustrated in Figure 11.2.

11.6 Conclusion

Considering the broad-spectrum biomedical applications of nanotechnology and especially SNPs, it is believed that nanotechnology can offer better options for the treatment of various diseases. Nanotechnology is currently employed as a tool for various biomedical applications. SNPs can be considered to be a viable alternative to antibiotics and other antimicrobial agents, and appear to have high potential to solve the problem of the emergence of resistant microbial infection. In the field of medicine, there are still many possibilities to explore new monotherapies. SNPs alone, and their combination with antibiotics, can be efficient antibacterial agents. Thus, these can be utilized in drug formulations that may increase its therapeutic efficacy and reduce its dose concentration. The other benefit of the use of combination therapy is to improve the pharmacokinetics of antibiotics, thus to reduce the toxicity. Packing the antibiotics into nanoparticles can prolong the half-life of antibiotic payload and improve the therapeutic index.

References

Akhtar, M.S., Panwar, J., and Yun, Y.S. (2013). Biogenic synthesis of metallic nanoparticles by plant extracts. *ACS Sustainable Chemistry and Engineering* 1: 591–602.

Anasane, N., Golińska, P., Wypij, M. et al. (2016). Acidophilic actinobacteria synthesised silver nanoparticles showed remarkable activity against fungi-causing superficial mycoses in humans. *Mycoses* 59 (3): 157–166.

Araj, S.E.A., Salem, N.M., Ghabeish, I.H., and Awwad, A.M. (2015). Toxicity of nanoparticles against *Drosohila melanogaster* (Diptera: Drosophilidae). *Journal of Nanomaterials* 2015: 1–9.

Awwad, A.M., Salem, N.M., and Abdeen, A.O. (2015). Novel approach for synthesis sulfur (S-NPs) nanoparticles using *Albizia julibrissin* fruits extract. *Advanced Materials Letters* 6 (5): 432–435.

Bader, M.S., Loeb, M., and Brooks, A.A. (2016). An update on the management of urinary tract infections in the era of antimicrobial resistance. *Postgraduate Medicine* 129 (2): 242–258.

Barkauskas, J., Juskenas, R., Mileriene, V., and Kubilius, V. (2007). Effect of sulfur on the synthesis and modification of carbon nanostructures. *Materials Research Bulletin* 42: 1732–1739.

Buzayan, M.M. and El-Garbulli, F.R. (2014). Detection of ESBL and AmpC β- lactamases producing in uropathogen *Escherichia coli* isolates at Benghazi Center of Infectious Diseases and Immunity. *International Journal of Current Microbiology and Applied Sciences* 3: 145–153.

Chen, H., Dong, W., Ge, J. et al. (2013). Ultrafine sulfur nanoparticles in conducting polymer shell as cathode materials for high performance lithium/sulfur batteries. *Scientific Reports* 3: 1910.

Colomb-Cotinat, M., Lacoste, J., Brun-Buisson, C. et al. (2016). Estimating the morbidity and mortality associated with infections due to multidrug-resistant bacteria (MDRB), France, 2012. *Antimicrobial Resistant Infection Control* 5: 56.

Connell, S.R., Tracz, D.M., Nierhaus, K.H. et al. (2003). Ribosomal protection proteins and their mechanism of tetracycline resistance. *Antimicrobial Agents and Chemotherapy* 47: 3675–3681.

Daniel, M.C. and Astruc, D. (2004). Gold Nanoparticles: assembly, supramolecular chemistry, quantum-size-related properties, and applications toward biology, catalysis, and nanotechnology. *Chemical Review* 104: 293–346.

Davies, J. and Davies, D. (2010). Origins and evolution of antibiotic resistance. *Microbiology and Molecular Biology Review* 74: 417–433.

Deepak, V., Kalishwaralal, K., Pandian, S.R.K., and Gurunathan, S. (2011). An insight into the bacterial biogenesis of silver nanoparticles, industrial production and scale-up. In: Metal Nanoparticles in Microbiology (eds. M. Rai and N. Duran), 17–35. Berlin, Heidelberg: Springer.

Devanesan, S., AlSalhi, M.S., Vishnubalaji, R. et al. (2017). Rapid biological synthesis of silver nanoparticles using plant seed extracts and their cytotoxicity on colorectal cancer cell lines. *Journal of Cluster Science* 28: 595–605.

Elamawi, R.M., Al-Harbi, R.E., and Hendi, A.A. (2018). Biosynthesis and characterization of silver nanoparticles using *Trichoderma longibrachiatum* and their effect on phytopathogenic fungi. *Egypt Journal of Biological Pest Control* 28 (28): 1–11.

Fair, R.J. and Tor, Y. (2014). Antibiotics and bacterial resistance in the 21st century. *Perspectives in Medicinal Chemistry* 6: 25–64.

Fernández, L. and Hancock, R.E.W. (2012). Adaptive and mutational resistance: role of porins and efflux pumps in drug resistance. *Clinical Microbiology Review* 25: 661–681.

Frieri, M., Kumar, K., and Boutin, A. (2017). Antibiotic resistance. *Journal of Infection and Public Health* 10 (4): 369–378.

Gao, W., Thamphiwatana, S., Angsantikul, P., and Zhang, L. (2014). Nanoparticle approaches against bacterial infections. *Nanomedicine and Nanobiotechnology* 6: 532–547.

Gnanajobith, G., Paulkumar, K., Vanaja, M. et al. (2013). Fruit-mediated synthesis of silver nanoparticles using *Vitis vinifera* and evaluation of their antimicrobial efficacy. *Journal of Nanostructure in Chemistry* 3: 67.

Golińska, P., Wypij, M., Rathod, D. et al. (2016). Synthesis of silver nanoparticles from two acidophilic strains of *Pilimelia columellifera* sub sp. *pallida* and their antibacterial activities. *Journal of Basic Microbiology* 56 (5): 541–556.

Griffith, C.M., Woodrow, J.E., and Seiber, J.N. (2015). Environmental behaviour and analysis of agricultural sulfur. *Pest Management Science* 71: 1486–1496.

Gugala, N. and Turner, R.J. (2018). The potential of metals in combating bacterial pathogens. In: Biomedical Applications of Metals (eds. M. Rai, A. Ingle and S. Medici), 129–150. Cham: Springer.

Gupta, A.K. and Nicol, K. (2004). The use of sulfur in dermatology. *Journal of Drugs Dermatology* 3 (4): 427–431.

Halder, S., Yadav, K.K., Sarkar, R. et al. (2015). Alteration of zeta potential and membrane permeability in bacteria: a study with cationic agents. *Springer Plus* 4: 1–14.

Husseiney, M.I., El-Aziz, M.A., Badr, Y., and Mahmoud, M.A. (2007). Biosynthesis of gold nanoparticles using *Pseudomonas aeruginosa*. *Spectrochimica Acta A* 67: 1003–1006.

Ilardi, E.A., Vitaku, E., and Njardarson, J.T. (2014). Data-mining for sulfur and fluorine: an evaluation of pharmaceuticals to reveal opportunities for drug design and discovery. *Journal of Medicinal Chemistry* 57 (7): 2832–2842.

Jaiswal, L., Shankar, S., and Rhim, J.W. (2019). Carrageenan-based functional hydrogel film reinforced with sulfur nanoparticles and grapefruit seed extract for wound healing application. *Carbohydrate Polymer* 224: 115191.

Kale, R.D. and Jagtap, P. (2018). Biogenic synthesis of silver nanoparticles using *Citrus limon* leaves and its structural investigation. In: Advances in Health and Environment Safety, Springer Transactions in Civil and Environmental Engineering (eds. N. Siddiqui, S. Tauseef and K. Bansal), 11–20. Singapore: Springer.

Keri, J. and Shiman, M. (2009). An update on the management of acne vulgaris. *Clinical, Cosmetic and Investigational Dermatology* 2: 105–110.

Khan, I., Saeed, K., and Khan, I. (2017). Nanoparticles: Properties, applications and toxicities. *Arabian Journal of Chemistry* 12: 908–931.

Khan, A.U., Malik, N., Khan, M. et al. (2018). Fungi-assisted silver nanoparticle synthesis and their applications. *Bioprocess and Biosystems Engineering* 41: 1.

Kouzegaran, V.J. and Farhadi, K. (2017). Green synthesis of sulphur nanoparticles assisted by a herbal surfactant in aqueous solutions. *Micro and Nano Letters* 12: 329–334.

Kumar, A. and Schweizer, H.P. (2005). Bacterial resistance to antibiotics: active efflux and reduced uptake. *Advance Drug Delivery Review* 57: 1486–1513.

Kumari, M., Pandey, S., Giri, V.P. et al. (2017). Tailoring shape and size of biogenic silver nanoparticles to enhance antimicrobial efficacy against MDR bacteria. *Microbial Pathogenesis* 105: 346–355.

Lambert, P.A. (2005). Bacterial resistance to antibiotics: modified target sites. *Advance Drug Delivery Review* 57: 1471–1485.

Landecker, H. and Blackman, L. (2016). Antibiotic resistance and the biology of history. *Body Society* 22 (4): 19–52.

Lee, C.R., Cho, I.H., Jeong, B.C., and Lee, S.H. (2013). Strategies to minimize antibiotic resistance. *International Journal of Environmental Research in Public Health* 10 (9): 4274–4305.

Libenson, L., Hadley, F.P., Mcllroy, A.P. et al. (1953). Antibacterial effect of elemental sulfur. *Journal of Infectious Diseases* 93: 28–35.

Lim, C., Takahashi, E., Hongsuwan, M. et al. (2016). Epidemiology and burden of multidrug-resistant bacterial infection in a developing country. *eLife* 5: e18082.

Lin, A.N., Reimer, R.J., and Carter, D.M. (1988). Sulfur revisited. *Journal of the American Academy of Dermatology* 18 (3): 553–558.

Llor, C. and Bjerrum, L. (2014). Antimicrobial resistance: risk associated with antibiotic overuse and initiatives to reduce the problem. *Therapeutic Advances in Drug Safety* 5 (6): 229–241.

Ma, A.H. and Hughes, G.J. (2018). Updates in management of complicated urinary tract infections: a focus on multidrug-resistant organisms. *American Journal of Therapy* 25 (1): 53–66.

MacGowan, A. and Macnaughton, E. (2017). Antibiotic resistance. *Medicine* 45 (10): 622–628.

Manimaran, M. and Kannabiran, K. (2017). Actinomycetes-mediated biogenic synthesis of metal and metal oxide nanoparticles: progress and challenges. *Letters in Applied Microbiology* 64 (6): 401–408.

Manjunath, G.N., Prakash, R., Annam, V., and Shetty, K. (2011). Changing trends in the spectrum of antimicrobial drug resistance pattern of uropathogens isolated from hospitals and community patients with urinary tract infections in Tumkur and Bangalore. *International Journal of Biology and Medical Research* 2: 504–507.

McCallan, S.E.A. (1949). The nature of the fungicidal action of copper and sulfur. *Botanical Review* 15: 629–643.

Moghaddam, A.B., Namvar, F., Moniri, M. et al. (2015). Nanoparticles biosynthesized by fungi and yeast: a review of their preparation, properties, and medical applications. *Molecules* 20: 16540–16565.

Morones-Ramirez, J.R., Winkler, J.A., Spina, C.S. et al. (2013). Silver enhances antibiotic activity against gram-negative bacteria. *Science Translational Medicine* 5: 190. https://doi.org/10.1126/scitranslmed.3006276.

Nagajyothi, P.C., Minh, T.N., Sreekanth, T.V.M. et al. (2013). Green route biosynthesis: characterization and catalytic activity of ZnO nanoparticles. *Material Letters* 108: 60–163.

Nikaido, H. (2009). Multidrug resistance in bacteria. *Annual Review of Biochemistry* 78: 119–146.

Nisar, P., Ali, N., Rahman, L. et al. (2019). Antimicrobial activities of biologically synthesized metal nanoparticles: an insight into the mechanism of action. *Journal of Biological Inorganic Chemistry* 24: 929. https://doi.org/10.1007/s00775-019-01717-7.

Orton, D.G. (1974). Sulfur dyes. In: The Chemistry of Synthetic Dyes, vol. 7 (ed. K. Venkataraman), 1–34. New York: Academic press.

Paralikar, P. and Rai, M. (2017). Bio-inspired synthesis of sulphur nanoparticles using leaf extract of four medicinal plants with special reference to their antibacterial activity. *IET Nanobiotechnology* 12 (1): 25–31.

Paralikar, P., Ingle, A.P., Tiwari, V. et al. (2019). Evaluation of antibacterial efficacy of sulfur nanoparticles alone and in combination with antibiotics against multidrug-resistant uropathogenic bacteria. *Journal of Environmental Science and Health, Part A* 54 (5): 381–390.

Parcell, S. (2002). Sulfur in human nutrition and applications in medicine. *Alternative Medicine Review* 7 (1): 22–44.

Priyadarshini, S., Gopinath, V., Priyadharsshini, N.M. et al. (2011). Synthesis of anisotropic silver nanoparticles using novel strain, *Bacillus flexus* and its biomedical application. *Colloids and Surface B Biointerfaces* 102: 232–237.

Rai, M.K., Deshmukh, S.D., Ingle, A.P., and Gade, A.K. (2012). Silver nanoparticles: the powerful nanoweapon against multidrug-resistant bacteria. *Journal of Applied Microbiology* 112 (5): 841–852.

Rai, M., Ingle, A.P., Yadav, A. et al. (2015). Strategic role of selected noble metal nanoparticles in medicine. *Critical Reviews in Microbiology* 19: 1–24.

Rai, M., Ingle, A.P., Birla, S. et al. (2016). Strategic role of selected noble metal nanoparticles in medicine. *Critical Reviews in Microbiology* 42 (5): 696–719.

Rai, M., Nagaonkar, D., and Ingle, A.P. (2018). Metal nanoparticles as therapeutic agents: a paradigm shift in medicine. In: Metal Nanoparticles: Synthesis and Applications in Pharmaceutical Sciences (eds. S. Thota and D.C. Crans), 33–48.

Roy Choudhury, S., Ghosh, M., Mandal, A. et al. (2011). Surface-modified sulfur nanoparticles: an effective antifungal agent against *Aspergillus niger* and *Fusarium oxysporum*. *Applied Microbiology and Biotechnology* 90: 733–743.

Roy Choudhury, S., Roy, S., Goswami, A., and Basu, S. (2012). Polyethylene glycol-stabilized sulfur nanoparticles: an effective antimicrobial agent against multidrug-resistant bacteria. *Journal of Antimicrobial Chemotherapy* 67 (5): 1134.

Roy Choudhury, S., Mandal, A., Chakravorty, D. et al. (2013a). Evaluation of physicochemical properties, and antimicrobial efficacy of monoclinic sulfur-nanocolloid. *Journal of Nanoparticle Research* 15: 1491.

Roy Choudhury, S., Mandal, A., Ghosh, M. et al. (2013b). Investigation of antimicrobial physiology of orthorhombic and monoclinic nanoallotropes of sulfur at the interface of transcriptome and metabolome. *Applied Microbiology and Biotechnology* 97: 5965–5978.

Sadeghi, B., Rostami, A., and Momeni, S.S. (2015). Facile green synthesis of silver nanoparticles using seed aqueous extract of *Pistacia atlantica* and its antibacterial activity. *Spectrochimica Acta, Part A: Molecular and Biomolecular Spectroscopy* 134: 326–332.

Sajid, H., Joo, J., Kang, J. et al. (2018). Antibiotic-loaded nanoparticles targeted to the site of infection enhance antibacterial efficacy. *Nature Biomedical Engineering* 2(2): 95–103.

Salem, N.M., Albanna, L.S., and Awwad, A.M. (2016). Green synthesis of sulfur nanoparticles using Punica granatum peels and the effects on the growth of tomato by foliar spray applications. *Environmental Nanotechnology, Monitoring and Management* 6: 83–87.

Santiago, P., Carvajal, E., Mendoza, D., and Rendon, L. (2006). Synthesis and structural characterization of sulfur nanowires. *Microscopy and Microanalysis* 12 (02): 690–691.

Seil, J.T. and Webster, T.J. (2012). Antimicrobial applications of nanotechnology: methods and literature. *International Journal of Nanomedicine* 7: 2767–2781.

Shankar, S. and Rhim, J.W. (2018). Preparation of sulfur nanoparticles incorporated antimicrobial chitosan films. *Food Hydrocolloids* 82: 116–123.

Shayegan-Mehr, E., Sorbiun, M., Ramazani, A., and Fardood, S.T. (2018). Plant-mediated synthesis of zinc oxide and copper oxide nanoparticles by using ferulagoangulata (schlecht) boiss extract and comparison of their photocatalytic degradation of Rhodamine B (RhB) under visible light irradiation. *Journal of Materials Science: Materials in Electronics* 29: 1333.

Shukla, A.K. and Iravani, S. (2017). Metallic nanoparticles: green synthesis and spectroscopic characterization. *Environmental Chemistry Letters* 15: 223.

Siddiqi, K.S. and Husen, A. (2016). Fabrication of metal nanoparticles from fungi and metal salts: scope and application. *Nanoscale Research Letters* 11: 98. https://doi.org/10.1186/s11671-016-1311-2.

Suleiman, M., Masri, M.A., Ali, A.A. et al. (2015). Synthesis of nano-sized sulfur nanoparticles and their antibacterial activities. *Journal of Materials and Environmental Science* 6 (2): 513–518.

Sun, J., Deng, Z., and Yan, A. (2014). Bacterial multidrug efflux pumps: mechanisms, physiology and pharmacological exploitations. *Biochemical and Biophysical Research Communications* 453: 254–267.

Suryavanshi, P., Pandit, R., Gade, A. et al. (2017). *Colletotrichum* sp.- mediated synthesis of sulphur and aluminium oxide nanoparticles and its in vitro activity against selected food-borne pathogens. *LWT - Food Science and Technology* 81: 188–194.

Tenover, F.C. (2006). Mechanisms of antimicrobial resistance in bacteria. *The American Journal of Medicine* 116: 3–10.

Thakur, S., Barua, S., and Karak, N. (2015). Self-healable castor oil based tough smart hyperbranched polyurethane nanocomposite with antimicrobial attributes. *RSC Advances* 5: 2167–2176.

Tripathi, R.M., Pragadeeshwara Rao, R., and Tsuzuki, T. (2018). Green synthesis of sulfur nanoparticles and evaluation of their catalytic detoxification of hexavalent chromium in water. *RSC Advances* 8: 36345–36352.

Varier, K.M., Gudeppu, M., Chinnasamy, A. et al. (2019). Nanoparticles: antimicrobial applications and its prospects. In: Advanced Nanostructured Materials for Environmental Remediation, Environmental Chemistry for a Sustainable World, vol. 25 (eds. M. Naushad, S. Rajendran and F. Gracia). Cham: Springer.

Ventola, C.L. (2015). The antibiotic resistance crisis: part 1: causes and threats. *Pharmacy and Therapeutics* 40 (4): 277–283.

Vishveshvar, K., Krishnan, A.M., and Vishnuprasad, H.K.S. (2018). Green synthesis of copper oxide nanoparticles using *Ixira coccinea* plant leaves and its characterization. *BioNanoScience* 8 (2): 554–558.

Wang, L., Hu, C., and Shao, L. (2017). The antimicrobial activity of nanoparticles: present situation and prospects for the future. *International Journal of Nanomedicine* 12: 1227–1249.

Wright, G.D. (2005). Bacterial resistance to antibiotics: enzymatic degradation and modification. *Advance Drug Delivery Review* 57: 1451–1470.

Wypij, M., Golinska, P., Dahm, H., and Rai, M. (2017). Actinobacterial-mediated synthesis of silver nanoparticles and their activity against pathogenic bacteria. *IET Nanobiotechnology* 11 (3): 336–342.

Wypij, M., Świecimska, M., Czarnecka, J. et al. (2018). Antimicrobial and cytotoxic activity of silver nanoparticles synthesized from two haloalkaliphilic actinobacterial strains alone and in combination with antibiotics. *Journal of Applied Microbiology* 124 (6): 1411–1424.

Yadav, A., Kon, K., Kratosova, G. et al. (2015). Fungi as an efficient mycosystem for the synthesis of metal nanoparticles: progress and key aspects of research. *Biotechnology Letters* 37: 2099–2120.

Yong, Z., Wei, Z., Ping, Z. et al. (2007). Novel nanosized adsorbing composite cathode materials for the next generation lithium battery. *Journal of Wuhan University of Technology* 22 (2): 234–239.

Zhu, X., Radovic-Moreno, A.F., Wu, J. et al. (2014). Nanomedicine in the management of microbial infection – overview and perspectives. *Nano Today* 9 (4): 478–498.

12

Role of Nanotechnology in the Management of Indoor Fungi

Erasmo Gámez-Espinosa[1], *Leyanet Barberia-Roque*[1], *and Natalia Bellotti*[1,2]

[1] *Center for Research and Development in Painting Technology (CIDEPINT; CONICET-CICPBA), National University of La Plata, La Plata, Buenos Aires, Argentina*
[2] *Faculty of Natural Sciences and Museum, National University of La Plata, La Plata, Buenos Aires, Argentina*

12.1 Introduction

Concern over indoor fungal growth has increased over the last decades. Research that relates human health issues to the deterioration of indoor air quality due to microbiological pollutants have been accumulated (Rath et al. 2011; Weber 2012; Täubel and Hyvärinen 2015). In addition, since the 1980s, it has been registered that there is an increase in the occurrence of natural disasters such as floods and extreme rainfall, on a global scale, which has enhanced the problems related to biodeterioration in indoor buildings (Bloom et al. 2009; Chow et al. 2019; EASAC 2018). It is worth mentioning that these kinds of events are directly related to fungal growth since water is a key factor in the development of these microorganisms (Johansson et al. 2013; Møller et al. 2017). Considering the strong impact of indoor fungi on public health, several guidelines have been proposed from different European and North American institutions. These institutions include the University of Connecticut Health Center, which proposed Guidance for Clinicians on the Recognition and Management of Health Effects Related to Mold Exposure and Moisture Indoors (Storey et al. 2004); the United States Environmental Protection Agency (US EPA), which proposed Mold Remediation in Schools and Commercial Buildings (EPA 2008); the World Health Organization (WHO), which proposed guidelines for indoor air quality, dampness, and mold (WHO 2009); the Canadian National Collaborating Centre for Environmental Health Mould Remediation Recommendations (Palaty 2014); and the US Centers for Disease Control and Prevention (CDC), proposed Mold Clean-Up After Disasters (2018), among others (US CDC 2018). These guidelines proposed that there is a dose-effect relationship, where more visible mold showed more symptoms. The improvement in molecular techniques enhances the interest in the field of mycology because due to these techniques it becomes possible to understand the mechanisms by which fungi affect the health of exposed human being.

Considering these facts, a great deal of attention has been given on the development of antimicrobial materials, especially those based on nanotechnology (Mittal et al. 2013; Singh 2016a; Soliman 2017). Moreover, different strategies focusing on efficient association of nanoscale materials in the development of bioactive surfaces which can prevent biofilm formation in indoor environments

Nanobiotechnology in Diagnosis, Drug Delivery, and Treatment, First Edition. Edited by Mahendra Rai, Mehdi Razzaghi-Abyaneh, and Avinash P. Ingle.

have been intensively studied in the last two decades (Bellotti et al. 2015; Manjumeena 2017; Ghorbani et al. 2018; Han et al. 2019; Barberia-Roque et al. 2019).

The main aim of this chapter is to integrate current knowledge about indoor fungal deterioration and its control through the use of nanotechnology. Moreover, nanomaterials used as effective antifungal agents and possible mechanisms involved in the inhibition of fungal growth have been also described.

12.2 Indoor Fungal Deterioration

12.2.1 Indoor Mycobiota

The mycobiota of indoor environments contains about 150 species and it is considered that a high level of exposure for a building indoor occupant is one greater than 1000 CFU m^{-3} (Sedlbauer 2002; Crook and Burton 2010). Most species belong to the so-called anamorphic fungi, which include deuteromycetes, hyphomycetes, or fungi imperfecti (Yang and Heinsohn 2007). But most are in the ascomycota phylum. In addition to these micro-fungi, a number of basidiomycetes are also found in indoor environments, growing on wood in buildings, and are considered important degraders of wooden building material (Adan and Samson 2011). The important sources of indoor fungal spores include wood products, foodstuffs, vegetables, carpet dust, and fruits. It is important to point out that some fungi may come from more than one source. The fungal genera most frequently found in indoor environments are *Cladosporium*, *Penicillium*, *Aspergillus,* and *Stachybotrys* (Verdier et al. 2014).

After germination, the spores produce a mycelium which covers diverse materials such as textile, paper, wood, coatings, wall paper, and gypsum, among others, depending on the moisture availability (Grant et al. 1989; Annila et al. 2018). Many fungi can produce numerous spores or other propagules, and this explains why there can be high concentrations in the air. The spore of anamorphic fungi is called a conidium. The structure bearing conidia is known as a conidiophore. The formation of conidia varies between the different genera and their efficiency to produce airborne propagules is mainly determined by the mode of conidium formation (Cole and Samson 1979). For example, species of *Aspergillus* and *Penicillium* produce numerous dry conidia which easily become airborne, and this explains the presence of these fungi in an indoor environment (Guerra et al. 2019; Kavkler et al. 2015). In contrast, with blastic arthric conidiogenesis, *Cladosporium* species are among the most abundant fungi in outdoor and indoor air (Anaya et al. 2016; Bensch et al. 2018; Asif et al. 2019).

12.2.2 Factors Influencing Indoor Fungal Growth

Many environmental parameters can influence the growth of indoor fungi. Among some are biotic, and others are physical and chemical, or abiotic factors. Biotic factors include the presence of fungal propagules or spores, viability of spores, the nature of the fungal species, and competing fungi and other organisms. Abiotic factors include nutrients, temperature, moisture, pH, oxygen and carbon dioxide, relative humidity, and light (Crawford et al. 2015; Liu et al. 2018). The viability of fungal spores is associated with several factors, such as age, UV light, and extreme conditions (Johansson et al. 2005; Chen et al. 2017). The dead spores present may be allergenic and contain secondary metabolites, but they cannot germinate and grow and hence are unable to cause infections. Moreover, some viable spores may remain in dormancy. Dormant fungal spores are usually either physically or chemically restricted from germination. A physical barrier, such as a thickened spore wall, restricts the absorption of water required for spore germination (Madelin 1994; Carlile et al. 2001).

Indoor fungi adapt and grow in environments that are not favorable for bacteria with low water activity and nutrients (Webster and Weber 2007). Different species of fungi have different abilities to access and utilize simple or complex forms of carbohydrate, organics, and mineral nutrients. Decomposition and degradation of a substrate is due to enzyme activities. The types of enzymes required depend on the substrates. The primary food source for indoor fungi is cellulosic matter (Yang and Heinsohn 2007). Fungi usually grow in a wide range of temperatures. In this context, range, including minimum, optimum, and maximum temperatures, can be defined as a temperature profile. Each species has its own profile. Some have narrow and some have wide temperature profiles. Fungi that can grow in a wider temperature range may also have a competitive edge (Griffin 1994). The relative humidity is critically important in indoor fungal growth; it has a secondary effect on condensation and the hygroscopicity of materials (Lattab et al. 2012). In fact, most indoor fungal growth occurs as a result of dampness, but not just due to high relative humidity and condensation on indoor surfaces (Crook and Burton 2010; Täubel and Hyvärinen 2015).

Vegetative and reproductive functions of fungi from indoor environments, together with biotic and abiotic factors, are responsible for deteriorating materials and affecting the health of people who are in these environments (WHO 2009; Hurraß et al. 2017). For example, Ponizovskaya et al. (2019) described the complex of microfungi colonizing mineral building materials, limestone and plaster, in interiors of cultural heritage (Ponizovskaya et al. 2019). These species can actively develop in materials, penetrating for years into the substrates and causing their deterioration in conditions of considerably more moisture content. In this group, *Acremonium charticola* and *Lecanicillium* sp. were able to solubilize calcium carbonate ($CaCO_3$). Moreover, the identification and quantification of filamentous fungi in samples from different indoor and outdoor environments based on traditional microculture methods, DNA extraction, and molecular analyses have been also proposed (Guerra et al. 2019).

Physicochemical characterization analysis was performed to evaluate biological growth; the isolated species produce acids and metabolites capable of causing chemical alterations in mortar substrates and physical damages due to the growth of filamentous structures. Kavkler et al. (2015) studied the presence of indoor fungi on historical textiles, including the canvases of easel paintings stored in museums and religious institutions (churches and cloisters) in Slovenia. Initial observations revealed that such paintings contain relatively widespread fungal contamination (Kavkler et al. 2015). Moreover, examination of the structural and physical changes to the fibers on contaminated and non-contaminated objects showed the most pronounced structural changes on flax and other cellulosic fibers, while proteinaceous fibers (wool and silk) were generally not affected (Kavkler et al. 2015). Surfaces of building materials (plasterboard, mortar, bricks, etc.) are generally highly porous and rough. In damp environments, these materials can provide favorable conditions for the proliferation and growth of microorganisms. Sampling of microbial communities on building materials and in the air is necessary to evaluate its presence and proliferation in indoor environments (Verdier et al. 2014).

As discussed earlier, the most commonly found fungal genera on indoor building materials include *Cladosporium*, *Penicillium*, *Aspergillus,* and *Stachybotrys* (Gutarowska and Czyżowska 2009; Andersen et al. 2011), and various factors such as moisture content, chemical composition, pH, and the physical properties of surfaces play important roles in influencing microbial growth on such surfaces or materials (Adan and Samson 2011). The particular behavior of porous materials in terms of water sorption and the effect of water on microbial proliferation are of prime importance (Nielsen et al. 2004; Verdier et al. 2014).

As mentioned, not only building surfaces or materials, but also quality of air in indoor environments, especially in food production plants and where heritage documents are kept, also needs to be studied. In this context, De Clercq et al. (2014) studied whether the production environment and common ingredients of chocolate confectioneries could be potential sources of contamination with xerophilic fungal species. In this sense, the relevance of fungal spores for food microbiology has been discussed. The function of spores is to disperse fungi to new areas and to get them through difficult periods. A number of fungal species form sexual spores, which are exceptionally stress-resistant and survive pasteurization and other treatments (Dijksterhuis 2019).

Fungi play a considerable role in the deterioration of cultural heritage. Due to their enormous enzymatic activity and their ability to grow at low water activity (a_w) values, fungi are able to inhabit and decay paintings, textiles, paper, parchment, leather, oil, casein, glue, and other materials used for historical art objects (Allsopp et al. 2004). The weathering of stone monuments is significantly increased by epi- and endolithic fungi. In museums and their storage rooms, climate control, regular cleaning, and microbiological monitoring are essential in order to prevent fungal contamination (Sterflinger 2010; Paiva de Carvalho et al. 2018; Melo et al. 2019). Immunosuppressed people exposed to fungi with pathogenic potentials of indoor environments may suffer from mycosis, allergies, and asthma (Perez-Nadales et al. 2014; Sardi et al. 2014). The fungi produce mycotoxins and other biologically active metabolites when growing in buildings, so influence of environmental conditions on the production of these metabolites is intensely investigated (Nielsen 2003; Täubel and Hyvärinen 2015). It was shown that *Stachybotrys chartarum* produced a number of mycotoxins when growing in buildings; *Aspergillus versicolor* produced high quantities of the carcinogenic mycotoxin, sterigmatocystin; *Chaetomium globosum* produced high quantities of chaetoglobosins; whereas *Trichoderma* species did not produce detectable quantities of trichothecenes when growing on materials (Nielsen 2002, 2003).

12.3 Conventional Approach Used for the Control of Indoor Fungi

Various antimicrobial compounds such as disinfectants and biocides are commonly applied to control growth of indoor fungi and bacteria. The most commonly used indoor bioactive compounds can be classified according to their mechanisms of action into two major groups: electrophiles and membrane-active (Chapman 2003). The electrophiles react with nucleophilic groups from biomolecules such as enzymes and nucleic acids present in microbial cells. Some of the most commercially used biocides (electrophiles) in antimicrobial paints are: formaldehydes, formaldehyde releasers, isothiazolinones, carbamates, and metal salts of silver and cooper (Falkiewicz-Dulik et al. 2015a). Among electrophiles, oxidizers like sodium hypochlorite (bleach) have been used for a long time due to their low cost and efficiency (Pereira et al. 2015). These kind of compounds oxidize organic matter in general, but they are being questioned and limited in use due to their toxicity.

On the other hand, the membrane-actives react with the cell membrane leading to its disruption or have ability to change cytoplasmic conditions (Chapman 2003). This group generally includes compounds like alcohols, phenolic derivatives, quaternary ammonium salts and pH actives (e.g. organic acids, parabens, and pyrithiones). Moreover, the control of the indoor fungi can be achieved by maintaining the hygiene by the periodic cleaning with disinfectants and controlling the indoor conditions (temperature, humidity, ventilation, and water leaks) (Adan and Samson 2011; Weber and Rutala 2013). On the other hand, antimicrobial or hygienic coatings having active antifungal ingredients can be used to control indoor fungi and bacteria growth, and therefore biofilm development (Johns 2003; Stobie et al. 2010).

In recent decades, attention has increasingly been paid to the environmental effect of the biocides used, leading to the emergence of new legislation that seeks to restrict their use, especially in North America and Europe (Ribeiro et al. 2018). Some additives are no longer allowed to be used, such as phenylmercurials, however, some others have been restricted in relation to their concentration in formulations (Paulus 2004; Falkiewicz-Dulik et al. 2015a). Some examples of these are aromatic and halogenated derivatives. Conventional biocides commonly added in commercial formulations include 5-Chloro-2-(2,4-dichlorophenoxy) phenol (Triclosan), 2-octyl-4-isothiazolin-3-one (OIT), dichlorooctylisothiazolone (DCOIT), methyl-2-benzimidazolecarbamate (Carbebendazim), and 3-(3,4-dichlorophenyl)-1,1-dimethylurea (Diuron) (Allsopp et al. 2004).

Usually, combinations of biocides are used to improve the spectrum of antifungal activity. However, the efficacy of such combinations, which are commonly determined in terms of minimum inhibition concentration (MIC) varies depending on the type and concentration of combination used. Therefore, more effective ("booster") compositions of biocides have been investigated in order to reduce the concentration of the active ingredients that would be used separately (Bellotti et al. 2012; Falkiewicz-Dulik et al. 2015a). For example, 2-methyl-4-isothiazolin-3-one (MIT) has relatively low antimicrobial activity, but it has shown a significant synergistic activity when combined with 1,2-benzisothiazolin-3-one (BIT) (Karsa and Ashworth 2002).

The addition of biocides in paints during the dispersion process is often not satisfactory, due to the fact that biocidal activity can be lost before the end of life of the coating (Sørensen et al. 2010; Mardones et al. 2019). It has been recorded that paint films in buildings, which should maintain their biocidal functionality for more than 10 years, actually do it for less than two years in extreme conditions (Eversdijk et al. 2012). The reasons that limit the efficiency of antimicrobial paints and coatings are: loss of biocidal efficacy of the film on the surface due to leaching or engaged in reactions with resin, pigment, and additives; degradation of the bioactive component by environmental factors; incompatibility between biocides within the paint; and commercial organic biocides could be used as a nutrient source by some microorganism (Edge et al. 2001; Falkiewicz-Dulik et al. 2015a; Andersson-Trojer et al. 2015; Kakakhel et al. 2019). The architecture of the paint films must be considered for a better understanding of this issue. Waterborne acrylic paints, most commonly used in buildings, are constituted by aqueous dispersion of polymer lattices, which after drying leads to the emergence of macroscopic pores. Therefore, this porosity favors the release of the biocides which reside in the pores or are adsorbed on particles (Andersson-Trojer et al. 2015).

Considering the facts mentioned here, several efforts are currently seeking to improve the bioactive ingredients used in formulation to prolong their useful life, to replace the toxic conventional ones, to decrease the concentrations used, and prevent their loss from the film. In this context, the emergence of nanotechnology presents an extensive field of study for new functional materials with size-dependent properties. Some strategies are based on finding free nanoparticles with antimicrobial activity, while others are based in associate antimicrobial agents to nanomaterials to protect them and control their release.

12.4 Nanotechnology for the Control of Fungal Growth

Nanotechnology has developed as one of the most groundbreaking scientific fields in the last few decades, since it exploits the enhanced reactivity of materials at a nanoscale. Currently, most scientists believe that nanomaterials are one of the mainstays of developing science and technology in the twenty-first century.

Materials with at least one dimension lower than 100 nm are considered "nano," and these materials have different physical and chemical properties from those in the microscale or bulk form

(Haupert and Wetzel 2005; Singh 2016b). They have surface/volume ratio higher than bulk ones, which is reflected in the fact that the majority of the atoms are located on the surface (Morones et al. 2005; Mathiazhagan and Joseph 2011). A large number of public domain investigations proposed that properties of nanoparticles (NPs) fundamentally depend on their size and shape. Similarly, the antimicrobial activity of NPs can be changed if they are modified (Morones et al. 2005; Raza et al. 2016).

NPs can be prepared using two major approaches: top-down (reduce in size to a suitable material) and bottom-up (build them from elemental entities like atoms and molecules) (Mittal et al. 2013). The top-down approach uses physical or chemical methods, which frequently have high-energy demand and produce NPs with surface imperfections; milling is a typical example of this approach (Landge et al. 2017; Thakkar et al. 2010). Bottom-up approaches are based on chemical or biological synthesis and usually produce colloidal dispersions of the NPs with fewer defects and more homogeneous chemical composition (Cao and Sun 2009; Singh et al. 2016).

Antifungal activity of nanomaterials depends on their properties, such as surface charge, composition, size, shape, and partial oxidation capacity (Kumar et al. 2013; Mittal et al. 2013; Singh et al. 2016). Studies that attempt to explain the mechanism of action of NPs are focused on assays that intrude the inhibition of spore germination, radial mycelial growth, and aflatoxin synthesis (Kasprowicz et al. 2010; Kairyte et al. 2013; Mitra et al. 2017). Taking into account the chemical nature of the NPs that were probed to be active against fungal strains, they can be classified into three main groups: metal, non-metal, and hybrid (metal/non-metal). Table 12.1 shows various nanomaterials which can be effectively used in the management of different fungi. Figure 12.1 shows NPs adhered to silica filler after the green synthesis process (Figure 12.1a,b). Visible differences between the surfaces of the siliceous material decorated with NPs and the original are indicated in Figure 12.1 c1 and c2.

12.4.1 Metal Nanoparticles

Metal (and metal oxide) NPs can be fine-tuned with several chemicophysical properties, size, surface to volume ratio, structural stability, and target affinity, for better efficiency and to facilitate their application in different fields (Elbourne et al. 2017). The most widely used synthesis method for metal NPs is wet-chemical, where NPs are formed from respective metal ions using a liquid system which contains reducing (e.g. sodium borohydride, methoxy polyethylene glycol, or hydrazine) and stabilizing agents (e.g. sodium dodecyl benzyl sulfate, polyvinyl pyrrolidone, or citrate) (Badawy et al. 2010; Singh 2016b). The chemical methods have been questioned due to the use of toxic solvents and the generation of hazardous by-products, which has increased interest in eco-friendly alternatives framed in green syntheses like bioreduction or biological methods (Singh et al. 2016).

The most studied metal NPs with antimicrobial activity are: silver (Ag), copper (Cu), zinc oxide (ZnO), and titanium dioxide (TiO_2) (Ruffolo et al. 2010; El Saeed et al. 2016; Nguyen et al. 2019). Without a doubt, AgNPs are the best known for their antimicrobial activity and their synthesis typically occurs by reduction of soluble silver salts in the presence of reducing agents such as citrate, glucose, ethylene glycol, or sodium borohydride (Badawy et al. 2010; Singh et al. 2016). AgNPs with different shapes have been produced besides the most common spherical ones, including pyramids, rods, triangular prisms, and cubes (Pal et al. 2007; Raza et al. 2016). Silver is able to control various pathogens with relative safety, if compared to synthetic fungicides, and it displays multiple modes of inhibitory action to microorganisms (Ogar et al. 2015). Hitherto, AgNPs have been shown to be effective biocides against *Aspergillus niger*, *Alternaria alternata*, *Alternaria bras-*

Table 12.1 Details about antifungal efficacy of different nanoparticles against a variety of fungi (MIC: minimum inhibitory concentration).

Nanoparticles	Synthesis method	Size of NPs (nm)	Test method	Fungi tested	MIC	References
Metal-based nanoparticles						
AuNPs	Chemical: using $SnCl_2$ and $NaBH_4$	7–15	Broth dilution (CLSI/M27-A3)	*Candida albicans, Candida tropicalis,* and *Candida glabrata*	4–16 $\mu g\,ml^{-1}$	Ahmad et al. (2013)
ZnONPs	Biological: using *Artocarpus gomezianus* fruit extract	30–40	Agar diffusion	*Aspergillus niger*	—	Anitha et al. (2018)
SiO_2 NPs doped Fe_2O_3	Chemical: coprecipitation method	6–20	Agar diffusion with dilutions	*Candida parapsilosis* and *Aspergillus niger*	0.18–0.24 $mg\,l^{-1}$	(Arshad et al. 2019)
ZnNPs-doped SiO_2	Chemical: deposition precipitation	7	Disc diffusion	*C. parapsilosis* and *A. niger*	—	Arshad et al. (2018)
AgNPs, CuNPs, and FeNPs	Biological: using tea leaf extracts	10–20, 26–40 and 42–60, respectively	Oxford cup diffusion And broth dilution (CLSI, 2012)	*Aspergillus flavus* and *A. parasiticus*	8 (Ag)-32(Cu)-128(Fe) $\mu g\,ml^{-1}$	Asghar et al. (2018)
AgNPs, CuNPs, ZnONPs, and AuNPs	Biological: extracellular products from *Enterococcus faecalis*	9–130, 20–90, 16–96 and 20–70, respectively	Broth dilution (potato dextrose)	*C. albicans* MTCC 3017, *Candida neoformans* MTCC 1347, *A. niger* MTCC 282, and *Fusarium oxysporum* MTCC 284	8–64 (Ag) 8 to ≥128(Cu) 8 to ≥128 (ZnO) ≥128 (Au) $\mu g\,ml^{-1}$	Ashajyothi et al. (2016)
AgNPs	Biological: using *Cassia* spp. aqueous leaf extracts	15–30	Agar well diffusion	*A. niger*, *Aspergillus fumigatus*, *A. flavus*, *Penicillium* sp., *C. albicans*, *Rhizoctonia solani*, *F. oxysporum*, and *Curvularia* sp.	—	Balashanmugam et al. (2016)

(Continued)

Table 12.1 (Continued)

Nanoparticles	Synthesis method	Size of NPs (nm)	Test method	Fungi tested	MIC	References
AgNPs	Biological: using *Selaginella* *Bryopteris* leaf extracts	15–30	Agar well diffusion	*A. niger*	1 mg/100 μL	Dakshayani et al. (2019)
AgNPs	Biological: using extracellular products from *Fusarium oxysporum*	93	Agar diffusion (M51-A2) and broth dilution (M38-A2) of CLSI	*A. flavus* MCT 00335, *Aspergillus nomius* MCT 00328, *A. parasiticus* MCT 00336, *Aspergillus melleus* MCT 00144, and *Aspergillus ochraceus* MCT 00435.	MIC_{50}: 4–8 $\mu g\,ml^{-1}$	Bocate et al. (2019)
FeNPs doped ZnO	Biological: using *Hibiscus Rosa* leaf extracts	15–170	Agar diffusion	*C. albicans*	—	Chai et al. (2019)
ZrO_2NPs	Chemical: calcination	17	Mycelium growth	*R. solani*	>100 $\mu g\,l^{-1}$	Derbalah et al. (2019)
AgNPs and AgClNPs	Biological: using *Malva sylvestris* leaf extracts	10–50	Broth microdilution according to CLSI (M7-A7)	*C. glabrata* CCUG 35267, *Candida orthopsilosis* ATCC 20503, *C. parapsilosis* ATCC 22019, and *C. tropicalis* CCUG 34274	7.8–62 $\mu g\,ml^{-1}$	Feizi et al. (2018)
AgNPs	Biological: using the culture supernatants of *Cryptococcus laurentii* and *Rhodotorula glutinis*	35–400 and 15–220 respectively	Broth microdilution	*B. cinerea*, *P. expansum*, *A.niger*, *Alternaria* sp. and *Rhizopus* sp.	2–4 $mg\,l^{-1}$	Fernández et al. (2016)

AgNPs	Chemical: coated with dodecanethiol		Antibiofilm resistance	*C. albicans* ATCC 18804	—	Ferreira et al. (2019)
CuONPs	Chemical: precipitation and autoclaved at 120 °C for 2 h	20–100	Agar diffusion	*Penicillium* sp.	—	Ghorbani et al. (2018)
TiO_2NPs and ZnONPs	Chemical synthesis	50–100	Agar diffusion	*C. albicans* ATCC 10231	—	Haghighi et al. (2011)
MgNPs doped ZnONPs	Chemical: coprecipitation	43–63	Agar diffusion Broth dilution	*C. albicans* ATCC 10231	2000 $\mu g\,ml^{-1}$	Haja Hameed et al. (2015)
ZnONPs	Commercial	~20	Agar dilution	*Botrytis cinerea* *Penicillium expansum*	>12 $mmol\,l^{-1}$ (>9.8 $\mu g\,ml^{-1}$)	He et al. (2011)
CuONPs and Fe_2O_3NPs	Biological: using *Euphorbia helioscopia* leaf extracts	7–10	Agar well diffusion	*Cladosporium herbarum*	—	Henam et al. (2019)
MnNPs doped ZnS	Chemical synthesis	~2.7	Agar dilution	*A. niger, A. fumigatus, A. flavus, Aspergillus terreus, Trichophyton mentagrophytes, Trichophyton rubrum, Trichophyton gypsum, Microsporum audouinii,* *Microsporum canis, F. oxysporum*	>30% p/v	Ibrahim et al. (2017)
AuNPs	Biological: using *Abelmoschus esculentus* aqueous extract	62	Agar well diffusion	*Puccinia graminis tritci*, *A. flavus*, *A. niger* and *C. albicans*	—	Jayaseelan et al. (2013)
ZnONPs	Commercial aqueous dispersion	195	Agar dilution with light exposition (400 nm lamp)	*Botrytis cinerea*	$>5\times10^{-3}$ M	Kairyte et al. (2013)

(Continued)

Table 12.1 (Continued)

Nanoparticles	Synthesis method	Size of NPs (nm)	Test method	Fungi tested	MIC	References
CuNPs	Chemical synthesis: with isopropyl alcohol	3–10	Agar disc diffusion	*Phoma destructiva* DBT-66, *Curvularia lunata* MTCC 2030, *Alternaria alternata* MTCC 6572, and *F. oxysporum* MTCC 1755	—	Kanhed et al. (2014)
AgNPs	Biological: using the culture filtrate of *Fusarium chlamydosporum* NG30 and *P. chrysogenum*	6–26 and 9–17 respectively	Broth microdilution (germination inhibitory effect)	*A. flavus* NRRL 3145 and *Aspergillus ochraceus* ATCC 22947	45–51 $\mu g\,ml^{-1}$	Khalil et al. (2019)
AgNPs, TiO_2NPs, ZnONPs, and SiO_2 NPs	Chenical synthesis	5, 11, 25, and 35 respectively	Broth microdilution	*C. albicans* ATCC 10145	—	Khan et al. (2018)
ZnONPs	Biological: using *Medicago sativa* aqueous extract	~14	Broth microdilution according to CLSI	*C. albicans* ATCC10231	9.31 $\mu g\,ml^{-1}$	Król et al. (2019)
AuNPs	Biological: using *Croton Caudatus Geisel* aqueous extract	20–50	Agar diffusion with dilutions	*A. niger* MTCC 281, *A. flavus* MTCC 277, *A. terreus* MTCC 1782, *F. oxysporum* MTCC 284, and *C. albicans* MTCC 227.	50–150 μg/well	Vijaya Kumar et al. (2019)
PtNPs	Biological: using *Xanthium strumarium* aqueous extract	22	Agar well diffusion with dilutions	*A. niger* MTCC 281, *A. flavus* MTCC 277, *C. tropicalis, C. parapsilosis,* and *C. albicans* (MTCC 227)	50 μg/well	P. V. Kumar et al. (2019a)
AgNPs	Biological: using the culture filtrate of *Trichoderma viride* MTCC 5661		Agar dilution	*Alternaria brassicicola* and *F. oxysporum*	5 $\mu g\,ml^{-1}$	Kumari et al. (2019)

AgNPs/Zeolitet	Chemical: ion exchange and thermal treatmen	2–34	Agar diffusion	*Trichoderma harzianum* Rifai	—	Machado et al. (2019)
ZnONPs	Biological: using *Pithecellobium dulce* peel extract	30	Broth dilution (biomass weighing)	*A. flavus* and *A. niger*	>1000 ppm	Madhumitha et al. (2019)
AgNPs	Biological: using lignin	10–15	Agar diffusion with dilutions	*A. niger*	0.3 μg ml^{-1}	Marulasiddeshwara et al. (2017)
NiONPs coupled CdO	Chemical: presipitation with ammonia and calcined at 200 °C		Agar well diffusion	*A. terreus*	—	Nallendran et al. (2019)
AgNPs	Commercial	~55	Agar dilution	*Penicillium brevicompactum, Aspergillus fumigatus, Cladosporium cladosporoides, Chaetomium globosum, Stachybotrys chartarum, and Mortierella alpina*	>100 ppm	Ogar et al. (2015)
Fe_2O_3NPs	Chemical: using tannic acid solution	10–30	Agar diffusion Broth dilution	*Trichothecium roseum, C. herbarum, P. chrysogenum, A. alternata* and *A. niger*	0.016–0.063 mg ml^{-1}	Parveen et al. (2018)
AuNPs	Chemical: with carbosilane–synthesis with $NaBH_4$	~3	Broth microdilution according to CLSI (M7-A7)	*C. albicans* ATCC 10231 and *C. glabrata* ZMF40	47.6 ppm	Peña-González et al. (2017)
Ag-TiO_2NPs	Chemical: electrochemically deposited of Ag on commercial TiO_2 NPs	20	Agar diffusion	*A. niger, Aspergilus terreus, A. flavus, Chaetominum globusum, Mirothecium verrucaria, Paecilomyces varioti, Aureobasidium pullulans, Penicilium cyclopium, Penicilium funiculosum, Penicilium glaucum, T. viride.*	—	Petica et al. (2019)

(Continued)

Table 12.1 (Continued)

Nanoparticles	Synthesis method	Size of NPs (nm)	Test method	Fungi tested	MIC	References
Non-metal and metal/non-metal (hybrid) nanoparticles						
β-d-glucan	—	60	Agar diffusion	*Pythium aphanidermatum*	—	Anusuya and Sathiyabama (2014)
PLGA loaded with Amphotericin B	Chemical: emulsion solvent evaporation	˂200	Broth microdilution (M27-A2) of CLSI.	*C. albicans* ATCC 64548 and *C. neoformans* ATCC 90112.	0.05–0.5 $\mu g\,ml^{-1}$	Moraes Moreira Carraro et al. (2017)
Chitosan loaded with clove essential oil	Chemical: emulsionic gelation technique	40–100	Agar dilution	*A. niger*	1.5 $mg\,ml^{-1}$	Hasheminejad et al. (2019)
Luliconazole	Nanocrystal	263–611	Agar well diffusion	*C. albicans* MTCC No 183	—	M. Kumar et al. (2019b)
Chitosan-gellan gum loaded with Ketoconazole	Synthesis by nanosuspension	~156	Mycelium growth inhibition	*A. niger* NICM 590	>100 $\mu g\,ml^{-1}$	Kumar et al. (2016)
Chitosan loaded with *Schinus molle* L. essential oil	Chemical: ionotropic gelation	~500	Broth microdilution	*A. parasiticus* ATCC 16992	>500 $\mu g\,ml^{-1}$	López-Meneses et al. (2018)
Chitosan/Ag	Chemical synthesis	373	Broth dilution Mycelium growth	*F. oxysporumin*	100 $\mu g\,ml^{-1}$ >1000 $\mu g\,ml^{-1}$	Dananjaya et al. (2017)
Chitin nanofibers/Ag	Physical: UV light reduction	10	Antibiofilm resistance	*A. alternata*, *Alternaria brassicae*, *Alternaria brassicicola*, *Bipolaris oryzae*, *Botrytis cinerea*, and *Penicillium digitatum*	—	Ifuku et al. (2015)
Chitosan	Chemical: by emulsion	~173	Broth microdilution	*C. albicans*	>75 $\mu g\,ml^{-1}$	Pan et al. (2019)
Sodium hyaluronate/TiO_2	Chemical: sol-gel method and calcined at 650 °C	23	Agar diffusion	*A. niger*	—	Safaei and Taran (2017)

CuO/C	Chemical: precipitation with sucrose and calcined at 500 °C	25–30	Agar dilution	*A. niger* and *A. flavus*	>1000 ppm	Roopan et al. (2019)
Alginate/CuO	precipitation, mixed, and dried	37	Agar diffusion	*A. niger*	—	Safaei et al. (2019)
Chitosan/Ag loaded with fluconazole	Mixed	30–50	Agar diffusion	*C. albicans* MTCC 4748 and *T. rubrum* MTCC 7859	—	Samrat et al. (2016)
Chitosan loaded with anionic proteins from *Penicillium oxalicum*	Mixed	10–30	Agar diffusion	*F. oxysporum*, *Pyricularia grisea*, and *Alternaria solani*	—	Sathiyabama and Parthasarathy (2016)
Clotrimazole/Ca $(OH)_2$	Mixed	294	Agar dilution	*A. niger* and *P. corylophilum*	4%	Sequeira et al. (2017)
Pectin/Ag	Mixed	5–50	Agar diffusion	*Aspergillus japonicus*	—	Su et al. (2019)
5-amino-2-mercapto benzimidazole/Ag_3O_4	Chemical: presipitation, mixed, and dried	18–29	Agar diffusion with dilution	*A. niger*	20 µl	Suresh et al. (2016)
Thyme and dill oil/Cu	Chemical synthesis and mixed	100–250	Conidia germination Mycelium growth	*Colletotrichum nymphaeae*	90–600 ppm	Weisany et al. (2019)
Chitosan loaded with *Origanum vulgare* oil	Electrospraying synthesis	290–483	Agar dilution	*A. alternata*	0.02–0.005% (w/v)	Yilmaz et al. (2019)
16-mercaptohexadecanoic acid/amino silano/magnetic particles	Chemical synthesis	—	Fungal cell viability in planktonic form	*C. albicans*	—	Wilczewska et al. (2019)

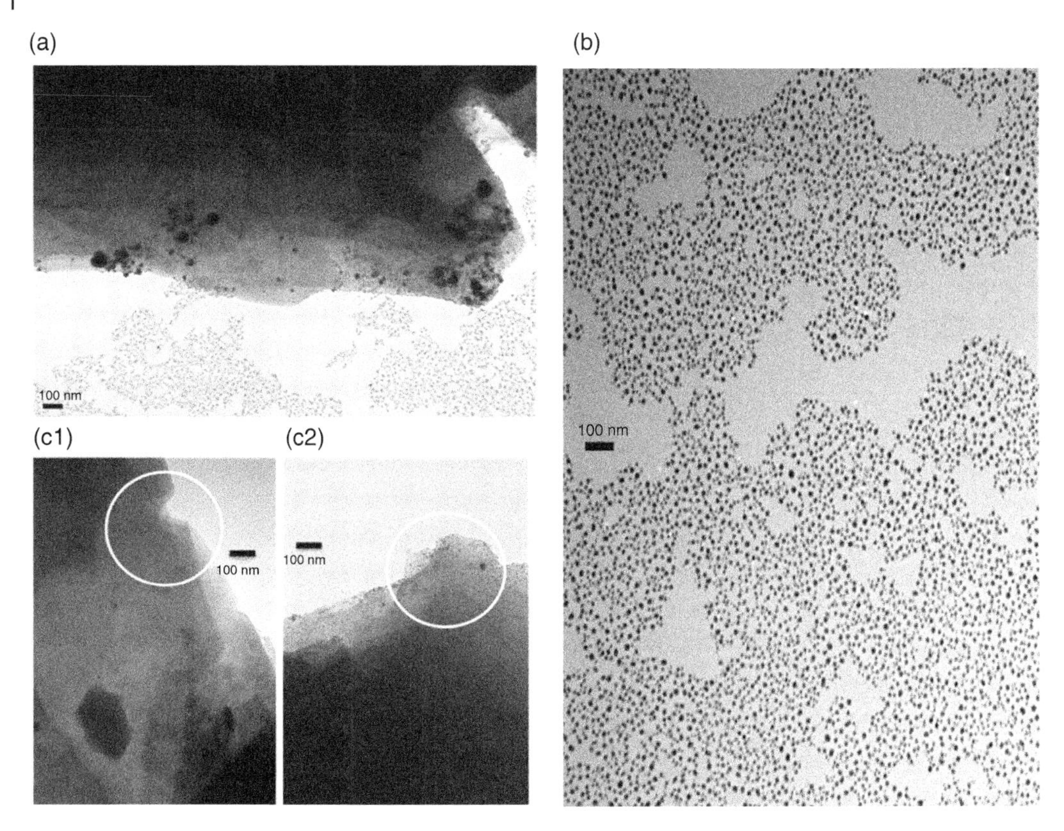

Figure 12.1 TEM micrographs of: (a) nanoparticles adhered to silica filler during the synthesis process; (b) free nanoparticles obtained by green synthesis; (c1) Siliceous filler before decorating with nanoparticles, and (c2) result of decorating with silver nanoparticles by green synthesis.

sicicola, *Alternaria solani, Botrytis cinerea, Fusarium oxysporum, Penicillium* spp., *Rhizoctonia solani*, and *Colletotrichum* spp. (Sardella et al. 2019).

AgNPs liberate silver ions in the fungal cells, which could attack the respiratory chain and cell division leading to cell death (Moritz and Geszke-Moritz 2013). In addition, they interact with the thiol groups of several enzymes, inactivating them and affecting processes such as nutrition (Chung and Toh 2014). Silver can also generate reactive oxygen species (ROS) that have high cytotoxic activity and can cause cell death (Morones et al. 2005; Singh et al. 2016). Silver ions can cause denaturation of proteins and DNA, which affects the replicative machinery in the fungal cell (Dananjaya et al. 2017). During the formation of the germ tube, the wall of the hypha is thinner and more fragile in the apical part; this may be the moment that allows the interaction of AgNPs in the cell wall, causing an increase in the permeability of the membrane, and with it an alteration in conidial viability (Jo et al. 2009; Mahmoud et al. 2014).

ZnONPs and their mechanism of action against two post-harvest pathogenic fungi *B. cinerea* and *Penicillium expansum*, were investigated (He et al. 2011). In this case, *P. expansum* was found to be more sensitive to treatment with ZnONPs than *B. cinerea*; NPs inhibited the growth by affecting cellular functions, which caused deformation in fungal hyphae, and prevented the development of conidiophores and conidia of *P. expansum* (He et al. 2011). In addition, Kairyte et al. (2013) obtained similar results against a *B. cinerea* strain when exposed to these NPs in suspension. Similarly, Sharma et al. (2010) studied the antifungal activity of ZnONPs against *Fusarium* sp. and proposed the fungal growth inhibition was due to the rupture of the cell membrane, resulting in the possible decrease in fungal enzymatic activity.

Iron oxide NPs were evaluated for their antifungal activity against the following: *Trichothecium roseum*, *Cladosporium herbarum*, *Penicillium chrysogenum*, *A. alternata*, and *A. niger*. The maximum inhibition in spore germination was caused against *T. roseum* (~88%) followed by *C. herbarum* (~85%) (Parveen et al. 2018). Iron oxide NPs cause oxidative stress through the generation of ROS and Fenton reaction. Since iron is a strong reducing agent, it induces the decomposition of functional groups in membrane proteins and lipopolysaccharides. Iron-based NPs also cause oxidation by intracellular oxygen, leading to oxidative damage via Fenton reaction. These NPs penetrate through disrupted membranes causing further damage and death of cells (Parveen et al. 2018).

Antimicrobial NPs obtained from aqueous plant extracts are reported to be very promising because these are accessible, effective, low cost, and eco-friendly (Mittal et al. 2013; Singh et al. 2016). AgNPs synthesized using aqueous extracts from different plants (*Schinus molle*, *Equisetum giganteum*, and *Ilex paraguariensis* Saint Hilaire) have been studied by Barberia et al. (2019) against fungal strains, i.e. *A. alternata* and *Chaetomium globosum*. These filamentous fungi are known to deteriorate indoor waterborne acrylic paints (Bellotti et al. 2013). Suspension with AgNPs from *E. giganteum* proved to be the most active, with a minimum inhibitory concentration of 3.3 and $67.5\,\mu g\,ml^{-1}$, respectively (Barberia-Roque et al. 2019). Biosynthesis of NPs is considered an economical and eco-friendly approach; moreover, it can be a novel substitute for NPs obtained by chemical synthesis (Xue 2016; Malkapur et al. 2017). The enhanced antifungal activity was reported for AgNPs biosynthesized by cell-free filtrate of *Trichoderma viride* (MTCC 5661) compared to chemically synthesized AgNPs of similar shape and size (Kumari et al. 2019). In this sense, biosynthesized AgNPs enhanced the reduction in dry weight by 20% and 48.8% of *F. oxysporum* and *A. brassicicola*, respectively, in comparison to their chemical counterparts; *A. brassicicola* revealed that osmotic imbalance and membrane disintegration are the major cause for fungal cell death after treatment with the biosynthesized AgNPs (Kumari et al. 2019).

12.4.2 Non-metal and Hybrid (Metal/Non-metal) Nanoparticles

Due to its structure, Kraft lignin formed by phenyl propanol and aryl-alkyl ether bonding can be a good source of polyols. The multiple hydroxyl groups present in the lignin's structure are essential raw materials for polyurethane production. Also, for polyolefins, polyethylene terephthalate (PET), and polycarbonate production, the plastics can be either replaced or enriched with bio-based components (Brodin et al. 2017). Considering sustainability concerns and the fact that petroleum products are commonly used in the polyurethane industry, bio-based polyols and lignopolyols could be an environmentally friendly solution (Mahmoud et al. 2014). Although a bioplastic is characterized as being produced from a renewable source, bioplastics are not necessarily biodegradable. As an example, biopolyethylene (BioPe) is similar to the fossil-based polyethylene and thus is not biodegradable. Hence, plastic biodegradability is determined by the chemical structure rather than origin (Brodin et al. 2017).

There are several examples of organic NPs obtained by encapsulation with various polymers such as chitosan, alginate, and poly(lactic-co-glycolic acid) (Pan et al. 2019; Safaei et al. 2019; Yilmaz et al. 2019). Bioactive biogenic compounds obtained from plants as essential oils or some of their components have been used. For example, chitosan nanoparticles (ChNPs) loaded with clove essential oil (CEO) were developed with the emulsion ionic gelation technique to improve the antifungal efficacy of CEO (Hasheminejad et al. 2019). The ChNPs demonstrated a superior performance against *A. niger*, isolated from spoiled pomegranate, compared with free CEO being active with a minimum concentration of 1.5 mg ml^{-1}. Similarly, López-Meneses et al. (2018) prepared polymeric ChNPs loaded with *S. molle* essential oil (SEO). These NPs have been studied as a possible substitute for fungicides against *Aspergillus parasiticus* and showed significant inhibition on spore germination (>80%) and aflatoxin production (>59%) at a concentration of 500 mg ml^{-1}. Antifungal agents such as ketoconazole and amphotericin B loaded in chitosan-gellan gum and poly(lactide-co-glycolide) NPs, respectively, were assessed against *A. niger* (Kumar et al. 2016; Moraes Moreira Carraro et al. 2017).

A large number of nanofiber fabrication methods have been reported, including template synthesis, molecular self-assembly, and hydrothermal methods. Spinning methods, including electrospinning, blow spinning, centrifugal spinning, and draw spinning, allow researchers to fabricate nanofibers from a precursor solution. A variety of polymeric nanofibers can be obtained by spinning methods. Compared with other techniques, spinning allows easier integration into industrial large-scale production (Huang et al. 2019). Polymer nanofibers fabricated via the facile electrospinning technique, mainly biopolymers, have ease of processing, excellent biocompatibility, and non-toxicity (Ambekar and Kandasubramanian 2019). Chitin nanofibers with AgNPs have been synthesized by Ifuku et al. (2015) and tested against 11 fungal strains. These hybrid AgNPs/chitin films showed an inhibition of spore germination >90% with 8 of the strains used.

Roopan et al. (2019) synthesized the bioactive hybrid CuO/C nanocomposite using sucrose as a capping agent. The antifungal activity of CuO/C nanocomposite was tested against *A. niger* and *A. flavus* at 1000 ppm and about 70% and 90% of inhibition, respectively, was reported. The authors proposed that this nano-complex causes interrupted transmembrane e^- transport, cell membrane disruption, mitochondria damage, and cytoplasm leakage.

Other examples are metal-organic framework (MOF) nanosheets which have attracted great attention due to their distinctive characteristics such as nanoscale and tunable thickness, high-aspect-ratio, large surface area, more exposed accessible active site, favorable mechanical flexibility, and optical transparency (Li et al. 2019). Recently, nanostructured MOFs, as a kind of crystalline material, were also constructed by the diversified interconnection of the organic linkers and metal nodes. These features endow MOF nanosheets with enhanced applications in gas separation, catalysts, sensing, energy storage and transfer, and enzyme inhibition.

12.4.3 Nanotechnological Management of Indoor Fungi

As mentioned, nanotechnology in general and NPs (nanomaterials) in particular play a key role in the control of growth of various fungi-causing infections in humans, plants, and other life forms. Colonization of fungi in indoor environments is considered a major concern because they have ability to cause many health-related issues. Therefore, management of indoor fungi is extremely important to avoid the ill effects caused by them. As discussed earlier, conventional approaches are available for the control of indoor fungi, but they have certain limitations. In this context, considering the antifungal potential of various NPs as discussed earlier, it is believed that such NPs can be effectively used in the management of indoor fungi. It is most unfortunate that very few reports are available on the management of indoor fungi using nanotechnological solutions. However, available reports revealed that NPs can be used as novel, effective, and eco-friendly alternative antifungal agents to chemical fungicides. Some of the nanotechnological applications that have been reported to the management of indoor fungi include the adoption of nano-enabled disinfectants, surface biocides, air filters, packaging, and rapid detection methods for contaminants (Vance et al. 2015; He and Hwang 2016; Chen et al. 2017).

Although the fate and potential toxicity of nanomaterials are not fully understood at this time and scientific risk assessments are required, it is evident that there have been significant advances in their applications (Brincat et al. 2016; King et al. 2018; Jogee and Rai 2020). Peanuts are vulnerable to fungal infections during long-term storage. Fungi infecting peanuts are toxigenic and cause health hazards. Further, the antifungal potential of AgNPs was evaluated and showed ability to inhibit fungal growth. *Cymbopogon citratus* leaf extract-mediated AgNPs were found to have prominent antifungal potential against all test fungi and its MIC was found to be $20\,\mu g\,ml^{-1}$ (Jogee et al. 2017). Pokhum et al. presented a facile and cost-effective approach to remove airborne microbes from indoor air by employing silver (Ag) and zinc oxide (ZnO) to decorate fibrous air filters (Pokhum et al. 2018). This method successfully led to homogeneous coating of active nanomaterials on the filter's surface. The developed Ag/ZnO air filter reduced the airborne psychrotrophic germ concentrations by ~50% and its efficiency increased to ~70% when combined with UVA illumination. Based on these results, a simple and low-cost ZnO/Ag air filter was successfully introduced as an effective strategy for removal of psychrotrophic microbes from indoor air.

Household cleaning products have been incorporating antimicrobial agents for past several decades, and they have achieved overwhelming success in gaining the confidence of the consumers. The staggering customer demands have motivated the industrial sector to constantly look for new effective antimicrobial products. For example, Microban® is a combination of polyvinylidene difluoride coating matrix along with silver as an active ingredient. Microban® technology offers protection from the deterioration of the coating from mold and mildew. The silver-based particles, on contact with microbes, do not allow the reproduction of microbes by interfering with metabolism and disrupting/damaging the cell walls. Moreover, the active ingredient interferes with the conversion of nutrients into energy, thereby inhibiting the reproductive process. This product, when used in waterborne or solvent-borne paint or coating, provides excellent protection in both indoor and outdoor environments (Tiwari and Chaturvedi 2018).

12.5 Hygienic Coatings and Nanotechnology

In order to inhibit or prevent the growth of microorganisms, including fungi, on building materials, the disruption of their vital processes is required. Figure 12.2 shows microscopic pictures of three mold strains commonly found in indoor spoilage materials growing on coatings in controlled conditions.

Hygienic paints are important tools to avoid indoor biological colonization and prevent biodeterioration which creates health problems in people and pets (Stobie et al. 2010; Falkiewicz-Dulik et al. 2015b). These functional paints can be used for painting in dwellings and hospitals. They can be also used in the food industry because in this sector they must deal with microbial growth as one of the most critical issues affecting production, processing, transport, and storing. Several applications of metal NPs in food safety are currently available (e.g. packaging material, air filter coatings) (Souza and Fernando 2016). Nanotechnology applied for the design of antimicrobial surfaces can eliminate pathogens in close proximity to the surface, preventing biofilm formation (Kaiser et al. 2013). The precise biocidal mechanism arising from these materials is complex in nature and is dependent on both the microbe and nanomaterial used (Bapat et al. 2018).

Additive paints and coatings with antimicrobial NPs have been studied (Kumar et al. 2008; Jo et al. 2009; Zielecka et al. 2011; Holtz et al. 2012; Dominguez-Wong et al. 2014; Barberia-Roque et al. 2019; Machado et al. 2019). There are two possible ways to incorporate NPs into a paint formulation: free or supported in other material. The direct use of metal NPs such as Ag, Cu, and ZnO in waterborne paints (latex type) can result in the decrease of their antimicrobial activity due to their reactivity with other components present in the formulation or their agglomeration (Zielecka et al. 2011; Bellotti et al. 2015; Arreche et al. 2017). Taking this into account, there are several

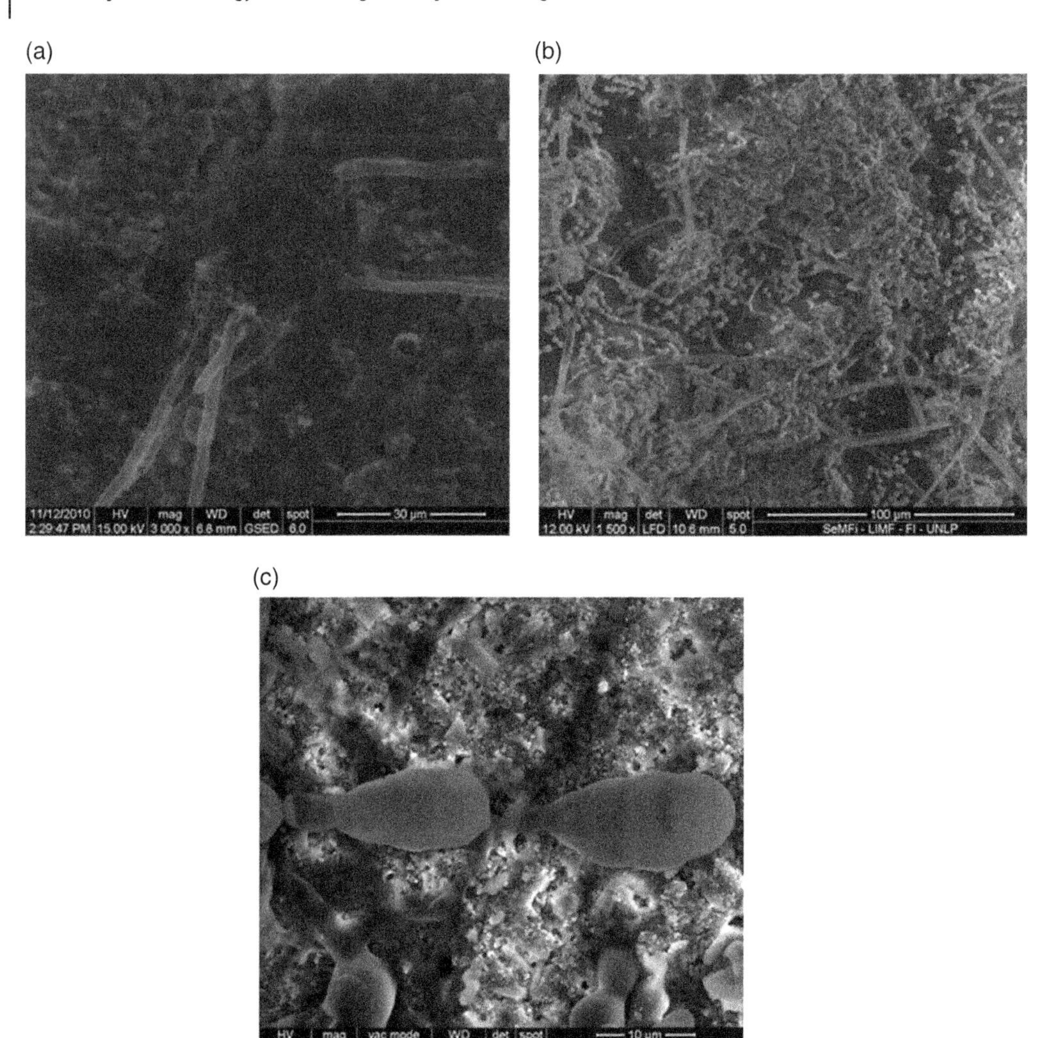

Figure 12.2 Active fungal growth on coatings: (a) *Chaetomium globosum* (KU936228), (b) *Aspergillus versicolor* (MG725821), and (c) *Alternaria alternata* (KU936229).

works performed that showed the efficient incorporation of bioactive nano-additives (in supported or immobilized form) in other materials to be applied in paints (Zielecka et al. 2011; Arreche et al. 2019; Machado et al. 2019).

Often the biocidal activity of the organic compounds ends long before the lifetime of the coating due to its low retention or degradation (Edge et al. 2001; Mardones et al. 2019). Therefore, usually they are loaded in natural or synthetic NPs that act as carriers (Hendessi et al. 2016; Kamtsikakis et al. 2017; Nguyen-Tri et al. 2018). However, some nanostructures have been developed which can be used as carriers in organic or inorganic matrix associated to the bioactive compound by electrostatic or covalent bonds (Hendessi et al. 2016; Kamtsikakis et al. 2017). For example, carvacrol, the active agent of essential thyme oil, has been loaded in halloysite nanotubes as a natural carrier to be applied in paints and coatings (Hendessi et al. 2016; Alkan Tas et al. 2019).

In the specific case of the paints, nanofunctionalized components commonly used, such as resins, pigments, fillers, and additives, have been reported (Kumar et al. 2008; Stobie et al. 2010; Dominguez-Wong et al. 2014; Fernández and Bellotti 2017; Machado et al. 2019). In this sense, conventional pigments such as TiO_2 and $CaCO_3$ have been modified at nanoscale level to gain antimicrobial functionality (Ferreira et al. 2013; Dominguez-Wong et al. 2014). Another example of a paint nano-functionalized component is acrylic resin associated to ZnONPs which have both anti-electrostatic and antibacterial functionalities at a concentration of ~ 5 wt% (Xu and Xie 2003). Siliceous matrix has been used in coating technology by the application of natural clays such as halloysite nanotubes; other aluminosilicates intensively studied are zeolites, which were associated with Ag and Zn to incorporate in waterborne acrylic formulations and probed to be efficient against fungal growth (Pereyra et al. 2014; Machado et al. 2019). On the other hand, synthetic matrix based in sol-gel technology have been incorporated to architectural paints (Arreche et al. 2017; Arreche et al. 2019). The synthesized nano-spheres by sol-gel method with Ag and Cu NPs were assessed in controlled conditions, showing a broader spectrum of antimicrobial activity (Zielecka et al. 2011).

12.6 Conclusion

The search for alternatives to control fungi in indoor environments has been nourished by the great impulse that nanotechnology has shown in recent decades. Antifungal nanoparticles that seek to replace commercial active compounds have led to the production of a large number of published works, but fewer publications are found in relation to specific applications of these. In relation to articles that deal specifically with this topic, it can be observed that there is diversity in assessment methods (e.g. agar diffusion, dilution in solid or liquid cultured medium, antibiofilm test, resistance to biodeterioration of films) and fungal strains selected as target. Mostly, the tests performed are carried out in controlled laboratory conditions.

The development of "smart" surfaces in nanotechnology, capable of responding to microbial cell interaction and avoid the biofilm development, is still a challenge. Eco-friendly biogenic compounds are intensely investigated; their incorporation in paint formulation largely requires the application of nanotechnology to the design of the nanostructured carriers. It is worth mentioning that it would be interesting in the future to count research works that faced the application and the assessment of these materials in more realistic conditions.

References

Adan, O.C.G. and Samson, R.A. (2011). *Fundamentals of Mold Growth in Indoor Environments and Strategies for Healthy Living*, 1e (eds. O.C.G. Adan and R.A. Samson). The Netherlands: Wageningen Academic Publishers.

Ahmad, T., Wani, I.A., Lone, I.H. et al. (2013). Antifungal activity of gold nanoparticles prepared by solvothermal method. *Materials Research Bulletin* 48 (1): 12–20.

Alkan Tas, B., Sehit, E., Erdinc Tas, C. et al. (2019). Carvacrol loaded halloysite coatings for antimicrobial food packaging applications. *Food Packaging and Shelf Life* 20: 100300.

Allsopp, D., Seal, K.J., Gaylarde, C.C. et al. (2004). Introduction to biodeterioration. *Introduction to Biodeterioration*: xiii–xiv.

Ambekar, R.S. and Kandasubramanian, B. (2019). Advancements in nanofibers for wound dressing: a review. *European Polymer Journal* 117: 304–336.

Anaya, M., Borrego, S.F., Gámez, E. et al. (2016). Viable fungi in the air of indoor environments of the National Archive of the Republic of Cuba. *Aerobiologia* 32 (3): 513–527.

Andersen, B., Frisvad, J.C., Søndergaard, I. et al. (2011). Associations between fungal species and water-damaged building materials. *Applied and Environmental Microbiology* 77 (12): 4180–4188.

Andersson-Trojer, M., Nordstierna, L., Bergek, J. et al. (2015). Use of microcapsules as controlled release devices for coatings. *Advances in Colloid and Interface Science* 222: 18–43.

Anitha, R., Ramesh, K.V., Ravishankar, T.N. et al. (2018). Cytotoxicity, antibacterial and antifungal activities of ZnO nanoparticles prepared by the Artocarpus gomezianus fruit mediated facile green combustion method. *Journal of Science: Advanced Materials and Devices* 3 (4): 440–451.

Annila, P.J., Lahdensivu, J., Suonketo, J. et al. (2018). Need to repair moisture- and mould damage in different structures in finish public buildings. *Journal of Building Engineering* 16, no. November 2017: 72–78.

Anusuya, S. and Sathiyabama, M. (2014). Preparation of β-d-glucan nanoparticles and its antifungal activity. *International Journal of Biological Macromolecules* 70: 440–443.

Arreche, R., Bellotti, N., Deyá, C., and Vázquez, P. (2017). Assessment of waterborne coatings formulated with sol-gel/Ag related to fungal growth resistance. *Progress in Organic Coatings* 108: 36–43.

Arreche, R.A., Igal, K., Bellotti, N. et al. (2019). Functionalized zirconia compounds as antifungal additives for hygienic waterborne coatings. *Progress in Organic Coatings* 128: 1–10.

Arshad, M., Qayyum, A., Shar, G.A. et al. (2018). Zn-doped SiO2 nanoparticles preparation and characterization under the effect of various solvents: antibacterial, antifungal and photocatlytic performance evaluation. *Journal of Photochemistry and Photobiology B: Biology* 185: 176–183.

Arshad, M., Abbas, M., Ehtisham-ul-Haque, S. et al. (2019). Synthesis and characterization of SiO_2 doped Fe_2O_3 nanoparticles: Photocatalytic and antimicrobial activity evaluation. *Journal of Molecular Structure* 1180: 244–250.

Ashajyothi, C., Prabhurajeshwar, C., Handral, H.K., and Kelmani, C.R. (2016). Investigation of antifungal and anti-mycelium activities using biogenic nanoparticles: an eco-friendly approach. *Environmental Nanotechnology, Monitoring and Management* 5: 81–87.

Asif, A., Zeeshan, M., and Jahanzaib, M. (2019). Assessment of indoor and outdoor microbial air quality of cafeterias of an educational institute. *Atmospheric Pollution Research* 10 (2): 531–536.

Badawy, A.M., Luxton, T.P., Silva, R.G. et al. (2010). Impact of environmental conditions (pH, ionic strength, and electrolyte type) on the surface charge and aggregation of silver nanoparticles suspensions. *Environmental Science and Technology* 44 (4): 1260–1266.

Balashanmugam, P., Balakumaran, M.D., Murugan, R. et al. (2016). Phytogenic synthesis of silver nanoparticles, optimization and evaluation of in vitro antifungal activity against human and plant pathogens. *Microbiological Research* 192: 52–64.

Bapat, R.A., Chaubal, T.V., Joshi, C.P. et al. (2018). An overview of application of silver nanoparticles for biomaterials in dentistry. *Materials Science and Engineering: C* 91: 881–898.

Barberia-Roque, L., Gámez-Espinosa, E., Viera, M., and Bellotti, N. (2019). Assessment of three plant extracts to obtain silver nanoparticles as alternative additives to control biodeterioration of coatings. *International Biodeterioration and Biodegradation* 141: 52–61.

Bellotti, N., Deyá, C., del Amo, B., and Romagnoli, R. (2012). "Quebracho" tannin derivative and boosters biocides for new antifouling formulations. *Journal of Coatings Technology and Research* 9 (5): 551–559.

Bellotti, N., Salvatore, L., Deyá, C. et al. (2013). The application of bioactive compounds from the food industry to control mold growth in indoor waterborne coatings. *Colloids and Surfaces B: Biointerfaces* 104: 140–144.

Bellotti, N., Romagnoli, R., Quintero, C. et al. (2015). Nanoparticles as antifungal additives for indoor water borne paints. *Progress in Organic Coatings* 86: 33–40.

Bensch, K., Groenewald, J.Z., Meijer, M. et al. (2018). *Cladosporium* species in indoor environments. *Studies in Mycology* 89: 177–301.

Bloom, E., Grimsley, L.F., Pehrson, C. et al. (2009). Molds and mycotoxins in dust from water-damaged homes in New Orleans after hurricane Katrina. *Indoor Air* 19 (2): 153–158.

Bocate, K.P., Reis, G.F., de Souza, P.C. et al. (2019). Antifungal activity of silver nanoparticles and simvastatin against toxigenic species of *Aspergillus*. *International Journal of Food Microbiology* 291: 79–86.

Brincat, J.P., Sardella, D., Muscat, A. et al. (2016). A review of the state-of-the-art in air filtration technologies as may be applied to cold storage warehouses. *Trends in Food Science and Technology* 50: 175–185.

Brodin, M., Vallejos, M., Opedal, M.T. et al. (2017). Lignocellulosics as sustainable resources for production of bioplastics – a review. *Journal of Cleaner Production* 162: 646–664.

Cao, Z. and Sun, Y. (2009). Polymeric N-halamine latex emulsions for use in antimicrobial paints. *ACS Applied Materials and Interfaces* 1 (2): 494–504.

Carlile, M.J., Watkinson, S.C., and Gooday, G.W. (2001). Spores, dormancy and dispersal. In: *The Fungi* (eds. M.J. Carlile and S.C. Watkinson), 185–243. Elsevier.

Chai, H.Y., Lam, S.M., and Sin, J.C. (2019). Green synthesis of magnetic Fe-doped ZnO nanoparticles via *Hibiscus rosa*-sinensis leaf extracts for boosted photocatalytic, antibacterial and antifungal activities. *Materials Letters* 242: 103–106.

Chapman, J.S. (2003). Biocide resistance mechanisms. *International Biodeterioration and Biodegradation* 51 (2): 133–138.

Chen, G., Luo, Q., Guo, X. et al. (2017). Study on mould germination risk in hydroscopic building wall. *Procedia Engineering* 205: 2712–2719.

Chow, N.A., Toda, M., Pennington, A.F. et al. (2019). Hurricane-associated mold exposures among patients at risk for invasive mold infections after hurricane Harvey-Houston, Texas, 2017. *MMWR. Morbidity and Mortality Weekly Report* 68 (21): 469–473.

Chung, P.Y. and Toh, Y.S. (2014). Anti-biofilm agents: recent breakthrough against multi-drug resistant *Staphylococcus aureus*. *Pathogens and Disease* 70 (3): 231–239.

Cole, G. and Samson, R.A. (1979). *Patterns of Development in Conidial Fungi*. London, UK: Pitman.

Crawford, J.A., Rosenbaum, P.F., Anagnost, S.E. et al. (2015). Indicators of airborne fungal concentrations in urban homes: understanding the conditions that affect indoor fungal exposures. *Science of the Total Environment* 517: 113–124.

Crook, B. and Burton, N.C. (2010). Indoor moulds, sick building syndrome and building related illness. *Fungal Biology Reviews* 24 (3–4): 106–113.

Dakshayani, S.S., Marulasiddeshwara, M.B., Sharath, S.K. et al. (2019). Antimicrobial, anticoagulant and antiplatelet activities of green synthesized silver nanoparticles using Selaginella (Sanjeevini) plant extract. *International Journal of Biological Macromolecules* 131: 787–797.

Dananjaya, S.H.S., Erandani, W.K.C.U., Kim, C. et al. (2017). Comparative study on antifungal activities of chitosan nanoparticles and chitosan silver nano composites against *Fusarium oxysporum* species complex. *International Journal of Biological Macromolecules* 105: 478–488.

De Clercq, N., Van Coillie, E., Van Pamel, E. et al. (2014). Detection and identification of xerophilic fungi in Belgian chocolate confectionery factories. *Food Microbiology* 46: 322–328.

Derbalah, A., Elsharkawy, M.M., Hamza, A., and El-Shaer, A. (2019). Resistance induction in cucumber and direct antifungal activity of zirconium oxide nanoparticles against *Rhizoctonia solani*. *Pesticide Biochemistry and Physiology* 157: 230–236.

Dijksterhuis, J. (2019). Fungal spores: highly variable and stress-resistant vehicles for distribution and spoilage. *Food Microbiology* 81: 2–11.

Dominguez-Wong, C., Loredo-Becerra, G.M., Quintero-González, C.C. et al. (2014). Evaluation of the antibacterial activity of an indoor waterborne architectural coating containing Ag/TiO2 under different relative humidity environments. *Materials Letters* 134: 103–106.

EASAC 2018, Extreme weather events in Europe. Preparing for climate change adaptation: an update on EASAC's 2013 study, pp. 1–8.

Edge, M., Allen, N.S., Turner, D. et al. (2001). The enhanced performance of biocidal additives in paints and coatings. *Progress in Organic Coatings* 43 (1–3): 10–17.

Elbourne, A., Crawford, R.J., and Ivanova, E.P. (2017). Nano-structured antimicrobial surfaces: from nature to synthetic analogues. *Journal of Colloid and Interface Science* 508: 603–616.

EPA 2008, Mold Remediation in Schools and Commercial Buildings Guide, United States Environmental Protection Agency, no. September, pp. 1–56.

Eversdijk, J., Erich, S.J.F., Hermanns, S.P.M. et al. (2012). Development and evaluation of a biocide release system for prolonged antifungal activity in finishing materials. *Progress in Organic Coatings* 74 (4): 640–644.

Falkiewicz-Dulik, M., Janda, K., and Wypych, G. (2015a). Analytical methods in biodegradation, biodeterioration, and biostabilization. In: *Handbook of Material Biodegradation, Biodeterioration, and Biostablization* (eds. M. Falkiewicz-Dulik, K. Janda and G. Wypych), 377–393. Canada: ChemTec Publishing.

Falkiewicz-Dulik, M., Janda, K., and Wypych, G. (2015b). Industrial Biocides. In: *Handbook of Material Biodegradation, Biodeterioration, and Biostablization* (eds. M. Falkiewicz-Dulik, K. Janda and G. Wypych), 33–65. Canada: ChemTec Publishing.

Feizi, S., Taghipour, E., Ghadam, P., and Mohammadi, P. (2018). Antifungal, antibacterial, antibiofilm and colorimetric sensing of toxic metals activities of ecofriendly, economical synthesized Ag/AgCl nanoparticles using *Malva sylvestris* leaf extracts. *Microbial Pathogenesis* 125: 33–42.

Fernández, M.A. and Bellotti, N. (2017). Silica-based bioactive solids obtained from modified diatomaceous earth to be used as antimicrobial filler material. *Materials Letters* 194: 130–134.

Fernández, J.G., Fernández-Baldo, M.A., Berni, E. et al. (2016). Production of silver nanoparticles using yeasts and evaluation of their antifungal activity against phytopathogenic fungi. *Process Biochemistry* 51 (9): 1306–1313.

Ferreira, C., Pereira, A.M., Pereira, M.C. et al. (2013). Biofilm control with new microparticles with immobilized biocide. *Heat Transfer Engineering* 34 (8–9): 712–718.

Ferreira, F.V., Mariano, M., Lepesqueur, L.S.S. et al. (2019). Silver nanoparticles coated with dodecanethiol used as fillers in non-cytotoxic and antifungal PBAT surface based on nanocomposites. *Materials Science and Engineering C* 98: 800–807.

Ghorbani, H.R., Alizadeh, V., Mehr, F.P. et al. (2018). Preparation of polyurethane/CuO coating film and the study of antifungal activity. *Progress in Organic Coatings* 123: 322–325.

Grant, C., Hunter, C.A., Flannigan, B., and Bravery, A.F. (1989). The moisture requirements of moulds isolated from domestic dwellings. *International Biodeterioration* 25: 259–284.

Griffin, D.H. (1994). *Fungal Physiology*, 2e. New York: Wiley, Wiley-Liss.

Guerra, F.L., Lopes, W., Cazarolli, J.C. et al. (2019). Biodeterioration of mortar coating in historical buildings: microclimatic characterization, material, and fungal community. *Building and Environment* 155: 195–209.

Gutarowska, B. and Czyżowska, A. (2009). The ability of filamentous fungi to produce acids on indoor building materials. *Annals of Microbiology* 59 (4): 807–813.

Haghighi, N., Abdi, Y., and Haghighi, F. (2011). Light-induced antifungal activity of TiO_2 nanoparticles/ZnO nanowires. *Applied Surface Science* 257 (23): 10096–10100.

Haja Hameed, A.S., Karthikeyan, C., Senthil Kumar, V. et al. (2015). Effect of Mg^{2+}, Ca^{2+}, Sr^{2+} and Ba^{2+} metal ions on the antifungal activity of ZnO nanoparticles tested against *Candida albicans*. *Materials Science and Engineering C* 52: 171–177.

Han, W., Wu, Z., Li, Y., and Wang, Y. (2019). Graphene family nanomaterials (GFNs)—promising materials for antimicrobial coating and film: a review. *Chemical Engineering Journal* 358: 1022–1037.

Hasheminejad, N., Khodaiyan, F., and Safari, M. (2019). Improving the antifungal activity of clove essential oil encapsulated by chitosan nanoparticles. *Food Chemistry* 275: 113–122.

Haupert, F. and Wetzel, B. (2005). *Polymer Composites from Nano- to Macro-Scale* (eds. F. Klaus, F. Stoyko and Zhong), 368. US: Springer-Verlag.

He, X. and Hwang, H.M. (2016). Nanotechnology in food science: functionality, applicability, and safety assessment. *Journal of Food and Drug Analysis* 24 (4): 671–681.

He, L., Liu, Y., Mustapha, A., and Lin, M. (2011). Antifungal activity of zinc oxide nanoparticles against *Botrytis cinerea* and *Penicillium expansum*. *Microbiological Research* 166 (3): 207–215.

Henam, S.D., Ahmad, F., Shah, M.A. et al. (2019). Microwave synthesis of nanoparticles and their antifungal activities. *Spectrochimica Acta - Part A: Molecular and Biomolecular Spectroscopy* 213: 337–341.

Hendessi, S., Sevinis, E.B., Unal, S. et al. (2016). Antibacterial sustained-release coatings from halloysite nanotubes/waterborne polyurethanes. *Progress in Organic Coatings* 101: 253–261.

Holtz, R.D., Lima, B.A., Filho, A.G.S. et al. (2012). Nanostructured silver vanadate as a promising antibacterial additive to water-based paints. *Nanomedicine: Nanotechnology, Biology, and Medicine* 8 (6): 935–940.

Huang, Y., Song, J., Yang, C. et al. (2019). Scalable manufacturing and applications of nanofibers. *Materials Today* 28: 98–113.

Hurraß, J., Heinzow, B., Aurbach, U. et al. (2017). Medical diagnostics for indoor mold exposure. *International Journal of Hygiene and Environmental Health* 220 (2): 305–328.

Ibrahim, I.M., Ali, I.M., Dheeb, B.I. et al. (2017). Antifungal activity of wide band gap thioglycolic acid capped ZnS:Mn semiconductor nanoparticles against some pathogenic fungi. *Materials Science and Engineering C* 73: 665–669.

Ifuku, S., Tsukiyama, Y., Yukawa, T. et al. (2015). Facile preparation of silver nanoparticles immobilized on chitin nanofiber surfaces to endow antifungal activities. *Carbohydrate Polymers* 117: 813–817.

Jayaseelan, C., Ramkumar, R., Rahuman, A.A., and Perumal, P. (2013). Green synthesis of gold nanoparticles using seed aqueous extract of *Abelmoschus esculentus* and its antifungal activity. *Industrial Crops and Products* 45: 423–429.

Jo, Y.K., Kim, B.H., and Jung, G. (2009). Antifungal activity of silver ions and nanoparticles on Phytopathogenic fungi. *Plant Disease* 93 (10): 1037–1043.

Jogee, P. and Rai, M. (2020). Application of nanoparticles in inhibition of mycotoxin-producing fungi. In: *Nanomycotoxicology*, 239–250. UK: Elsevier.

Jogee, P.S., Ingle, A.P., and Rai, M. (2017). Isolation and identification of toxigenic fungi from infected peanuts and efficacy of silver nanoparticles against them. *Food Control* 71: 143–151.

Johansson, P, Samuelson, I, Ekstrand-tobin, A, et al. 2005, Microbiological growth on building materials – critical moisture levels. Swedish National Testing and Research Institute, Paper A41-T4-S-05-3.

Johansson, P., Bok, G., and Ekstrand-Tobin, A. (2013). The effect of cyclic moisture and temperature on mould growth onwood compared to steady state conditions. *Building and Environment* 65: 178–184.

Johns, K. (2003). Hygienic coatings: the next generation. *Surface Coatings International Part B: Coatings Transactions* 86 (2): 101–110.

Kairyte, K., Kadys, A., and Luksiene, Z. (2013). Antibacterial and antifungal activity of photoactivated ZnO nanoparticles in suspension. *Journal of Photochemistry and Photobiology B: Biology* 128: 78–84.

Kaiser, J.P., Zuin, S., and Wick, P. (2013). Is nanotechnology revolutionizing the paint and lacquer industry? A critical opinion. *Science of the Total Environment* 442: 282–289.

Kakakhel, M.A., Wu, F., Gu, J.D. et al. (2019). Controlling biodeterioration of cultural heritage objects with biocides: a review. *International Biodeterioration and Biodegradation* 143: 104721.
Kamtsikakis, A., Kavetsou, E., Chronaki, K. et al. (2017). Encapsulation of antifouling organic biocides in poly(lactic acid) nanoparticles. *Bioengineering* 4 (4): 81.
Kanhed, P., Birla, S., Gaikwad, S. et al. (2014). In vitro antifungal efficacy of copper nanoparticles against selected crop pathogenic fungi. *Materials Letters* 115: 13–17.
Karsa, D.R. and Ashworth, D. (2002). *Industrial Biocides Selection and Application, Royal Society of Chemistry*, 173. UK: Royal Society of Chemistry Publisher.
Kasprowicz, M.J., Kozioł, M., and Gorczyca, A. (2010). The effect of silver nanoparticles on phytopathogenic spores of *Fusarium culmorum*. *Canadian Journal of Microbiology* 56 (3): 247–253.
Kavkler, K., Gunde-Cimerman, N., Zalar, P., and Demšar, A. (2015). Fungal contamination of textile objects preserved in Slovene museums and religious institutions. *International Biodeterioration and Biodegradation* 97: 51–59.
Khalil, N.M., Abd El-Ghany, M.N., and Rodríguez-Couto, S. (2019). Antifungal and anti-mycotoxin efficacy of biogenic silver nanoparticles produced by *Fusarium chlamydosporum* and *Penicillium chrysogenum* at non-cytotoxic doses. *Chemosphere* 218: 477–486.
Khan, S.T., Ahmad, J., Ahamed, M., and Jousset, A. (2018). Sub-lethal doses of widespread nanoparticles promote antifungal activity in *Pseudomonas protegens* CHA0. *Science of the Total Environment* 627: 658–662.
King, T., Osmond-McLeod, M.J., and Duffy, L.L. (2018). Nanotechnology in the food sector and potential applications for the poultry industry. *Trends in Food Science and Technology* 72: 62–73.
Król, A., Railean-Plugaru, V., Pomastowski, P., and Buszewski, B. (2019). Phytochemical investigation of *Medicago sativa* L. extract and its potential as a safe source for the synthesis of ZnO nanoparticles: the proposed mechanism of formation and antimicrobial activity. *Phytochemistry Letters* 31: 170–180.
Kumar, A., Vemula, P.K., Ajayan, P.M., and John, G. (2008). Silver-nanoparticle-embedded antimicrobial paints based on vegetable oil. *Nature Materials* 7 (3): 236–241.
Kumar, A., Chisti, Y., and Chand, U. (2013). Synthesis of metallic nanoparticles using plant extracts. *Biotechnology Advances* 31 (2): 346–356.
Kumar, S., Kaur, P., Bernela, M. et al. (2016). Ketoconazole encapsulated in chitosan-gellan gum nanocomplexes exhibits prolonged antifungal activity. *International Journal of Biological Macromolecules* 93: 988–994.
Kumar, M., Shanthi, N., Mahato, A.K. et al. (2019a). Preparation of luliconazole nanocrystals loaded hydrogel for improvement of dissolution and antifungal activity. *Heliyon* 5 (5): e01688.
Kumar, P.V., Jelastin, S.M., and Prakash, K.S. (2019b). Green synthesis derived Pt-nanoparticles using *Xanthium strumarium* leaf extract and their biological studies. *Journal of Environmental Chemical Engineering* 7 (3): 103–146.
Kumari, M., Giri, V.P., Pandey, S. et al. (2019). An insight into the mechanism of antifungal activity of biogenic nanoparticles than their chemical counterparts. *Pesticide Biochemistry and Physiology* 157: 45–52.
Landge, S., Ghosh, D., and Aiken, K. (2017). Solvent-free synthesis of nanoparticles. In: *Green Chemistry: An Inclusive Approach*, 609–646. Elsevier.
Lattab, N., Kalai, S., Bensoussan, M., and Dantigny, P. (2012). Effect of storage conditions (relative humidity, duration, and temperature) on the germination time of *Aspergillus carbonarius* and *Penicillium chrysogenum*. *International Journal of Food Microbiology* 160 (1): 80–94.
Li, Y., Fu, Z., and Xu, G. (2019). Metal-organic framework nanosheets: preparation and applications. *Coordination Chemistry Reviews* 388: 79–106.

Liu, Z., Ma, S., Cao, G. et al. (2018). Distribution characteristics, growth, reproduction and transmission modes and control strategies for microbial contamination in HVAC systems: a literature review. *Energy and Buildings* 177: 77–95.

López-Meneses, A.K., Plascencia-Jatomea, M., Lizardi-Mendoza, J. et al. (2018). *Schinus molle* L. essential oil-loaded chitosan nanoparticles: preparation, characterization, antifungal and anti-aflatoxigenic properties. *LWT – Food Science and Technology* 96: 597–603.

Machado, G.E., Pereyra, A.M., Rosato, V.G. et al. (2019). Improving the biocidal activity of outdoor coating formulations by using zeolite-supported silver nanoparticles. *Materials Science and Engineering C* 98 (257): 789–799.

Madelin, T.M. (1994). Fungal aerosols: a review. *Journal of Aerosol Science* 25 (8): 1405–1412.

Madhumitha, G., Fowsiya, J., Gupta, N. et al. (2019). Green synthesis, characterization and antifungal and photocatalytic activity of *Pithecellobium dulce* peel-mediated ZnO nanoparticles. *Journal of Physics and Chemistry of Solids* 127: 43–51.

Mahmoud, M.A., Eifan, S.A., and Majrashi, M. (2014). Application of silver nanoparticles as antifungal and antiaflatoxin B1 produced by *Aspergillus flavus*. *Journal of Nanomaterials and Biostructures* 9 (1): 151–157.

Malkapur, D., Devi, M.S., Rupula, K., and Sashidhar, R.B. (2017). Biogenic synthesis, characterization and antifungal activity of gum kondagogu-silver nano bio composite construct: assessment of its mode of action. *Nanobiotechnology* 11 (7): 866–873.

Manjumeena, R. (2017). Imparting potential antibacterial and antifungal activities to water-based interior paint using phytosynthesized biocompatible nanoparticles of silver as an additive-a green approach. In: *Handbook of Antimicrobial Coatings*. Elsevier Inc.

Mardones, L.E., Legnoverde, M.S., Monzón, J.D. et al. (2019). Increasing the effectiveness of a liquid biocide component used in antifungal waterborne paints by its encapsulation in mesoporous silicas. *Progress in Organic Coatings* 134: 145–152.

Marulasiddeshwara, M.B., Dakshayani, S.S., Sharath Kumar, M.N. et al. (2017). Facile-one pot-green synthesis, antibacterial, antifungal, antioxidant and antiplatelet activities of lignin capped silver nanoparticles: a promising therapeutic agent. *Materials Science and Engineering C* 81: 182–190.

Mathiazhagan, A. and Joseph, R. (2011). Nanotechnology – a new prospective in organic coating – review. *International Journal of Chemical Engineering and Applications* 2 (4): 225–237.

Melo, D., Sequeira, S.O., Lopes, J.A., and Macedo, M.F. (2019). Stains versus colourants produced by fungi colonising paper cultural heritage: a review. *Journal of Cultural Heritage* 35: 161–182.

Mitra, C., Gummadidala, P.M., Afshinnia, K. et al. (2017). Citrate-coated silver nanoparticles growth-independently inhibit aflatoxin synthesis in *Aspergillus parasiticus*. *Environmental Science and Technology* 51 (14): 8085–8093.

Mittal, A.K., Chisti, Y., and Banerjee, U.C. (2013). Synthesis of metallic nanoparticles using plant extracts. *Biotechnology Advances* 31 (2): 346–356.

Møller, E.B., Andersen, B., Rode, C., and Peuhkuri, R. (2017). Conditions for mould growth on typical interior surfaces. *Energy Procedia* 132: 171–176.

Moraes Moreira Carraro, T.C., Altmeyer, C., Maissar Khalil, N., and Mara Mainardes, R. (2017). Assessment of in vitro antifungal efficacy and in vivo toxicity of amphotericin B-loaded PLGA and PLGA-PEG blend nanoparticles. *Journal de Mycologie Medicale* 27 (4): 519–529.

Moritz, M. and Geszke-Moritz, M. (2013). The newest achievements in synthesis, immobilization and practical applications of antibacterial nanoparticles. *Chemical Engineering Journal* 228: 596–613.

Morones, J.R., Elechiguerra, J.L., Camacho, A. et al. (2005). The bactericidal effect of silver nanoparticles. *Nanotechnology* 16 (10): 2346–2353.

Nallendran, R., Selvan, G., and Balu, A.R. (2019). NiO coupled CdO nanoparticles with enhanced magnetic and antifungal properties. *Surfaces and Interfaces* 15: 11–18.
Nguyen, T.N.L., Do, T.V., Nguyen, T.V. et al. (2019). Antimicrobial activity of acrylic polyurethane/ Fe_3O_4-Ag nanocomposite coating. *Progress in Organic Coatings* 132: 15–20.
Nguyen-Tri, P., Nguyen, T.A., Carriere, P., and Ngo Xuan, C. (2018). Nanocomposite coatings: preparation, characterization, properties, and applications. *International Journal of Corrosion* 2018: 1–19.
Nielsen, K. (2002). *Mould Growth on Building Materials Secondary Metabolites, Mycotoxins and Biomarkers*. Lyngby, Denmark: Technical University of Denmark.
Nielsen, K. (2003). Mycotoxin production by indoor molds. *Fungal Genetics and Biology* 39 (2): 103–117.
Nielsen, K.F., Holm, G., Uttrup, L.P., and Nielsen, P.A. (2004). Mould growth on building materials under low water activities. Influence of humidity and temperature on fungal growth and secondary metabolism. *International Biodeterioration and Biodegradation* 54 (4): 325–336.
Ogar, A., Tylko, G., and Turnau, K. (2015). Antifungal properties of silver nanoparticles against indoor mould growth. *Science of the Total Environment* 521–522: 305–314.
Paiva de Carvalho, H., Mesquita, N., Trovão, J. et al. (2018, 2017). Fungal contamination of paintings and wooden sculptures inside the storage room of a museum: are current norms and reference values adequate? *Journal of Cultural Heritage* 34: 268–276.
Pal, S., Tak, Y.K., and Song, J.M. (2007). Does the antibacterial activity of silver nanoparticles depend on the shape of the nanoparticle? A study of the gram-negative bacterium *Escherichia coli*. *Applied and Environmental Microbiology* 73 (6): 1712–1720.
Palaty, C 2014, Mould Remediation Recommendations, available at: http://www.ncceh.ca/sites/default/files/Mould_Remediation_Evidence_Review_March_2014.pdf, pp. 1–7.
Pan, C., Qian, J., Fan, J. et al. (2019). Preparation nanoparticle by ionic cross-linked emulsified chitosan and its antibacterial activity. *Colloids and Surfaces A: Physicochemical and Engineering Aspects* 568: 362–370.
Parveen, S., Wani, A.H., Shah, M.A. et al. (2018). Preparation, characterization and antifungal activity of iron oxide nanoparticles. *Microbial Pathogenesis* 115: 287–292.
Paulus, W. (2004). Introduction to microbicides. *Directory of Microbicides for the Protection of Materials*: 1–8.
Peña-González, C.E., Pedziwiatr-Werbicka, E., Martín-Pérez, T. et al. (2017). Antibacterial and antifungal properties of dendronized silver and gold nanoparticles with cationic carbosilane dendrons. *International Journal of Pharmaceutics* 528 (1–2): 55–61.
Pereira, S.S.P., de Oliveira, H.M., Turrini, R.N.T., and Lacerda, R.A. (2015). Disinfection with sodium hypochlorite in hospital environmental surfaces in the reduction of contamination and infection prevention: a systematic review. *Revista da Escola de Enfermagem* 49 (4): 675–681.
Pereyra, A.M., Gonzalez, M.R., Rosato, V.G., and Basaldella, E.I. (2014). A-type zeolite containing Ag+/Zn2+ as inorganic antifungal for waterborne coating formulations. *Progress in Organic Coatings* 77 (1): 213–218.
Perez-Nadales, E., Almeida Nogueira, M.F., Baldin, C. et al. (2014). Fungal model systems and the elucidation of pathogenicity determinants. *Fungal Genetics and Biology* 70: 42–67.
Petica, A., Florea, A., Gaidau, C. et al. (2019). Synthesis and characterization of silver-titania nanocomposites prepared by electrochemical method with enhanced photocatalytic characteristics, antifungal and antimicrobial activity. *Journal of Materials Research and Technology* 8 (1): 41–53.
Pokhum, C., Intasanta, V., Yaipimai, W. et al. (2018). A facile and cost-effective method for removal of indoor airborne psychrotrophic bacterial and fungal flora based on silver and zinc oxide nanoparticles decorated on fibrous air filter. *Atmospheric Pollution Research* 9 (1): 172–177.

Ponizovskaya, V.B., Rebrikova, N.L., Kachalkin, A.V. et al. (2019). Micromycetes as colonizers of mineral building materials in historic monuments and museums. *Fungal Biology* 123 (4): 290–306.

Rath, B., Young, E.A., Harris, A. et al. (2011). Adverse respiratory symptoms and environmental exposures among children and adolescents following hurricane Katrina. *Public Health Reports* 126 (6): 853–860.

Raza, M., Kanwal, Z., Rauf, A. et al. (2016). Size- and shape-dependent antibacterial studies of silver nanoparticles synthesized by wet chemical routes. *Nanomaterials* 6 (4): 74.

Ribeiro, M., Simões, L.C., and Simões, M. (2018). Biocides. In: *Reference Module in Life Sciences* (ed. T. Schmidt), 1–13. Elsevier.

Roopan, S.M., Devi Priya, D., Shanavas, S. et al. (2019). CuO/C nanocomposite: synthesis and optimization using sucrose as carbon source and its antifungal activity. *Materials Science and Engineering C* 101: 404–414.

Ruffolo, S.A., La Russa, M.F., Malagodi, M. et al. (2010). ZnO and $ZnTiO_3$ nanopowders for antimicrobial stone coating. *Applied Physics A* 100 (3): 829–834.

Saeed, A.M.E., Abd El-Fattah, M., Azzam, A.M. et al. (2016). Synthesis of cuprous oxide epoxy nanocomposite as an environmentally antimicrobial coating. *International Journal of Biological Macromolecules* 89: 190–197.

Safaei, M. and Taran, M. (2017). Fabrication, characterization, and antifungal activity of sodium hyaluronate-TiO_2 bionanocomposite against *Aspergillus niger*. *Materials Letters* 207: 113–116.

Safaei, M., Taran, M., and Imani, M.M. (2019). Preparation, structural characterization, thermal properties and antifungal activity of alginate-CuO bionanocomposite. *Materials Science and Engineering C* 101: 323–329.

Samrat, K., Nikhil, N.S., Karthick Raja Namasivamyam, S. et al. (2016). Evaluation of improved antifungal activity of fluconazole - silver nanoconjugate against pathogenic fungi. *Materials Today: Proceedings* 3 (6): 1958–1967.

Sardella, D., Gatt, R., and Valdramidis, V.P. (2019). Metal nanoparticles for controlling fungal proliferation: quantitative analysis and applications. *Current Opinion in Food Science* 30: 49–59.

Sardi, J.D.C.O., Pitangui, N.D.S., Rodríguez-Arellanes, G. et al. (2014). Highlights in pathogenic fungal biofilms. *Revista Iberoamericana de Micologia* 31 (1): 22–29.

Sathiyabama, M. and Parthasarathy, R. (2016). Biological preparation of chitosan nanoparticles and its in vitro antifungal efficacy against some phytopathogenic fungi. *Carbohydrate Polymers* 151: 321–325.

Sedlbauer, K. (2002). Prediction of mould growth by hygrothermal calculation. *Journal of Thermal Envelope and Building Science* 25 (4): 321–336.

Sequeira, S.O., Laia, C.A.T., Phillips, A.J.L. et al. (2017). Clotrimazole and calcium hydroxide nanoparticles: a low toxicity antifungal alternative for paper conservation. *Journal of Cultural Heritage* 24: 45–52.

Sharma, D., Rajput, J., Kaith, B.S. et al. (2010). Synthesis of ZnO nanoparticles and study of their antibacterial and antifungal properties. *Thin Solid Films* 519 (3): 1224–1229.

Singh, A.K. (2016a). Experimental methodologies for the characterization of nanoparticles. In: *Engineered Nanoparticles* (ed. A.K. Singh), 125–170. Elsevier.

Singh, A.K. (2016b). Structure, synthesis, and application of nanoparticles. In: *Engineered Nanoparticles* (ed. A.K. Singh), 19–76. Elsevier.

Singh, P., Kim, Y.J., Wang, C. et al. (2016). The development of a green approach for the biosynthesis of silver and gold nanoparticles by using *Panax ginseng* root extract, and their biological applications. *Artificial Cells, Nanomedicine and Biotechnology* 44 (4): 1150–1157.

Soliman, G.M. (2017). Nanoparticles as safe and effective delivery systems of antifungal agents: achievements and challenges. *International Journal of Pharmaceutics* 523 (1): 15–32.

Sørensen, G., Nielsen, A.L., Pedersen, M.M. et al. (2010). Controlled release of biocide from silica microparticles in wood paint. *Progress in Organic Coatings* 68 (4): 299–306.
Souza, V.G.L. and Fernando, A.L. (2016). Nanoparticles in food packaging: biodegradability and potential migration to food—a review. *Food Packaging and Shelf Life* 8: 63–70.
Sterflinger, K. (2010). Fungi: their role in deterioration of cultural heritage. *Fungal Biology Reviews* 24 (1–2): 47–55.
Stobie, N., Duffy, B., Colreavy, J. et al. (2010). Dual-action hygienic coatings: benefits of hydrophobicity and silver ion release for protection of environmental and clinical surfaces. *Journal of Colloid and Interface Science* 345 (2): 286–292.
Storey, BE, Dangman, K, Schenck, P, et al. 2004, Guidance for Clinicians on the Recognition and Management of Health Effects Related to Mold Exposure and Moisture Indoors, available at: https://www.epa.gov/mold/guidance-clinicians-recognition-and-management-health-effects-related-mold-exposure-and.
Su, D.-l., Li, P.-j., Ning, M. et al. (2019). Microwave assisted green synthesis of pectin based silver nanoparticles and their antibacterial and antifungal activities. *Materials Letters* 244: 35–38.
Suresh, S., Karthikeyan, S., Saravanan, P. et al. (2016). Comparison of antibacterial and antifungal activity of 5-amino-2-mercapto benzimidazole and functionalized Ag_3O_4 nanoparticles. *Karbala International Journal of Modern Science* 2 (2): 129–137.
Täubel, M. and Hyvärinen, A. (2015). Occurrence of Mycotoxins in indoor environments. In: *Environmental Mycology in Public Health*, 299–323. Elsevier.
Thakkar, K.N., Mhatre, S.S., and Parikh, R.Y. (2010). Biological synthesis of metallic nanoparticles. *Nanomedicine: Nanotechnology, Biology, and Medicine* 6 (2): 257–262.
Tiwari, A. and Chaturvedi, A. (2018). Antimicrobial coatings-technology advancement or scientific Myth. In: *Handbook of Antimicrobial Coatings*, 1–5. Elsevier.
US CDC 2018, Mold Clean-Up After Disasters: When to Use Bleach, National Center for Environmental Health, https://www.cdc.gov/mold/mold-cleanup-bleach.html
Vance, M.E., Kuiken, T., Vejerano, E.P. et al. (2015). Nanotechnology in the real world: redeveloping the nanomaterial consumer products inventory. *Beilstein Journal of Nanotechnology* 6: 1769–1780.
Verdier, T., Coutand, M., Bertron, A., and Roques, C. (2014). A review of indoor microbial growth across building materials and sampling and analysis methods. *Building and Environment* 80: 136–149.
Vijaya Kumar, P., Mary Jelastin Kala, S., and Prakash, K.S. (2019). Green synthesis of gold nanoparticles using Croton Caudatus Geisel leaf extract and their biological studies. *Materials Letters* 236: 19–22.
Weber, R.W. (2012). Allergen of the month-*Stachybotrys chartarum*. *Annals of Allergy, Asthma and Immunology* 108 (6): A9.
Weber, D.J. and Rutala, W.A. (2013). Self-disinfecting surfaces: review of current methodologies and future prospects. *American Journal of Infection Control* 41 (5): S31–S35.
Webster, J. and Weber, R. (2007). *Introduction to Fungi*, 3e. New York: Cambridge University Press.
Weisany, W., Samadi, S., Amini, J. et al. (2019). Enhancement of the antifungal activity of thyme and dill essential oils against *Colletotrichum nymphaeae* by nano-encapsulation with copper NPs. *Industrial Crops and Products* 132: 213–225.
Wilczewska, A.Z., Kosińska, A., Misztalewska-Turkowicz, I. et al. (2019). Magnetic nanoparticles bearing metallocarbonyl moiety as antibacterial and antifungal agents. *Applied Surface Science* 487: 601–609.
World Health Organization (2009). *WHO Guidelines for Indoor Air Quality: Dampness and Mould*, 228.

Xu, T. and Xie, C.S. (2003). Tetrapod-like nano-particle ZnO/acrylic resin composite and its multi-function property. *Progress in Organic Coatings* 46 (4): 297–301.

Xue, B. (2016). Biosynthesis of silver nanoparticles by the fungus *Arthroderma fulvum* and its antifungal activity against genera of *Candida*, *Aspergillus* and *Fusarium*. *International Journal of Nanomedicine*: 1899–1906.

Yang, C.S. and Heinsohn, P.A. (2007). *Sampling and Analysis of Indoor Microorganisms*, 277. Wiley.

Yilmaz, M.T., Yilmaz, A., Akman, P.K. et al. (2019). Electrospraying method for fabrication of essential oil loaded-chitosan nanoparticle delivery systems characterized by molecular, thermal, morphological and antifungal properties. *Innovative Food Science and Emerging Technologies* 52: 166–178.

Zielecka, M., Bujnowska, E., Kępska, B. et al. (2011). Antimicrobial additives for architectural paints and impregnates. *Progress in Organic Coatings* 72 (1–2): 193–201.

13

Nanotechnology for Antifungal Therapy

Jacqueline Teixeira da Silva and Andre Correa Amaral

Laboratory of Nanobiotechnology, Institute of Tropical Pathology and Public Health, Federal University of Goiás, Goiânia, Brazil

13.1 Introduction

The ability of microorganisms to colonize so many habitats is related to their metabolism being capable of adapting to different environmental conditions by producing primary and secondary metabolites that guarantee their survival. They do not live isolated, being found close to other organisms, sharing chemical signals and causing diseases (Netzker et al. 2018) which can be contagious and are easily transferred from a sick to a healthy individual (Zhu et al. 2014). For decades, microorganisms including fungi have developed considerable resistance to conventional drugs (Voltan et al. 2016). In the 1970s, the diversity, accelerated growth, and the burden of antimicrobial resistance were already being evidenced (Tang et al. 2014). The numbers in cases of microbial resistance have been growing, representing a worldwide public health challenge, because of the emerging cases of resistant infections that can occur in both hospital and community settings, associated with high morbidity and mortality (Netzker et al. 2018).

Since 1960, few new classes of antibiotics have been introduced for clinical use (Netzker et al. 2018). Then, the available choices for the development of new therapies are limited. Microorganisms have been resistant to one or more classes of antibiotics, and conventional drug options are less effective or ineffective (Brooks and Brooks 2014; Khameneh et al. 2016). Because of the increase of resistance mechanisms by pathogenic microorganisms, research had failed to find new drugs (Tang et al. 2014); the problem starts from the similarity of mechanisms and structures of antibiotics already available, making the speed of research development slow (Khameneh et al. 2016).

The term resistance can be understood in different ways (Pfaller 2012): microbiological resistance, which occurs when an antibiotic concentration higher than the usual range is used to inhibit the growth of a pathogen; clinical resistance, where the microorganism is inhibited by a concentration greater than could be safely achieved at normal antibiotic dosage; and compound resistance, where the microorganism is not inhibited by concentrations in the usual range or achievable antibiotic concentrations.

Many factors contribute to resistant and emergent cases. Among them, large-scale drug usage, without knowledge or medical prescription, represents an important contribution to aggravating this scenario (Zhu et al. 2014). The molecular basis for antimicrobial resistance is mediated by elements

Nanobiotechnology in Diagnosis, Drug Delivery, and Treatment, First Edition. Edited by Mahendra Rai, Mehdi Razzaghi-Abyaneh, and Avinash P. Ingle.

such as plasmids, transposons, and integrons. Molecular alterations with genetic element transfers, efflux pump overexpression, and biofilm formation are the most important resistance mechanisms (Khameneh et al. 2016).

Microbial resistance can be classified as innate or acquired. Resistance may occur through horizontal transfer or mutations, and adaptive resistance through genetic alterations. The related mechanisms are drug alteration by enzymatic modification and inactivation, alteration in target or in cell permeability and efflux, or by resistance mechanisms as late drug penetration, genetic material change, cellular metabolism alteration, and super mutations (Brooks and Brooks 2014; Khameneh et al. 2016). These changes occur during exposure to antimicrobial agents in a process of adaptive evolution (Pippi et al. 2015). Resistance mechanisms have spread across many classes of antibiotics, and have developed into at least one bacterial species per year of new drug approval, but with resistance levels appearing within months (Brooks and Brooks 2014).

Just like bacteria, fungi can also gain resistance when exposed to different types of drugs. The ability to overcome the inhibitory action of the drug may be influenced by its action on the fungal cell membrane, disorganization in the cell wall, interference in the synthesis of DNA and RNA, or influenced by environmental factors leading to colonization or replacement of susceptible species with resistant species (Pfaller 2012). Microorganisms can also use other ways to escape inhibition and death processes by evading host immune recognition. The multiplicity of evasion strategies used by a single organism demonstrate multiple levels of defensive attack and escape in which during evolution, pathogens were able to exploit the host immune system for their own benefit, seeking survival and causing damage to the host (Zipfel et al. 2007).

Multi-resistant microorganisms are more complexly treated using conventional methods and some cases may be clinically intractable (Zhu et al. 2014). Multidrug resistance thus far surpasses new therapeutic options and alternatives, demonstrating the need for changes in current therapies, targeting local microbial attack strategies that have microbiota protection methods and elimination of resistance cases (Brooks and Brooks 2014).

The development of microorganism biofilms are found in the hospital environment. These structures are composed of polymeric matrices, made of polysaccharides, proteins, and lipids. They form a resistance barrier in instruments and environments that prevent the passage of antibiotics, xenobiotics, and hinder the patient's immune system activity, once again hindering eradication of infections because of the expression of virulence factors and the resistance gene mutations that relate to biofilms (Khameneh et al. 2016; Ong et al. 2017; Gucwa et al. 2018). Antimicrobial resistance is an arduous challenge that can severely affect patients, increasing virulence, causing delays and limitations in drug therapy (Gastmeier 2010), causing higher mortality, and increasing treatment costs and risks in surgery, chemotherapy and transplantation. These affect not only low-income countries, but in the post-antibiotic era, it affects the entire world without distinction (Tang et al. 2014).

Resistant microorganisms found in communities or hospitals are shared in various locations and facilities, and there are cases of patients who may be colonized or infected with resistant strains before hospital admission. Increasing numbers of immunocompromised individuals, and the susceptibility of the elderly and children to infections, contribute to the disease and infection spread, accompanied by the use of more drug therapies. The constant use of these treatments increases the risk factors and the dispersal of strains in health facilities, community, and even among animals (Gastmeier 2010). Several approaches have been researched and used in an attempt to eliminate cases of resistance. The aim of reducing emergent resistance is essential for drug treatment to work as expected and for available therapy options. Understanding resistance mechanisms will assist in deciding the appropriate therapy in each risk group (Khameneh et al. 2016).

13.2 Basic Aspects of Nanotechnology in Medicine

In the beginning of the 1990s, nanoparticle research began to be carried out, aiming to be used for the transport of active ingredients with pharmacological actions and to carry molecules such as proteins and peptides, allowing greater therapeutic efficacy. These nanostructured drug delivery systems present in different shapes and can be synthesized from different types of materials to compose the systemic matrix, which is fundamental for defining the purpose of the system (Do Nascimento et al. 2016). Table 13.1, describes a few examples of nanoformulated antifungal drugs.

Considering the problem of microbial resistance, medical therapies must face arduous challenges. The search for even more effective treatments requires the frequent adaptation of microorganisms to new arsenals developed (Zhu et al. 2014). In the field of nanotechnology, nanocarriers have been used to assist in the defence against resistant microorganisms, by ensuring a therapeutic increase, providing efflux pump alteration activities, and presenting an anti-biofilm activity and increased penetration, protection against degradation and inactivation by enzymes, resulting in the pathogen death (Khameneh et al. 2016).

Nanomaterials have been applied to treat microbial infections by compromising resistance mechanisms (Zhu et al. 2014). Nanoparticles are used for a diversity of applications and have been ideal for drug delivery because they offer greater drug stability, penetration, and absorption, increased surface area, and prevention of enzymatic degradation. Formulations using natural products offer a broad potential for efficacy (Ong et al. 2017), overcoming resistance mechanisms and causing fewer side effects (Zhu et al. 2014).

Several approaches have been researched and used in an attempt to eliminate cases of resistance. The objective of reducing emergent resistance is essentially for drug treatment to be as effective as possible, for available therapy options, and understanding resistance mechanisms will assist in deciding the appropriate therapy in each risk group (Khameneh et al. 2016). Although nanotechnology brings a variety of changes in the field of antimicrobial treatment, its clinical development still faces testing challenges and clinical safety assurance, yet they are easily achievable (Zhu et al. 2014).

Nanoparticles may assist in the problem of antimicrobial resistance, related to increased absorption, decreased efflux pumps from microbial cells, biofilm formation, and drug delivery to the target site (Voltan et al. 2016). Its applications involve the development of particles through organic and inorganic nanomaterials, and the amount of new commercially available nanotechnology-based materials are considerable and indicated for diagnosis and treatment according to their specificity and sensitivity characteristics in relation to various conditions, in which their use is high. An advantage in antibiotic treatment involves the indication of simultaneous resistance to decreased development, leading to different delivery strategies, and increased drug concentration at the target site (Zhu et al. 2014).

The role of nanotechnology in the development of new therapeutic systems creates a landscape of new products, vaccine adjuvants, or target site delivery, helping the pharmaceutical and

Table 13.1 Examples of commercially available antifungal nanoformulated drugs.

Brand name	Composition	Company	Approval year
Abelcet	Liposomal amphotericin B lipid complex	Sigma-tau	1995
Ambisome	Liposomal amphotericin B	Gilead sciences	1997
Cresemba®	Isavuconazolium sulfate	Astellas	2015

biotechnological sectors to have a major impact on potential hospital infection. The potential of nanoparticles for microbial infections has often been demonstrated in clinical drug testing (Zhu et al. 2014). Laboratory methodologies have been updated since the advent of nanotechnology. Techniques such as enzyme-linked immunosorbent assay (ELISA) and PCR (polymerase chain reaction) that have a high level of sensitivity and reproducibility will be overcome with the opportunities that nanotechnology will bring: faster, sensitive, specific, and cost-effective techniques compared to other techniques for diagnosing microbial infections such as the use of gold, magnetic, or fluorescently labeled nanoparticles (Zhu et al. 2014).

Applications in medicine have involved the use of nanoparticles for certain types of treatment. Magnetic nanoparticles, for example, have been used as contrast agents in magnetic resonance. Functionalized gold nanoparticles have several advantages which increase their options for biological application, such as in the identification of pathogens in genomic verification and hybridization, as well as fluorescent nanoparticles (Zhu et al. 2014).

Polymeric matrix nanoparticles and liposomes are the most widely used for biological applications, because of their many advantages such as biocompatibility, efficacy, biodegradability, controlled release and non-toxicity, present stability, and solubility, thus assisting for an effective therapy (Do Nascimento et al. 2016; Ong et al. 2017). As an example, β-lactam antibiotics have low bioavailability and rely on porins to enter the cell, thus nanoparticles can increase drug solubility by increasing absorption. It can also minimize unnecessary drug exposure and organ toxicity injury (Tang et al. 2014).

13.3 Nanoparticulate Drug Delivery Systems for Refined Antifungal Therapy

To date various nanomaterials such as metallic nanoparticles, liposomes, polymeric nanoparticles, etc. have been successfully exploited as drug delivery systems for antifungal therapy. These nanomaterials are briefly discussed here.

13.3.1 Metallic Nanoparticles

Metal salts, when reduced to a zero atomic dimension, synthesized from a certain number of atoms or bound molecules, form metallic nanoparticles (Figure 13.1) (Krishnaswamy and Orsat 2017). Their physical, chemical, optical, and electronic properties differ not only in size but also in morphology and structure. Its synthesis can follow different methodologies such as microemulsions, chemical reduction, or thermal decomposition of metal salts, and thus, different types of particles can be found, such as magnetic nanoparticles, titanium and zinc oxide nanoparticles, gold, iron, silver, copper, cadmium, and platinum (Krishnaswamy and Orsat 2017).

Magnetic nanoparticles are made from iron and can range in size from 2 to 100 nm. Its size has influence on properties such as redox potential of metal salts and reducing agent, bonding forces between metals, nature of solvent, type of stabilizing agent used, preparation conditions, and temperature (Krishnaswamy and Orsat 2017). The stability of these nanoparticles can be achieved when particles are added to an inert environment, or when exposed to surface protective agents (Krishnaswamy and Orsat 2017). In this way magnetic fluids are formed which allow greater stability as well as a functionalization for these nanoparticles.

The synthesis of metal nanoparticles prepared from silver nitrate ($AgNO_3$) can also be obtained by environmentally friendly methods using catalysts of natural origin. These catalysts may be

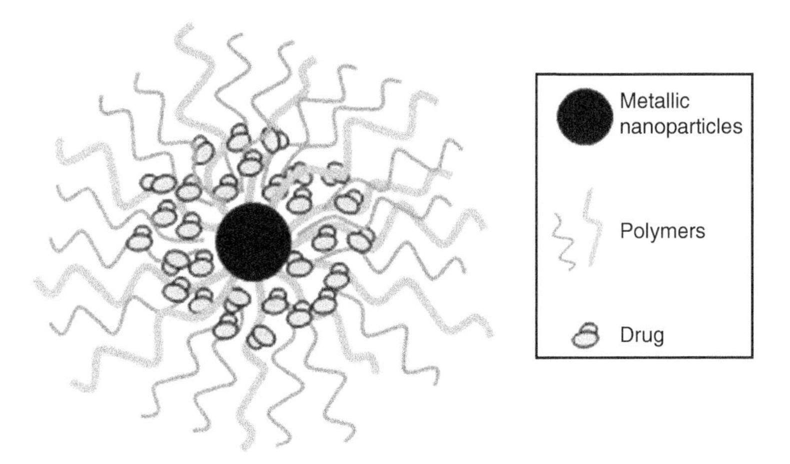

Figure 13.1 Schematic representation of metallic nanoparticles as a drug delivery system.

microorganisms such as acidophilic actinobacterium *Pilimelia columellifera* subsp. *pallida* S19 (Wypij et al. 2017) or plant extracts such as marine mangrove *Rhizophoa mucronata* (Singh et al. 2013). Through these preparation methods, nanoparticles are obtained ranging in size from 5.8 to 12.7 nm. In the case of *R. mucronata* plant extract, it acts as a matrix to reduce and stabilize the synthesized nanoparticles.

The use of magnetic nanoparticles has been applied to thermal ablation in tumors, in addition to radiological imaging, vaccination, and antiangiogenic and proangiogenic function tests (Sharma 2019). Understanding the capture, clearance, accumulation, and pharmacokinetic systems that are involved in the nanoparticle synthesis process is necessary to understand the toxicity and forms of biological acceptance of the nanoparticle, as well as the understanding of oxidation, reduction, and dissolution data (Singh et al. 2017).

Studies indicate that gold nanoparticles are effective both in enhancing photosensitizers for photodynamic-based antifungal therapy and exhibiting *in vitro* inhibitory activities against spore germination of dermatophyte fungi. Based on these characteristics, Tawfik et al. (2016) developed a therapy composed of methylene blue photosensitizer and gold nanoparticles to be tested in the experimental rabbit model for onychomycosis caused by *Trichophyton mentagrophytes*. The results were promising, as 96% of the animals treated using this strategy achieved cure. The combination of gold nanoparticles and another antifungal was also evaluated against *Candida albicans* (Rahimi et al. 2019). In the study, the researchers investigated the anticandida effects of the gold nanoparticle-conjugated indolicidin peptide. In the *in vitro* assays, the authors observed a fourfold reduction in conjugate when compared to isolated peptide. Reduction of toxicity in NIH3T3 fibroblasts was also observed.

The conjugation of silver and conventional antifungal nanoparticles improves the therapeutic activity of both when compared with the compounds tested alone. The synergistic effect of silver nanoparticles with amphotericin B, ketoconazole, and fluconazole was observed in assays with different fungi, all demonstrating improved antifungal activity (Wypij et al. 2017). Another ecologically friendly synthesis method for metallic nanoparticles was achieved by its chemical reduction with cysteine, whereby nanoparticles with a size of around 19 nm were obtained (Bonilla et al. 2015). When tested against *Candida* isolates, they showed good antifungal activity, suggesting their application as proof of concept for more efficient fungal therapy. Conjugation of silver nanoparticles with conventional antifungals significantly increases the potential of the drug. This effect may be related to the particularity of nanoparticles penetrating the fungus cell (Lara et al. 2015). Disruption of the integrity of the fungal cell wall or membrane causes structural weakness that not only allows the passage of ions to the cell but also allows the penetration of a higher amount of the antifungal.

13.3.2 Liposomes

Liposomes are microscopic spherical colloidal structures with the presentation of an aqueous center surrounded by a lipid bilayer (Singh et al. 2017; Pund and Joshi 2017). This type of particle created from the binding of synthetic and/or natural phospholipids is known to form spherical carriers for drug and substance transport between the lipid bilayer (Figure 13.2), and can form varieties of aqueous phase forms, ranging in size from 400 nm to 2.5 μm (Jeevanandam et al. 2017; Sharma 2019). They are generally composed of phospholipids, which are amphiphilic molecules formed by a three-carbon glycerol backbone and a hydrophilic polar head linked by a double tail of fatty acids, which may be either from natural or synthetic origins (Jeevanandam et al. 2017).

The hydrophobic portion of these molecules may give a rigid or malleable architecture to these structures and may change the way the drug will be released. These structures are similar to biological membranes, forming a bilayer which, depending on the physical and chemical characteristics of the drug, may be located within or between the bilayers. Liposomes can confer efficient administration to the drug since the interaction between this structure and the target action can occur by membrane fusion and the consequent release of the drug to both cell and pathogen (Jeevanandam et al. 2017; Sharma 2019).

Polar ends indicate the inner and outer location of the particle, thus forming an interface with the aqueous phase, where hydrophilic substances can be encapsulated within the particle (Jeevanandam et al. 2017). Lipophilic substances are protected against enzymes, immunogens, and chemicals, and prevent metabolic degradation of the drug prior to reaching the target site, and this action is important for increasing the therapeutic index (Jeevanandam et al. 2017; Sharma 2019).

Biodegradability, biocompatibility, non-immunogenicity, structural versatility, amphiphilic nature, and possibility of change in lipid composition make this nanoparticle a great option for specific target delivery (Pund and Joshi 2017). This delivery system modifies the kinetics and

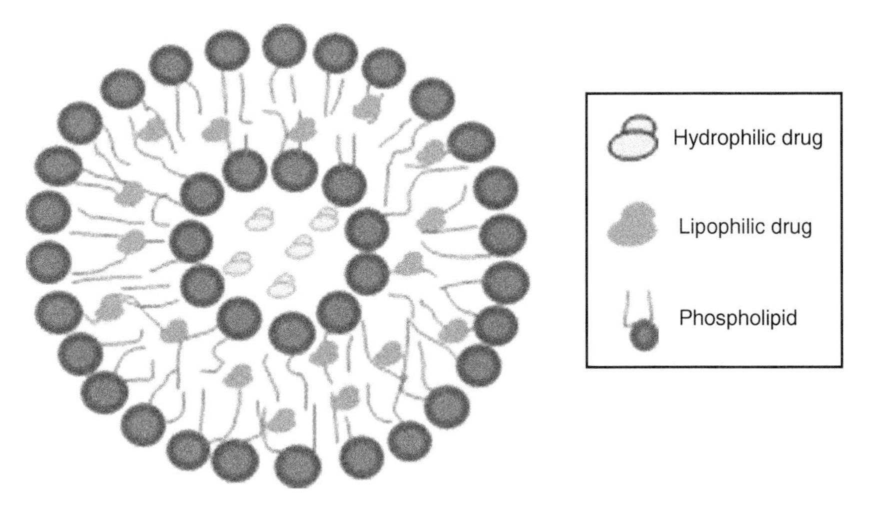

Figure 13.2 Schematic representation of a liposome as a drug delivery system.

distribution of cytostatic agents and leads to increased drug release (Singh et al. 2017). Liposome types are developed as virosomes, magnetic liposomes, and photodynamic liposomes, to name a few (Pund and Joshi 2017). Cationic liposomes have high affinity for cell membranes, which are used as delivery vehicles for various materials such as antimicrobial substances, vaccines, and genetic materials (Jeevanandam et al. 2017).

Amphotericin B has an excellent clinical spectrum and fungicidal activity through interaction with sterols in the plasma membrane forming transmembrane channels, which cause a disordered passage of ions culminating in cell death (Gruszecki et al. 2003). Although the chemical structure of this drug favors interaction with ergosterol present in fungi, it can also interact with cholesterol present in humans (Yilma et al. 2007). Thus, the conventional formulation for amphotericin B with sodium deoxycholate causes severe patient side effects such as nephrotoxicity and hepatoxicity, limiting its use. Thus, the incorporation of amphotericin B into lipid complexes aims at reducing toxicity and may increase the concentration of the drug in certain organs.

Of the three lipid formulations for amphotericin B, Abelcet® (lipid complex), AmBisome®, and Amphotec® (Kleinberg 2006), Ambisome® is the considered to be a true liposomal formulation consisting of unilamellar liposomes composed of hydrogenated phosphatidylcholine, soybean, distearoyl phosphatidylglycerol, and cholesterol (Adler-Moore and Proffitt 2002). These lipid formulations in clinical use have shown interesting therapeutic effects. Lipid complex formulation for amphotericin B has been shown to be effective in treating severe cases of paracoccidioidomycosis, leading to cure in 28 patients (Peçanha et al. 2016). The authors reported that only three patients had renal impairment, the main side effect for amphotericin B, but without the need to discontinue treatment. In a descriptive study from January 2005 to December 2014, 69 patients with severe cases of coccidioidomycosis were found to improve significantly when treated with either liposomal amphotericin B or lipid complex (Sidhu et al. 2018).

These liposomal formulations allow the drug to be administered directly to the airway through nebulization (Hanada et al. 2014; Mihara et al. 2014; Godet et al. 2017). In certain uncommon infections, such as in lung infection caused by *Hormographiella aspergillata* for which there is no standard therapy, alternatives should be evaluated. For this case, the aerosolized form was used for Ambisome resulting in therapeutic success for the patient, even in resistant cases for non-aerosolized forms for this drug (Godet et al. 2017). In another case, in which the patient did not respond to intravenous treatment with vorizonazole to treat *Aspergillus* empyema, a therapeutic success was achieved switched to aerosolized liposomal amphotericin B (Hanada et al. 2014).

The strategy of using liposomes was also applied to the antifungal itraconazole. For this study, the authors obtained liposomes using soy phosphatidylcholine, cholesterol, and sterylamine along with itraconazole (Leal et al. 2015). In animal experiments with the fungus *Aspergillus flavus* inducing fungal keratitis with endophthalmitis, the authors reported better antifungal activity than animals treated with the conventional form for itraconazole. Following drug repositioning, this strategy for a liposomal formulation for itraconazole was also evaluated in inhibiting the Hedgehog pathway related to different cancers (Pace et al. 2019).

Still in experimental stages, other drugs are being incorporated into liposomes such as flucitocin (Salem et al. 2016) and miconazole nitrate (Pandit et al. 2014). In the first case, the flucytosine antifungal was formulated in liposomes containing gold nanoparticles. The formulation was evaluated in the experimental model of intraocular endophthalmitis fungal in the rabbit's cornea (Salem et al. 2016). The data indicated that the higher the zeta potential of the formulation, the better the positive treatment results. In addition, because of the gold nanoparticles used in the formulation, it was possible to track the particles in the anterior portion of the animals' eye. In the second work, nitrate and miconazole were formulated in ultraflexible liposomes (Pandit et al. 2014) formed with soya phosphatidylcholine. The authors concluded that through this type of formulation, greater liposome flexibility occurs and thus improves permeability. This conclusion was reached after animal testing to mimic cutaneous candidiasis caused by *C. albicans*.

13.3.3 Polymeric Nanoparticles

Polymeric nanoparticles (Figure 13.3) have generated interest in research development sectors because of their varied properties and therapeutic potency, which is the representation of controlled drug release technology (Schaffazick et al. 2003; Troiano et al. 2016) that encompasses a group of formed particles by polymers and/or polymer conjugations to nanoparticles (Vasile 2019). The use of polymers in the preparation of nanoparticles has been sought because of their ease of manipulation and ability to control their preparation parameters. Their length scale and broad rate of incorporable chemical functionalities, with characteristics of good bioavailability, biocompatibility, and lower cellular toxicity, are explored for use in research and development of new controlled release products (Li and Huck 2002; Singh et al. 2017).

Nanoparticles may be formed of polymers of natural or synthetic origin. Among the natural polymers, the most used are alginate, chitosan, gelatin, and albumin. Synthetic polymers, non-biodegradable and biodegradable, are often found as poly (D, L-lactic-*co*-glycolic acid; PLGA), poly (D, L-lactic acid; PLA) and poly (ethylene glycol; PEG), among others. They can be used in the formation of micelles, dendrimers, capsules, or colloidal solutions to prevent protein adsorption (Pund and Joshi 2017; Sharma 2019; Vasile 2019). Their distribution and structural organization have been demonstrated in two conformations, known as nanocapsules and nanospheres (Schaffazick et al. 2003; Vasile 2019).

Nanocapsules are particles formed by a polymeric shell surrounding an oily core, wherein the drug may be adsorbed within the core or disposed of in the particle-forming capsule. In the nanosphere conformation, it differs by the constitution of the shell, forming a massive matrix of polymers, in which size and charge, plus surface properties, participate in the cellular interaction, providing the binding of the drug in the polymeric or adsorbed wall and complexed in the polymeric matrix (Schaffazick et al. 2003; Sharma 2019; Vasile 2019). Its preparation can be accomplished through a variety of techniques such as solvent evaporation, spontaneous emulsification,

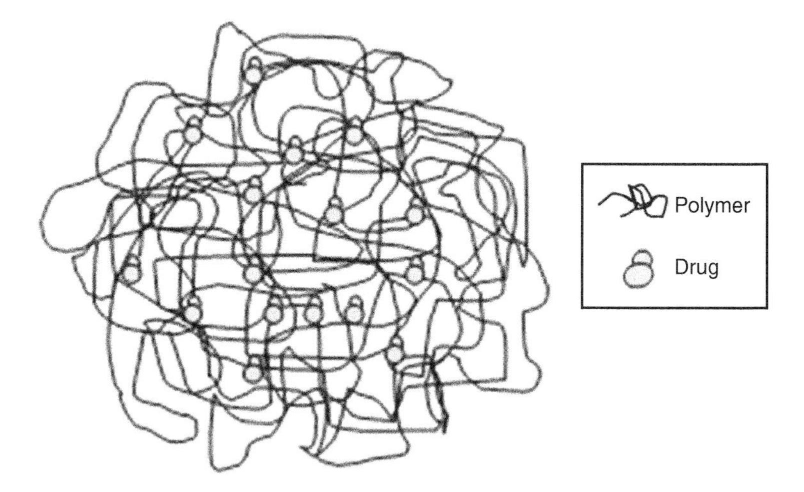

Figure 13.3 Schematic representation of a polymeric nanoparticle as a drug delivery system.

solvent diffusion, polymerization, and ionic gelation, promoting chemical bonding with drug molecules on the nanoparticle polymeric surface (Pund and Joshi 2017; Singh et al. 2017).

The nanometer scale classification, which varies in size from 10 nm to 1 μm (Sharma 2019), is dependent on its method of preparation and synthesis conjugation according to the polymer structure. This directly affects the physical and chemical characteristics of the nanoparticle, influencing size and shape distribution, chemical area, porosity, coating parameters, stability, pH, release, adsorption rate, and drug solubility (Pund and Joshi 2017; Singh et al. 2017). Features such as particle surface functionalization and active release rate provide information on drug delivery capability at its target site (Singh et al. 2017), and its ability to control release that is dependent of the particles' chemical conformation provided by degradation of the polymeric matrix or diffusion of the drug through the polymer (Vasile 2019).

The importance of polymeric nanoparticles has been demonstrated by the expanding range of their applications. The use in the treatments of metabolic and neurological diseases, such as diabetes, schizophrenia, and bipolar disorder, extending to cancer treatment, strengthens the idea of functionalities for this type of drug delivery system (Troiano et al. 2016). Unlike conventional treatments, where pharmacological administration involves the administration of up to two doses per day, the new models transform intervals, increasing the administration period from one week to six months, in addition to decreasing side effects (Troiano et al. 2016).

Although this system has several advantages in its use, disadvantages can be verified from the use of synthetic polymers, with a varying accumulation rate in organs and phagocyte mononuclear system, which can lead to situations of toxic degradation, toxic monomer aggregation, and residual material. However, these disadvantages can be avoided by controlling the developed particle, which requires understanding the effects of surface size, shape, and morphology, and their functionalities (Singh et al. 2017).

Polymeric nanoparticles have been successfully applied as proof of concept for alternative therapies to treat fungal infections as both therapy and prophylaxis (Amaral et al. 2009; Ribeiro et al. 2013). The versatility of these nanostructures allows the incorporation of a larger amount of drug capable of slowly releasing over time and thus reducing the number of doses. As with other nanostructured drug delivery systems, it is possible to functionalize these nanoparticles for target direction in the body. In paracoccidioidomycosis fungal infection, which affects the lungs, PLGA nanoparticles were prepared to carry the amphotericin B drug and functionalized with dimercaptosuccinic acid, which has lung tropism (Amaral et al. 2009). This strategy allowed the experimental model for this mycosis to administer one dose every three days, because the formulation contained three times the recommended dose to treat the infection. It was possible to investigate the biodistribution of the formulation that demonstrated its tropism to the lungs, probably by macrophage phagocytosis (Souza et al. 2015).

This ability to be easily phagocytized by macrophages was used to administer a P10 immunomodulatory peptide as a strategy to promote therapeutic improvement in animals infected with the fungus *Paracoccidioides brasiliensis* (Amaral et al. 2010). This peptide was incorporated into PLGA and administered in combination with the recommended antifungal for the treatment of this mycosis. Therapeutic improvement was observed by enhancing the peptide compared to its free administration with Freund's adjuvant with fourfold reduction of the peptide. This fact may be correlated with the best immunomodulatory response promoted by the formulation. This was observed when the use of another scFv peptide also incorporated within PLGA allowed a reduction of fungal burden in mice (Jannuzzi et al. 2018). The type of immunomodulatory response triggered by this Th1-type copolymer promotes the animal's ability to develop mechanisms to reverse fungal infection.

Another interesting feature of polymeric nanoparticles is their ability to be mucoadhesive. Thus, drug delivery strategies for the treatment of mucosal-affecting fungal infections, such as vulvovaginal candidiasis, have been explored (Lucena et al. 2018; Amaral et al. 2019; Costa et al. 2019). Through this strategy, different compounds can be co-associated, promoting synergism of molecules and thereby increasing the effectiveness of therapy. Mucoadhesive polymers can be used to increase the retention time of the formulation in the vaginal mucosa, promoting slow release of compounds such as antifungal drugs.

13.4 Conclusions and Final Remarks

Launching new drugs may be of little interest to the pharmaceutical industries as it involves a long development period and is costly. These factors directly influence the adoption of strategies for the use of drugs already available in clinical practice for the treatment of resistant microorganisms.

In general, antifungal therapies present adverse side effects, ineffectiveness against some fungal species, and resistance development, and some have high toxicity and high cost in treating the patient. Thus, there is a need for alternative therapies and for stimulating the discovery of new chemical compounds with biological potential against isolated strains and biofilms. Different therapeutic options can be considered through the analysis of data from epidemiological studies, crossing with regional and local information, elaborating a bias resistance tracking system based on data searches, and contributing to an empirical treatment.

An interesting strategy for circumventing the problem of resistance might be the use of a combination therapy of the conventional drug and molecules presenting antimicrobial potential in alternative drug delivery systems. For this strategy, using nanotechnology to prepare drug formulations has been shown to be an interesting option for the more efficient treatment of infectious diseases, contributing to reversing the problem of antimicrobial resistance. Applying the nanotechnology principles can be attractive to develop rapid, sensitive, specific, and cost-effective techniques for the diagnosis of infectious diseases.

Because of their physicochemical and biological characteristics, nanoparticles are considered promising sources of new diagnostic applications to advance the therapeutic techniques employed in the biomedical field. The development of formulations aimed at transporting assets to the target site, with specific binding between cell ligands, forms a new therapeutic category and has ensured more effective responses in diagnostic therapies. Drug administration and medical treatments such as stem cell research and regenerative medicine, as well as the production of antibacterial vaccines including protein isolates, polysaccharides, and free DNA, are some of the key sectors that can be enhanced by innovations in nanotechnology.

References

Adler-Moore, J. and Proffitt, R. (2002). AmBisome: liposomal formulation, structure, mechanism of action and pre-clinical experience. *Journal of Antimicrobial Chemotherapy* 49 (Suppl. S1): 21–30.

Amaral, A.C., Bocca, A.L., Ribeiro, A.M. et al. (2009). Amphotericin B in poly(lactic-co-glycolic acid) (PLGA) and dimercaptosuccinic acid (DMSA) nanoparticles against paracoccidioidomycosis. *Journal of Antimicrobial Chemotherapy* 63 (3): 526–533.

Amaral, A.C., Marques, A.F., Muñoz, J.E. et al. (2010). Poly(lactic-glycolic acid) nanoparticles markedly improve immunological protection provided by peptide P10 against murine paracoccidioidomycosis. *Brazilian Journal of Pharmacology* 159 (5): 1126–1132.

Amaral, A.C., Saavedra, P.H.V., Oliveira Souza, A.C. et al. (2019). Miconazole loaded chitosan-based nanoparticles for local treatment of vulvovaginal candidiasis fungal infections. *Colloids and Surfaces B Biointerfaces* 174: 409–415.

Bonilla, J.J.A., Guerrero, D.J.P., Suárez, C.I.S. et al. (2015). In vitro antifungal activity of silver nanoparticles against fluconazole-resistant *Candida* species. *World Journal of Microbiology and Biotechnology* 31: 1801–1809.

Brooks, B.D. and Brooks, A.E. (2014). Therapeutic strategies to combat antibiotic resistance. *Drug Delivery* 78: 14–27.

Costa, A.F., Evangelista, A.D., Santos, M.C. et al. (2019). Development, characterization, and in vitro-in vivo evaluation of polymeric nanoparticles containing miconazole and farnesol for treatment of vulvovaginal candidiasis. *Medical Mycology* 57 (1): 52–62.

Do Nascimento, T.G., Da Silva, P.F., Azevedo, L.F. et al. (2016). Polymeric nanoparticles of Brazilian red propolis extract: preparation, characterization, antioxidant and leishmanicidal activity. *Nanoscale Research Letters* 11: 301.

Gastmeier, P. (2010). Healthcare-associated versus community-acquired infections: a new challenge for science and society. *International Journal of Medical Microbiology* 300: 342–345.

Godet, C., Cateau, E., Rammaert, B. et al. (2017). Nebulized liposomal amphotericin B for treatment of pulmonary infection caused by *Hormographiella aspergillata*: case report and literature review. *Mycopathologia* 182: 709–713.

Gruszecki, W.L., Gagos, M., Herec, M., and Kernen, P. (2003). Organization of antibiotic amphotericin B in model lipid membranes. A mini review. *Cellular & Molecular Biology Letters* 8: 161–170.

Gucwa, K., Kusznierewicz, B., Milewski, S. et al. (2018). Antifungal activity and synergism with azoles of polish propolis. *Pathogens* 7: 56.

Hanada, S., Uruga, H., Takaya, H. et al. (2014). Nebulized liposomal amphotericin B for treating *Aspergillus* empyema with bronchopleural fistula. *American Journal of Respiratory and Critical Care Medicina* 189 (5): 607–608.

Jannuzzi, G.P., Souza, N.A., Françoso, K.S. et al. (2018). Therapeutic treatment with scFv-PLGA nanoparticles decreases pulmonary fungal load in a murine model of paracoccidioidomycosis. *Microbes and Infection* 20 (1): 48–56.

Jeevanandam, J., Aing, Y.S., Chan, Y.S. et al. (2017). Nanoformulation and application of phytochemicals as antimicrobial agents. In: *Antimicrobial Nanoarchitectonics* (ed. A.M. Grumezescu), 61–82. Elsevier.

Khameneh, B., Diab, R., Ghazvini, K., and Bazzaz, B.S.F. (2016). Breakthroughs in bacterial resistance mechanisms and the potential ways to combat them. *Microbial Pathogenesis* 95: 32–42.

Kleinberg, M. (2006). What is the current and future status of conventional amphotericin B? *International Journal of Antimicrobial Agents* 27 (Suppl. 1): 12–16.

Krishnaswamy, K. and Orsat, V. (2017). Sustainable delivery systems through green nanotechnology. In: *Nano- and Microscale Drug Delivery Systems* (ed. A.M. Grumezescu), 17–32. Elsevier.

Lara, H.H., Romero-Urbina, D.G., Pierce, C. et al. (2015). Effect of silver nanoparticles on *Candida albicans* biofilms: an ultrastructural study. *Journal of Nanobiotechnology* 13 (91): 1–12.

Leal, A.F.G., Leite, M.C., Medeiros, C.S.Q. et al. (2015). Antifungal activity of a liposomal itraconazole formulation in experimental *Aspergillus flavus* keratitis withi endophthalmitis. *Mycopathologia* 179: 225–229.

Li, H. and Huck, W.T.S. (2002). Polymers in nanotechnology. *Current Opinion in Solid State and Materials Science* 6: 3–8.

Lucena, P.A., Nascimento, T.L., Gaeti, M.P.N. et al. (2018). In vivo vaginal fungal load reduction after treatment with itraconazole-loaded polycaprolactone-nanoparticles. *Journal of Biomedical Nanotechnology* 14 (7): 1347–1358.

Mihara, T., Kayeya, H., Izumikawa, K. et al. (2014). Efficacy of aerosolized liposomal amphotericin B against murine invasive pulmonary mucormycosis. *Journal of Infection Chemotherapy* 20: 104–108.

Netzker, T., Flak, M., Krespach, M.K.C. et al. (2018). Microbial interactions trigger the production of antibiotics. *Current Opinion in Microbiology* 45: 117–123.

Ong, T.H., Chitra, E., Ramamurthy, S. et al. (2017). Chitosan-propolis nanoparticle formulation demonstrates anti-bacterial activity against *Enterococcus faecalis* biofilms. *PLoS One* 12: 3.

Pace, J.R., Jog, R., Burgess, D.J., and Hadden, M.K. (2019). Formulation and evaluation of itraconazole liposomes for Hedgehog pathway inhibition. *Journal of Liposome Research*: 1–7.

Pandit, J., Garg, M., and Jain, N.K. (2014). Miconazole nitrate bearing ultraflexible liposomes for the treatment of fungal infection. *Journal of Liposome Research* 24 (2): 163–169.

Peçanha, P.M., Souza, S., Falqueto, A. et al. (2016). Amphotericin B lipid complex in the treatment of severe paracoccidioidomycosis: a case series. *International Journal of Antimicrobial Agents* 48: 428–430.

Pfaller, M.A. (2012). Antifungal drug resistance: mechanisms, epidemiology, and consequences for treatment. *The American Journal of Medicine* 125: S3–S13.

Pippi, B., Lana, A.J.D., Moraes, R.C. et al. (2015). In vitro evaluation of the acquisition of resistance, antifungal activity and synergism of Brasilian red propolis with antifungal drugs on *Candida* spp. *Journal of Applied Microbiology* 118: 839–850.

Pund, S. and Joshi, A. (2017). Nanoarchitectures for neglected tropical protozoal diseases: challenges and state of the art. In: *Nano- and Microscale Drug Delivery* (ed. A.M. Grumezescu), 439–480. Elsevier.

Rahimi, H., Roudbarmohammadi, S., Delavari, H., and Roudbary, M. (2019). Antifungal effects of indolicidin-conjugated gold nanoparticles against fluconazole-resistant strains of *Candida albicans* isolated from patients with burn infection. *International Journal of Nanomedicine* 14: 5323–5338.

Ribeiro, A.M., Souza, A.C., Amaral, A.C. et al. (2013). Nanobiotechnological approaches to delivery of DNA vaccine against fungal infection. *Journal of Biomedical Nanotechnology* 9 (2): 221–230.

Salem, H.F., Ahmed, S.M., and Omar, M.M. (2016). Liposomal flucytosine capped with gold nanoparticles formulations for improved ocular delivery. *Drug Design, Development and Therapy* 10: 277–295.

Schaffazick, S.R., Guterres, S.S., Freitas, L.L., and Pohlmann, A.R. (2003). Caracterização e Estabilidade Físico-Química de Sistemas Poliméricos Nanoparticulados para Administração de Fármacos. *Química Nova* 26: 726–737.

Sharma, M. (2019). Transdermal and intravenous nano drug delivery systems: present and future. In: *Applications of Targeted Nano Drugs and Delivery Systems* (eds. S.S. Mohapatra, S. Ranjan, N. Dasgupta, et al.), 499–550. Elsevier.

Sidhu, R., Lash, D.B., Heidari, A. et al. (2018). Evaluation of amphotericin B lipid formulations for treatment of severe coccidioidomycosis. *Antimicrobial Agents and Chemotherapy* 62 (7): e02293-17.

Singh, M., Kumar, M., Kalaivani, R. et al. (2013). Metallic silver nanoparticle: a therapeutic agent in combination with antifungal drug against human fungal pathogen. *Bioprocess and Biosystems Engineering* 36: 407–415.

Singh, N., Joshi, A., Toor, A.P., and Verma, G. (2017). Drug delivery: advancements and challenges. In: *Nanostructures for Drug Delivery* (eds. E. Andronescu and A.M. Grumezescu), 865–886. Elsevier.

Souza, A.C., Nascimento, A.L., de Vasconcelos, N.M. et al. (2015). Activity and in vivo tracking of amphoreticin B loaded PLGA nanoparticles. *European Journal of Medicinal Chemistry* 95: 267–276.

Tang, S.S., Apisarnthanarak, A., and Hsu, L.Y. (2014). Mechanisms of β-lactam antimicrobial resistance and epidemiology of major community-and healthcare-associated multidrug-resistance bacteria. *Advanced Drug Delivery Reviews* 78: 3–13.

Tawfik, A.A., Noaman, I., El-Elsayyad, H. et al. (2016). A study of the treatment of cutaneous fungal infection in animal model using photoactivated composite of methylene blue and gold nanoparticles. *Photodiagnosis and Photodynamic Therapy* 15: 59–69.

Troiano, G., Nolan, J., Parson, D. et al. (2016). A quality by design approach to developing and manufacturing polymeric nanoparticle drug products. *The AAPS Journal* 18: 1354–1365.

Vasile, C. (2019). Polymeric nanomaterials: recent developments, properties and medical applications. In: *Polymeric Nanomaterials in Nanotherapeutics* (ed. C. Vasile), 1–66. Elsevier.

Voltan, A.R., Quindós, G., Alarcón, K.P.M. et al. (2016). Fungas diseases: could nanostructured drug delivery systems be a novel paradigm for therapy? *International Journal of Medicine*: 3715–3730.

Wypij, M., Czarnecka, J., Dahm, H. et al. (2017). Silver nanoparticles from *Pilimelia columellifera* subsp. *pallida* S19 strain demonstrated antifungal activity against fungi causing superficial mycoses. *Journal of Basic Microbiology* 57: 793–800.

Yilma, S., Liu, N., Samoylov, A. et al. (2007). Amphotericin B channels in phospholipid membranes-coated nanoporous silicon surfaces: implications for photovoltaic driving of ions across membranes. *Biosensors Bioelectronics* 22 (8): 1605–1611.

Zhu, X., Radovic-Moreno, A.F., Wu, J. et al. (2014). Nanomedicine in the management of microbial infection – overview and perspectives. *Nano Today* 384: 321.

Zipfel, P.F., Wurzner, R., and Sherka, C. (2007). Complement evasion of pathogens: common strategies are shared by diverse organisms. *Molecular Immunology* 44: 3850–3857.

14

Chitosan Conjugate of Biogenic Silver Nanoparticles: A Promising Drug Formulation with Antimicrobial and Anticancer Activities

Smitha Vijayan and Jisha Manakulam Shaikmoideen

School of Bioscience, Mahatma Gandhi University, Kottayam, Kerala, India

14.1 Introduction

Silver nanoparticles (AgNPs) have unique properties and exhibit a number of medically important applications. Apart from all the applications and advantages, biological silver nanoparticles (Bio-AgNPs) still face some challenges in the industry. The main challenge they come across is their bioreduction mechanisms. The exact mechanism behind the synthesis is still not elucidated and while using a biological system the biomolecules involved in the reduction should be clearly squeezed out. The particle size, dispersion and surface morphology depend on the reducing and stabilizing agents involved in the synthesis. In order to produce the particles more effectively, it also needs a well-elucidated genetic system (Zhao et al. 2018).

Another major challenge associated with AgNPs is *in vitro* and *in vivo* cytotoxicity. Its toxicity toward the normal cells is a limiting factor for their medical applications. Chemically and physically synthesized AgNPs are toxic to both animals and the environment. This issue can be successfully resolved by synthesizing AgNPs with biological systems. Biological nanoparticles are less toxic to the normal cell lines and if toxicity occurs, it is only dose-dependent in nature. The toxicity of Bio-AgNPs can be studied through MTT (3-(4, 5-Dimethyl-2-thiazolyl)-2, 5-diphenyl-H-tetrazolium bromide) assay. In the case of normal L 929 fibroblast cells, they showed about 80% of cell viability at $50\,\mu g\,ml^{-1}$ concentrations (Vijayan et al. 2016a). *In vivo* toxicity studies revealed that they are toxic due to their affinity toward the SH (sulfhydral) or thiol group present in the proteins. Such affinity may adversely affect the normal cells of the body. *In vivo* studies on mice showed the accumulation of nanoparticles in the liver cells due to –SH silver precipitation (Zhao et al. 2018). Studies on mice liver cells after treatment with AgNPs of 20 nm size revealed that they are getting accumulated in Kupffer cells of liver but without causing any significant problem to the organ. There is no inflammation or impaired liver function. The level of albumin was measured to detect liver activity. This study states that the liver is not affected by the AgNPs (Kermanizadeh et al. 2014).

However, the toxicity of AgNPs can be reduced by certain methods. The important approach employed for the reduction of toxicity is their conjugation with a nontoxic natural polymer like chitosan, starch, gelatine, dextrin, dextran, etc. It was observed that conjugation of AgNPs with such polymers helps in the reduction of toxicity of the particles, and will also help in the stabilization of AgNPs. Still, extensive studies are required to evaluate the biocompatibility and stability of

Nanobiotechnology in Diagnosis, Drug Delivery, and Treatment, First Edition. Edited by Mahendra Rai, Mehdi Razzaghi-Abyaneh, and Avinash P. Ingle.

the conjugate in order to develop a potential conjugate. So far limited attempts have been made regarding the conjugation of Bio-AgNPs with natural polymers and the available studies revealed that it is a positive approach. Regiel et al. (2013) synthesized and explained the role of silver chitosan nanoconjugates (Ch-AgNP) and demonstrated their antimicrobial activity. Moreover, the enhanced antioxidant activity of Ch-AgNPs was studied by Minz (2015). Gelatin-conjugated nanoparticles and their enhanced activity was reported by Ahmad et al. (2003). A detailed and long term *in vivo* needs to be performed to get a detailed account of the toxicity of AgNPs and their conjugates. According to the available reports and studies AgNPs conjugates can be used as a better solution for the current situation of antibiotic resistance, including costly cancer treatment measures. Considering these facts, the present focus is on the role of chitosan-conjugated AgNPs in the management of various pathogens.

14.2 Conjugation of AgNPs with Natural Polymers

Different approaches have been proposed to enhance the stability and dispersion of nanoparticles; the most widely accepted, cost-effective approach is their conjugation with a polymer. As far as AgNPs are concerned, it was found that their encapsulation into a non-metal natural polymer is a reliable and eco-friendly approach (Wang et al. 2017). A large number of polymers are available which have the ability to enhance the stability and dispersion of metal nanoparticles, but at the same time it is necessary to assure that such conjugation should make the nanoparticles more potent. In this context, the compatibility between the polymer and the nanoparticles should be studied extensively. Various polymers which are often used for the conjugation of AgNPs include gelatine, starch, dextrin, liposome, polylactic acid, polyvinyl alcohol, etc. (Regiel et al. 2013). Among the different polymers available, biopolymers have attracted a great deal of attention due to their eco-friendly, renewable, sustainable, abundant, and nontoxic nature. The preparation of composite with AgNPs should be done with great care so that conjugation should increase the activity nanoparticles and decrease the toxicity. Bankura et al. (2012) studied the conjugation of AgNPs with dextran polymer. During the preparation of a composite, immense concern should be given for their uniform distribution, reduced toxicity, extended stability, and enhanced activity (Dananjaya et al. 2014). In the current world of nanotechnology, the frequency of using polymer-metal nanocomposite is increased because of its distinctive property and widened applications.

A variety of biomolecules can be conjugated with AgNPs to obtain AgNP-biomolecule hybrid or conjugate. Proteins, enzymes, oligopeptides, and polysaccharides can be used as a matrix for conjugation of Bio-AgNPs. The main mechanisms involved in the conjugation process are assumed to be electrostatic adsorption, direct chemisorption of thiol derivatives, covalent binding through bi-functional linkers, and specific affinity interactions. All these kinds of conjugation of AgNPs showed extended stability and activity (Regiel et al. 2013). Wang et al. (2015) evaluated the antifungal activity of Ch-AgNP conjugate and reported that antifungal activity of AgNPs was significantly increased after its conjugation with chitosan. Similarly, Kaur et al. (2013) demonstrated the antibacterial effect of Ch-AgNPs and found a significant enhancement in the activity against the tested pathogens.

Moreover, some of the researchers also studied the impact of Ch-AgNPs on viral and cancer cells, but only limited studies are available on the anticancer property of the ChBio-AgNP conjugate (Huang et al. 2004, 2007; Wei et al. 2009; Mori et al. 2013; Peng et al. 2017; Venkatesan et al. 2017). The studies of AgNP conjugate on MDA MB cell lines showed very significant cell death and apoptosis. The cell cycle arrest studied by flow cytometry showed that the arrest occurs at S phase only (Venkatesan et al. 2017). Gelatin-AgNPs and their antibacterial effects were studied to an extent by Singh et al. (2016). The pH stability of AgNPs was found to increase due to gelatine conjugation. This gelatine conjugate showed extensive bioactivity and stability in various pH ranges from 2 to 13 (Sivera et al. 2014). Darroudi et al. (2011) studied the characteristics and stability of green AgNPs conjugated with gelatine and found that thus-prepared conjugates are stable, but the bioactivity was not elucidated in the study.

Antibacterial, antibiofilm, and cytotoxicity of starch-stabilized AgNPs were studied and they exhibited potential antibacterial activity against both Gram-positive and Gram-negative bacteria with enhanced stability and no cytotoxicity to normal cells (Jena et al. 2012). Only few studies are available which demonstrate the antibacterial efficacy and stability of AgNPs-starch conjugates (AshaRani et al. 2009; Ghaseminezhad et al. 2012; Raji et al. 2012; Valencia et al. 2013, Cheviron et al. 2014; Khachatryan et al. 2016; Vasileva et al. 2017). All the studies performed in the past broadly proposed that conjugation of Bio-AgNPs with natural polymers is a promising approach to develop highly stable, biocompatible, economical, and nontoxic nanoconjugated complex.

So far, limited research attempts have been made on the conjugation of Bio-AgNPs with chitosan. Hence, this chapter focuses on ex situ conjugation of Bio-AgNPs with low molecular weight chitosan. The Bio-AgNP chitosan conjugates were widely employed for antimicrobial studies, especially against bacteria, but less concern is given on antifungal and anticancer activities.

14.3 Conjugation of Bio-AgNPs with Chitosan

Chitosan is a natural polymer which has been reported as a robust antimicrobial agent with many other advantages. There are reports regarding the use of chitosan as a polymer-based protective agent for silver, gold, and platinum nanoparticles. Moreover, conjugation of chitosan in various nanoparticles is reported to be very easy, biocompatible, biodegradable, and nontoxic (Ahmad et al. 2003). In one of the studies, authors synthesized Bio-AgNPs using the mycoendophyte *Colletotrichum gloeosporioides* (KX881911), isolated from the medicinal plant *Withania somnifera* (L.), and further these Bio-AgNPs were conjugated with the natural polymer chitosan to increase their stability and to enable the sustained release of Bio-AgNPs. However, previous studies on copper nanoparticle-chitosan conjugate revealed that conjugation with metal nanoparticles enhances the antimicrobial activity of chitosan (Du et al. 2009). Similarly, studies on conjugation of AgNPs with chitosan revealed the enhanced stability of the AgNPs, and it was also reported that there was an increase in antimicrobial as well as antioxidant activity of AgNPs. Such increased activity of this conjugate against Gram-positive and Gram-negative bacteria was also reported by Wei et al. (2009). Low molecular weight chitosan has more stability with AgNPs and the conjugates developed with such chitosan exerts more potent antibacterial efficacy (Richardson et al. 1999). Venkatesan et al. (2017) studied the conjugation of chitosan with AgNPs and alginate, where it showed high stability and increased activity. All the previous methods used for conjugation nanoparticles with chitosan, calcium chloride, or sodium hydroxide are only for regeneration of nanoparticles (Kaur et al. 2013; Venkatesan et al. 2017). Honary et al. (2011) studied the antibacterial effect of chitosan-AgNP conjugate and its application in wound dressing.

14.4 Methods for Conjugation

The conjugation can be performed either by in situ or ex situ approaches (Pecher and Mecking 2010). The in situ approach is the preparation of nanoparticles with the polymer, in which the polymer gets naturally conjugated with the nanoparticles. But here it is necessary to use a potential reducing agent to enhance the reaction. It is a one-step process for the preparation of nanocomposite. However, in the case of ex situ conjugation, already prepared nanoparticles are mixed with polymers to make a conjugate. The most important advantage in this approach is that no adverse chemicals are required to reduce the respective metal salt. The biologically prepared nanoparticles can be conjugated with a natural polymer to assure a completely green process. Conjugation of chitosan with chemically synthesized AgNPs through the in situ approach is an extensively studied area, whereas the ex situ mode of conjugation with Bio-AgNPs is much less explored. As discussed earlier, so far there are only a few studies which have been performed on conjugation of Bio-AgNPs with chitosan (Guo et al. 2014). Ex situ conjugation of Bio-AgNPs with low molecular weight chitosan was conducted by Vijayan et al. (2019). Thus, prepared AgNPs chitosan conjugates were widely employed for antimicrobial studies, especially against bacteria, but less focus has been given to evaluation of antifungal and anticancer activities. Generally, the ex situ approach for conjugation of chitosan with Bio-AgNPs is preferred because this is purely green, cost-effective, and environmentally reliable. Vijayan et al. (2019) performed the conjugation of chitosan derived from *Penaeus monodon* (Tiger prawns) (Divya et al. 2014) with biologically synthesized AgNPs. Moreover, the authors proposed that there is no involvement of any chemicals to ensure stabilization or reduction. Hence, authors claimed that the conjugate developed and proposed was cost effective, nontoxic, and environmentally friendly.

14.5 Techniques Used for the Characterization of ChBio-AgNPs

To establish the successful synthesis of ChBio-AgNP conjugate, it is very important to characterize the various parameters associated with it, such as size, shape, structure/crystallinity, concentration, topography, agglomeration, surface charge, and stability. There are many techniques used for the characterization of AgNPs, which can be employed for the characterization of ChBio-AgNPs. They mainly include UV-Visible spectroscopy, X-ray Diffraction pattern analysis (XRD), Transmission Electron Microscopy (TEM), Scanning Electron Microscopy (SEM), Atomic Force Microscopy (AFM), Dynamic Light Scattering analysis (DLS), Zeta potential analysis, and Fourier Transform Infrared Analysis (FTIR) (Wang et al. 2017).

14.5.1 UV-Visible Spectroscopy

It is an easy, reliable, simple, sensitive, and selective method for the characterization of nanoparticles and its conjugates. For AgNPs and their conjugates, the conduction band and valence band lie very close to each other, which allow the free movement of electrons. These free electrons are responsible for the surface plasmon resonance (SPR) absorption band of the AgNPs and its conjugates. This unique property makes the identification of AgNPs formation easy by means of UV spectroscopy (Desai et al. 2012). The SPR property varies according to the size, shape, and dispersity of the nanoparticles. Figure 14.1a shows the UV spectrum scan of ChBio-AgNPs. The stability of AgNPs and its conjugates can be detected by UV spectroscopy, where the same peak at the same wavelength indicates stability. It can be checked to a period of up to 12 months (Zhang et al. 2012).

14.5.2 X-Ray Diffraction (XRD) Analysis

X-ray diffraction analysis is a well-known scientific procedure which has been used for the examination of both the atomic and precious crystalline nature of many materials, including metal

(a)

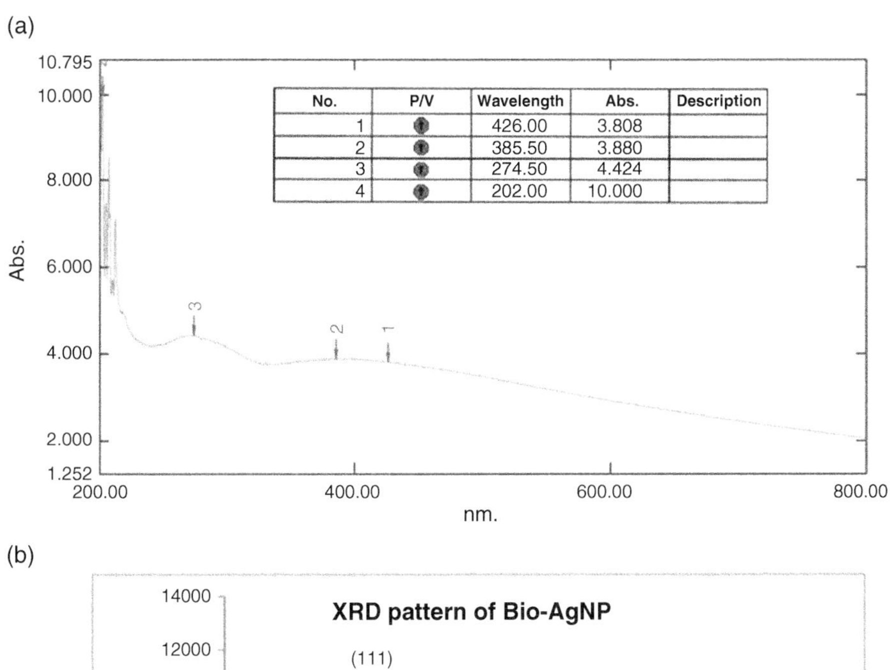

(b)

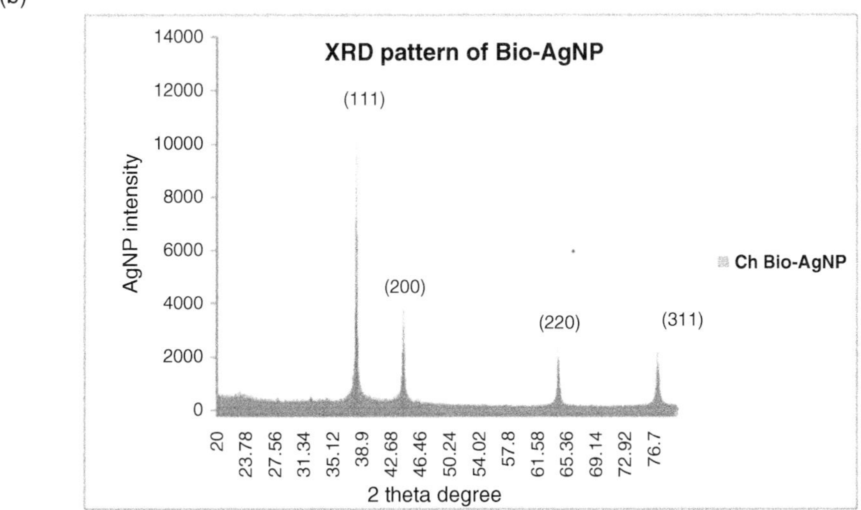

Figure 14.1 (a) UV-Visible spectrum analysis of ChBio-AgNPs conjugate showing a peak at 426 nm; (b) XRD pattern of ChBio-AgNPs showing strong peaks at 38.59, 44.72, 65.335, and 77.98 angles of 2theta indexed indicating fcc structure.

nanoparticles. XRD is excellent in its determination of crystallinity of a compound by observing the X-ray diffraction pattern. The diffraction pattern obtained can be compared with the Joint Committee on Powder Diffraction Standards (JCPDS) library. XRD is a technique with many advantages and fewer limitations. It works on the principle of Bragg's law. It also helps in the calculation of nanoparticle size with the help of the Debye-Scherrer equation (Waseda et al. 2011). The main advantage of this technique is that the size calculation is easy, and the results are accurate. On the other hand, the only drawback of this technique is that it can only be performed with powdered solid materials (Dey et al. 2009; Ivanisevic 2010; Cabral et al. 2013). Figure 14.1b shows the XRD pattern of ChBio-AgNPs which showed strong peaks at 38.59, 44.72, 65.335, and 77.98 angles of 2 theta indicating the with fcc structure.

14.5.3 Transmission Electron Microscopy (TEM)

Transmission electron microscopy is an efficient, sophisticated technique to study the dispersion, morphology, and size of nanoparticles. In nanotechnology, TEM plays a major role in the characterization of nanoparticles (Williams and Carter 2009). The selected area electron diffraction (SLAED) pattern is also coupled with TEM. This pattern helps in obtaining the details of electron diffraction of metal nanoparticles. This SLAED pattern can be compared with the XRD pattern to confirm the presence of metal (Lin et al. 2016). Similarly, for the characterization of ChBio-AgNPs, TEM and SLAED also play an important role. Moreover, energy dispersive spectroscopy (EDAX) is another technique associated with TEM; it helps in the quantification of silver metal ions present in the area. TEM analysis also helps to make histogram or particle distribution pattern of particles using TEM image and Image J software. Figure 14.2 shows the TEM, SLAED, and EDAX analysis of ChBio-AgNPs. However, TEM has certain disadvantages, such as: sample preparation is time-consuming; immense skill is required to operate it; and thin preparation of the sample required. The quality of the image depends on the condition of the equipment and accuracy in sample preparation. The high maintenance of vacuum conditions are necessary (Hall et al. 2007).

14.5.4 Scanning Electron Microscopy (SEM)

Scanning electron microscopy is a completely equipped surface imaging strategy. It helps to identify distinctive molecule sizes, and estimate disseminations, nanomaterial shapes, and the surface morphology of nanoparticles and its conjugates. Figure 14.3a shows the SEM micrograph for

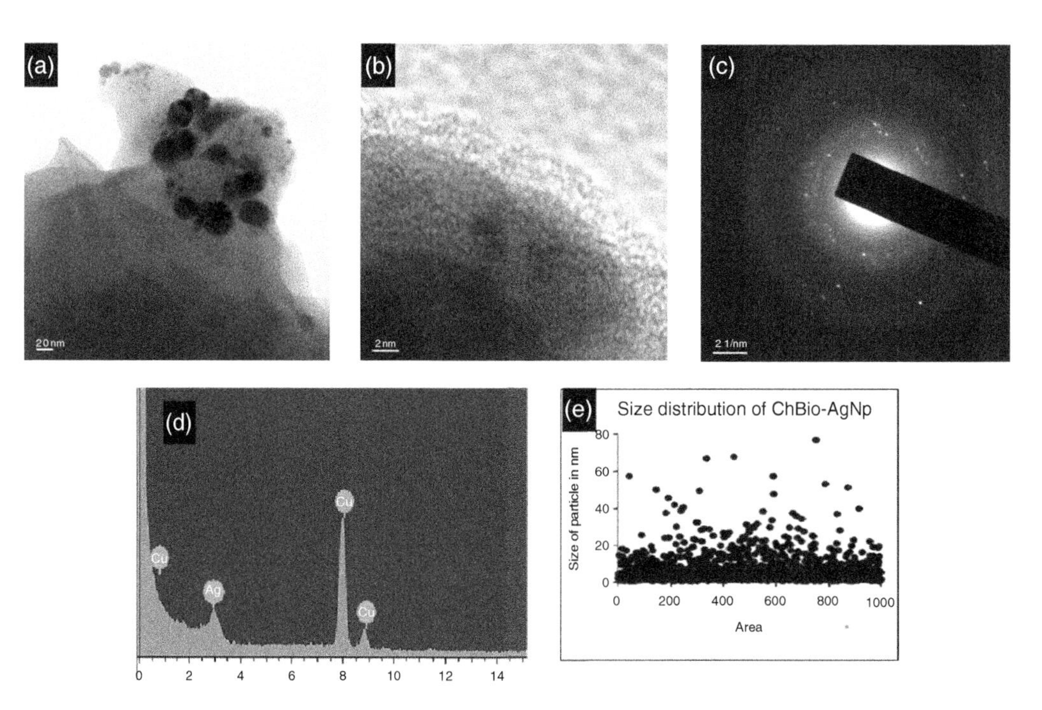

Figure 14.2 TEM micrographs of ChBio-AgNPs showing (a) particles inside the matrix; (b) high resolution images of spherical ChBio-AgNPs; (c) SLAED pattern of silver; (d) EDAX pattern showing the presence of silver; (e) Particle size distributions calculated with TEM micrograph using Image J software showing the distribution of particles below 50 nm.

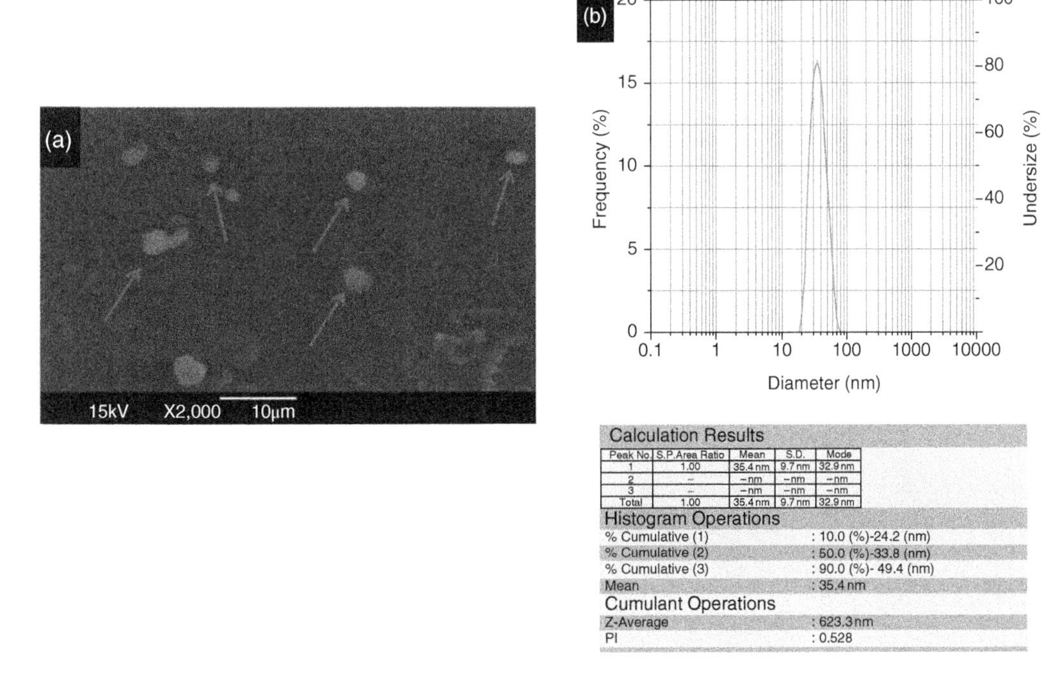

Calculation Results

Peak No.	S.P.Area Ratio	Mean	S.D.	Mode
1	1.00	35.4 nm	9.7 nm	32.9 nm
2	–	–nm	–nm	–nm
3	–	–nm	–nm	–nm
Total	1.00	35.4 nm	9.7 nm	32.9 nm

Histogram Operations

% Cumulative (1)	: 10.0 (%)-24.2 (nm)
% Cumulative (2)	: 50.0 (%)-33.8 (nm)
% Cumulative (3)	: 90.0 (%)- 49.4 (nm)
Mean	: 35.4 nm

Cumulant Operations

Z-Average	: 623.3 nm
PI	: 0.528

Figure 14.3 (a) SEM micrographs of ChBio-AgNPs showing well dispersed spherical particles (highlighted with arrows); (b) DLS analysis for distribution of size of ChBio-AgNPs.

spherical well-dispersed ChBio-AgNPs. The disadvantage of SEM is that it cannot resolve the issue of observing the internal structure of the particles. Modern high-resolution SEM is now available to observe the small particles of 10 nm sizes. EDAX is an electron diffraction pattern available with SEM helping in the analysis of purity of nanoparticle and observing the crystallinity of the particles (Fissan et al. 2014).

14.5.5 Dynamic Light Scattering (DLS) and Zeta Potential

Dynamic light scattering is helpful to determine the size of nanoparticles with the light diffraction of particles in a solution. This depends on the concentration of particles in the solution, dilution pattern, and presence of stabilizing or capping molecules (Zhang et al. 2016). Because of the Brownian movement, the size obtained from DLS is always higher than TEM. It has sample specific limitations and it mainly measures the particle size by the pattern of laser light scattering on it. It utilizes laser wavelength of 633 nm (He–Ne) and a scattering angle of 173°, a measurement temperature of 25 °C, a medium viscosity of 0.8872 m Pas, a medium refractive index of 1.330, and material refractive index of 0.200 (Tomaszewska et al. 2013). Figure 14.3b represents the DLS analysis for size distribution of ChBio-AgNPs.

Moreover, the zeta sizer available with DLS gives the stability output of the AgNPs by measuring the net surface charge of the molecules. Zeta potential between +30 and −30 mV is considered as stable. But this value varies with the dilution pattern, temperature, and pH of the medium of dispersion in case of AgNPs and its conjugates (Haider and Mehdi 2014).

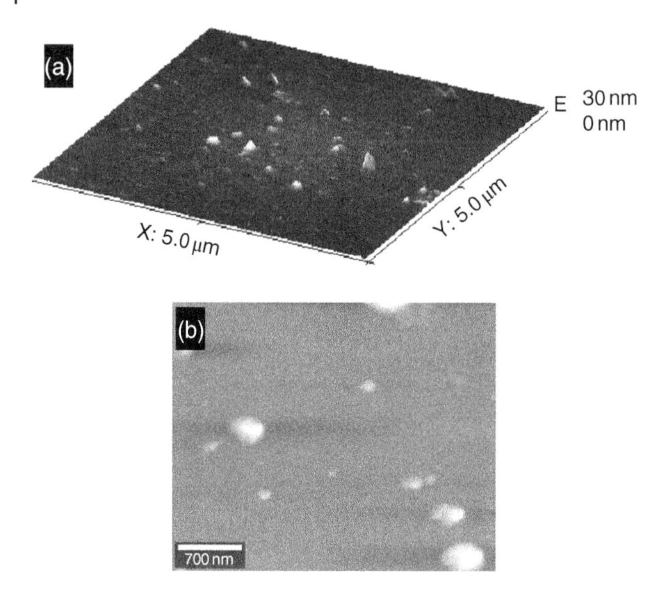

Figure 14.4 (a) AFM topology of ChBio-AgNPs showing well distributed particles of various sizes; and (b) AFM micrograph showing well dispersed spherical ChBio-AgNPs.

14.5.6 Atomic Force Microscopy (AFM)

AFM is another important tool for measuring nanoscale materials. It uses a cantilever with a sharp probe to scan the surface of the specimen. Cantilever works under Hooke's law in three different modes: contact mode, non-contact mode, and intermittent sample contact mode through which surface topology, size, and distribution of particles can be performed AFM gives a very brilliant idea about surface topology and shape of the particles without any surface coating or special treatments. It is a very good characterization technique for measuring the sub-nanometer scale in aqueous fluids (Zhang et al. 2016). The exact configuration and homogenous size distribution can be brilliantly elucidated by AFM analysis. A skilled person is necessary to avoid erroneous measurements and to select the correct mode (Abo-State and Partila 2015). Sample preparation is a very crucial thing in AFM analysis (Singh et al. 2016). Figure 14.4a,b show the AFM micrographs of the distribution and shape of ChBio-AgNPs, respectively.

14.5.7 Fourier Transform Infrared (FTIR) Spectroscopy

FTIR helps in detecting the absorbance in an order of 10^{-3} and can identify the small absorption difference of a residue which may occur by means of molecule stretching, wagging, and bending. It is widely used in case of biologically synthesized nanoparticles to identify the presence of biomolecules in the reaction solution. It is counted with advantages such as easy, cost effective, no sample heating, sample to error ratio is less, and results are accurate. It is a noninvasive technique (Banu and Rathod 2011; Zhang et al. 2016).

14.6 Bioactivities of ChBio-AgNP Conjugate

ChBio-AgNPs are reported to have potential biological activities, which mainly include antimicrobial and antioxidant activities. All these bioactivities are briefly discussed in the following section.

14.6.1 Antibacterial Activity of ChBio-AgNPs

AgNPs are widely studied for their antibacterial activity. It is proposed that antibacterial efficacy of AgNPs depends on various factors like size, shape, zeta potential, etc. However, in case of the conjugate, the polymer used and the pH of the solution are also very crucial. So far, there are a large number of studies performed which clearly reveal the potential of AgNPs as antibacterial agents (Rekha and Vedpriya 2013; Fatima et al. 2015). After conjugation, comparative assessment of the antibacterial efficacy was elucidated by well diffusion assay and it revealed enhanced antibacterial activity of ChBio-AgNPs with increased zone size. The enhanced activity may be due to the synergistic effect of chitosan and AgNPs. Chitosan is well known for its antibacterial effect (Jayakumar et al. 2010; Palem et al. 2018). The increased antibacterial activity of ChBio-AgNP conjugate was also reported by Palem et al. (2018). The large number of hydroxyl groups in the polymer helps in formation of good conjugate with metal nanoparticles and this further provides them with increased stability and antibacterial activity (Jena et al. 2012).

Not only antibacterial efficacy, but antibiofilm activity of ChBio-AgNPs was also found to increase, and they showed biofilm inhibition from the minimum concentration, i.e. from 25 $\mu g\,ml^{-1}$ onwards. It was suggested that the increased antibiofilm activity may be due to the mucoadhesive property of chitosan and inhibition of exopolysaccharide synthesis by Bio-AgNPs. Moreover, the presence of amino and hydroxyl groups in chitosan exhibit such strong properties to Bio-AgNPs (Jena et al. 2012; Regiel et al. 2013).

The modes of action of the ChBio-AgNPs are not yet well elucidated but it was proposed that the possible mechanisms for the action of AgNPs can be applicable for ChBio-AgNPs also. The mucoadhesive property of chitosan and the interaction mediated by the electrostatic forces mediated by the protonated NH_3^+ groups of chitosan enhance the activity of AgNPs (Goy et al. 2016). Chitosan also executes blockage of some essential nutrient flow. ChBio-AgNPs bind with the thiol group of cysteine-containing protein in the bacterial cell wall and in the respiratory systems. This will lead to the disintegration of membrane and impairment in the respiratory system. The same will lead to the destruction of succinate dehydrogenase and NADH dehydrogenase. Moreover, antibacterial action was also depicted by membrane integrity analysis, where it showed leakage of cell inclusions and the formation of pits on the cell membrane. The polycationic activity of chitosan helps in the creation of membrane disintegration with Bio-AgNPs. The TEM analysis also explained the activity of ChBio-AgNPs on membrane disruption and the presence of nanoparticles on the membrane also confirms the same (Soman and Ray 2016).

In this context, different studies have been performed against Gram-positive and Gram-negative bacteria; extensively studied strains include *Escherichia coli* and *Staphylococcus aureus* (Ruparelia et al. 2008; Vijayan et al. 2016b). It is proven that these particles are efficient in the biocidal effect against both Gram-positive and Gram-negative cells. These composites are responsible for creating uncontrolled tyrosine phosphorylation inside the cell and they affect the total system, including gene expression and regulation which eventually lead to cell death (Shrivastava et al. 2007; Aerle et al. 2013). The antibacterial activity of AgNPs against multidrug resistant (MDR) *E. coli* and *S. aureus* was effectively studied by Singh et al. (2014). The results obtained showed a good inhibition pattern at 20 μl concentration. ChBio-AgNPs is also reported to be potential antibiofilm agents and their efficacy was tested on common biofilm-forming bacteria (*S. aureus*) of catheters by Thomas et al. (2015). Their efficacy in preventing biofilm formation makes them potential coating agents for hospital devices. The capacity to prevent microbial attachment is very important and it also helps to disseminate the medicine successfully (Thomas et al. 2015; Ramachandran and Sangeetha 2017). Overall, from all the reports available it is confirmed that ChBio-AgNPs exhibited broad-spectrum antibacterial activity against both Gram-positive and Gram-negative bacteria.

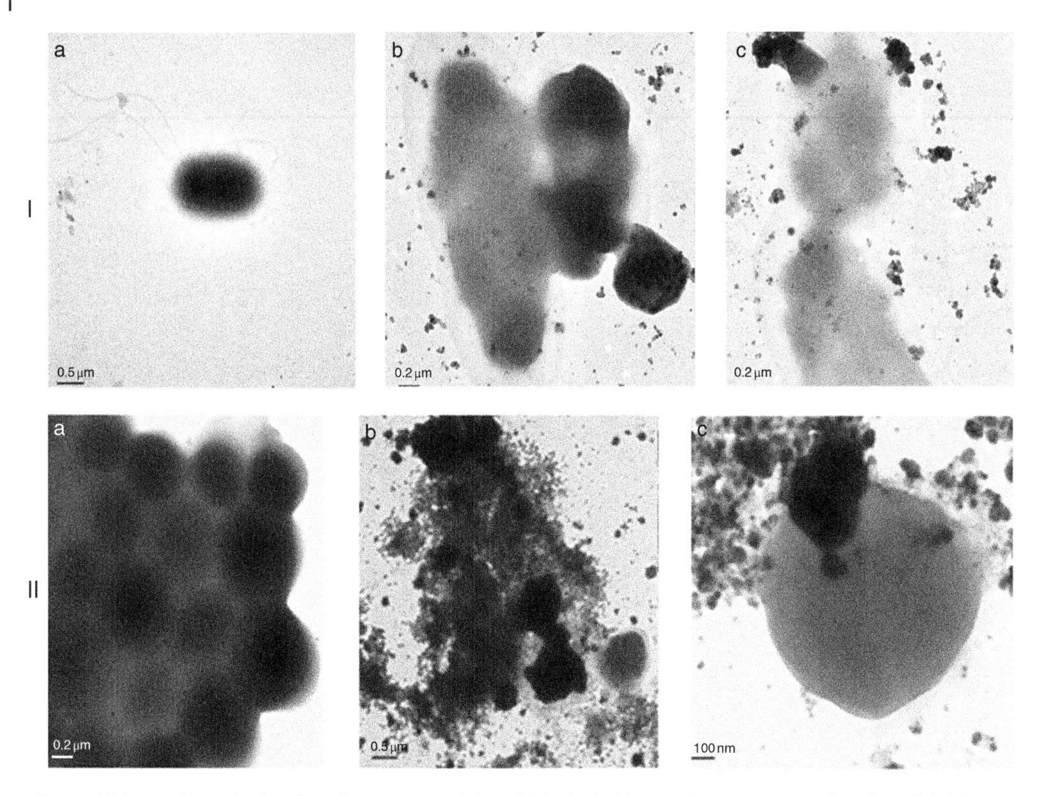

Figure 14.5 TEM analysis of antibacterial activity of ChBio-AgNPs conjugates. [I] against *E. coli:* (a) intact *E coli* cell; (b) Entry of nanoparticles into the cell and distortion of cell structure; (c) Formation of pits on the cell surface. [II] against *S. aureus:* (a) intact *S. aureus* cluster; (b) destroyed cluster showing extensive damage of cells; (c) Single cell with membrane damage and presence of nanoparticles inside and outside of the cells.

Interestingly, until now no incident of antimicrobial resistance against ChBio-AgNPs has been reported. Figure 14.5 shows the antibacterial activity of ChBio-AgNPs against *E. coli* and *S. aureus.*

14.6.2 Antifungal Activity of ChBio-AgNPs

The antifungal activity of these conjugates was studied using different methods, like well diffusion assay and minimum inhibitory concentration (MIC) assay. One of the studies performed recorded the MIC of 25 $\mu g\,ml^{-1}$ against the human pathogenic *Candida* species obtained from the Lakeshore hospital in Cochin, Kerala, India. AgNPs were reported to have inhibitory efficacy against both yeast and mold simultaneously (Panacek et al. 2009). Ishida (2017) demonstrated that Bio-AgNPs showed potential efficacy against all the tested *Candida* species at a concentration below 13 $\mu g\,ml^{-1}$. It is believed that the antifungal activities of AgNPs are mainly due to membrane damage. The particles penetrate the cell and create pits and pores in the membrane. The membrane disintegration leads to the leakage of trehalose and glucose, which is necessary for the cells to maintain stability which eventually leads to cell death (Xue et al. 2016). Xue et al. (2016) studied the antifungal activity of AgNPs against *Candida* sp. and the MIC for the activity was found in between 0.125 and 4.00 $\mu g\,ml^{-1}$. *Candida albicans* and *Candida tropicalis* were successfully inhibited by AgNPs at a MIC of 0.42 and 0.84 $mg\,l^{-1}$, respectively (Panacek et al. 2009).

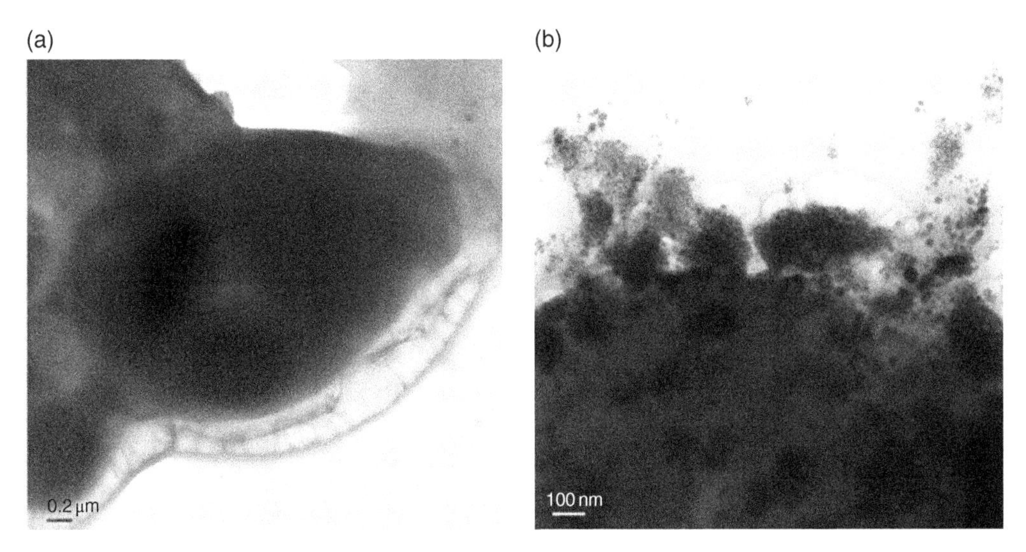

Figure 14.6 TEM analysis of antifungal activity of ChBio-AgNPs conjugates. (a) Intact cell of *Candida* sp.; (b) cell disruption of *Candida* sp. by ChBio-AgNPs.

The polycationic activity of chitosan helps in binding the fungal cells more firmly. Bio-AgNPs and chitosan together cause membrane damage due to reactive oxygen species (ROS) generation and also by means of disrupting the membrane stabilizing proteins by binding on the sulfur-containing groups and precipitating them (Wang et al. 2015). *In vitro* ROS generation is the main deal in antifungal activity; it causes severe damage to the cell. The irreversible changes occurred in the membrane of fungal cells and it leads the cell to leakage and death (Khan et al. 2018). ChBio-AgNPs act together to inhibit the bud formation in the *Candida* species and they are also capable of blocking the spore germination. The attachment of the AgNPs with the fungal cells was enhanced by the conjugated chitosan. The presence of particles inside the cells is the prime evidence for the mechanism of action (Figure 14.6) (Mori et al. 2013; Zhang et al. 2016; Wang et al. 2017). A study conducted by the authors revealed the extensive antifungal activity of ChBio-AgNPs against various human pathogenic *Candida* species (*C. albicans, C. tropicalis, C. dubliniensis, C. krusei*, and *C. glabrata*) with concentrations of 1–7 $\mu g\,ml^{-1}$.

14.6.3 Antioxidant Activity of ChBio-AgNPs

Chitosan also showed excellent antioxidant activities which have been studied by many researchers (Minz 2015; Majekodunmi 2016). The DPPH (2,2-diphenyl-1-picryl-hydrazyl-hydrate) assay of chitosan showed excellent radical scavenging activity, but the evaluation of antioxidant activity of ChBio-AgNPs conjugate is a less explored field hence, very few studies are conducted related to this field. There are reports on the anticancerous study of chitosan and AgNP conjugates; the anticancerous effect can be related to the antioxidant activity (Li and Li 2014). Li and Li (2014) studied the efficacy of these conjugates using various assays such as DPPH assay, 2,2-Azino-bis(3-ethylbenzthiazoline-6-sulfonic) acid (ABTS) assay, hydroxyl radical scavenging assay, reducing power assay, and super oxide radical scavenging assay. The IC_{50} value of standard and sample vary according to the different parameters of the experiment, so that various assays are necessary to be carried out at the same concentrations to establish the results. DPPH assay was found to be rapid and dependable radical scavenging assay, where authors reported IC_{50} value of $1.976 \pm 0.2959\,\mu g\,ml^{-1}$

for the ChBio-AgNPs and $2.561 \pm 0.4084\,\mu g\,ml^{-1}$ for standard. These results clearly depict the antioxidant nature of ChBio-AgNP conjugates. These results were found to be promotable and can be supported with the findings of various previous studies (Manivasagan et al. 2014; Minz 2015; Majekodunmi 2016). ABTS is another very rapid and potent antioxidant assay; it can be used at various pH flexibility. This compound is very stable and has the ability to accept an electron from ChBio-AgNPs, and showed very promising rate of inhibition. IC_{50} values reported for standard (ascorbic acid) and ChBio-AgNPs are 2.561 ± 0.4084 and $1.976 \pm 0.2959\,\mu g\,ml^{-1}$, respectively (Veerapur et al. 2009). Moreover, IC_{50} calculated from hydroxyl radical scavenging activity was found to be $1.778 \pm 0.2325\,\mu g\,ml^{-1}$ for standard ($R^2 = 0.99$) and $1.732 \pm 0.2386\,\mu g\,ml^{-1}$ for the ChBio-AgNPs. It is a good scavenger of radicals and it can be employed as a natural antioxidant agent (Ishida 2017). The reducing power assay depicts the increased reducing capacity of ChBio-AgNPs with respect to the increase in concentration. The efficiency of reduction depends on the number of reductants present in the sample. Therefore, from these assays it is revealed that ChBio-AgNPs have potential antioxidant capacity (Loganayaki et al. 2013; Sivera et al. 2014; Bhakya et al. 2016). Another study demonstrated that ChBio-AgNPs showed considerably good radical scavenging activity in superoxide radical scavenging assay. The IC_{50} recorded for standard and ChBio-AgNPs were 3.066 ± 0.4866 and $2.22 \pm 0.4022\,\mu g\,ml^{-1}$, respectively (Nimmi and George 2012; Linn 2015).

14.6.4 Anticancer Efficacy of ChBio-AgNPs

The chitosan is a natural polymer usually derived from the exoskeleton of shrimp shells, which is a waste product of the seafood industry. Chitosan is not soluble in water but can be dissolved in weak acids like acetic, citric, or tartaric acid. When it dissolves in dilute acidic solution, the free amino groups undergo protonation and ionizable $R\text{-}NH^{3+}$ are formed (Sugano et al. 1988). It is commonly available in high and low molecular weight forms which determine their activity, especially during the preparation of conjugate and nanoparticles. It is a therapeutic agent widely used as a cholesterol reducing, wound dressing substance. As discussed earlier it also has antiulcer and antimicrobial properties (Ahmed and Aljaeid 2016). Chitosan–metal nanoparticle conjugates can be better drug carriers because of their biocompatibility, biodegradability and nontoxic nature. The mucoadhesive property of chitosan is due to the interaction between $R\text{-}NH^{3+}$ and the negative charge on the mucosal membrane. This property makes them suitable carriers for drug delivery and a better polysaccharide to conjugate with metal nanoparticles (Sugano et al. 1988; Rabea et al. 2005). The size of Bio-AgNPs used for the preparation of conjugate is very important; if the size is small it will be more effective. Chitosan has the capability to increase cell membrane permeability and can act as an absorption enhancer. It enhances the absorption across the intestinal epithelia and also increases the residence time of drugs in the system. The mucoadhesive property of chitosan is helpful for using it as a conjugating agent (Schmitt et al. 1998; Roldo et al. 2004; Sarmento et al. 2007; Mizrahy and Peer 2012).

So far, few studies have been performed about the efficacy of ChBio-AgNPs on human lung carcinoma (A_{549}), hepatocarcinoma (Hep G_2), and MCF (Michigan Cancer Foundation)-7 (breast cancer) cell line which have been significant (Wang et al. 2017). The ChBio-AgNP conjugate is very advantageous as the AgNP used is biogenic. The toxicity study on normal cells found them to be nontoxic. Anticancerous studies on various tumor cell lines using AgNPs by MTT assay were carried out in various attempts (Zhao et al. 2018). Han et al. (2014) demonstrated that chemically synthesized nanoparticles are less efficient in cancer treatment compared to Bio-AgNPs.

Minz (2015) reported the promising anticancerous efficacy of Ch-AgNP (chemically synthesized) conjugates. Similarly, Abaza et al. (2018) studied the anticancerous efficacy of chitosan/poly (vinyl alcohol) blend doped with gold and AgNPs on prostate cancer cell lines and it showed significant impact on the death of cancer cell lines. In one of the studies, the anticancerous efficacy of ChBio-AgNPs on MDA MB and Si Ha cell lines was evaluated. Further, the MTT assay performed showed excellent cytotoxicity toward both cell lines. The IC_{50} value for Bio-AgNPs and ChBio-AgNPs were found to be 4.346 ± 0.6381 and $0.9851 \pm 0.0065\,\mu g\,ml^{-1}$, respectively, for MDA MB. However, the IC_{50} value for Bio-AgNPs and ChBio-AgNPs were reported to be 24.35 ± 1.390 and $2.086 \pm 0.319\,\mu g\,ml^{-1}$, respectively, for Si Ha cell lines (Vijayan et al. 2019). These values stated that ChBio-AgNPs are a potent and highly effective antitumor agent.

The efficacy of ChBio-AgNPs as an antitumor agent was described by Vijayan et al. (2019) by the acridine orange-ethidium bromide (AO/EtBr) double staining method. It showed early and late apoptosis and it was later supported with a lactate dehydrogenase assay (LDH), Caspase7/9 assay, and DNA fragmentation. The mode of action was also established by *in vitro* ROS generation. Both cell lines are efficient in RO (reactive oxygen) generation. Cell cycle arrest was studied by flow cytometry and revealed that the arrest occurs at G_1/M phase for Si Ha cells and G_2/S for MDA MB cells. Vijayan et al. (2019) conducted a gene expression study by real-time polymerase chain reaction (RT-PCR) and showed upregulation of p53 and p38 genes. These studies clearly explain the mode of action of ChBio-AgNPs on the tumor cells at the genetic level. Hence, this formulation can be suggested to the pharmaceutical industry, which is always in search of a new therapeutic formulation to defeat cancer.

In the two cell lines (Si HA and MDA MB), it showed early and late apoptosis with morphological changes such as membrane blubbing, shrinkage, and chromatin condensation. The aforementioned approaches are widely accepted and employed for the detection of apoptosis, which is the key feature in antitumour/anticancerous activity detection (Fani et al. 2016; Bendale et al. 2017). This apoptosis can be again supported well with caspase, LDH, and DNA fragmentation assay.

Caspases are a group of cysteine-containing potent proteases. They play a crucial role in the anticancer activity by initiating apoptosis and thereafter DNA fragmentation. Caspases are involved in two categories such as initiating and execution caspases. Caspase 7 and 3 are redundant (Gurunathan et al. 2015). An optimal peptide recognition sequence is usually shared by Caspase 7 and 3, which make them redundant. They are generally activated by caspase 8 and 9. All these enzymes play a major role in antitumour activity by inducing apoptosome formation. On treatment with ChBio-AgNPs it induces the formation of caspases. This was noticed in the case of MDA MB and Si Ha cells. The increased level of caspase 7 and 9 were obtained after ChBio-AgNP treatment (Lamkanfi and Kanneganti 2010). The role of caspase in apoptosis and antitumour activity was also elucidated in many reports (Shahbazzadeh et al. 2011; Gurunathan et al. 2015; Elshawy et al. 2016; Yuan et al. 2017; Khan et al. 2018).

Moreover, release of LDH from the cytosol is a good indicator of necrosis and apoptosis. It is not a standalone assay; the results of LDH should always be concluded with other assay and findings (Gurunathan et al. 2015). Yuan et al. (2017) demonstrated the efficacy of Ch-AgNPs in inducing the release of LDH during apoptosis, and the release is gradual and directly proportional to the dose of the Ch-AgNPs and also the time of interaction.

Similarly, fragmentation or laddering of DNA is an evidence of apoptosis. During apoptosis, the large fragments of DNA are cleaved by the nucleases and create fragments of DNA and can be well studied by agarose gel electrophoresis. The characteristic event in apoptosis is DNA fragmentation and the two cell lines (MDA MB and Si Ha) exhibited the fragmentation after treatment with ChBio-AgNPs. Previous studies on DNA fragmentation of various tumor cell lines after treatment with chitosan–metal nanoparticle conjugates revealed that these particles can initiate apoptosis and thereafter can cause DNA fragmentation (Ishida et al. 2013; Moghaddam et al. 2015; Elshawy et al. 2016; Ishida 2017; Yuan et al. 2017).

In an antitumor study, it is very important to identify the stage at which the cell cycle arrest takes place. Flow cytometry is a very sophisticated, reliable technique to depend on the analysis. Identification of cell cycle arrest point help in the identification of the enzyme present in the particular anticancer activity. There are many reports available which deal with the cell cycle arrest caused by the AgNPs, but very few in the case of ChBio-AgNPs. In most of the cells, it is reported that the cell cycle arrest occurs at the G_2/M phase. In MDA MB cells the previous reports suggest that the cell cycle arrest at G_2/M phase and the gene p38 and p53 are usually studied (Fani et al. 2016; Azizi et al. 2017). In the case of ChBio-AgNPs cell cycle arrest also occurs at G_1/M phase for Si Ha cells and G_2/S for MDA MB cells.

In flow cytometry, the DNA is stained with a fluorescence dye called propidium iodide. The cells in S phase contain more DNA will produce more bright fluorescence and G2 are supposed to be brighter than G1 (Murad et al. 2016). Based on this aspect the DNA from the ChBio-AgNP-treated Si Ha cells and MDA MB cells were measured and correlated with the cell population. Choudhari et al. (2013) studied the cell cycle arrest in Si Ha cells and it was reported to be at G_1/S phase. Pumiputavon et al. (2017) also studied the cell cycle arrest of Si Ha cells and found to be at G_1/S phase.

RT-PCR is a well-established method to study gene expression. It is widely used and is a sophisticated technique with a high rate of accuracy. Gene expression studies on ChBio-AgNPs treated MDA MB and Si Ha cells revealed the upregulation of the gene (Fani et al. 2016). The relative rate of PCR products was calculated by the comparative analysis of Ct ($2-\Delta\Delta$Ct). The fold change in the target gene was compared with the average of the ß-actin gene. Both target and reference genes were measured in the samples and referenced. GenEx software was used for the measurements, and Data Assist software, version 3, from Applied Biosystems (USA), was used for the RNA fold alterations calculation. The previous studies on p53 gene expression include the studies performed by Yuan et al. (2017), Mocan (2013), and Ishida (2017).

14.7 Cytotoxicity Analysis

14.7.1 RBC Lysis Assay

RBC lysis is a primary method to depict the cytotoxic nature of compounds. It is a method used by many researchers due to its high sensitivity, potency, easy, and economical nature. AgNPs have a tendency to cause membrane damage and chitosan is a polycationic polymer; these characters are enough to cause membrane damage in cells (Oves et al. 2013). To evaluate the ability to cause damage in blood cells, hemolytic assay is usually carried out. But in the case of normal blood cells, the ChBio-AgNPs showed no toxicity at low concentrations (Table 14.1). RBC lysis occurred only at very high concentration of ChBio-AgNPs, and it clearly indicates that the formulation is safe for treatment at the IC_{50} concentration (Vijayan et al. 2016a,b; Khan et al. 2018; Bilal et al. 2019).

14.7.2 MTT Assay

The MTT assay is important to determine the cytotoxicity to normal fibroblast L929 cells helped to ascertain the noncytotoxic nature of ChBio-AgNPs. It was reported that only at high concentration the particles are cytotoxic in nature. The effective concentration which showed characteristic antimicrobial and antifungal activity does not show any cytotoxicity (Składanowski et al. 2016). The cytotoxicity of nanosilver (AgNPs) on normal human cell lines such as HF2 and M cells showed

Table 14.1 Dose-dependent RBC lysis assay of ChBio-AgNPs.

Concentration in μg ml^{-1}	Absorbance at 576 nm for ChBio-AgNPs	Hemolysis (%)
25	0.033 ± 0.008	−0.0485
50	0.042 ± 0.002	0.169
75	0.065 ± 0.008	0.728
100	0.074 ± 0.003	0.946
125	0.096 ± 0.002	1.496
150	0.104 ± 0.002	2.11

Data represented as mean ± SD, n = 3.

concentration-dependent cytotoxicity and it also revealed that nanosilver has no potential toxicity on normal cell line at low concentrations. This low concentration is well enough for the treatment of some diseases (Shahbazzadeh et al. 2011). In an attempt to study cytotoxicity of ChBio-AgNPs, human fibroblasts, HS27 cells were treated with ChBio-AgNPs and found that at lowest concentration 98% of cells are viable (Vijayan et al. 2019). Huang et al. (2016) reported that ChBio-AgNPs are a low-cytotoxic alternate for antimicrobial drugs. Moreover, Regiel et al. (2013) suggested that conjugation of AgNPs with natural polymers like chitosan will decrease the toxicity of AgNPs.

14.8 Conclusion

The overuse of antibiotics and expanded drug resistance is a major issue in this century. Bio-AgNPs and their chitosan conjugate can be used as a better solution for the issue. They need further in silico and *in vivo* studies for their administration as a drug formulation. Scale up is an important field in this aspect as far as the commercial production of the drug is concerned. ChBio-AgNPs showed excellent anticancer activity and hence, this conjugate can be used as effective nanomedicine for cancer treatment. As cancer treatment is very costly, any approach to decrease the cost is appreciable. Moreover, available cancer treatments cause many side effects to the patients in a systemic manner, but the available reports suggest that ChBio-AgNPs are nontoxic and/or have fewer side effects. However, further extensive studies are required involving animal models to reach at any prominent conclusion.

Acknowledgment

This work was supported by Mahatma Gandhi University (Section Order no: 529/A6/2/JRF2018-2019/Academic), Kottayam, Kerala, India. I also acknowledge Business Innovation and Incubation Centre (BIIC), Mahatma Gandhi University, for Patent Facilitation and Start-up award.

References

Abaza, A., Hegazy, E.A., Mahmoud, G.A., and Elsheikh, B. (2018). Characterization and antitumor activity of chitosan/poly (vinyl alcohol) blend doped with gold and silver nanoparticles in treatment of prostatic cancer model. *Journal of Pharmacy and Pharmacology* 6: 659–673.

Abo-State, M.A.M. and Partila, A.M. (2015). Microbial production of silver nanoparticles by *Pseudomonas aeruginosa* cell free extract. *International Journal of Environmental Research and Public Health* 3 (3): 91–98.

Aerle, R., Lange, A., Moorhouse, A. et al. (2013). Molecular mechanisms of toxicity of silver nanoparticles in zebrafish embryos. *Environmental Science and Technology* 47: 8005–8014.

Ahmad, A., Senapati, S., Islam, M.K. et al. (2003). Extracellular biosynthesis of monodisperse gold nanoparticles by a novel extremophilic actinomycete *Thermomonospora* Sp. *Langmuir* 19 (8): 3550–3553.

Ahmed, T.A. and Aljaeid, B.M. (2016). Preparation, characterization, and potential application of chitosan, chitosan derivatives, and chitosan metal nanoparticles in pharmaceutical drug delivery. *Drug Design, Development and Therapy* 10: 483.

AshaRani, P.V., Mun, G.L., Hande, K., and Valiyaveettil, M.S. (2009). Cytotoxicity and genotoxicity of silver nanoparticles in human cells. *ACS Nano* 3 (2): 279–290.

Azizi, M., Ghourchian, H., Yazdian, F. et al. (2017). Anti-cancerous effect of albumin coated silver nanoparticles on MDA-MB 231 human breast cancer cell line. *Scientific Reports* 7 (1): 5178.

Bankura, K.P., Maity, D., Mollick, M.M.R. et al. (2012). Synthesis, characterization and antimicrobial activity of dextran stabilized silver nanoparticles in aqueous medium. *Carbohydrate Polymers* 89 (4): 1159–1165.

Banu, A. and Rathod, V. (2011). Synthesis and characterization of silver nanoparticles by *Rhizopus Stolonier*. *International Journal of Biomedical and Advance Research* 2 (5): 121–130.

Bendale, Y., Bendale, V., and Paul, S. (2017). Evaluation of cytotoxic activity of platinum nanoparticles against normal and cancer cells and its anticancer potential through induction of apoptosis. *Integrative Medicine Research* 6 (2): 141–148.

Bhakya, S., Muthukrishnan, S., Sukumaran, M., and Muthukumar, M. (2016). Biogenic synthesis of silver nanoparticles and their antioxidant and antibacterial activity. *Applied Nanoscience* 6 (5): 755–766.

Bilal, M., Zhao, Y., Rasheed, T. et al. (2019). Biogenic nanoparticle-chitosan conjugates with antimicrobial, antibiofilm and anticancer potentialities: development and characterization. *International Journal of Environmental Research and Public Health* 19: 1–14.

Cabral, M., Pedrosa, F., Margarido, F., and Nogueira, C.A. (2013). End-of-life Zn–MnO_2 batteries: electrode materials characterization. *Environmental Technology* 34 (10): 1283–1295.

Cheviron, P., Gouanvé, F., and Espuche, E. (2014). Green synthesis of colloid silver nanoparticles and resulting biodegradable starch/silver nanocomposites. *Carbohydrate Polymers* 108 (1): 291–298.

Choudhari, A.S., Suryavanshi, S.A., and Kaul-Ghanekar, R. (2013). The aqueous extract of *Ficus religiosa* induces cell cycle arrest in human cervical cancer cell lines SiHa (HPV-16 positive) and apoptosis in HeLa (HPV-18 positive). *PLoS One* 8 (7): e70127.

Dananjaya, S.H.S., Godahewa, G.I, Jayasooriya, R.G.P.T, et al (2014). Chitosan silver nanocomposites (CAgNCs) as potential antibacterial agent to control *Vibrio tapetis*. Journal of Vetinary Science Technology 5:209.

Darroudi, M., Ahmad, M.B., Abdullah, A.H., and Ibrahim, N.A. (2011). Green synthesis and characterization of gelatin-based and sugar-reduced silver nanoparticles. *International Journal of Nanomedicine* 6: 569–574.

Desai, R., Mankad, V., Gupta, S., and Jha, P. (2012). Size distribution of silver nanoparticles: UV-visible spectroscopic assessment. *Nanoscience and Nanotechnology Letters* 4: 30–34.

Dey, A., Mukhopadhyay, A.K., Gangadharan, S. et al. (2009). Characterization of microplasma sprayed hydroxyapatite coating. *Journal of Thermal Spray Technology* 18 (4): 578–592.

Divya, K., Rebello, S., and Jisha, M.S. (2014). A simple and effective method for extraction of high purity chitosan from shrimp shell waste. In: *Proceeding of International Conference on Advances in*

Applied Science and Environmental Engineering, Institute of Research engineers and doctors, IRED Headquarters, 141–145.

Du, W.L., Niu, S.S., Xu, Y.L. et al. (2009). Antibacterial activity of chitosan tripolyphosphate nanoparticles loaded with various metal ions. *Carbohydrate Polymers* 75 (3): 385–389.

Elshawy, O.E., Helmy, E.A., and Rashed, L.A. (2016). Preparation, characterization and in vitro evaluation of the antitumor activity of the biologically synthesized silver nanoparticles. *Advances in Nanoparticles* 5: 149–166.

Fani, S., Kamalidehghan, B., Lo, K.M. et al. (2016). Anticancer activity of a monobenzyltin complex C1 against MDA-MB-231 cells through induction of apoptosis and inhibition of breast cancer stem cells. *Scientific Reports* 6 (1): 38992.

Fatima, F., Preethi, B., and Neelam, B. (2015). Antimicrobial and immunomodulatory efficacy of extracellularly synthesized silver and gold nanoparticles by a novel phosphate solubilizing fungus *Bipolaris tetramera*. *BMC Microbiology* 15: 52. https://doi.org/10.1186/s12866-015-0391-y.

Fissan, H., Ristig, S., Kaminski, H. et al. (2014). Comparison of different characterization methods for nanoparticle dispersions before and after aerosolization. *Analytical Methods* 6 (18): 7324.

Ghaseminezhad, S.M., Hamedi, S., and Shojaosadati, S.A. (2012). Green synthesis of silver nanoparticles by a novel method: comparative study of their properties. *Carbohydrate Polymers* 89 (2): 467–472.

Goy, R.C., Morais, T.B.S., and Assis, B.G.S. (2016). Evaluation of the antimicrobial activity of chitosan and its quaternized derivative on *E. coli* and *S. aureus* growth. *Revista Brasileira de Farmacognosia* 26 (1): 122–127.

Guo, Q.G., Reza, W., Thomas, A. et al. (2014). Comparison of in situ and ex situ methods for synthesis of two-photon polymerization polymer nanocomposites. *Polymers* 6 (7): 2037–2050.

Gurunathan, S., Park, J.H., Han, J.W., and Kim, J.H. (2015). Comparative assessment of the apoptotic potential of silver nanoparticles synthesized by *Bacillus tequilensis* and *Calocybe indica* in MDA-MB-231 human breast cancer cells: targeting p53 for anticancer therapy. *International Journal of Nanomedicine* 10: 4203–4223.

Haider, M.J. and Mehdi, M.S. (2014). Study of morphology and zeta potential analyzer for the silver nanoparticles. *International Journal of Scientific and Engineering Research* 5 (7): 381–387.

Hall, J.B., Dobrovolskaia, M.A., Patri, A.K., and McNeil, S.E. (2007). Characterization of nanoparticles for therapeutics. *Nanomedicine* 2: 789–803.

Han, J.W., Gurunathan, S., and Jeong, J.K. (2014). Oxidative stress mediated cytotoxicity of biologically synthesized silver nanoparticles in human lung epithelial. *Nanoscale Research Letters* 9: 459–473.

Honary, S., Ghajar, K., Khazaeli, P., and Shalchian, P. (2011). Preparation, characterization and antibacterial properties of silver-chitosan nanocomposites using different molecular weight grades of chitosan. *Tropical Journal of Pharmaceutical Research* 10 (101): 69–69.

Huang, H., Yuan, Q., and Yang, X. (2004). Preparation and characterization of metal–chitosan nanocomposites. *Colloids and Surfaces B: Biointerfaces* 39 (1–2): 31–37.

Huang, J., Li, Q., Sun, D. et al. (2007). Biosynthesis of silver and gold nanoparticles by novel sundried Cinnamomum camphora leaf. *Nanotechnology* 18 (10): 105104.

Huang, H., Lai, W., Cui, M. et al. (2016). An evaluation of blood compatibility of silver nanoparticles. *Scientific Reports* 6 (1): 25518.

Ishida, T. (2017). Anticancer activities of silver ions in cancer and tumor cells and DNA damages by Ag+ – DNA base- pairs reactions. *MOJ Tumor Research* 1 (I): 8–16.

Ishida, K., Cipriano, T.F., and Rocha, G.M. (2013). Silver nanoparticle production by the fungus *Fusarium oxysporum*: nanoparticle characterization and analysis of antifun- gal activity against pathogenic yeasts. *Memórias do Instituto Oswaldo Cruz* 109: 220–228.

Ivanisevic, I. (2010). Physical stability studies of miscible amorphous solid dispersions. *Journal of Pharmaceutical Sciences* 99 (9): 4005–4012.

Jayakumar, R., Menon, D., Manzoor, K. et al. (2010). Biomedical applications of chitin and chitosan based nanomaterials-a short review. *Carbohydrate Polymers* 82: 227–232.

Jena, P., Mohanty, S., and Mallick, R. (2012). Toxicity and antibacterial assessment of chitosancoated silver nanoparticles on human pathogens and macrophage cells. *International Journal of Nanomedicine* 7: 1805–1818.

Kaur, P., Choudhary, A., and Thakur, T. (2013). Synthesis of chitosan-silver nanocomposites and their antibacterial activity. *International Journal of Scientific and Engineering Research* 4 (4): 869–872.

Kermanizadeh, A., Chauché, C., Balharry, D. et al. (2014). The role of Kupffer cells in the hepatic response to silver nanoparticles. *Nanotoxicology* 8 (sup1): 149–154.

Khachatryan, K., Khachatryan, G., and Fiedorowicz, M. (2016). Silver and gold nanoparticles embedded in potato starch gel films. *Journal of Materials Science and Chemical Engineering* 4 (02): 22.

Khan, S.H., Tawfik, A., Wahab, A.S., and Khan, M.H.U. (2018). Nanosilver: new ageless and versatile biomedical therapeutic scaffold. *International Journal of Nanomedicine* 13: 733–762.

Lamkanfi, M. and Kanneganti, T.D. (2010). Caspase-7: a protease involved in apoptosis and inflammation. *International Journal of Biochemistry and Cell Biology* 42 (1): 21–24.

Li, Q. and Li, X. (2014). Nanosilver particles in medical applications: synthesis, performance, and toxicity. *International Journal of Nanomedicine* 2014: 2399–2407.

Lin, W.P., Wang, J.T., Chang, S.C. et al. (2016). The antimicrobial susceptibility of *Klebsiella pneumoniae* from community settings in Taiwan, a trend analysis. *Scientific Reports* 6: 36280.

Linn, C.Q. (2015). Investigation on free radical scavenging activity of bigenic silver nanoparticle. *International Journal of Pharma and Bio Sciences* 6 (2): 349–353.

Loganayaki, N., Siddhuraju, P., and Manian, M. (2013). Antioxidant activity and free radical scavenging capacity of phenolic extracts from *Helicteres isora* L. and *Ceiba pentandra* L. *Journal of Food Science and Technology* 50 (4): 687–695.

Majekodunmi, S.O. (2016). Current development of extraction, characterization and evaluation of properties of chitosan and its use in medicine and pharmaceutical industry. *American Journal of Polymer Science* 6 (3): 86–91.

Manivasagan, P., Venkatesan, J., Sivakumar, K., and Kim, S.K. (2014). Actinobacteria mediated synthesis of nanoparticles and their biological properties: a review. *Critical Reviews in Microbiology* 42 (12): 1–13.

Minz, AP 2015, Evaluation of antioxidant and anticancer efficacy of chitosan based nanoparticles, unpublished thesis, National Institute of Technology, Rourkela, pp. 413.

Mizrahy, S. and Peer, D. (2012). Polysaccharides as building blocks for nanotherapeutics. *Chemical Society Reviews* 41: 2623–2640.

Mocan, T. (2013). Hemolysis as expression of nanoparticles-induced cytotoxicity in red blood cells. *Biotechnology, Molecular Biology and Nanomedicine* 1 (1): 7–12.

Moghaddam, A.B., Namvar, F., Moniri, M. et al. (2015). Nanoparticles biosynthesized by fungi and yeast: a review of their preparation, properties, and medical applications. *Molecules* 20 (9): 16540–16565.

Mori, Y., Ono, T., Miyahira, Y. et al. (2013). Antiviral activity of silver nanoparticle/chitosan composites against H1N1 influenza a virus. *Nanoscale Research Letters* 8 (1): 93.

Murad, H., Hawat, M., Ekhtiar, A. et al. (2016). Induction of G1-phase cell cycle arrest and apoptosis pathway in MDA-MB-231 human breast cancer cells by sulfated polysaccharide extracted from Laurencia papillosa. *Cancer Cell International* 16 (1): 39.

Nimmi, O.S. and George, P. (2012). Evaluation of the antioxidant potential of a newly developed polyherbal formulation for antiobesity. *International Journal of Pharmacy and Pharmaceutical Sciences* 4 (Suppl.3): 505–510.

Oves, M., Khan, M.S., and Zaidi, A. (2013). Antibacterial and cytotoxic efficacy of extracellular silver nanoparticles biofabricated from chromium reducing novel OS4 strain of *Stenotrophomonas maltophilia*. *PLoS One* 8 (3): e59140.

Palem, R.R., Ganesh, S.D., Kronekova, Z. et al. (2018). Green synthesis of silver nanoparticles and biopolymer nanocomposites: a comparative study on physico-chemical, antimicrobial and anticancer activity. *Bulletin of Materials Science* 41 (2): 55.

Panacek, A., Kolar, M., Vecerova, R. et al. (2009). Antifungal activity of silver nanoparticles against *Candida* spp. *Biomaterials* 30 (31): 6333–6340.

Pecher, J. and Mecking, S. (2010). Nanoparticles of conjugated polymers. *Chemical Reviews* 110 (10): 6260–6279.

Peng, Y., Song, C., Yang, C. et al. (2017). Low molecular weight chitosan-coated silver nanoparticles are effective for the treatment of MRSA-infected wounds. *International Journal of Nanomedicine* 12: 295–304.

Pumiputavon, K., Chaowasku, T., Saenjum, C. et al. (2017). Cell cycle arrest and apoptosis induction by methanolic leaves extracts of four Annonaceae plants. *BMC Complementary and Alternative Medicine* 17 (1): 294.

Rabea, E.I., Badawy, M.E., and Steurbaut, W. (2005). Fungicidal effect of chito- san derivatives containing an N-alkyl group on grey mould botryti77s cinerea and rice leaf blast *Pyricularia grisea*. *Communications in Agricultural and Applied Biological Sciences* 70 (3): 219–223.

Raji, V.M., Chakraborty, and Parikh, P.A. (2012). Synthesis of starch-stabilized silver nanoparticles and their antimicrobial activity. *Particulate Science and Technology* 30 (6): 565–577.

Ramachandran, R. and Sangeetha, D. (2017). Antibiofilm efficacy of silver nanoparticles against biofilm forming multidrug resistant clinical isolates. *The Pharma Innovation International Journal* 6 (11): 36–43.

Regiel, A., Kyzioł, A., and Arruebo, M. (2013). Chitosan-silver nanocomposites: modern antibacterial materials. *CHEMIK* 67 (8): 683–692.

Rekha, J.K. and Vedpriya, A. (2013). Biological syntheis of silver nanoparticles from aqueous extract of endophytic fungus *Aspergillus terrus* and its antibacterial activity. *International Journal of Nanomaterials and Biostructures* 3 (2): 35–37.

Richardson, S.C., Kolbe, H.V., and Duncan, R. (1999). Potential of low molecular mass chitosan as a DNA delivery system: biocompatibility, body distribution and ability to complex and protect DNA. *International Journal of Pharmaceutics* 178 (2): 231–243.

Roldo, M., Hornof, M., Caliceti, P., and Bernkop-Schnürch, A. (2004). Mucoadhesive thiolated chitosans as platforms for oral controlled drug delivery: synthesis and in vitro evaluation. *European Journal of Pharmaceutics and Biopharmaceutics* 57: 115–121.

Ruparelia, J.P.S., Chatterjee, A.K., Duttaqupta, S.P., and Mukherji, S. (2008). Specificity in antimicrobal activity of silver and copper nanoparticles. *Acta Biomaterialia* 4: 707–716.

Sarmento, B., Ribeiro, A.J., Veiga, F. et al. (2007). Insulin-loaded nanoparticles are prepared by alginate ionotropic pre-gelation followed by chitosan polyelectrolyte complexation. *Journal of Nanoscience and Nanotechnology* 7: 2833–2841.

Schmitt, C., Sanchez, C., Desobry-Banon, S., and Hardy, J. (1998). Structure and technofunctional properties of protein-polysaccharide complexes: a review. *Critical Reviews in Food Science and Nutrition* 38: 689–753.

Shahbazzadeh, D., Ahari, H., and Motalebi, A.A. (2011). In vitro effect of nanosilver toxicity on fibroblast and mesenchymal stem cell lines. *Iranian Journal of Fisheries Sciences* 10 (3): 487–496.

Shrivastava, S., Bera, T., Roy, A. et al. (2007). Characterization of enhanced antibacterial effects of novel silver nanoparticles. *Nanotechnology* 18 (22): 1–14.

Singh, D., Rathod, V., Ninganagouda, S. et al. (2014). Optimization and characterization of silver nanoparticle by endophytic fungi *Penicillium* sp. isolated from *Curcuma longa* (Turmeric) and application studies against MDR *E. coli* and *S. aureus*. *Bioinorganic Chemistry and Applications* 2014: 408021.

Singh, A.K., Singh, D., and Rathod, V. (2016). Nanosilver coated fabrics show antimicrobial property. *World Journal of Pharmacy and Pharmaceutical Sciences* 5 (2): 1023–1035.

Sivera, M., Kvitek, L., and Soukupova, J. (2014). Silver nanoparticles modified by gelatin with extraordinary pH stability and long-term antibacterial activity. *PLoS One* 9 (8): e103675.

Składanowski, M., Golinska, P., and Rudnicka, K. (2016). Evaluation of cytotoxicity, immune compatibility and antibacterial activity of biogenic silver nanoparticles. *Medical Microbiology and Immunology* 205 (6): 603–613.

Soman, S. and Ray, J.G. (2016). Silver nanoparticles synthesized using aqueous leaf extract of *Ziziphus oenoplia* (L.) Mill: characterization and assessment of antibacterial activity. *Journal of Photochemistry and Photobiology B: Biology* 163: 391–402.

Sugano, M., Watanabe, S., Kishi, A. et al. (1988). Hypocholes- terolemic action of chitosans with different viscosity in rats. *Lipids* 23 (3): 187–191.

Thomas, R., Soumya, K.R., Mathew, J., and Radhakrishnan, E.K. (2015). Inhibitory effect of silver nanoparticle fabricated urinary catheter on colonization efficiency of Coagulase Negative Staphylococci. *Journal of Photochemistry and Photobiology B: Biology* 149: 68–77.

Tomaszewska, E., Soliwoda, K., Kadziola, K. et al. (2013). Detection limits of DLS and UV-vis spectroscopy in characterization of polydisperse nanoparticles colloids. *Journal of Nanomaterials* 2013: 1–13.

Valencia, A., Germán, L.C.O.V., Ferrari, R., and Vercik, A. (2013). Synthesis and characterization of silver nanoparticles using water-soluble starch and its antibacterial activity on *Staphylococcus aureus*. *Starch* 65 (11–12): 931–937.

Vasileva, P., Alexandrova, T., and Karadjova, I. (2017). Application of starch-stabilized silver nanoparticles as a colorimetric sensor for mercury (II) in 0.005 Mol/L nitric acid. *Journal of Chemistry* 2017: 6897960. https://doi.org/10.1155/2017/6897960.

Veerapur, V.P., Prabhakar, K.R., and Parihar, V.K. (2009). *Ficus racemosa* stem bark extract: a potent antioxidant and a probable natural radioprotector. *Evidence-based Complementary and Alternative Medicine: Ecam* 6 (3): 317–324.

Venkatesan, J., Lee, J.Y., Kang, D.S. et al. (2017). Antimicrobial and anticancer activities of porous chitosan-alginate biosynthesized silver nanoparticles. *International Journal of Biological Macromolecules* 98: 515–525.

Vijayan, S., Divya, K., George, T.K., and Jisha, M.S. (2016a). Biogenic synthesis of silver nanoparticles using endophytic fungi *Fusarium oxysporum* isolated from *Withania Somnifera* (L.), its antibacterial and cytotoxic activity. *Journal of Bionanoscience* 10 (5): 369–376.

Vijayan, S., Koilaparambil, D., George, T.K., and Jisha, M.S. (2016b). Antibacterial and cytotoxicity studies of silver nanoparticles synthesized by endophytic *Fusarium Solani* isolated from *Withania somenifera* (L.). *Journal of Water and Environmental Nanotechnology* 1 (2): 91–103.

Vijayan, S., Divya, K., and Jisha, M.S. (2019). In vitro anticancer evaluation of chitosan/biogenic silver nanoparticle conjugate on Si Ha and MDA MB cell lines. *Applied Nanoscience* https://doi.org/10.1007/s13204-019-01151-w.

Wang, L.S., Wang, C.Y., Yang, C.H. et al. (2015). Synthesis and anti-fungal effect of silver nanoparticles–chitosan composite particles. *International Journal of Nanomedicine* 10: 2685–2696.

Wang, C., Gao, X., Zhongqin, C. et al. (2017). Preparation, characterization and application of polysaccharide-based metallic nanoparticles: a review. *Polymers* 9 (12): 1–34.

Waseda, Y., Matsubara, E., and Shinoda, K. (2011). *X-Ray Diffraction Crystallography: Introduction, Examples and Solved Problems*, 67–106. Germany: Springer Verlag Berlin.

Wei, D.S., Wuyong, Q., Weiping, Y. et al. (2009). The synthesis of chitosan-based silver nanoparticles and their antibacterial activity. *Carbohydrate Research* 344: 2375–2382.

Williams, D.B. and Carter, C.B. (2009). *The Transmission Electron Microscope*, 3–22. MA: Springer Verlag Boston.

Xue, B., He, D., Gao, S. et al. (2016). Biosynthesis of silver nanoparticles by the fungus *Arthroderma fulvum* and its antifungal activity against genera of *Candida*, *Aspergillus* and *Fusarium*. *International Journal of Nanomedicine* 11: 1899–1906.

Yuan, Y.G., Peng, Q.L., and Gurunathan, S. (2017). Effects of silver nanoparticles on multiple drug-resistant strains of *Staphylococcus aureus* and *Pseudomonas aeruginosa* from mastitis-infected goats: an alternative approach for antimicrobial therapy. *International Journal of Molecular Sciences* 18 (3): 569.

Zhang, H., Smith, J.A., and Oyanedel-Craver, V. (2012). The effect of natural water conditions on the anti-bacterial performance and stability of silver nanoparticles capped with different polymers. *Water Research* 46 (3): 691–699.

Zhang, X.F., Liu, Z.G., Shen, W., and Gurunathan, S. (2016). Silver nanoparticles: synthesis, characterization, properties, applications, and therapeutic approaches. *International Journal of Molecular Sciences* 17 (9): 1534.

Zhao, X., Zhou, L., Rajoka, M.S.R. et al. (2018). Fungal silver nanoparticles: synthesis, application and challenges. *Critical Reviews in Biotechnology* 38 (6): 817–835.

15

Leishmaniasis: Where Infection and Nanoparticles Meet

Mohammad Imani and Azam Dehghan

Novel Drug Delivery Systems Department, Iran Polymer and Petrochemical Institute, Tehran, Iran

15.1 Introduction

Leishmaniasis, among the other TriTryp diseases, is a nearly neglected tropical disease in terms of new drug development. This neglect can be attributed to several factors, including its geographical endemicity profile, lack of financial support, and lack of an integrated network of academicians and research and development (R & D) people from industry organized in multidisciplinary centers, among the many other factors like low mean income of the people in the affected countries and so on (Alcântara et al. 2018). The disease is caused by protozoan parasites from more than 20 *Leishmania* species. The disease mostly affects tropical (mostly developing) countries. The parasites are transmitted to humans after bites of the infected female *Phlebotomine* sandfly, i.e. a 2–3 mm long insect vector (Islan et al. 2017).

At least two thousand years ago, cutaneous lesions due to leishmaniasis were first described in people of the Old World. Illustrations of the lesions have been found on human faces decorating pottery created 1500 years ago in Peru and Ecuador. Leishmania lesions were first described in modern medicine by Russell in 1756. Subsequently, Leishman reported the identification of the parasite in India in 1900 (Mukbel et al. 2005). The disease, with a current mortality rate of 50 000 deaths per year, threatens approximately 350 million people in more than 90 countries all over the world, and affects 0.7–1.2 million more cases annually in its cutaneous form (WHO 2019a). The number of new cases suffering from visceral leishmaniasis has now dropped to less than 0.1 million persons per year thanks to more efficient dosage forms developed using previously known drug entities. The type of disease caused by the parasite depends on the infecting species and the host's immune response (Islan et al. 2017), but three main forms occur: cutaneous leishmaniasis (CL), mucocutaneous leishmaniasis (MCL), and visceral leishmaniasis (VL), also called kala-azar. Cutaneous manifestations can be subdivided into localized, diffuse, leishmaniasis recidivans, and post-kala-azar dermal leishmaniasis (PKDL) (Shaw and Carter 2014; Gutiérrez et al. 2016).

The currently available therapies are based on old pentavalent antimony agents with a known toxic profile, amphotericin B (AmB), miltefosine, and a few other agents. The shortcomings in the drug development sector force compensation paid by developing more delivery options not only aimed to improve the safety of the currently used therapeutic agents but also to provide more efficacious treatment when possible. It worth noting that the *Leishmania* parasite lives in the host

Nanobiotechnology in Diagnosis, Drug Delivery, and Treatment, First Edition. Edited by Mahendra Rai, Mehdi Razzaghi-Abyaneh, and Avinash P. Ingle.

immune cells, i.e. macrophages. Pharmacokinetically, this makes the process of drug therapy more challenging than other conventional infectious diseases because it adds another barrier to be passed by the therapeutic agent to reach the target parasite. In this specific case, nanotechnology may be helpful clinically due to the dependency of pharmacokinetic parameters on the particle size of particulate drug delivery systems (DDS). In another aspect, nanoparticles (NPs) may enhance the delivery of drugs to target cells – macrophages in the case of leishmaniasis – upon their enhancement in some of the physiological characteristics of the human body, like macrophage activation after administration of some polyelectrolytes, e.g. chitosan or enhanced permeation and retention phenomenon in the reticuloendothelial system (RES).

Here, in this chapter, we briefly introduce leishmaniasis as a disease and describe its clinical features, including epidemiological profile, clinical forms, the life cycle of the parasite in the human body, and its transmission routes. Clinical features of the disease will be considered, including the available strategies for the management of the disease. The most dominant part of this chapter is devoted to describing how nanotechnology may be helpful to provide safer options and provide a bulk of literature for the reader to find the most relevant reports presented in the form of tables.

15.2 Clinical Forms

Cutaneous leishmaniasis is the most common form of the disease. It usually produces ulcers on the exposed parts of the body, such as the face, arms, and legs. There may be a large number of lesions, sometimes up to 200, which can cause serious disability. When the ulcers heal, they invariably leave permanent scars, which are often the cause of social problems. In MCL, the lesions can lead to the partial or total destruction of the mucous membranes of the nose, mouth, and throat cavities also the surrounding tissues (Table 15.1).

Visceral leishmaniasis, also known as kala-azar, is characterized by irregular bouts of fever, substantial weight loss, swelling of the spleen and liver, and anemia. If the disease is not treated, the fatality rate in developing countries can be as high as 100% within twoyears. PKDL is a complication of VL in areas where *Leishmania donovani* is endemic. It is characterized by a hypopigmented macular, maculopapular, and nodular rash, usually in patients who have recovered from VL. It usually appears six months to one or more years after apparent cure of the disease but may occur earlier or even concurrently with VL. PKDL heals spontaneously in the majority of cases in Africa, but rarely in patients in India. It is considered to have an important role in maintaining and contributing to the transmission of the disease, particularly in interepidemic periods of VL, acting as a reservoir for parasites (WHO 2019b). The clinical manifestation of different types of leishmaniasis is depicted in Figure 15.1.

15.3 Epidemiology

Recurrent epidemics of VL in East Africa (Ethiopia, Kenya, Sudan, and South Sudan) have caused high morbidity and mortality in the affected communities. Likewise, major epidemics of CL have affected different parts of Afghanistan and Syria. In 2017, 20 792 out of 22 145 (94%) new cases reported to the World Health Organization (WHO) occurred in seven countries: Brazil, Ethiopia, India, Kenya, Somalia, South Sudan, and Sudan. In the WHO South-East Asia Region (Figure 15.2a), the kala-azar elimination program is progressing satisfactorily, so countries such as Bangladesh that reported more than 9000 cases in 2006 reported 255 and 192 new cases in 2016 and 2017, respectively.

Table 15.1 Clinical and epidemiological characteristics of the main *Leishmania* species.

	Subgenus	Clinical form	Main clinical features	Natural progression	Risk groups	Main reservoir	High-burden countries (regions)	Estimated annual worldwide incidence
Leishmania donovani[a]	Leishmania	VL and PKDL	VL: Persistent fever, splenomegaly, weight loss, anemia. PKDL: Multiple painless macular, popular, or nodular lesions.	VL: Fatal within 2 yrs. PKDL: Lesions self-heal in up to 85% of cases in Africa but rarely in Asia	VL: Predominantly adolescents and young adults. PKDL: Young children in Sudan with no clearly established risk factors.	Humans	India, Bangladesh, Ethiopia, Sudan, and South Sudan	VL: 50 000–90 000 cases. PKDL: Unknown
Leishmania tropica[a]	Leishmania	CL, LR, rarely VL	Ulcerating dry lesions, painless, and frequently multiple.	CL lesions often self-heal within 1 yr.	Not well-defined.	Humans but zoonotic foci exist.	Eastern Mediterranean, the Middle East, and northeastern and southern Africa	CL: 200 000–400 000
Leishmania aethiopica[a]	Leishmania	CL, DCL, DsCL, and oronasal CL	Localized cutaneous nodular lesions; occasionally oronasal; rarely ulcerates.	Self-healing, except for DCL, within 2–5 yr.	Limited evidence; adolescents.	Hyraxes	Ethiopia and Kenya	CL: 20 000–40 000
Leishmania major[a]	Leishmania	CL	Rapid necrosis, multiple wet sores, and severe inflammation	Self-healing in >50% of cases within 2–8 months; multiple lesions slow to heal, and severe scarring	No well-defined risk groups.	Rodents	Iran, Saudi Arabia, north Africa, the Middle East, central Asia, and west Africa	CL: 230 000–430 000
Leishmania infantum[a]	Leishmania	VL and CL	CL: Typically single nodules and minimal inflammation. VL: Persistent fever and splenomegaly.	CL: Lesions self-heal within 1 yr and confers individual immunity VL: Fatal within 2 yr.	CL: Older children and young adults. VL: Children under 5 yr and immunocompromised adults.	Dogs, hares, and humans	CL: Central America. VL: China, southern Europe, Brazil, and South America.	CL: Unknown VL: 6200–12 000 (Old World) and 4500–6800 (New World)

(Continued)

Table 15.1 (Continued)

	Subgenus	Clinical form	Main clinical features	Natural progression	Risk groups	Main reservoir	High-burden countries (regions)	Estimated annual worldwide incidence
Leishmania mexicana[b]	Leishmania	CL, DCL, and DsCL	Ulcerating lesions, single or Multiple.	Often self-healing within 3–4 months.	No well-defined risk groups.	Rodents and marsupials	South America	Limited number of cases, included in the 187 200–300 000 total cases of New World CL[c]
Leishmania amazonensis[b]	Leishmania	CL, DCL, and DsCL	Ulcerating lesions, single or multiple.	Not well-described.	No well-defined risk groups.	Possums and rodents	South America	Limited number of cases, included in the 187 200–300 000 total cases of New World CL[c]
Leishmania braziliensis[b]	Viannia	CL, MCL, DCL, and LR	Ulcerating lesions can progress to mucocutaneous form; local lymph nodes are palpable before and early in the onset of the lesions.	Might self-heal within 6 months; 2–5% of cases progress to MCL.	No well-defined risk groups.	Dogs, humans, rodents, and horses	South America	Majority of the 187 200–300 000 total cases of New World CL[c]
Leishmania guyanensis[b]	Viannia	CL, DsCL, and MCL	Ulcerating lesions, single or multiple that can progress to mucocutaneous form; palpable lymph nodes.	Might self-heal within 6 months.	No well-defined risk groups.	Possums, sloths, and anteaters	South America	Limited number of cases, included in the 187 200–300 000 total cases of New World CL[c]

VL, Visceral leishmaniasis; PKDL, Post-kala azar dermal leishmaniasis; CL, Cutaneous leishmaniasis; LR, Leishmaniasis recidivans; DCL, Diffuse cutaneous leishmaniasis; DsCL, Disseminated cutaneous leishmaniasis; MCL, Mucocutaneous leishmaniasis.
[a] Old World leishmaniasis.
[b] New World leishmaniasis.
[c] Estimates are of all New World leishmaniases with *Leishmania braziliensis* comprising the vast majority of these cases.
Source: Adapted from Burza et al. (2018) with permission.

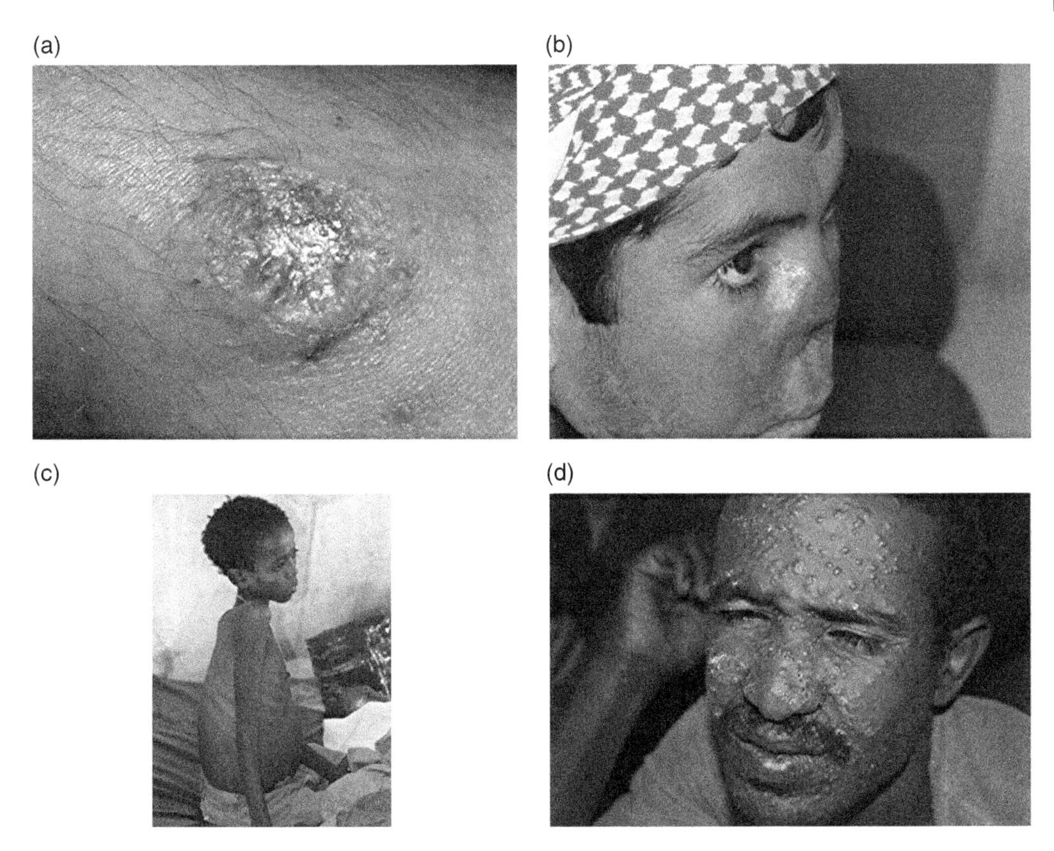

Figure 15.1 Clinical manifestations of leishmaniases including (a) cutaneous, (b) mucocutaneous, (c) visceral, and (d) post-kala-azar dermal leishmaniasis. *Source:* Adopted from WHO (2019b) with permission.

The majority of CL cases occur in Afghanistan, Algeria, Brazil, Colombia, Iran, Pakistan, Peru, Saudi Arabia, and Syria. Anthroponotic cutaneous leishmaniasis is predominantly urban and periurban, and shows patterns of spatial clustering similar to those of anthroponotic visceral leishmaniasis in southeast Asia (Figure 15.2b). The disease is usually characterized by large outbreaks in densely populated cities, especially in war and conflicted zones, refugee camps, and in settings where there is large-scale migration of populations. The epidemiology of CL in the South American continent is complex, with intra- and interspecific variation in transmission cycles, reservoir hosts, sandfly vectors, clinical manifestations, response to therapy, and multiple circulating *Leishmania* species in the same geographical area. Almost 90% of MCL cases occur in Bolivia, Brazil, and Peru (WHO 2019a).

15.4 Life Cycle and Transmission

Leishmania parasites exist in two morphological forms, amastigotes and promastigotes, which are found in the mammalian host and sandfly, respectively. Amastigotes entering into the sandfly midgut after a blood meal initially transform into procyclic promastigotes. Procyclic promastigotes are non-infective and rapidly divide, then transform into infective metacyclic promastigotes within 4–7 days, which migrate to the pharynx of the sandfly. The six developmental forms of

(a)

Status of endemicity of cutaneous leishmanisis worldwide, 2016

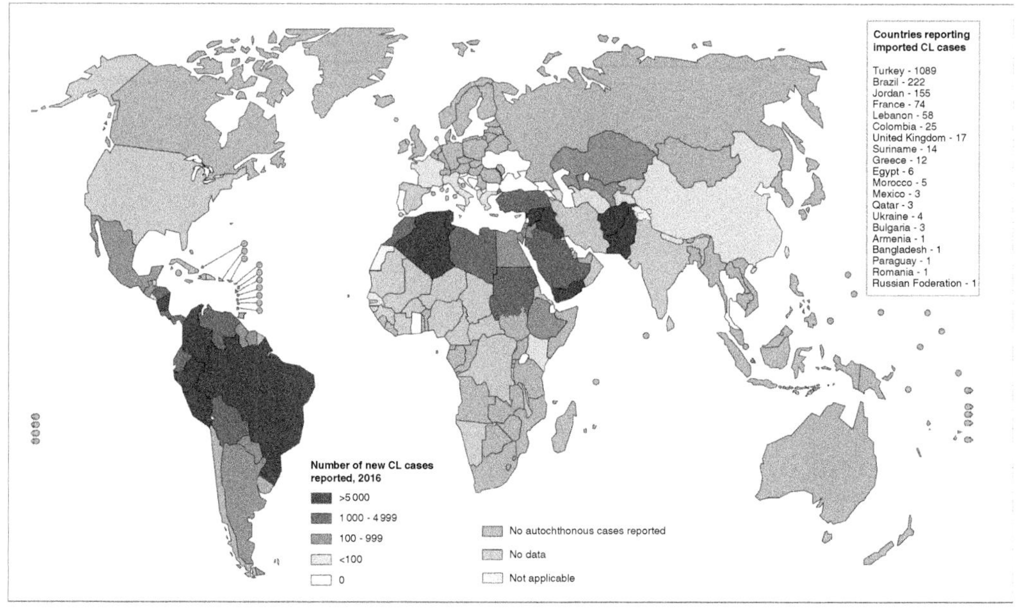

Data Source: World Health Organization
Map Production: Control of Neglected Tropical Diseases (NTD)
World Health Organization

(b)

Status of endemicity of visceral leishmanisis worldwide, 2016

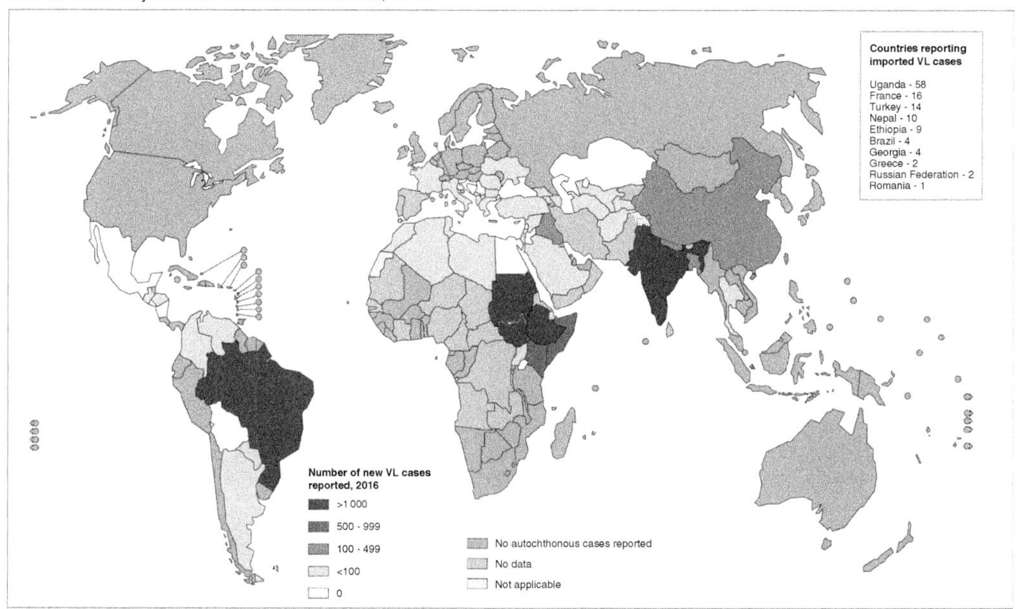

The boundaries and names shown and the designations used on this map do not imply the expression of any opinion whatsoever on the part of the World Health Organization concerning the legal status of the country, territory, city or area or of its authorities, or concerning the delimitation of its frontiers or boundaries. Dotted lines on maps represent approximate border lines for which there may not yet be full agreement. © WHO 2018. All rights reserved

Data Source: World Health Organization
Map Production: Control of Neglected Tropical Diseases (NTD)
World Health Organization

World Health Organization

Figure 15.2 Worldwide endemicity of (a) cutaneous and (b) visceral leishmaniasis in 2016. *Source:* Adopted from WHO (2019a) with permission.

promastigotes distinguished morphologically include procyclic, nectomonad, leptomonad, haptomonad, paramastigote, and metacyclic promastigotes. The process is schematically depicted in Figure 15.3.

Following the bite of an infected sandfly, *Leishmania* metacyclic promastigotes undergo phagocytosis by the host macrophages or dendritic cells through binding to various receptors, including the complement receptors and F_C receptors. Once a promastigote is inside a phagosome, fusion with lysosomes and endosomes occurs, forming a phagolysosome in which promastigotes transform into amastigotes. The multiplication of amastigotes within macrophages causes eventual rupture of the macrophage and allows the spread of infection to other macrophages. The *Leishmania* life cycle is completed when a sandfly feeds on the blood of an infected host taking up the amastigotes. The endemic transmission cycle of *Leishmania* typically occurs through the sandfly and mammalian reservoirs including rodents and canines, however humans can be incidental hosts. More recently with HIV infection, a new mode of transmission was recorded via needle between HIV infected people (Mukbel et al. 2005).

15.5 Diagnosis, Detection, and Surveillance

Efficient case management based on early diagnosis and treatment is the key to limiting morbidity and preventing mortality. Laboratory diagnosis of leishmaniasis can be made by microscopic examination of the stained specimen, *in vitro* culture, animal inoculation, detection of parasite DNA in tissue samples, detection of a parasite antigen, or specific antileishmanial antibodies (Kumar 2013). Direct visualization of amastigotes in clinical specimens is the diagnostic gold standard in regions where tissue aspiration is feasible and microscopy and technical skill are available. Diagnostic sensitivity for splenic, bone marrow, and lymph node aspirate smears is >95%, 55–97%, and 60%, respectively

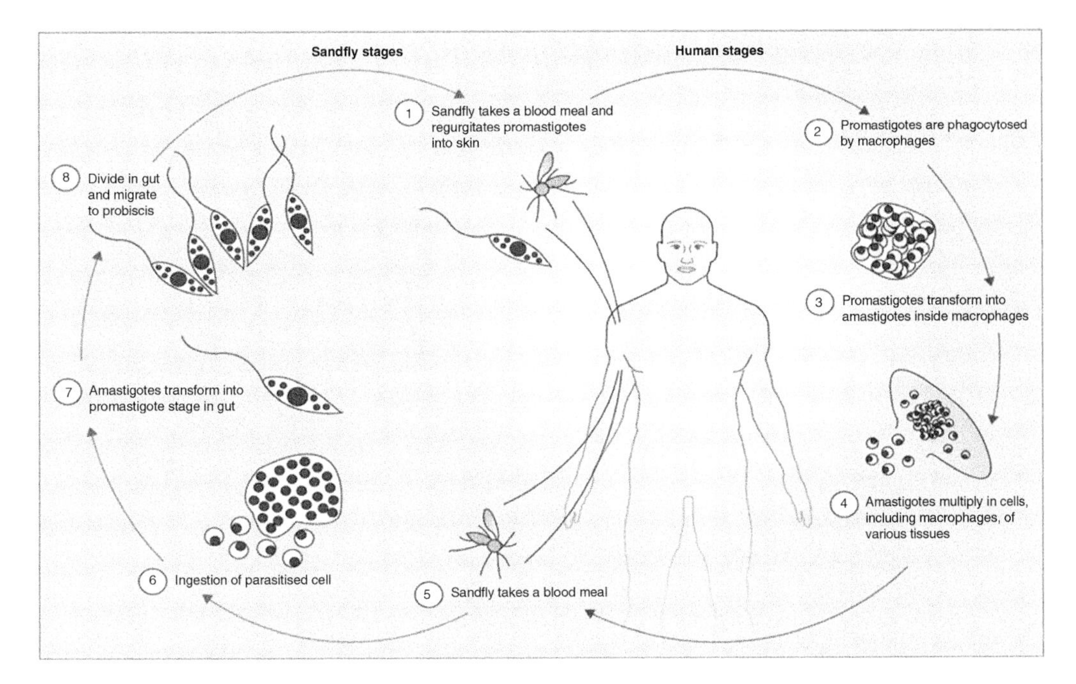

Figure 15.3 Life cycle of *Leishmania*. *Source:* Adapted from Burza et al. (2018) with permission from Elsevier.

(Sundar 2003). Otherwise, serum antileishmanial immunoglobulin G in high titer is the diagnostic standard, primarily with direct agglutination tests or other laboratory-based serological assays (Asfaram et al. 2018).

Freeze-dried antigen and rapid detection of the anti-K39 antibody with finger stick blood in an immune chromatographic strip test is also an alternative serodiagnostic method. Testing urine for leishmanial antigen or antibody is a new approach (Sundar 2003; Islam et al. 2004). Various immunodiagnostic tests include antibody detection, complement fixation test, immunodiffusion test, countercurrent immunoelectrophoresis, indirect hemagglutination, indirect immunofluorescent assay, direct agglutination test, enzyme-linked immunosorbent assay (ELISA) with crude soluble antigen, ELISA with the fucose-mannose ligand, ELISA with rK39 antigen, and rapid strip test with rK39. DNA detection by polymerase chain reaction with *L. donovani* primer using blood, bone marrow, and skin samples are currently underused. Different DNA sequences in the genome of *Leishmania* have been documented in the diagnosis and prognosis of visceral leishmaniasis (Kumar 2013).

15.6 Treatment Strategies for Leishmaniasis

Available drugs against *Leishmania* can be quite toxic and costly, and there may be some parasitic resistance, which explains how the current situation is complicated for treatment of leishmaniasis. During the last seven decades, pharmacotherapy of leishmaniasis has been dependent on toxic antimonial compounds. Now, the situation has not changed a lot with a limited number of drugs available in the arsenal for treatment of leishmaniasis (Table 15.2).

The old-fashioned pentavalent antimony-based drugs, i.e. glucantime® and pentostam, were developed and introduced as antileishmanial drugs during the 1950s (Gutiérrez et al. 2016). These are the standard first-line medicines in most parts of the world with a higher than 90% overall cure rate. Initial treatment of VL should be based on a daily injection of $20\,\mathrm{mg\,kg^{-1}}$ body weight of Sb(V), and injections are usually given for 28–30 days. The mechanism of action of the said compounds is believed, in one model, to be based on the reductive bioconversion of Sb(V) to Sb(III), by the parasite or by the infected host cells, to create an active agent Sb(III)-trypanothione conjugate. In a second model, Sb(V) is the active species itself, which directly exerts activity against *Leishmania*. Sb(V) can act selectively against the topoisomerase of the promastigote rather than against that of the monocyte. Sb(V) is also reported to bind ribonucleosides in an environment similar to that of lysosomes. Unfortunately for the first-line antimonials, the development of resistance is a primary concern. The resistance mechanism has been proven to be multifactorial and is principally due to reduced Sb(III) uptake or/and increased cell efflux/sequestration through an abnormally high level of trypanothione and increased expression of the metalloid-thiol pump. Resistance to therapeutics has also been imputed to different efflux pumps or ATP-binding cassette transporters, such as P-glycoprotein and multidrug resistance protein (Bruni et al. 2017).

In a second instance, the macrolide antifungal agent AmB is a good alternative to the former antimony compounds. The high molecular weight and polar nature of AmB prevent its oral absorption, so it is administered as a deoxycholate salt (Fungizone®) by single or repeated intravenous infusion. AmB, a highly effective antileishmanial agent, cures more than 90% of the cases of VL and is the drug of choice for antimonial-resistant cases. AmB is used against cutaneous or mucosal leishmaniasis and is effective for treating immunocompromised patients. Lipid preparations of the drug have reduced toxicity, but the cost of the drug and the difficulty of administration remain a problem in endemic regions. The mechanism of action of AmB is based on its binding to the ergosterol in the parasite cell membrane rather than to human cholesterol. AmB complexes with ergosterol precursors form pores

Table 15.2 Chemical structure and selected properties of existing antileishmanial therapeutic agents.

Drug	Chemical structure	Route of administration	Dose regimen	Clinical efficacy	Major side effects
Meglumine Antimoniate (Glucantime)	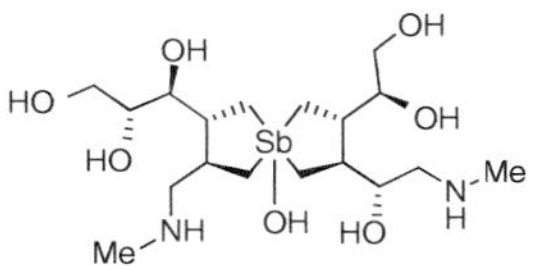	Slow IVb infusion or IM	20 mg kg^{-1} d^{-1} 30 d	35–95% depending on area, resistance in India	Painful injection; nephrotoxicity; cardiotoxicity and pancreatitis
Sodium stibogluconate (Pentostam®)	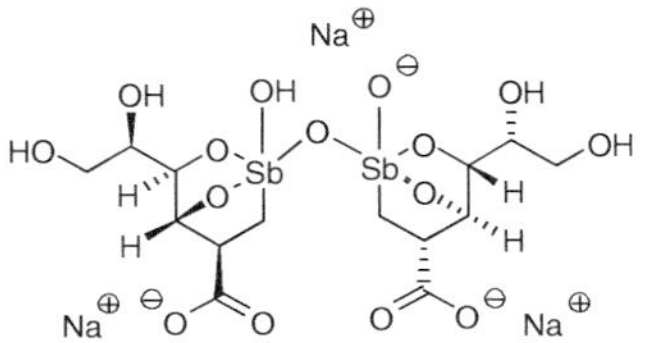	Slow IV infusion or IM	20 mg kg^{-1} d^{-1} 30 d	35–95% depending on area, resistance in India	Painful injection; cardiotoxicity and pancreatitis
Amphotericin B	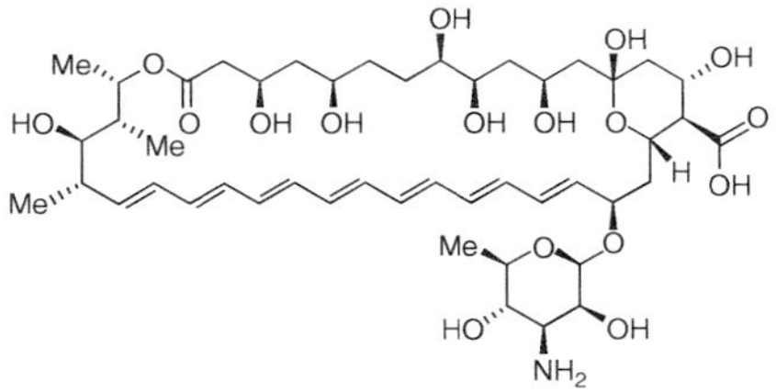	IV infusion	10–30 mg kg^{-1} d^{-1} total dose over 1–10 d	>95% efficacy in India, variable response in Africa	Rigors and chills; nephrotoxicity; hypokalemia; anaphylaxis
Miltefosine	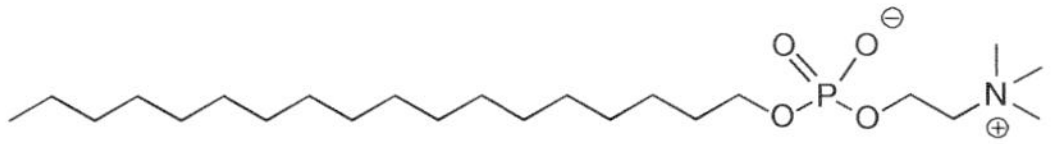	PO	1.5–2.0 mg kg^{-1} d^{-1} 28 d	94–97%	Teratogenicity; gastrointestinal side effects; nephrotoxicity and hepatotoxicity

(Continued)

Table 15.2 (Continued)

Drug	Chemical structure	Route of administration	Dose regimen	Clinical efficacy	Major side effects
Paromomycin sulfate	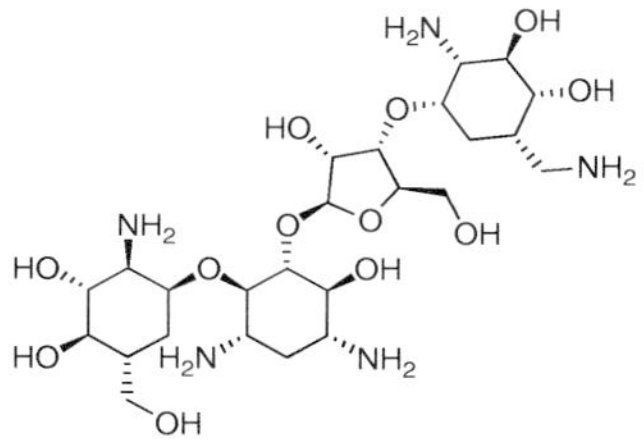	IM	15–20 mg kg^{-1} d^{-1} 21 d	93–95% India, 64–85% Africa	Painful injections; reversible nephrotoxicity; hepatotoxicity and ototoxicity
Pentamidine	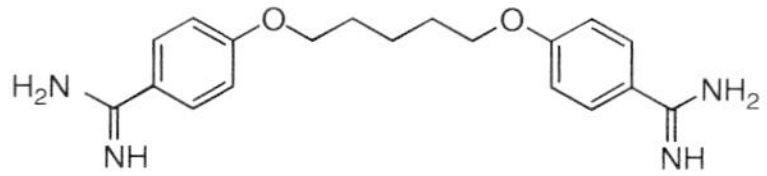	IV or IM	4 mg kg^{-1} d^{-1} 3 doses	55%	Diabetes

Source: Adapted from Mowbray (2018) with modifications.

through the cell membrane allowing ions influx (Wetzel and Phillips 2018). In a clinical trial in India, Fungizone was 99% effective when administered daily or on alternate days by IV (intravenous) slow infusion at a dose of 0.75–1.0 $mg\,kg^{-1}\,d^{-1}$ for 15–20 doses. Different AmB formulations have been developed and are currently available, including Abelcet®, Amphocil®, Amphotec® (lipid complexes), AmBisome® (liposomes), and Ampholip®, in order to reduce the severe side effects including nephrotoxicity and hematotoxicity. In contrast to AmB effectiveness against VL, it is extremely expensive to develop in poor countries and requires intravenous administration (Bruni et al. 2017).

Miltefosine (Impavido®) is an alkyl phosphocholine derivative which was first synthesized as an antineoplastic drug, but it is now considered as a promising alternative to pentavalent antimony-based drugs. The antiprotozoal activity of miltefosine was discovered in the 1980s as it was being evaluated for cancer chemotherapy. In 2002, it was approved in India as the first orally administered drug for VL. It is highly curative against VL but it is also effective against CL at the same time. Teratogenicity is the main drawback for miltefosine and its mechanism of action is not clearly understood. Studies suggest that miltefosine may alter ether-lipid metabolism, cell signaling, or glycosylphosphatidylinositol anchor biosynthesis. The recommended dose for oral miltefosine in both visceral and cutaneous leishmaniasis is 150 $mg\,kg^{-1}\,d^{-1}$ for 28 days, given in three divided doses to adults over 45 kg and 100 $mg\,kg^{-1}\,d^{-1}$, given in two divided doses for patients weighing 30–45 kg. Due to its hemolytic activity, miltefosine cannot be given intravenously (Wetzel and Phillips 2018).

Many other compounds are considered as second-line drugs against leishmaniasis, including paromomycin (PA) and pentamidine, as well as other agents present at different stages of the approval process (Mowbray 2018). Paromomycin (Aminosidine®), an aminoglycoside antibiotic, is used as an oral agent to treat *Entamoeba histolytica* infection, cryptosporidiosis, and giardiasis. Topical formulations have been used to treat trichomoniasis and CL. Parenteral administration has been tried to treat CL, both alone and in combination with antimony compounds. Paromomycin shares the same mechanism of action as neomycin and kanamycin, i.e. binding to the 30S ribosomal subunit with the same spectrum of antibacterial activity.

Pentamidine, a positively charged aromatic diamidine compound, is an alternative agent for the treatment of CL. It is a broad-spectrum agent active against several species of pathogenic protozoa and some fungi, with an unknown mechanism of action. The compounds display multiple effects on any given parasite and act by disparate mechanisms in different parasites. Pentamidine can be used in doses of 2–3 $mg\,kg^{-1}$ (IV or IM) daily or once every two days for 4–7 doses to treat CL. This compound provides an alternative to antimonials, lipid formulations of AmB, or miltefosine, but overall it is the least well-tolerated agent (Wetzel and Phillips 2018).

The efficacy of oral azithromycin (500 $mg\,d^{-1}$ for 20 days) against CL from two endemic regions of Brazil was compared to the standard treatment using IM injections of pentavalent antimonials, i.e. glucantime 20 $mg\,kg^{-1}\,d^{-1}$ for 20 days. Fexinidazole, a 2 substituted 5-nitroimidazole formulated for oral administration, is in phase II clinical trial for oral treatment of adult patients affected by VL (Clinical trials 2019).

In the search for more effective drugs to treat leishmaniasis, researchers have chosen to follow different approaches; among them, natural products or privileged structures are without a doubt excellent starting points. Throughout the history of medicinal chemistry, natural products have played and continue to play an invaluable role in drug discovery. The compounds representing potential prototypes for drug design studies against *Leishmania* can be broadly classified as alkaloids, terpenoids, saponins, phenylpropanoids, flavonoids, lignoids, naphthoquinones, and iridoids (Lago and Tempone 2018).

In attempts to overcome the limitations of this small set of antileishmanial drugs, researchers have also explored the potential of combination therapies to increase efficacy, reduce toxicity, improve tolerance, and shorten treatment durations, with the added benefit of perhaps reducing the risk of resistance. These efforts have met with some success, providing improved options for some patient populations, such as the combination of sodium stibogluconate (SSG) with paromomycin for VL in East Africa; antimonials with allopurinol and paromomycin (Mowbray 2018); meglumine antimoniate (MA) and pentoxifylline for CL; WR 279 396 (topical paromomycin and gentamicin) in non-complicated, non-severe CL; oral miltefosine with topical imiquimod for CL; allopurinol vs. glucantime vs. allopurinol/glucantime in Brazilian patients with CL; topical immucillin DI4G with IV meglumine antimoniate; and finally imiquimod (Aldara®) and glucantime in patients affected by Peruvian CL. 18-methoxyoronaridine as a candidate for tegumentary leishmaniasis treatment is in randomized phase II clinical trial (Clinical trials 2019).

Further improvements for leishmaniasis treatment may be possible through optimization of combination treatments, but providing a major step forward with efficacious, safe, well-tolerated, and convenient-to-use treatments which are genuinely adapted to the needs of patients will need a new generation of drugs designed for this purpose (Mowbray 2018) or improved delivery of the currently approved drugs using nanotechnology-based drug delivery approaches.

15.7 Nanotechnological Approach to Leishmaniasis Treatment

Nanoparticles have been used in the diagnosis and treatment of many diseases inasmuch as they benefit from large specific surface area and special chemical, physical, and mechanical properties in comparison with larger particles (Qasim et al. 2014; Want et al. 2016). Leishmaniasis spreads rapidly and it is presently a major public health concern in developing countries. The major challenge in the treatment of leishmaniasis is the fact that the parasite infects the macrophages, therefore traditional antileishmanial drugs face difficulties to find and penetrate inside the infected macrophages. For example, in VL, a drug must target parasites within infected macrophages in the spleen, liver, and bone marrow, whereas in CL the drug must reach parasites in the cutaneous lesion(s). Once at the site of infection, any drug must first permeate the infected macrophage membrane before crossing the membrane of the parasitophorous vacuole and finally cross the plasma membrane of the parasite (Figure 15.4) (Shaw and Carter 2014).

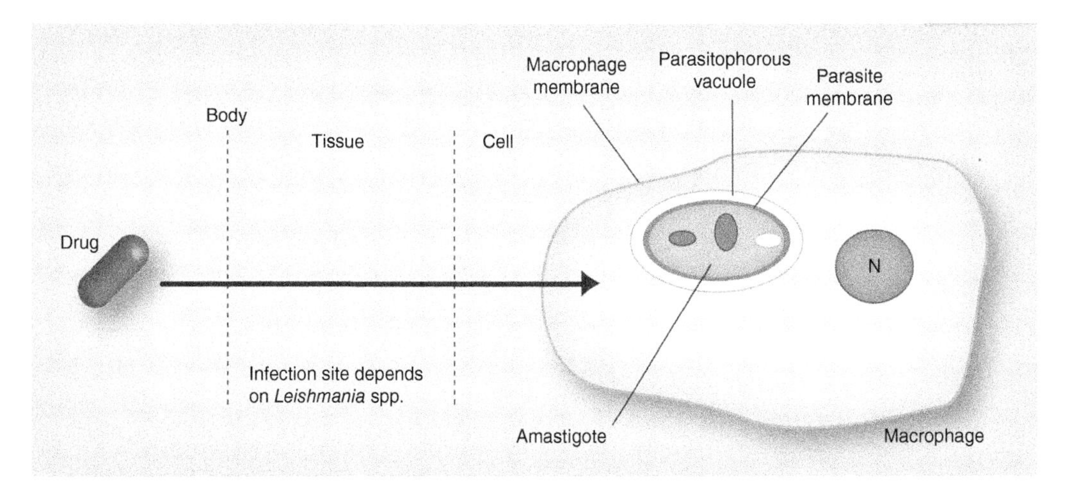

Figure 15.4 The way that a drug must pass to access intracellular leishmania amastigotes within macrophages. A drug enters the body by the route administered and has to reach the sites where infected macrophages reside, and as a consequence has to cross multiple membranes to enter the parasite. N: Nucleus of the cell. *Source:* Adapted from Shaw and Carter (2014); an open article.

Recently, the combination of antileishmanial drugs with nanocarriers has been emerging as a promising approach in the treatment of leishmaniasis. These nanocarriers can penetrate into macrophages and release the drug inside the cell, leading to a locally high concentration of the therapeutic agent and ultimately killing the protozoa. In this context, the main strategy in the treatment of leishmaniasis is to target the drugs directly to macrophages by using nanocarriers, which have the ability to overcome the biological barriers involved. Moreover, the use of said nanocarriers would permit to reduce the drug toxicity, enhance treatment efficacy, improve selectivity, modulate the drug pharmacokinetics, increase drug solubilization, decrease drug-resistance potential, protect the drug from degradation, and promote a sustained drug delivery directly in the target site (Gutiérrez et al. 2016). Another advantage is the possibility to design the nanocarriers to carry more than one drug, allowing the combination therapy that might have a pharmacologically synergistic effect. Nanocarriers allow a wide range of surface modifications that may be further investigated to increase parasite selectivity (Singh et al. 2016).

Targeted drug delivery to the macrophages appears to be an attractive proposition to improve therapeutic efficacy of enclosed drug. Thus, macrophages can be exploited as Trojan horses for targeted drug delivery, i.e. macrophage-mediated drug delivery approach. Nanocarriers can migrate across the different membrane barriers and release their drug cargo at sites of infection (Jain et al. 2013).

Immunological response against foreign moiety is the natural phenomenon of the body which can be utilized for the passive targeting mediated by the intravasation of the drug and drug carrier systems based on their physicochemical properties. Particle uptake by the cells of the RES is an excellent example of passive targeting. The endeavor potential of macrophages for rapid recognition and clearance of foreign particles has provided a rational approach to macrophage-specific targeting with nanocarriers. Passive capture of colloidal carriers by macrophages offers therapeutic opportunities for the delivery of antileishmanial drugs since it involves macrophage cells of the RES. Passive targeting is being exaggerated by pathophysiological factors (inflammation/infection and enhanced permeation and retention effect) as well as physicochemical factors (particle size, surface charge, and molecular weight) of the DDS.

Active targeting redefines the biofate or natural distribution pattern of the drug carrier system with modification or manipulation of the surface of carriers so that it can be identified by specific cells. Binding of drug-loaded carriers to target cells is facilitated by the use of ligands or engineered homing devices and thus enhances the receptor-mediated localization of drug. The leishmania parasite resides in intracellular phagolysosomes and hence there is a need to target the DDS intracellularly. Intracellular targeting is hierarchically the third order of targeting and involves receptor-based, ligand-mediated entry of a drug complex into a cell by endocytosis followed by lysosomal degradation of the carrier leading to the release of the drug. The uptake of leishmania promastigote by macrophages is a receptor-mediated process that involves the expenditure of energy by the macrophage, but not by the parasite. Due to the obligate intracellular nature of the pathogen, this organism expresses several different ligands on its surface that can interact with a variety of different macrophage receptors, to ensure its uptake by phagocytic cells. These include the receptors for complement, fibronectin, sugars (such as the mannose-fucose, galactosyl receptor), and so on. Different receptors bind carrier molecules with different avidity (Chowdhary et al. 2016).

Moreover, finding new uses for existing drugs – so-called drug repurposing – and different new target-based approaches are interesting and fruitful methods for discovering and developing new drugs for neglected tropical diseases, including leishmaniasis (Martinez and Gil 2018). Conceptually, many of the unlikely properties of conventional antileishmanial drugs or the poor immunogenicity of subcellular compounds of leishmania could be improved through the use of DDS and nano-devices provided by pharmaceutical technology. DDS could improve the solubility of poorly water-soluble drugs, e.g. AmB or protecting of antigenic proteins, DNA, or RNA from rapid degradation and drug targeting to the site of action, i.e. macrophages (Jain et al. 2013). As leishmania parasites are also mainly confined in macrophages, DDS could improve the therapeutic index of antileishmanial drugs, decreasing the effective dose and the off-target toxic effects produced by an inadequate biodistribution.

Various novel nano-sized DDS are currently being experimented and utilized against leishmaniasis such as dendrimers (monodisperse polymers), solid lipid nanoparticles (SLNs) and nano-structured lipid nanoparticles (NLPs), ultradeformable lipid matrices, and various nano-structured polyelectrolyte complexes. Dendrimers are highly stable, water-soluble, unimolecular micelles capable of forming complexes with hydrophobic drugs in their inner hydrophobic pockets and can be administered by the oral route. SLNs and NLPs have solid hydrophobic cores of variable crystallinity, stabilized by amphipathic surfaces. Drugs loaded in the particles are retained and released in a controlled manner, as a function of core phase transitions in response to external stimuli such as changes in humidity, heat, light, or mechanical stress exerted by the surrounding medium. They can be administered by oral and topical routes. The ultradeformable lipid matrices are those vesicles that are capable of experiencing spontaneous locomotion and penetration to deeper layers across water nano-channels in the stratum corneum. Ultradeformable liposomes do not fuse on the surface of the stratum corneum, and penetrate without being destroyed. Ultradeformable liposomes could efficiently transport low- or high-molecular-weight hydrophilic drugs across thickened lesions that represent an additional barrier to absorption in CL (Bhatia and Goli 2016).

15.7.1 Polymeric Nanoparticles

15.7.1.1 Biodegradable Polymers

Abu-Ammar and colleagues prepared AmB-loaded nanoparticles for local treatment of CL. Poly(lactic-*co*-glycolic acid) (PLGA) nanoparticles loaded with AmB were produced to reduce drug toxicity and facilitate localized delivery over a prolonged time. The *in vitro* release experiments revealed the sustained activity of the AmB-releasing nanoparticles and inhibition of promastigote growth over 15 days. This profile is expected to minimize drug toxicity associated with exposure to high drug concentrations caused by rapid release after usage of traditional dosage forms, and to enhance efficacy (Abu Ammar et al. 2018).

Asthana et al. (2015a) prepared a lactoferrin-appended AmB-loaded nano-reservoir. The biodistribution study illustrated preferential accumulation of the nano-reservoirs in the liver and spleen (Figure 15.5a). The augmented antileishmanial activity was shown by the nanoparticles along with a significant reduction (~88%) in splenic parasite burden of the infected hamsters, compared to the conventional formulations. Superior efficacy, desired stability, and reliable safety of the cost-effective lactoferrin-appended AmB-loaded nano-reservoirs suggest its potential for leishmaniasis therapeutics (Asthana et al. 2015a).

Halder et al. reported lactoferrin-modified betulinic acid-loaded PLGA nanoparticles (Lf-BANPs) (Lf-BANPs). The amastigotes count in macrophages was more effectively reduced by Lf-BANP than betulinic acid (BA) and betulinic acid nanoparticles (BANP) (Figure 15.5b). Lf-BANPs reduced the pro-parasitic, anti-inflammatory cytokine IL-10, but increased nitric oxide (NO) production in *L. donovani*-infected macrophages, indicating that Lf-BANP possesses significant anti-leishmanial activity (Halder et al. 2018).

Moreover, Nahar and Jain (2009) also reported on the fabrication and evaluation of AmB-loaded, engineered PLGA nanoparticles. PLGA was conjugated to mannose via direct coupling (M-PLGA) and via PEG spacer (MPEG- PLGA) and engineered PLGA nanoparticles (M-PNPs and M-PEG-PNPs)

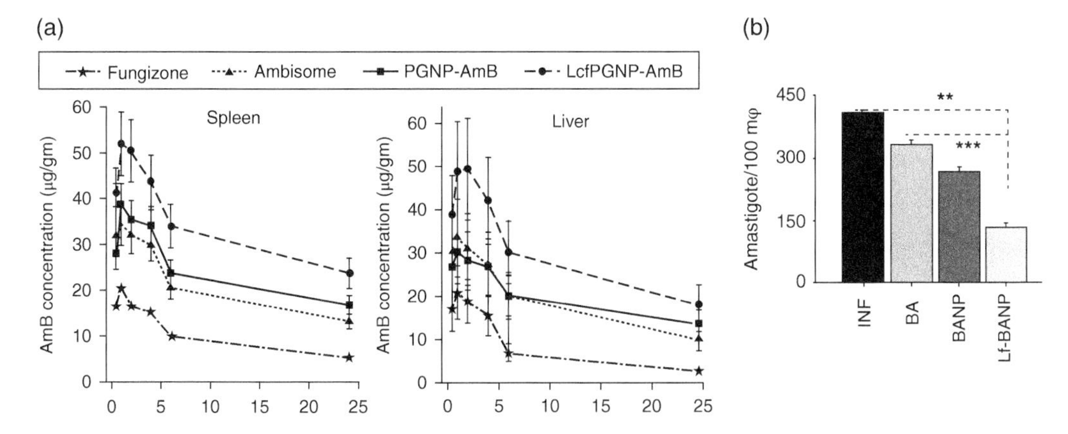

Figure 15.5 Preferential biodistribution of AmB over time in liver and spleen after administration of lactoferrin-appended nanoreservoir. *Source:* Adapted from Asthana et al. (2015a); an open access article. (b) Lf-BANP induces leishmanicidal activity in BALB/c-derived peritoneal macrophages. *L. donovani*-infected macrophages were treated with betulinic acid (BA) BA was loaded onto PLGA nanoparticles (BANPs) and coated with Lactoferrin (Lf-BANPs). The macrophages were infected with *L. donovani* promastigotes at a ratio of 1:10 for 6 h at 33 °C and BA (2.5 µg/ml), BANP (2.5 µg/ml) and Lf-BANP (2.5 µg/ml) were added to the cultures. After 72 h, the macrophages were fixed, Giemsa-stained, and the number of amastigotes per 100 infected macrophages were counted under a microscope. Lf-BANP treatment reduced infection significantly. *Source:* Adapted from Halder A et al. (2018).

which were prepared from respective conjugates. Engineered polymeric nanoparticles (PNPs) with spacer showed enhanced uptake, potential antileishmanial activity, and higher disposition in macrophage-rich organs, suggesting improved macrophage targeting. The results suggest that the engineering of nanoparticles could lead to the development of efficient carriers for macrophage targeting (Nahar and Jain 2009).

In another study, the authors developed an oral nanoparticle-based formulation of curcumin PLGA. In combination with miltefosine, it exhibited a synergistic effect on both promastigotes and amastigotes under *in vitro* conditions. The combination also exhibited increased lymphocyte proliferation. The present study thus establishes the possible use of nano-curcumin as an adjunct to antileishmanial chemotherapy. The increase in inhibition of parasite burden by combination therapy compared to CNP monotherapy was modest (14%) only, but the difference was statistically significant (Tiwari et al. 2017). Moreover, Table 15.3 shows details of PNPs made from biodegradable polymers intended for the treatment of leishmaniasis.

15.7.1.2 Micellar Polymeric Nanoparticles

Tavares et al. (2019) evaluated a Pluronic® F127-based polymeric micellar system containing Clioquinol (ICHQ). Immunological and parasitological analyses showed that this composition was highly effective in chronically infected mice, based on the significant reductions in the parasite burden analyzed in different organs as shown in Figure 15.6. This fact can be considered relevant, since ICHQ-containing polymeric micelles are cheaper to produce, with an estimated value of 10 times less in comparison to the production of AmB-containing liposomal formulations (Tavares et al. 2019).

Mendonça et al. (2019) incorporated a naphthoquinone derivate (Flau-A) into a Pluronic F127-based polymeric micellar system (Flau-A/M) to be effective against *Leishmania amazonensis* infection. According to their findings, AmB, AmBisome, Flau-A, or Flau-A/M-treated animals presented significantly lower average lesion diameter and parasite burden in tissue and organs

Table 15.3 A survey of reports on polymeric nanoparticles made from biodegradable polymers intended for treatment of leishmaniasis.

Polymer	Technique	Drug	Comment	References
PLGA	Nanoprecipitation (modified)	AmB	AMB NPs exhibited significant reduction in the lesion size compared to control, while the remaining groups showed inadequate efficacy.	Abu Ammar et al. (2018)
PLGA	Nanoprecipitation	AmB	Significant increase in *in vitro* (with ~82% parasite inhibition) and *in vivo* antileishmanial activity for 3-O-sn-Phosphatidyl-L-serine anchored PLGA nanoparticles.	Singh et al. (2018)
PLGA	O/W emulsification/ solvent evaporation	AmB	Lactoferrin-appended AmB bearing showed antileishmanial activity by significantly reducing (~88%) splenic parasite burden in infected hamsters.	Asthana et al. (2015a)
PLGA	Ultrafiltration	AmB	Significant reduction in liver parasite burden.	Italia et al. (2012)
PLGA	Nanoprecipitation	AmB	The relative oral bioavailability of the AmB-NP was found to be ~800% as compared to Fungizone.	Italia et al. (2009)
PLGA	O/W emulsification/ solvent evaporation	AmB		Nahar et al. (2009)
PLGA	O/W emulsification/ solvent evaporation	AmB	Designed mannose-anchored PLGA nanoparticles for improved macrophage targeting	Nahar and Jain (2009)
PLGA	Double (W/O/W) emulsification/solvent evaporation	AmB	AmB-loaded, mannose-modified PLGA nanoparticles were found more efficacious in the treatment of VL in both *in vitro* and *in vivo* than unmodified nanoformulations.	Ghosh et al. (2017)
PLGA	Solvent displacement	Artemisinin	Positive delayed-type hypersensitivity reaction enhancement in pro-inflammatory cytokines i.e. IFN-γ and IL-2	Want et al. (2015)
PLGA	O/W emulsification/ solvent evaporation (modified)	Betulinic acid	Lactoferrin-modified PLGA nanoparticles reduced the pro-parasitic, anti-inflammatory cytokine IL-10	Halder et al. (2018)
PLGA	O/W emulsification/ solvent evaporation	Caffeic acid phenethyl ester	Superior antileishmanial activity on both forms of parasites	Abamor (2017)
PLGA	O/W emulsification/ solvent evaporation	Curcumin and miltefosine	Significant leishmanicidal activity both *in vitro* and *in vivo*	Tiwari et al. (2017)

Table 15.3 (Continued)

Polymer	Technique	Drug	Comment	References
PLGA	Double (W/O/W) emulsification/solvent evaporation	Doxorubicin	Superior performance *in vivo*, i.e. enhanced bioavailability and lower toxicity	Kalaria et al. (2009)
PLGA	Nanoprecipitation	Itraconazole	A significant reduction in *Leishmania braziliensis* amastigotes from 54.9% to 28% and 21.1% for PLGA-ITZ-mannose NPs and PLGA-ITZ NPs, respectively.	Biswaro et al. (2019)
PLGA	Sonication	Paromomycin	Mannosylated thiolated paromomycin-loaded PLGA nanoparticles reduced parasitic burden.	Afzal et al. (2019)
PLGA	Double (W/O/W) emulsification/solvent evaporation	Saponin β-aescin		Van de Ven et al. (2011)
PLGA	Double (W/O/W) emulsification/solvent evaporation (modified)	AmB		Asthana et al. (2014)
PLGA	Nanoprecipitation	AmB	Improved oral bioavailability, lower hemolysis, and nephrotoxicity.	Italia et al. (2009)
PLGA-chitosan	Nanoprecipitation	Doxorubicin-AmB	Macrophage targeting by chitosan-anchored PLGA nanoparticles loaded with AmB and Dox.	Singh et al. (2016)
Poly(D,L-lactide)	Nanoprecipitation	Primaquine	The 50% lethal dose of primaquine-loaded nanoparticles was significantly reduced.	Rodrigues Junior et al. (1995)
PCL	Double (W/O/W) emulsification/solvent evaporation (modified)	Meglumine antimoniate	Improved efficacy	Sousa-Batista et al. (2019)
PCL	O/W emulsification/ solvent evaporation	*Nigella sativa* oil	Decreased infection indexes of macrophages	Abamor et al. (2018)
PCL-Pluronic	Nanoprecipitation	Red propolis extract		do Nascimento et al. (2016)
PCL-Pluronic®	Nanoprecipitation	Red propolis extract		Farias Azevedo et al. (2018)
PCL	O/W emulsification/ solvent evaporation	Quercetin	Reduced viability of promastigotes c. 20 times and of amastigotes c. 5 times compared to the control group.	Abamor (2018)

PLGA, Poly(lactic-co-glycolic acid); PCL, Poly(ε-caprolactone).

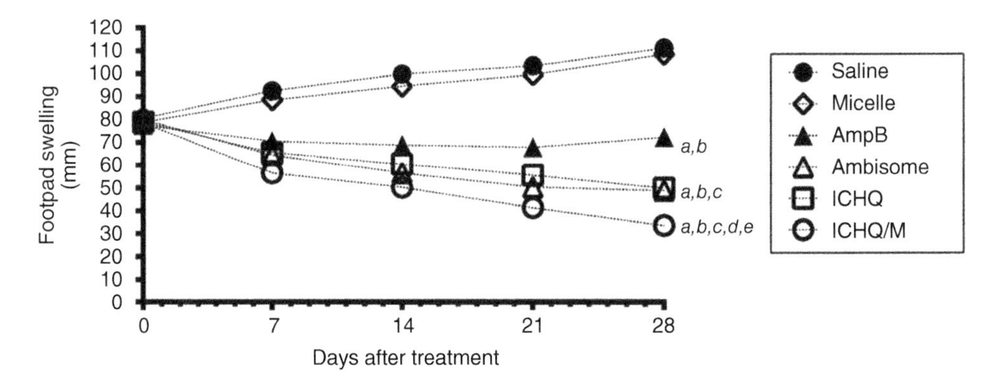

Figure 15.6 Lesion average diameter (mean ± SD) in chronically infected mice animal model after treatment with a Pluronic F127-based polymeric micellar system containing Clioquinol. *Source:* Adapted from Tavares et al. (2019) with permission from Elsevier.

compared to the control (saline and drug-free micelle) groups. Flau-A or Flau-A/M-treated mice were those presenting the most significant reductions in the parasite burden when compared to the others. These animals also developed a more polarized antileishmanial Th1 immune response (Mendonça et al. 2019). The list of micellar drug delivery devices made from copolymers is mentioned in Table 15.4.

15.7.1.3 Natural Polymers

Zadeh-Mehrizi et al. (2018a) prepared AmB-releasing, nanosized chitosan formulations (AK) with considerable improvement against *Leishmania major* adopting the phase-separation method. Complete wound healing and parasite inhibition were achieved by using AK 10 $mg\,kg^{-1}$ dose in terms of improvement of the treatment indicators (Figure 15.7a). Increasing the therapeutic dose of AK to 10 $mg\,kg^{-1}$ provided the successful treatment of *L. major*'s pathological effects *in vitro* and *in vivo* (Zadeh-Mehrizi et al. 2018a). Moreover, chitosan–chondroitin sulfate nanodelivery vehicle for specific delivery of AmB to *Leishmania*-infected macrophage was studied. Taking the advantage of *Leishmania* being highly auxotrophic for heme, hemoglobin was attached on the surface of the delivery device for specifically targeting the *leishmania*-infected macrophages. None of the NP formulations decreased the cell viability significantly, however free AmB was found to be significantly toxic. LD50 concentration (0.28 $\mu g\,ml^{-1}$) of free AmB was used for the comparative study of toxicity (Figure 15.7b) (Bose et al. 2016).

Chaubey et al. (2017) prepared mannose-conjugated curcumin-chitosan nanoparticles (Cur-MCN) (Figure 15.7c). *In vivo* antileishmanial activity in hamsters was shown to be significantly higher in terms of suppression of parasite replication in the spleen using Cur-MCN than unconjugated chitosan nanoparticles. *In vitro* cytotoxicity studies against the J774A.1 cell line demonstrated its comparative nontoxicity toward the macrophage cells. The potential of Cur-MCN was also confirmed by minimal cytotoxicity observed in studies performed in an animal model.

Tripathi et al. (2015) developed 4-sulfated N-acetyl galactosamine (4-SO_4GalNAc) anchored chitosan nanoparticles (SCNPs) for efficient chemotherapy in leishmaniasis by using the dual strategy of targeting. Modified chitosan nanoparticles (AmB–SCNPs) provided significantly higher localization of AmB in liver and spleen compared to unmodified chitosan nanoparticles (AmB–CNPs) after IV administration. The study stipulates that 4-SO_4GalNAc assures targeting of resident macrophages. Highly significant anti-leishmanial activity ($P < 0.05$ compared to AmB–CNPs) was observed with AmB–SCNPs, causing 75.30 ± 3.76% inhibition of splenic parasitic burdens (Figure 15.7d) (Tripathi et al. 2015).

Table 15.4 Micellar drug delivery devices made from copolymers as appearing in the literature.

Copolymer	Technique	Drug(s)	Comment	References
Cashew gum-*g*-PLA	Nanoprecipitation with subsequent spontaneous emulsification	AmB	Process efficiency for incorporating of AmB in the emulsion was 21–47%	Richter et al. (2018)
Chitosan-PLGA	Osmosis-based methodology	Paromomycin	36-fold lower IC50 encapsulation efficiency: 67.16 ± 14%	Afzal et al. (2019)
Chitosan-Pluronic® F127	Thin-film hydration	AmB		Singh et al. (2017)
MPEG-*b*-PLA	Nanoprecipitation	Doxorubicin and mitomycin C	Decreased cellular toxicity in mouse macrophages	Shukla et al. (2012)
PCL-*b*-PDMAEMA PEG-*b*-PCL	Nanoprecipitation	AmB	Nanoparticles were seen to be 10 times less cytotoxic than free AmB	Shim et al. (2011)
PEG-*b*-PCL	Nanoprecipitation	Bisnaphthalimidopropyl	1.02 ± 0.41 μM and 0.73 ± 0.06 μM antileishmanial activity for THP-1 and J774 macrophages	Costa Lima et al. (2012)
PEG-*b*-PLA	Solvent-displacement	Primaquine	Enhanced leishmanicidal activity of primaquine; Decreased primaquine toxicity	Heurtault et al. (2001)
PEG-*b*-PLGA	Doubling the amount	AmB	Improved bioavailability from 1.5% to 10.5%; Significant oral absorption	Radwan et al. (2017)
PEG-*b*-PLGA	Nanoprecipitation	AmB	Significant inhibition of amastigotes in the splenic tissue (93.02 + 6.6)	Kumar et al. (2014b)
PEG-*b*-PLGA	Nanoprecipitation	Miltefosine	Significant nanoencapsulated miltefosine (23.21 ± 23)	Kumar et al. (2016a)
PEO(5000)-*b*-PCL(5000)	Nanoprecipitation	4-Nitrobenzaldehyde thiosemicarbazone	Significant antiprotozoal activity; Low cytotoxicity	Britta et al. (2016)
PEO-*b*-PDMAEMA	Micellar structures	Miltefosine and TCAN-26	The increase of the quantity of the drug up to 5% w/w	Karanikolopoulos et al. (2007)

(*Continued*)

Table 15.4 (Continued)

Copolymer	Technique	Drug(s)	Comment	References
PEO-*b*-PPO-*b*-PEO (Poloxamer® 407)	Nanogel formulation	Paromomycin	Cutaneous drug con.: 31.652 $\mu g g^{-1} cm^{-2}$	Brugués et al. (2015)
Pluronic F127	Cold method	8-hydroxyquinoline	Significant reduction in the average lesion diameter; Higher levels of parasite-specific IFN-γ, IL-12, and GM-CSF	Lage et al. (2016)
Pluronic F127	Cold method	AmB	Effective in stimulating the production of Th1 cytokines and inhibiting the development of Th2 cytokines	Mendonça et al. (2018)
Pluronic F127	Cold method	AmB	The cytotoxic concentration (CC50) values were ca. 119.5 ± 9.6 and 134.7 ± 10.3 μM, respectively.	Mendonça et al. (2016)
Pluronic F127	Cold method	Clioquinol		Tavares et al. (2019)
Pluronic F127	Cold method	Flau-A		Mendonça et al. (2019)
Pluronic F-127 and P-123	Solid dispersion thin-film method		Noninvasive and selective for a target region	Oyama et al. (2019)
Polysorbate 80	Simulation	AMB – Nystatin	Coarse-grained (CG) molecular dynamics (MD) simulation	Mobasheri et al. (2015)

The only exception based on a non-ionic surfactant is cited in the last row.
PLGA, Poly(lactide-co-glycolic acid); MPEG, Methoxy Poly(ethylene glycol); PEG, Poly(ethylene glycol); PLA, Poly(lactide); PEO, Poly(ethylene oxide); PCL, Poly(ε-caprolactone); PPO, Poly(propylene oxide); PDMAEMA, Poly(N,N-dimethylamino-2-ethyl methacrylate); O/W, oil-in-water; ATRP, Atom transfer radical polymerization.

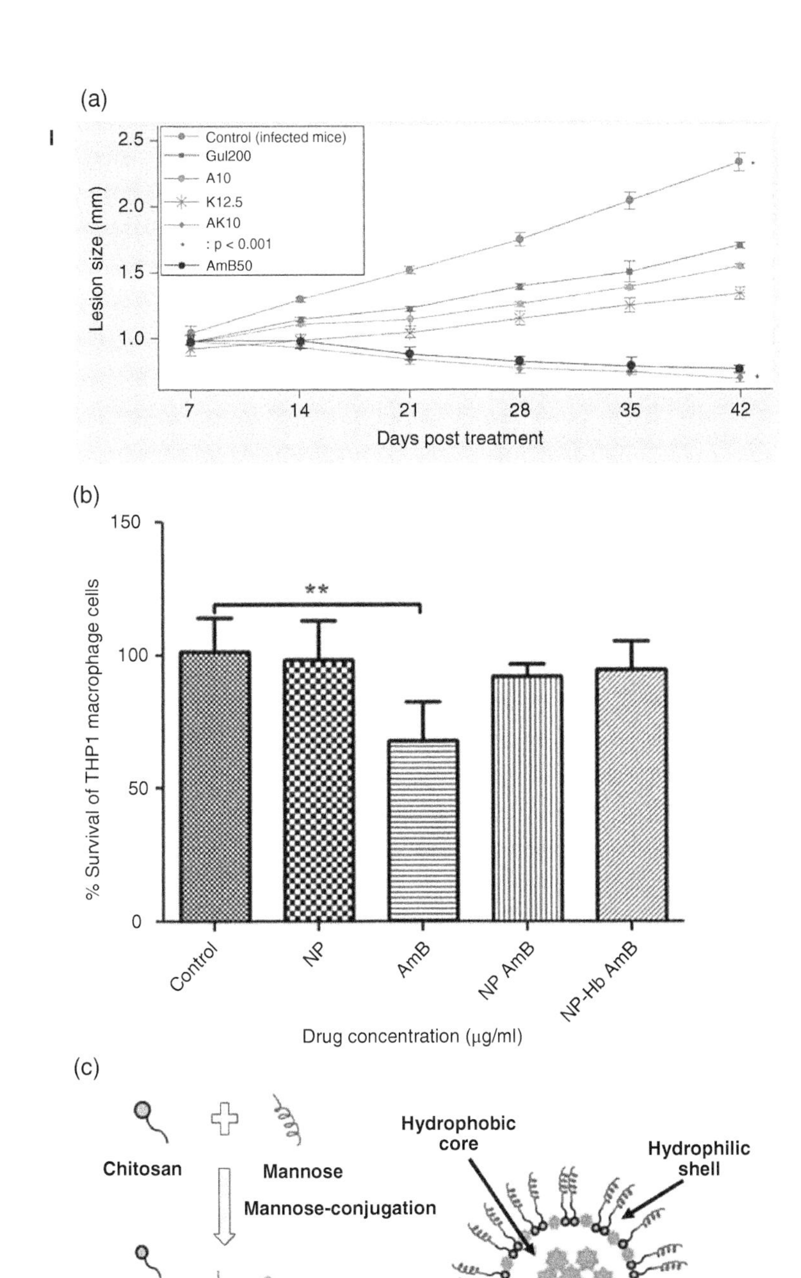

Figure 15.7 (a) Lesion size change in *Leishmania major*-infected Balb/c mice administered after administration of various formulations. Nanochitosan (K), Nanodrug Amp-K (AK), glucantime (GUL). *Source:* Adapted from Zadeh-Mehrizi et al. (2018a); an open access article. (b) Cytotoxicity profile for free AmB, chitosan–chondroitin sulfate nanoparticles (NPs), AmB-loaded chitosan–chondroitin sulfate nanoparticles (NP-AmB) and hemoglobin attached AmB-loaded chitosan–chondroitin sulfate nanoparticles (NP-Hb AMB). *Source:* Adapted from Bose et al. (2016) with permission from Elsevier. (c) Schematic representation of mannose-conjugated chitosan nanoparticle. *Source:* Adapted from Chaubey et al. (2017) with permission from Elsevier. (d) Anti-leishmanial activity of AmB formulations against established infection of *Leishmania donovani* in hamsters (unmodified chitosan nanoparticles: CNPs, modified chitosan nanoparticles: SCNPs). *Source:* Adapted from Tripathi et al. (2015) with permission from Elsevier.

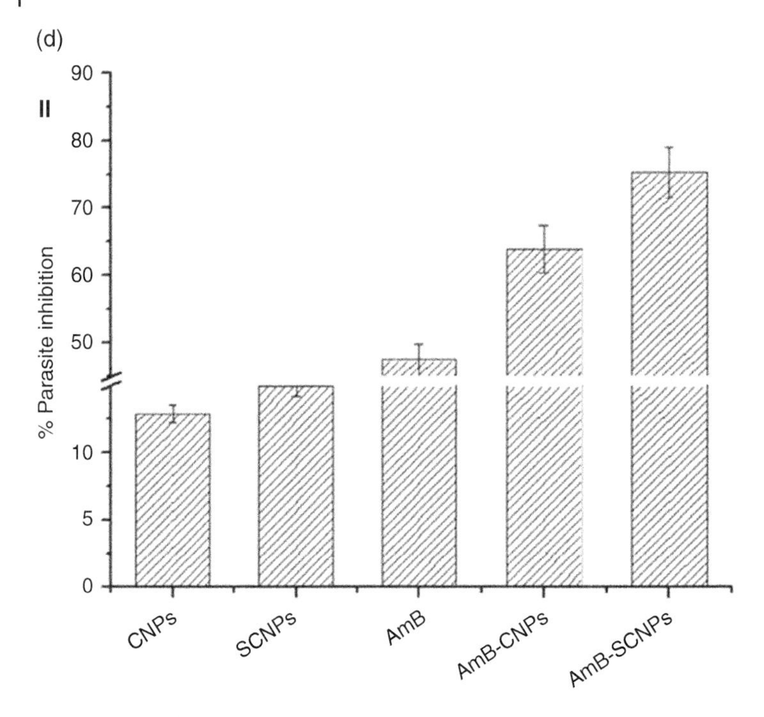

Figure 15.7 (Continued)

During recent years, protein-based nanoparticles have shown great potential as drug carriers. The great advantage in the application of proteins as nanoparticles are due to their biocompatibility, biodegradation, and low immunogenicity, besides their extraordinary binding capacity for various drugs, long-circulating properties in the body, and stability. Bovine serum albumin (BSA) has been widely used as a nanodrug delivery vehicle due to its tolerance by the living organism and low cost, as well as its simple preparation techniques (Casa et al. 2018).

Nahar et al. (2010) evaluated the potential of surface-functionalized gelatin nanoparticles (f-GNPs) for efficient macrophage-specific delivery of AmB in the treatment of VL. Further, the effect of a spacer for macrophage targeting was also evaluated. Gelatin was functionalized either through conjugation to mannose via direct coupling (mGelatin) or via PEG spacer (m-Gelatin). m-GNPs showed a 5.4-fold reduction in IC50 in comparison with the plain AmB suggesting significant enhancement of antileishmanial activity. Table 15.5 shows the nanodrug delivery devices proposed for the treatment of leishmanial infection using naturally occurring polymers or their semisynthetic counterparts.

15.7.1.4 Bioconjugates

One of the studies demonstrated the development of a macrophage-mediated drug targeting delivery system by conjugating the anti-leishmanial drug pentamidine (Pent) with the biocompatible polymer hyaluronic acid (HA). Later, the bioconjugate was employed as a delivery platform and targeting scaffold. Biological assays against *L. major* amastigote-infected macrophages and primary bone marrow-derived macrophages (BMDM) confirmed the validity of the strategy as the resulting bioconjugate increased both the potency and the selectivity index of the drug (Micale et al. 2015).

Similarly, Sokolsky-Papkov et al. (2006) prepared dextran-AmB conjugate. The antifungal and antileishmanial dextran-AmB imine conjugate, which contains unreacted aldehyde groups,

Table 15.5 Nanodrug delivery devices intended for treatment of leishmanial infection using naturally occurring polymers or their semisynthetic counterparts.

Polymer	Technique	Drug	Comment	References
Bovine serum albumin	Desolvation	AmB	Nanoparticles showed a significant decrease in the lesion thickness at the infected footpad.	Casa et al. (2018)
Bovine serum albumin	Precipitation	Meglumine antimoniate	The highest antileishmanial activity was observed for the carriers made with glutaraldehyde ($10\,\mu l\,ml^{-1}$).	Barazesh et al. (2018)
Bovine serum albumin (maleylated)		Methotrexate	Rapid and effective killing of intracellular parasites	Chaudhuri et al. (1989)
Bovine serum albumin (mannosylated)		Methotrexate	Reduced spleen parasite burden	Chakraborty et al. (1990)
Chitosan	Deposition	AmB	*In vitro* antileishmanial activity and *in vivo* pro-inflammatory mediator's expression were observed with AMB entrapped, mannose-grafted chitosan nanocapsules, led to significant reduction (~90%) in splenic parasite burden.	Asthana et al. (2015b)
Chitosan	Emulsification/ solvent evaporation	AmB		Asthana et al. (2013)
Chitosan	In situ gelation	AmB	The nanoparticles elicited 90% macrophage viability and 71-fold enhancement in drug uptake compared to the native drug.	Shahnaz et al. (2017)
Chitosan	Ionic gelation	AmB	Inhibition of splenic parasitic burden: $75.30 \pm 3.76\%$	Tripathi et al. (2015)
Chitosan	Phase separation	AmB	Drug loading efficiency: 90%; Cellular uptake: 98.6% Sustained drug-release	Zadeh-Mehrizi et al. (2018a)
Chitosan	Polyelectrolyte complexation (PEC) technique	AmB	Oral bioavailability of AmB-encapsulated N,N-dimethyl-N,N,N-trimethyl-6-O-glycol chitosan nanoparticles: 24%.	Serrano et al. (2015)
Chitosan		Curcumin	Mannose-conjugated chitosan nanoparticles CC50 for Cur-MCN: 26 ± 0.60	Chaubey et al. (2017)
Chitosan	High-pressure homogenization	Miltefosine-AmB		Tripathi et al. (2017)

(Continued)

Table 15.5 (Continued)

Polymer	Technique	Drug	Comment	References
Chitosan	Ionotropic gelation	Rifampicin	Ex vivo uptake of mannose-conjugated chitosan nanoparticles was 2.31 times higher compared to unconjugated chitosan nanoparticles.	Chaubey and Mishra (2014)
Chitosan	Continuous stirring	Ursolic acid	Orally effective ursolic acid-loaded nanoparticles suppressed the parasite burden to 98.75% *in vivo*.	Das et al. (2017)
Chitosan (mannosylated)	Ionic gelation technique	Curcumin	Significant improvement in the mean resident time (39.38 h) compared to free curcumin solution (0.30 h) according to the pharmacokinetic study of optimized nanoparticles	Chaubey et al. (2014)
Chitosan glycol stearate	Ionotropic complexation	AmB		Gupta et al. (2015b)
Chitosan–chondroitin sulfate	Polyelectrolyte complexation (PEC) technique	AmB	Hemoglobin-guided nanocarrier	Bose et al. (2016)
Chitosan–chondroitin sulfate	Polyelectrolyte complexes	AmB	Significant reductions in the lesion size and in the parasite burden	Ribeiro et al. (2014)
Chitosan-PLGA	Nanoprecipitation technique	Doxorubicin-AmB	Chitosan-anchored PLGA nanoparticles loaded with AmB and Dox	Singh et al. (2016)
Chitosan-Pluronic F127	Thin-film hydration	AmB	Excellent antileishmanial activity: *in vitro* (1.82-fold compared to AmB suspension) *in vivo* ($75.84 \pm 7.91\%$ parasitic inhibition).	Singh et al. (2017)
Gelatin	Electrospinning	AmB		Nanda and Mishra (2011)
Gelatin	Precipitation	AmB	A 5.4-fold reduction in IC50 compared to plain AmB	Nahar et al. (2010)
β- (1,3)-D-glucan	Solvent evaporation	4-[(2E)-N′-(2,2′-bithienyl-5-methylene) hydra-zinecarbonyl]-6,7-dihydro-1-phenyl-1Hpyrazolo[3,4-d] pyridazin-7-one	Significant decrease in J774A1 macrophage toxicity ($CC50 \geq 18.53\,\mu g\,ml^{-1}$)	Volpato et al. (2018)

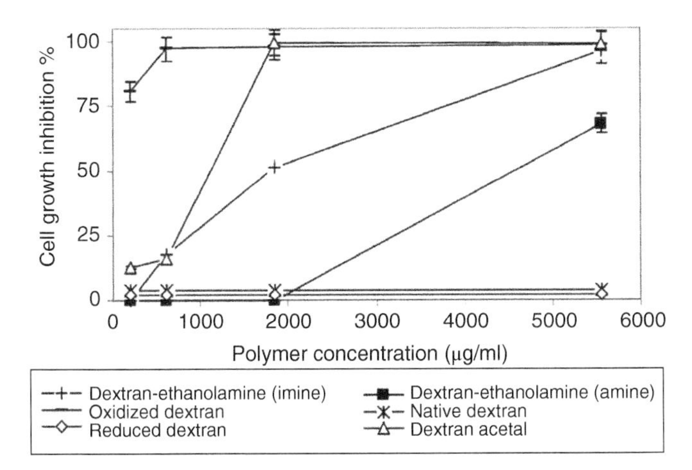

Figure 15.8 Cytotoxicity of dextran-AmB bioconjugates against *Leishmania*. *Source:* Adapted from Sokolsky-papkov et al. (2006) with permission from American Chemical Society.

was modified with ethanolamine and compared to dextran-AmB amine and imine conjugates. The dextran-AmB-ethanolamine conjugate was at least 15 times less hemolytic than the free AmB. The imine conjugates released free AmB while the amine conjugate did not. It was concluded that aldehyde groups may contribute to cell toxicity. The toxicity is reduced by converting the aldehyde groups into imine conjugates with ethanolamine (Figure 15.8). The results have direct implications for the safety of AmB-polysaccharide conjugates used against fungal and leishmanial infections (Sokolsky-papkov et al. 2006). Moreover, some selected reports on using bioconjugate drug delivery devices for the treatment of leishmanial infection are enlisted in Table 15.6.

15.7.2 Inclusion Compounds

Inclusion compounds made from host–guest complexation of antileishmanial agents with ligands (mostly cyclodextrins and their derivatives) are suitable candidates, considering the efficacy of this methodology, for enhancing the solubility of the active agents, enhancing their pharmacokinetics profile, and reducing the inherent toxicity of the active agents used. Physicochemical and *in vitro* biological properties of β-cyclodextrin-furazolidone (FZD) inclusion complexes were investigated by Carvalho et al. against *L. amazonensis*. Biological *in vitro* evaluations demonstrated that free FZD and the FZD:β-CD complexes presented significant leishmanicidal activity against *L. amazonensis* with IC50 values of 6.16 and 1.83 $\mu g\, ml^{-1}$ for the complexes prepared by kneading and lyophilization methods, respectively. The data showed that these complexes reduced the survival of promastigotes and presented no toxicity for the cells examined. The results indicated that the compounds could be a cost-effective alternative for use in the pharmacotherapy of leishmaniasis in dogs infected with *L. amazonensis* (Carvalho et al. 2018).

Demicheli et al. (2004) investigated the ability of β-cyclodextrin to enhance the oral absorption of antimony and to promote the oral efficacy of meglumine antimoniate against experimental cutaneous leishmaniasis. The effectiveness of the complex given orally was equivalent to that of meglumine antimoniate given intraperitoneally at a twofold-higher antimony dose. The antileishmanial efficacy of the complex was confirmed by the significantly lower parasite load in the lesions of the treated animals than in the saline-treated control group (Demicheli et al. 2004).

Table 15.6 Some selected reports on using bioconjugate drug delivery devices for treatment of leishmanial infection.

Polymer	Drug	Comment	References
Hyaluronic acid	Pentamidine	Effective in the biological assay against *Leishmania major* amastigote-infected macrophages ≈ 2-fold increase in the potency of the drug	Micale et al. (2015)
Dextran Imine	AmB	The bioconjugate was at least 15 times less hemolytic than free AmB.	Sokolsky-papkov et al. (2006)
Cell-Penetrating Tat (48-60)	Paromomycin	CPP-mediated delivery of PMM increased biodistribution into deeper layers of the Leishmania ulcer CPP acting as a skin vehicle for the drug	Defaus et al. (2017)
Pectin	AmB		Kothandaraman et al. (2017)
β-Cyclodextrin	Meglumine Antimoniate	Pharmacokinetic studies of in dogs showed a prolongation of the serum mean residence time of Sb from 4.1 to 6.8 h.	Ribeiro et al. (2010)
Gum arabic	Primaquine		Nishi and Jayakrishnan (2007)
Arabinogalactan	AmB		Ehrenfreund-kleinman et al. (2004)
Polypeptide	Methotrexate	≈ 60 times less hemolytic against sheep erythrocytes than the free drug	Kóczán et al. (2002)
Neoglycoprotein	Doxorubicin	≈ 12.5 times more elimination of the intracellular amastigotes of *Leishmania donovani* in peritoneal macrophages than the free drug	Sett et al. (1993)
Diphenylalanine nanotube	Folic acid		Castillo et al. (2014)
N-(2-hydroxypropyl) methacrylamide (HPMA)	AmB		Nicoletti et al. (2010)
N-(2-hydroxypropyl) methacrylamide (HPMA)	8-Aminoquinoline	The targeted conjugates were significantly more effective *in vivo* (67–80% inhibition) than non-targeted conjugate (47% inhibition).	Nan et al. (2004)

Frézard et al. (2008) reported enhanced oral delivery of antimony from meglumine antimoniate/β cyclodextrin nanoassemblies. This study demonstrated that the heated meglumine antimoniate (MA) and β cyclodextrin (β-CD) mixture still produced significantly lower serum Sb levels when compared to the MA/β-CD composition (Figure 15.9), indicating that the freeze-drying process is required for achieving a high absorption of Sb by oral route. Another important observation was

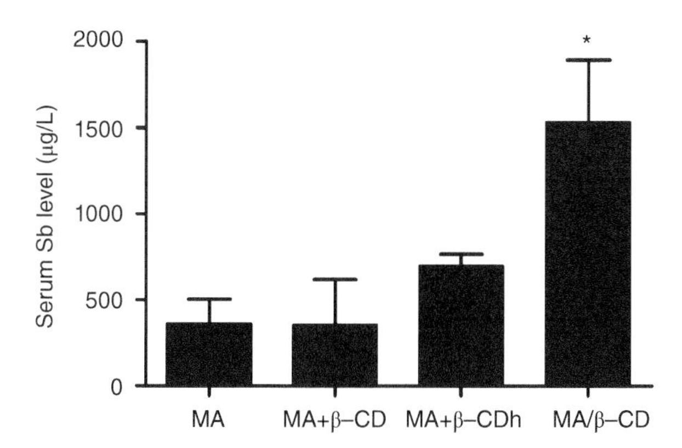

Figure 15.9 Sb concentration in the serum of Swiss mice 3h after oral administration of different compositions of MA at 100 mg Sb/kg. MA: meglumine antimoniate freshly prepared in water at 0.05 mol/L Sb and 25 °C; MA+ β-CD: equimolar mixture of MA and β cyclodextrin freshly prepared in water at 0.05 mol/L Sb and 25 °C; MA+ β -CDh: equimolar mixture of MA and β -CD in water at 0.05 mol/L Sb and 25 °C pre-incubated for 48 h at 55 °C; MA/ β -CD: freeze-dried MA+ β - CDh freshly reconstituted in water at 0.05 mol/L Sb and 25 °C. *Source:* Adopted from Frézard, F et al (2008) with permission from Elsevier.

the ability of the MA/β-CD composition to act as a sustained release system of the antimonial drug MA, suggesting that this property may result in the change of the drug absorption site in the gastrointestinal tract (Frézard et al. 2008). Table 15.7 shows the various inclusion compounds used for treatment of leishmanial infection.

15.7.3 Dendrimers

A novel strategy for targeted delivery of AmB to macrophages with muramyl dipeptide (MDP) conjugated multimeric poly(propyleneimine) (PPI) dendrimers have been developed. Highly significant reduction in toxicity was observed in terms of the hemolytic toxicity and cytotoxicity in erythrocytes and J774A.1 macrophage cells, respectively. MDP-PIA formulation showed appreciable macrophage-targeting potential in cell uptake studies (Figure 15.10a). The developed formulation showed higher or equivalent anti-parasitic activity against parasite-infected macrophage cell lines and *in vivo* infection in BALB/c mice (Figure 15.10b). These results suggest the developed MDP-conjugated dendrimeric formulation of AmB as a promising immunostimulant targeted DDS and a safer alternative to marketed formulations (Jain et al. 2015). Table 15.8 shows the drug delivery devices developed for antileishmanial agents based on dendrimers.

15.7.4 Liposomes

Liposomes are self-assembling, closed colloidal structures composed of lipid bilayers having a spherical morphology in which an outer lipid bilayer surrounds a central aqueous phase. The bilayer structure of liposomes have provided them a great potential to carry hydrophilic therapeutic agents in the core aqueous phase, as well as hydrophobic ones in the lipophilic shell phase. Targeting of liposomal delivery systems can not only be achieved using their specific characteristics like particle size or surface charge but also by conjugating specific ligands to the liposome surface for active targeting. In this way, these bilayered vesicles act as efficient carriers to deliver drugs, vaccines, diagnostic agents, and other bioactive agents to their specific site of action, leading to rapid advancement in the field of liposomal drug delivery to target tissues.

Table 15.7 Some selected reports on using inclusion compounds for treatment of leishmanial infection.

	Technique	Drug	Comment	References
β-cyclodextrin	Kneading and lyophilization	Furazolidone	The data showed that these complexes reduced the survival of promastigotes and presented no toxicity for tested cell.	Carvalho et al. (2018)
2-hydroxypropyl-α-cyclodextrin (HPαCD) Methyl-β-cyclodextrin (MβCD) 2-hydroxypropyl-β-cyclodextrin (HPβCD)	Coevaporation	5,6-dichloro-2-(trifluoromethyl)-1H-benzimidazole (G2)		Rojas-Aguirre et al. (2012)
Cyclodextrin/poly(anhydride)	Solvent displacement	Atovaquone	These nanoparticle formulations showed more adequate physicochemical properties in terms of size (<260 nm), drug loading (17.8 and 16.9 $\mu g\ mg^{-1}$, respectively), and yield (>75%).	Calvo et al. (2011)
β-cyclodextrin		Meglumine antimoniate		Frézard et al. (2008)
β-cyclodextrin		Meglumine antimoniate	The effectiveness of the complex given orally was equivalent to that of meglumine antimoniate given intraperitoneally at a twofold-higher antimony dose.	Demicheli et al. (2004)

Alving was a pioneer to demonstrate the capabilities of liposomes in the treatment of experimental visceral leishmaniasis in a hamster animal model (Alving 1983). Accumulation of drug-loaded liposomes in the liver and spleen was also observed and could be visualized within infected Kupffer cells (Heath et al. 1984). To this end, liposomes have been proposed as effective carriers to target drugs at macrophages infected in VL (Alving 1986).

AmB, a very potent antifungal and antileishmanial agent, suffers from significant renal toxicity. Liposomal DDS for AmB increases the efficacy of the drug while decreasing its toxicity at the same time (Alving 1986). Liposomal AmB is very effective not only against the systemic fungal diseases but also against the leishmanial infections. Now, it is established that AmB is significantly more active in its liposomal form compared to its non-liposomal form, perhaps due to the reduced drug toxicity rather than the altered drug distribution at the site of infection (Ahmad et al. 1991). A case of VL, unresponsive to several courses of treatment with standard drugs, was successfully cured by a 21-day course (50 $mg\ d^{-1}$) of AmBisome (Gilead Sciences, Inc.) liposomal AmB formulation. The efficacy of liposomal AmB formulation is further supported by experimental studies on BALB/c mice animal model for *L. donovani* infection where ED50 values for AmBisome was determined *eq.* to 0.15–0.25 $mg\ kg^{-1}$ compared to 0.95–4.9 $mg\ kg^{-1}$ for conventional AmB parenteral formulation. A lack of toxicity for AmBisome formulation was noted in both studies (Croft et al. 1991).

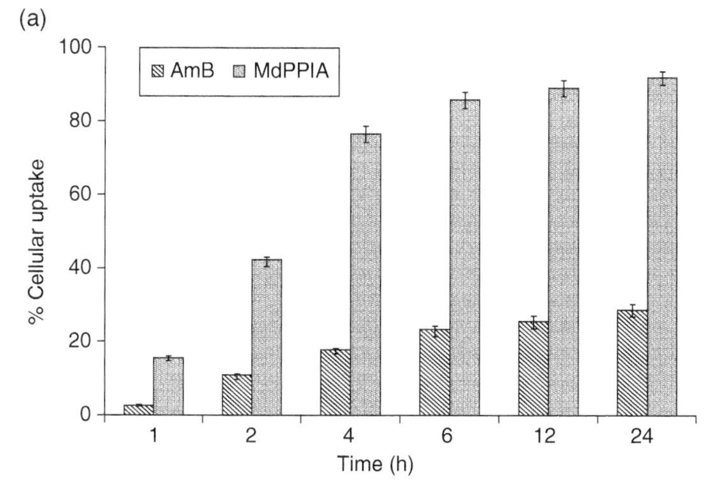

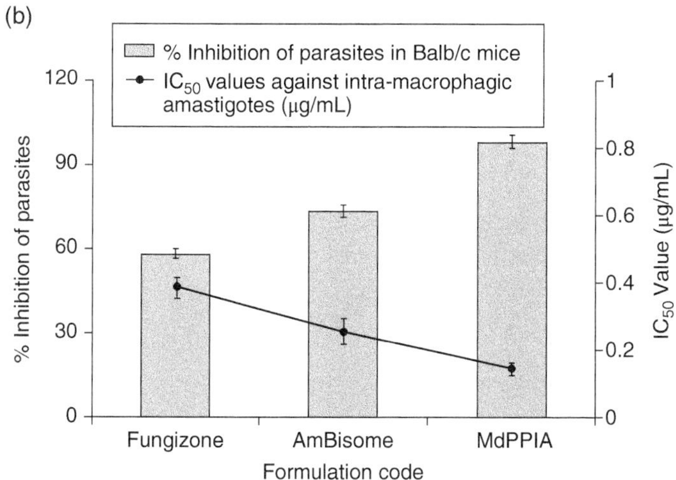

Figure 15.10 (a) Cellular uptake of AmB and muramyl dipeptide (MdP) conjugated multimeric poly(propyleneimine) (PPI) dendrimers (MdPPIA) by J774A.1 cells (b) along with their antileishmanial activity in BALB/c mice animal model of leishmaniasis. *Source:* Both figures are adapted from Jain et al. (2015) with permission from Elsevier. Values represent mean ± S.D. (n = 6).

Table 15.8 Drug delivery devices developed for antileishmanial agents based on dendrimers.

Dendrimer	Technique	Drug	Comment	References
Polyethylene glycol		AmB	The drug solubility rate was increased by 478-fold.	Zadeh-Mehrizi et al. (2018b)
Maleimido-poly(ethylene glycol)		Doxorubicin	PG-DOX(pH)-PEG is actively internalized through the acidic endocytic pathway and co-localized surrounding the amastigotes.	Gutierrez-Corbo et al. (2017)
Poly(propylene imine)		AmB		Jain et al. (2015)
PAMAM	Coprecipitation	AmB		Kumar et al. (2014a)

Agrawal et al. (2002) reported on the higher rate and amount of tuftsin-bearing liposomal AmB uptake from the hamsters' circulation compared to the conventional liposomal AmB formulation. Tuftsin (Thr–Lys–Pro–Arg), as well as grafting on of Liposome-AmB, significantly alters its distribution in the macrophage-rich organs of infected animals. It may thus be suggested that the higher efficacy of Tuft-Lip-AmB observed here could be due to both the increased drug tolerance and enhanced uptake of the liposomized drug by the macrophage-rich organs of the infected mice. Tuftsin-bearing liposomes also increase the drug accessibility to areas like bone marrow which are otherwise inaccessible to free AmB, and are the main cause of relapse of the leishmanial infection. Moreover, as observed *in vitro*, the difference between the anti-leishmanial activities of Lip-AmB and Tuft-Lip-AmB was much more prominent at lower AmB dose ($0.5\,mg\,kg^{-1}$), compared to a higher dose ($1.0\,mg\,kg^{-1}$) (Agrawal et al. 2002).

In another study, Rathore et al. (2011) demonstrated that mannosylated liposomes showed a maximum reduction in parasite load (i.e. $78.8 \pm 3.9\%$). The biodistribution study clearly showed the higher uptake of mannosylated liposomes in the liver and spleen and hence the active targeting to the RES, which in turn would provide a direct attack of the drug to the site where the pathogen resides, rendering the other organs free and safe from the toxic manifestations of the drug. The hepatosplenic quantitative uptake of cationic and mannosylated liposomes was analyzed from the biodistribution study.

AmB, the most effective drug against leishmaniasis, shows serious toxicity. As *Leishmania* species are obligate intracellular parasites of antigen-presenting cells (APC), an immunopotentiating APC-specific AmB nanocarrier would be ideally suited to reduce the drug dosage and regimen requirements in leishmaniasis treatment. Researchers developed a nanocarrier that can be used for effective treatment of CL in a mouse animal model, while also enhancing *L. major*-specific T-cell immune responses in the infected host. They used a Pan-DR-binding epitope (PADRE)-derivatized-dendrimer (PDD), complexed with liposomal amphotericin B (LAmB) in an *L. major* mouse animal model, and analyzed the therapeutic efficacy of low-dose PDD/LAmB vs. full dose LAmB. PDD was shown to escort LAmB to APCs *in vivo*, enhanced the drug efficacy by 83% and drug APC targeting by 10-fold, and significantly reduced parasite burden and toxicity. Fortuitously, the PDD immunopotentiating effect significantly enhanced parasite-specific T-cell responses in immunocompetent infected mice. PDD reduced the effective dose and toxicity of LAmB and resulted in elicitation of strong parasite-specific T-cell responses. A reduced effective therapeutic dose was achieved by selective LAmB delivery to APC, bypassing bystander cells, reducing toxicity, and inducing anti-parasite immunity (Daftarian et al. 2013).

Perez et al. (2016) prepared ultradeformable liposomes containing amphotericin B (AmB-UDL) for improving the topical delivery of AmB in CL. AmB-UDL showed 100% and 75% anti-promastigote and anti-amastigote activity, respectively, on *Leishmania braziliensis* (Figure 15.11a). The total accumulation of AmB in the skin was 40 times higher when applied as AmB-UDL than as AmBisome.

Surface modification of liposomes with targeting ligands is known to improve the efficacy with reduced untoward effects in treating infective diseases like VL. Patere et al. (2016) designed modified ligands by modifying polysaccharide with a long chain lipid that was incorporated in liposomes to target AmB to RES and macrophages. AmB was encapsulated in surface-modified liposomes (SML). SML with 3% w/w of ML retained the vesicular nature with a particle size of $\sim$205 nm, encapsulation efficiency of $\sim$95%, and good stability. SML showed increased cellular uptake in RAW 264.7 cells which could be attributed to receptor-mediated endocytosis. Pharmacokinetics and biodistribution studies revealed high $t_{1/2}$, area under the curve $(AUC)_{0-24}$, reduced clearance, and prolonged retention in liver and spleen with AmB SML compared to other formulations. In promastigote and amastigote models, AmB SML showed enhanced performance with a low 50% inhibitory

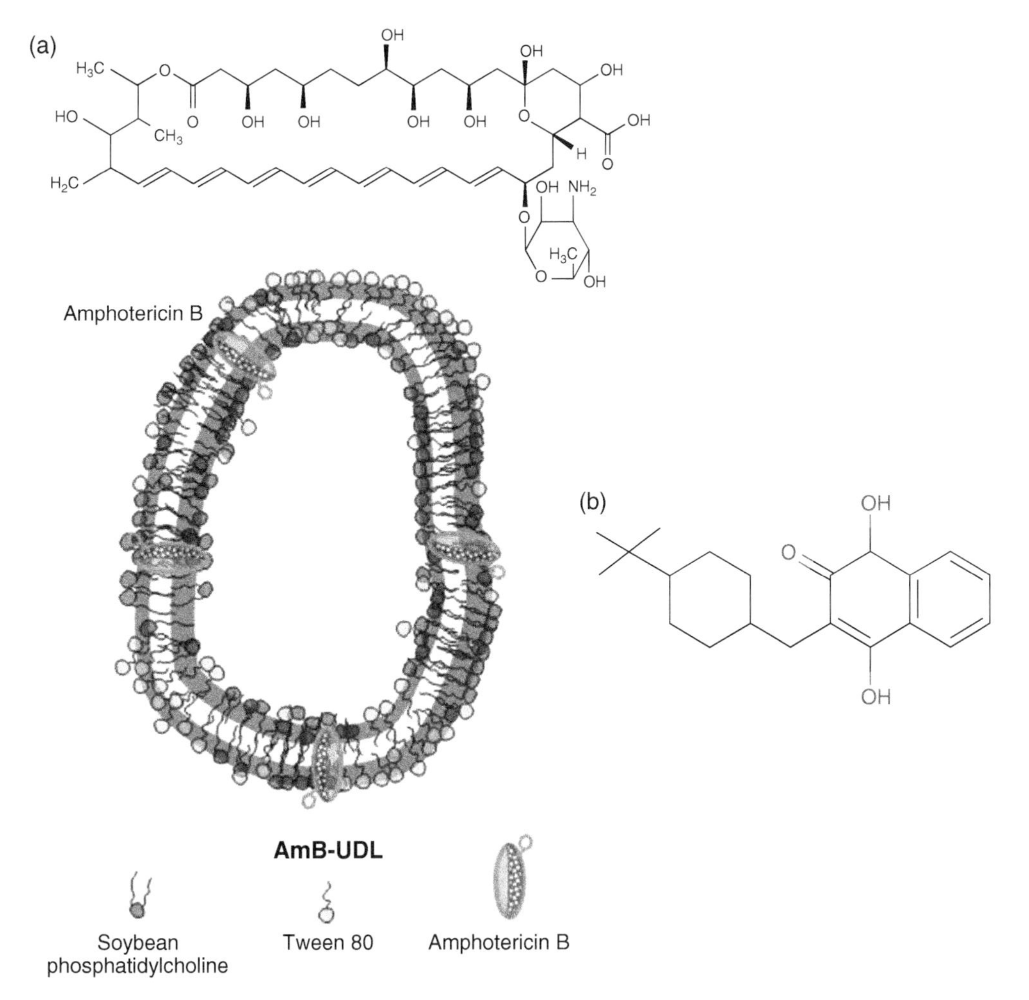

Figure 15.11 (a) Schematic representation of ultradeformable liposomes (UDL) contacting AmB. *Source:* Aadapted from Perez et al. (2016) with permission from Elsevier. (b) Chemical structure of Buparvaquone.

concentration (IC50) compared to AmB solution and AmB CL. Thus, due to the targeting ability of modified ligands, SML has the potential to achieve enhanced efficacy in treating visceral leishmaniasis (Patere et al. 2016).

Buparvaquone (BPQ) (Figure 15.11b), a veterinary drug used to treat theileriosis, has shown promising activities against protozoan parasites, including *Leishmania* spp., *Plasmodium falciparum*, and *Cryptosporidium parvum*. The poor distribution and low bioavailability of BPQ contributed to the limited *in vivo* efficacy in animal models of VL and CL.

Similarly, Reimão et al. (2012) developed a novel liposomal formulation, containing phosphatidylserine (PS), of BPQ to evaluate its *in vivo* effectiveness in *Leishmania infantum chagasi*-infected hamsters. The activity of BPQ was evaluated against both the promastigote forms of different *Leishmania* species and the intracellular amastigotes of *L. infantum chagasi*. BPQ was significantly ($P < 0.05$) selective against *L. infantum chagasi* intracellular amastigotes, with an IC50 value of $1.5\,\mu M$; no *in vitro* mammalian cytotoxicity could be detected. These results showed the potential of PS-liposome formulations for the successful targeted delivery of BPQ in visceral leishmaniasis.

Phosphatidylserine-liposomes (PS-liposomes) have been an efficient tool to deliver drugs to Leishmania-infected macrophages. The association of the negatively charged phosphatidylserine into liposomes was demonstrated to improve the selective uptake by macrophages via scavenger receptors, resulting in higher *in vitro* and *in vivo* efficacies against Leishmania parasites. Costa-Silva et al. (2017) entrapped BPQ in a liposomal formulation containing 20% (molar ratio) of the negatively charged phosphatidylserine and a safe and effective regimen of administration was investigated using the hamster animal model. This could be a result of a differential bioavailability of the liposomal BPQ when administered via subcutaneous route. Herein, they reported that BPQ has an immunomodulatory effect in the host cells and when entrapped into PS-liposomes, the drug was highly effective at low doses to eliminate the *L. infantum* parasites in the hamster animal model. The phosphatidylserine-liposomes used in this study were demonstrated to be a useful tool to deliver a BPQ to *Leishmania*-infected macrophages in the liver and spleen, as well as the bone marrow, an important site for parasite persistence (Costa-Silva et al. 2017).

Pentavalent antimonial agents, such as SSG or meglumine antimoniate (MA), have remained the mainstay of treatment for all forms of leishmaniasis for more than 70 years. Roychoudhury et al. investigated the efficacy of SSG in phosphatidylcholine (PC)–stearylamine-bearing liposomes (PC-SA-SSG), PC-cholesterol liposomes (PC-Chol-SSG) and free AmB against SSG-resistant *L. donovani* strains in 8-week infected BALB/c mice. Unlike free and PC-Chol-SSG, PC-SA-SSG was effective in curing mice infected with two differentially originated SSG-unresponsive parasite strains at significantly higher levels than AmB. Successful therapy correlated with complete suppression of disease-promoting IL-10 and TGF-b, upregulation of Th1 cytokines, and expression of macrophage microbicidal NO (Figure 15.12a). The design of this single-dose combination therapy with PC-SA-SSG for visceral leishmaniasis, having reduced toxicity and long-term efficacy irrespective of SSG sensitivity, may prove promising, not only to overcome SSG resistance in *Leishmania*, but also for drugs with similar resistance-related problems in other diseases (Roychoudhury et al. 2011).

Moosavian Kalat et al. (2014) developed a topical liposomal meglumine antimoniate (MA, Glucantime) (Lip-MA) formulation and evaluated the therapeutic effects of the preparation on lesions induced by *L. major* in BALB/c mice. The *in vitro* permeation data showed that almost 1.5% of formulations applied in the mouse skin were penetrated and the amount retained in the skin was about 65%. The results showed a significantly ($P < 0.001$) smaller lesion size in the treated groups of mice compared to the control groups which received either drug-free liposomes or phosphate-buffered saline (PBS) (Figure 15.12b). The spleen parasite burden was significantly ($P < 0.001$) lower in the treated groups compared to the control groups receiving either drug-free liposomes or PBS at the end of the treatment period.

Reis et al. (2017) evaluated the immune pathological changes using a mixture of conventional and PEGylated liposomes with meglumine antimoniate (MA). Histopathological analyses demonstrated that animals treated with liposomal formulation of MA (Lipo MA) showed a significant decrease in the inflammatory process and the absence of granulomas. The results confirmed that Lipo MA is a promising antileishmanial formulation able to reduce the inflammatory response, accompanied by a significant reduction of the parasite burden into hepatic and splenic compartments in treated animals (Reis et al. 2017). Moreover, in one of the recent studies, Dar et al. (2018) developed nano-deformable liposomes (NDLs) for the dermal delivery of SSG against CL. The ex vivo skin permeation study revealed that SSG-NDLs gel provided 10-fold higher skin retention toward the deeper skin layers, attained without use of classical permeation enhancers (Figure 15.12c). Moreover, *in vivo* skin irritation and histopathological studies verified the safety of the topically applied formulation. The anti-leishmanial activity on intramacrophage amastigote models of *Leishmania tropica* showed that

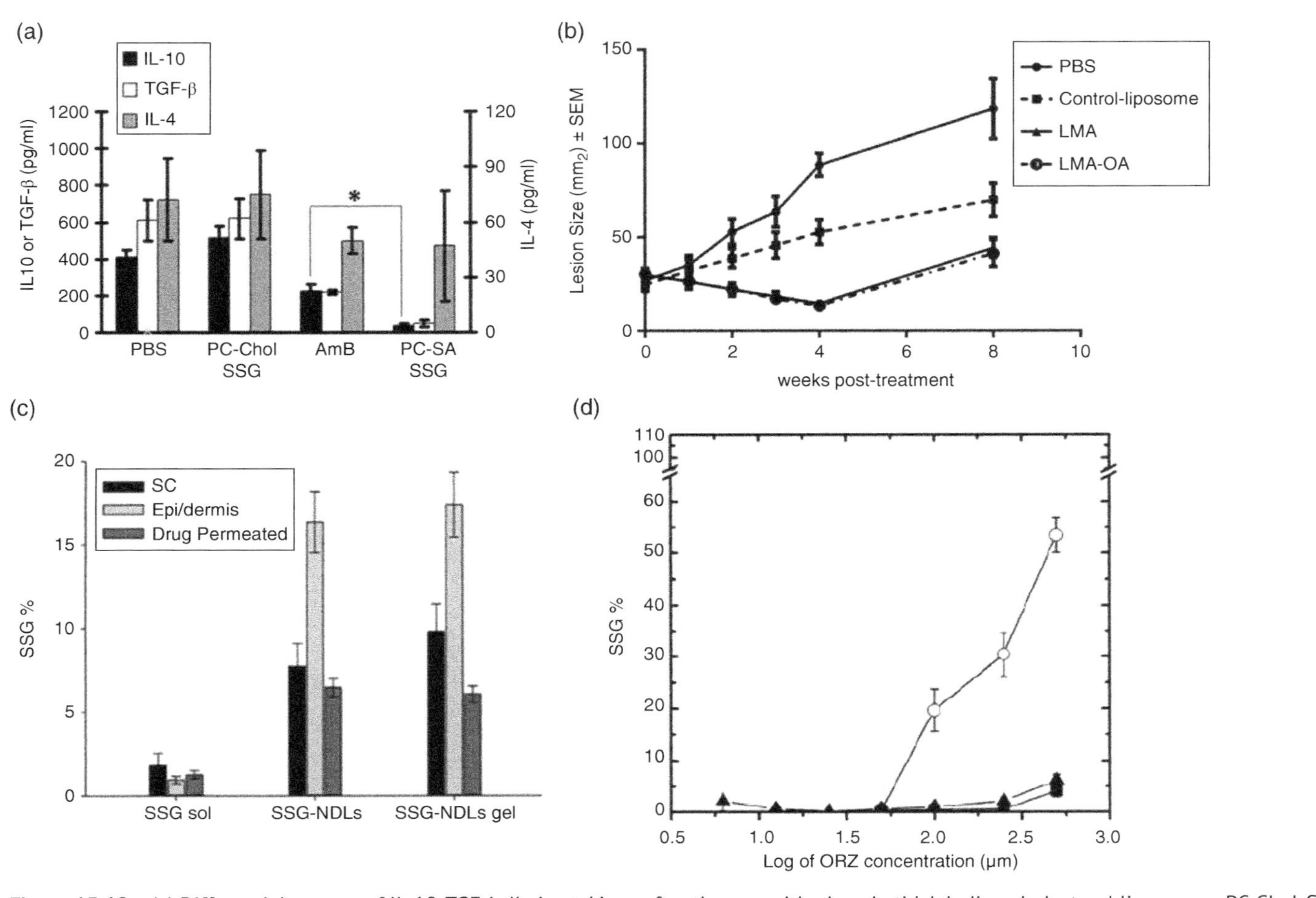

Figure 15.12 (a) Differential pattern of IL-10, TGF-b, IL-4 cytokines after therpay with phosphatidylcholine cholesterol liposomes: PC-Chol-SSG, phosphatidylcholine cholesterol stearylamine-bearing liposomes: PC-SA-SSG. *Source:* Adapted from Roychoudhury et al. (2011); an open access article. (b) Lesion size in the different groups of infected subcutaneously female six to eight weeks BALB/c mice treated using liposomal meglumine antimoniate (LMA) and liposomal meglumine antimoniate with oleic acid (LMA-OA). *Source:* Adapted from Moosavian Kalat et al. (2014) with permission from Elsevier. (c) Ex vivo permeation and retention studies of SSG in different formulations including in PBS of pH 7.4 and nano-deformable liposomes (NDLs). Stratum corneum: SC and epidermis/dermis layers of the skin: Epi/dermis. *Source:* Adapted from Dar et al. (2018); an open access article. (d) Haemolytic activity induced by liposomal and free Oryzalin (ORZ). RBCs were exposed to different concentrations of freeze-dried reconstituted liposomal formulations F1: (DMPC:DMPG 7:3 (■), F2: DMPC:DMPG (9:1) (▲) and of free ORZ (O)). *Source:* Adopted from Lopes et al., (2012a) with permission from Elsevier.

IC50 value of the SSG-NDLs was ≈fourfold lower than the plain drug solution, with marked increase in the selectivity index. The *in vivo* results displayed higher anti-leishmanial activity by efficiently healing lesions and successfully reducing parasite burden (Dar et al. 2018).

Borborema et al. (2018) developed a formulation of meglumine antimoniate encapsulated into phosphatidylserine liposomes (MA-LP) and analyzed the *in vitro* antileishmanial activity, physicochemical properties, and pharmacokinetic profile in a mouse animal model. MA-LP was 23-fold more effective against *L. infantum*-infected macrophages *in vitro* than the free drug, with a selectivity index higher than 220. The pharmacokinetic studies demonstrated that the liposomes increased the uptake of the drug by the liver and spleen and promoted sustained levels.

Miltefosine (Hexadecylphosphocholine) is currently on trial as a first-choice, orally active drug for the treatment of VL when resistance to organic pentavalent antimonials becomes epidemic. Another study presents new HePC liposomal formulations composed of HePC/egg yolk phosphatidylcholine (EPC)/stearylamine (SA) 10 : 10 : 0.1, 10 : 10 : 0.5, and 10 : 10 : 1 (molar ratio). *In vivo* results showed that liposomal HePC seemed to be less toxic than the free drug despite the absence of significant antitrypanosomal activity. Moreover, SA-bearing liposomes were found to be able to provoke an *in vitro* lysis of the bloodstream forms of *Trypanosoma brucei gambiense*, a human pathogen. These liposomes were proven to be accumulated at the cell surface damaging the plasma membrane (Papagiannaros et al. 2005).

Najafian et al. (2016) designed nanoliposomal miltefosine *in vitro* conditions against the Iranian strain of *L. major* (MRHO/IR/75/ER) for the first time. The results indicate that nano-liposomes are suitable carriers for the loading and transportation of HePC drug to eliminate *L. major* intracellular parasites. These features, along with the use of SA in lipid formulations, strengthen the antiparasitic effect of HePC against intracellular parasites, and reducing the size of liposomes and the production of nano-liposomes loaded with HePC has a key role in better penetration of HePC into the site of parasite reproduction. The topical application of nano-liposomal HePC can make faster treatment of CL and reduce the duration of therapy for the cutaneous lesions, thereby reducing the risk of disease relapse (Najafian et al. 2016).

Dinitroanilines like Trifluralin (TFL) and Oryzalin (ORZ) are widely used in herbicide formulations. Their herbicidal effect is due to an antimitotic activity that in turn is determined by the binding of the dinitroanilines to tubulins, the main structural component of microtubules.

Lopes et al. (2012a) optimized liposomal formulations containing ORZ by selection of the appropriated preparation method, lipid composition, and experimental conditions. These formulations are stable in different storage conditions and have demonstrated superior biopharmaceutical properties over free ORZ. Incorporation of ORZ in liposomes proved to be important in reducing the hemolysis of red blood cells and the cytotoxic activity in THP-1 cells observed for free ORZ, and showing that they are active against intracellular leishmanial infection (Figure 15.12d). *In vivo* studies demonstrated the efficacy of the liposomal formulation to passively target ORZ to the main organ of leishmanial infections (liver and spleen). These results seem to indicate that ORZ liposomes could be candidates as therapeutic agents against leishmaniasis.

In another study, Lopes et al. (2014) demonstrated the *in vitro* and *in vivo* therapeutic activity of these oryzalin nanoformulations, and established a systematic comparison of both systems. After oryzalin incorporation, suitable physicochemical properties for parenteral administration were obtained. Nano-formulations revealed reduced cytotoxicity and hemolytic activity when compared with free-oryzalin, while retaining the *in vitro* intracellular activity. Results demonstrate the superiority of both oryzalin nanoformulations on the reduction of parasitic burden in the liver and spleen compared to the control group (84–91%) and similar to Glucantime. A strong reduction in ED50 values (3–65 fold) as compared to free-oryzalin was also obtained, depending on the organ and nanoformulation used (Lopes et al. 2014).

Paromomycin (PA), an aminoglycoside antibiotic, is a highly polar molecule and relatively lipid insoluble. Its chemical structure is similar to that of neomycin. Paromomycin has a broad spectrum of activity, including activity against protozoa, such as *Leishmania* spp., bacteria, and cestodes. Ferreira et al. (2004) prepared liposomal formulations containing PA to investigate their potential as topical delivery systems of this antileishmanial. Improved skin permeation and retention across intact skin were obtained with encapsulation of PA in LUV liposomes in comparison with solution and empty LUV + PA. Controlled topical delivery across stripped skin was observed for PA entrapped in LUV liposomes (Ferreira et al. 2004).

Carneiro et al. (2010) evaluated the potential of liposomes loaded with paromomycin (PA), an aminoglycoside antibiotic associated with poor skin penetration, for the topical treatment of CL. PA-loaded liposomes enhanced *in vitro* drug permeation across stripped skin and improved the *in vivo* antileishmanial activity in experimentally infected mice. The results showed that PA-loaded liposomes provide enhanced topical delivery in pigskin models, a relevant model for human skin. Improved efficacy of liposomal PA, in experimentally infected mice, an animal model widely used for the evaluation of antileishmanial activity of drug formulations, is also reported. These findings suggest that liposomes represent a promising alternative for the topical treatment of cutaneous leishmaniasis with PA (Carneiro et al. 2010). Table 15.9 enlisted the selected liposomal DDS used for the treatment of leishmaniasis.

15.7.5 Other Vesicular Nanoparticles

Niosomes, bilayer vesicles obtained by mixing a non-ionic surfactant and cholesterol, are widely used as vehicles for drug delivery. They cannot only be prepared using a large number of surfactants available in low cost but also benefit from a number of advantages in many features, e.g. higher stability and reduction of drug degradation. High purity, excellent content uniformity, and increased drug bioavailability have attracted much attention toward niosomes as candidates for drug delivery to various organs of human body. Both lipophilic or hydrophilic drugs can be entrapped in the lipophilic membrane or aqueous cores of niosomes, respectively.

Nazari-Vanani et al. (2018) reported on miltefosine- and ketoconazole-loaded nano-niosomes. The nano-niosomes exhibited improved antileishmanial efficacy against CL, probably due to the enhanced and controlled dissolution of the drugs which in turn all affect drug release from the formulations. After incubating of the niosomes with *L. major* promastigote and amastigote forms for 48 hours, parasite growth was inhibited *in vitro*.

Niosomal form of Harmine, a β-carboline alkaloid isolated from *Peganum harmala* was prepared by Lala et al. (2004) and tested for its antileishmanial properties both *in vitro* and *in vivo*. Cell death could be attributed to necrosis due to the nonspecific membrane damage exerted by the DDS according to the confocal microscopy findings.

Gupta and Vyas (2007) developed AmB-bearing emulsomes for passive and active macrophage targeting. AmB was formulated in trilaurin-based, nanosized lipid particles (emulsomes) stabilized by soy phosphatidylcholine (PC) as a new intravenous DDS for macrophage targeting. Emulsomes were modified by coating them with a macrophage-specific ligand (O-palmitoyl mannan, OPM). OPM-coated emulsomes could fuse with the macrophages of liver and spleen due to ligand-receptor interactions and could target the contained bioactive agents inside them. The proposed plain and OPM-coated emulsome-based systems showed excellent potential for passive and active intra-macrophage targeting, respectively, and the approach could be a successful alternative to the currently available drug regimens recommended for visceral leishmaniasis and systemic fungal infections. Table 15.10 shows the details of various vesicular nanoparticles used for the delivery of antileishmanial bioactive agents.

Table 15.9 Selected liposomal drug delivery systems appearing in the literature for treatment of leishmaniasis.

Morphology	Lipid	Technique	Drug(s)	Comment	References
ML	Unsaturated soy phosphatidylcholine Phospholipon 90G Saturated soy phosphatidylcholine Phospholipon 90H	Thin-film hydration	AmB	Enhanced performance with low 50% inhibitory concentration (IC50); particle size of ~205 nm; E.E. of ~95% and good stability.	Patere et al. (2016)
UD	Soybean Phosphatidylcholine Sodium cholate	Thin-film hydration	AmB	100% and 75% anti-promastigote and anti-amastigote activity on *Leishmania braziliensis*	Perez et al. (2016)
ML	Soy phosphatidylcholine Cholesterol	Cast film	AmB	Maximum reduction in parasite load for mannose-coupled liposomes (i.e. 78.8 ± 3.9%) Higher uptake of mannosylated liposomes in the liver and spleen	Rathore et al. (2011)
ML	Hydrogenated soy phosphatidylcholine 1,2-Dimyristoyl-sn-glycero-3-phosphocholine 1,2-Dimyristoyl-sn-glycero-3-[phospho-rac-(1-glycerol)] 1,2-Dipalmitoyl-sn-glycero-3-phosphocholine 1,2-Dipalmitoyl-sn-glycero-3-[phospho-rac-(1-glycerol)] 1,2-Distearoyl-sn-glycero-3-phosphocholine 1,2-distearoyl-sn-glycero-3-[phospho-rac-(1-glycerol)] Cholesterol	Thin-film hydration	AmB	Formulation with superior *in vitro* antifungal activity Maximum tolerated intravenous doses (MTD) of less than 10 mg kg^{-1}	Iman et al. (2011)
N/M	Dimyristoyl phosphatidylcholine Soy phosphatidylcholine	Ethanol injection	AmB	Liposome with size (~100 nm), zeta (−43.3 ± 2.8 mV) and percent entrapment efficiency (>95%).	Singodia et al. (2012a)

N/M	Egg phosphatidylcholine Cholesterol	Solvent evaporation	AmB	The increased efficacy, encapsulation of AmB in the tuftsin-bearing liposomes Tuftsin-bearing liposomes as drug vehicles in treatment of the macrophage-based infections	Agrawal et al. (2002)
N/M	Cholesterol Soy phosphatidylcholine Egg phosphatidylcholine	Methanol injection	AmB	Significant reduction in parasite load in the spleens of infected animals	Ahmad et al. (1991)
N/M	Dimyristoyl phosphatidylcholine cholesterol	Ethanol injection	AmB	Enhanced internalization due to binding of 4-sulfated-N–acetylgalactosamine to extracellular domain	Singodia et al. (2012b)
N/M	Soy phosphatidylcholine	Lipid-film hydration	Miltefosine	Cell internalization of reactive oxygen species production, and toxicity/genotoxicity and transdermal delivery of formulation was studied	Escobar et al. (2018)
ML	1,2- dioleoyl-sn-glycero-3-phosphocholine Cholesterol	Lipid-film hydration Extrusion	Miltefosine	Nanoliposomes with 56.7 nm size and+15.5 mV ζ-potential High encapsulation efficiency (95.3%)	Najafian et al. (2016)
Small UL	L-α-dipalmitoyl phosphatidylcholine Cholesterol	Ethanol injection	Miltefosine		Ribeiro et al. (2016)
UL	Hexadecylphosphocholine Egg phosphatidylcholine	Reverse phase evaporation	Miltefosine		Papagiannaros et al. (2005)
ML	Hydrogenated phosphatidylcholine Phosphatidylserine	Film hydration	Meglumine antimoniate	Increased drug uptake by liver and spleen and promoted sustained levels of drug was observed.	Borborema et al. (2018)
N/M	Soy phosphatidylcholine	Thin-film hydration	Sodium stibogluconate	Formulation with 10-fold higher skin retention toward the deeper skin layers	Dar et al. (2018)

(Continued)

Table 15.9 (Continued)

Morphology	Lipid	Technique	Drug(s)	Comment	References
ML	Distearoylphosphatidylcholine Cholesterol Dicetylphosphate	Freeze-drying	Meglumine antimoniate	Antileishmanial formulation able to reduce the inflammatory response and induce a type 1 immune response	Reis et al. (2017)
N/M	Soy phosphatidylcholine Cholesterol	Fusion	Meglumine antimoniate	The spleen parasite burden was significantly lower in the treated groups compared to the control groups	Moosavian Kalat et al. (2014)
N/M	Phosphatidylcholine Cholesterol	Film hydration	Sodium stibogluconate	Reduced toxicity and long-term efficacy after single-dose combination therapy with phosphatidylcholine stearylamine-bearing liposomes for VL	Roychoudhury et al. (2011)
ML	Soy phosphatidylcholine Cholesterol	Film hydration	Meglumine antimoniate	Meglumine antimoniate containing liposomes were ≥10-fold more effective than the free drug	Borborema et al. (2011)
Small UL	Dicetylphosphate Cholesterol	Freeze-drying	Meglumine antimoniate	The targeting of antimony to the bone marrow was improved (approximately threefold) with the liposomal formulation	Schettini et al. (2006)
N/M	L-α-phosphatidylcholine Dipalmitoyl phosphatidylcholine Cholesterol Dicetyl phosphate	Hydrating	Sodium stibogluconate		Hunter et al. (1988)
N/M	Egg phosphatidylcholine Saturated egg phosphatidylserine Cholesterol	Hydrating	Buparvaquone	Safe and effective treatment of murine leishmaniasis by the formulation was achieved	da Costa-Silva et al. (2017)
N/M	Phosphatidylserine	Methanol injection	Buparvaquone	Liposome caused a significant reduction in the parasite burden at a 60-fold lower dose than did the free buparvaquone	Reimão et al. (2012)

N/M	Dimyristoylphosphatidylcholine Dimyristoylphosphoglycerol	Dehydration–rehydration	Oryzalin	A strong reduction in ED50 values (3 to 65-fold) as compared to free-oryzalin was observed.	Lopes et al. (2014)
N/M	Dipalmitoylphosphatidylcholine Dipalmitoylphosphatidylglycerol Dimyristoylphosphatidylcholine Dimyristoylphosphoglycerol	Dehydration-rehydration	Oryzalin	An increased Oryzalin (ORZ) delivery was observed in the main organs of leishmanial infection with a 9–13-fold higher accumulation compared to the free ORZ.	Lopes et al. (2012a)
Large UL	Soy phosphatidylcholine Cholesterol	Reverse-phase evaporation	Paromomycin		Carneiro et al. (2010)
Large UL	Soy phosphatidylcholine Cholesterol	Solvent evaporation	Paromomycin	The skin permeation across stripped skin was 1.55 ± 0.31%, 1.29 ± 0.40%, 0.20 ± 0.08%, and 0.50 ± 0.19% for PC LUV, PC:CH LUV, empty LUV + PA and aqueous solution, respectively	Ferreira et al. (2004)
N/M	Distearoyl phosphatidylethanolamine Dimiristoyl phosphatidylcholine Dimiristoyl phosphatidylglycerol Dipalmitoyl phosphatidylcholine Dipalmitoyl phosphatidylglycerol	Dehydration-rehydration	Paromomycin		Gaspar et al. (2015)
Small UL	Phosphatidylcholine Cholesterol	Dehydration–rehydration	Azithromycin	Better penetration of liposomal forms of drugs into macrophages was achieved	Rajabi et al. (2016)
ML	Phospholipids Dimyristoylphosphatidylcholine Dimyristoylphosphatidylglycerol	Thin-film hydration	2-((2,6-dinitro-4-trifluoromethyl-phenyl)-butylamino)-ethanol 4-(2,6-dinitro-4-trifluoro methyl-phenylamino)-phenol	Treatment of infected mice with liposomal formulation reduced the amastigote loads in the spleen up to 97%, compared to the loads for untreated controls	Carvalheiro et al. (2015)

(Continued)

Table 15.9 (Continued)

Morphology	Lipid	Technique	Drug(s)	Comment	References
N/M	Soy L-α-phosphatidylcholine (soy [hydrogenated]) 1,2-Distearoyl-sn-glycero-3-[phospho-rac-(1-glycerol)] Cholesterol	Freezing-thawing	Resiquimod	Increased production of γ-interferon and interleukin-10	Peine et al. (2014)
UL	Egg yolk phosphatidylcholine Phosphatidyl glycerol Phosphatidylethanolamine Cholesterol	Freeze-drying Double emulsion	Meglumine antimoniate Miltefosine Paromomycin	The effect of surface charge was significant on preparation procedure, size, and encapsulation efficiency.	Momeni et al. (2013)
ML	Saturated egg phosphatidylcholine Phosphatidylserine Cholesterol	Lipid hydration	Furazolidone	Liposomal furazolidone administered intraperitoneally reduced spleen (74%) and liver (32%) parasite burden at a 100-fold lower dose than the free drug	Tempone et al. (2010)
N/M	Saturated egg phosphatidylcholine Phosphatidylserine Cholesterol	Dehydration–rehydration	Pentavalent antimony	Liposome-entrapped antimony was 16-fold more effective (IC50 5 14.11 mM) than the free drug (IC50 5225.9 mM) against *Leishmania chagasi*-infected macrophages	Tempone et al. (2004)
ML	Phosphatidyl ethanolamine Cholesterol	Thin-film hydration	Quercetin	Highly potent in reducing the parasite burden in the spleen	Sarkar et al. (2002)
Small UL	Dipalmitoylphosphatidylcholine 1,2-Dipalmitoyl-phosphatidyl-glycerole 1,2-Distearoyl-sn-glycero-3-phosphoethanolamine	Thin-film hydration	Camptothecin	Treatment of infected mice with intraperitoneal injections of free and liposomal camptothecin significantly reduced the parasite loads in the livers by 43–55%.	Proulx et al. (2001)
ML	L-α-phosphatidylethanolamine Cholesterol	Thin film hydration	Doxorubicin	The results indicate that immunostimulation with IFN-γ facilitates leishmanicidal activity of a standard chemotherapeutic drug.	Kole et al. (1999)

VL, Visceral leishmaniasis; ML, Multilamellar; UD, Ultradeformable; UL, Unilamellar; PDT, Photodynamic therapy, N/M, not mentioned, PC, Soybean phosphatidylcholine, CH, cholesterol.

Table 15.10 Other vesicular nanoparticles including niosomes, emulsomes, and transfersomes reported for delivery of antileishmanial bioactive agent.

Structure	Surfactant(s)	Technique	Drug(s)	Comment	References
Emulsome	Soy phosphatidylcholine Trilaurin Cholesterol		AmB		Gupta et al. (2007)
Emulsome	Soy phosphatidylcholine Trilaurin Cholesterol	Cast film technique	AmB	A significantly higher drug concentration in the liver was estimated over a period of 24 h for OPM coated emulsomes.	Gupta and Vyas (2007)
Emulsome	Trilaurin Cholesterol	Thin-film dispersion	Silybin	Silybin emulsomes displayed a sustained-release profile.	Zhou and Chen (2015)
Niosome	L-α-phosphatidylcholine Dipalmitoyl phosphatidyl choline Dicetyl phosphate Cholesterol	Ether injection	Sodium stibogluconate	Higher drug levels in the infected reticuloendothelial system	Baillie et al. (1986)
Niosome	Sorbitan monopalmitate Cholesterol Dicetyl phosphate	Thin-film hydration	Miltefosine Ketoconazole	Growth of promastigote and amastigote forms of *Leishmania major* was inhibited *in vitro* after 48 h incubation.	Nazari-Vanani et al. (2018)
Niosome	Sorbitan monopalmitate Cholesterol	Thin-film hydration	Harmine	Significant reduction in spleen parasite load was observed	Lala et al. (2004)
Niosome	Sorbitan monopalmitate Cholesterol	Thin-film hydration	Quercetin	Nanocapsulated quercetin was found to be the most potent in reducing the parasite burden in the spleen.	Sarkar et al. (2002)
Niosome	Span40 Span60 Cholesterol	Thin-film hydration	Itraconazole	The mean number of amastigotes in each macrophage treated with itraconazole niosome (34.9 and 3.0) were significantly lower than those treated with itraconazole alone (62.0 and 3.8).	Khazaeli et al. (2014)
Niosome	Span40 Cholesterol	Thin-film hydration	14-Deoxy-11-oxoandrographolide	Max. efficacy of niosomal drug after six injections: 91%	Lala et al. (2003)

(Continued)

Table 15.10 (Continued)

Structure	Surfactant(s)	Technique	Drug(s)	Comment	References
Niosome	Triglycerol monostearate Hexaglycerol distearate Diethylene glycol mono *n*-hexadecyl ether Tetraethylene glycol mono *n*-hexadecyl ether Hexaethylene glycol mono *n*-hexadecyl ether Dicetyl phosphate Cholesterol	Thin-film hydration	Sodium stibogluconate	A reduction in splenic and bone marrow parasite burdens was achieved using large vesicles (mean diameter > 800 nm)	Williams et al. (1995)
Transfersome	Soy phosphatidylcholine Cholesterol	Fusion	Paromomycin sulfate	Significantly smaller lesion size in the mice Lower splenic parasite burden	Bavarsad et al. (2012)
Transfersome	Soy phosphatidylcholine Cholesterol	Hand shaking	AmB	Suitable system for transdermal application	Singodia et al. (2010)

15.7.6 Solid Lipid Nanoparticles (SLNs)

SLNs, as a colloidal drug carrier, combine the advantage of PNPs, fat emulsions, and liposomes. The small particle size and relatively narrow size distribution of SLNs provide biological opportunities for site-specific drug delivery along with the controlled release of active drug over a long period. Surface modification can easily be accomplished over SLNs, further enhancing site-specific drug delivery capabilities. Being a lipid-based nanoparticulate delivery system, the probability of cytotoxicity is expected to be minimal, mainly due to their more physiological acceptability compared to the PNPs previously discussed, and offers ease of industrial-scale production by hot dispersion techniques (Jain et al. 2014).

Jain et al. (2014) prepared AmB-loaded SLNs intended for the intervening of the experimental leishmaniasis. Chitosan-coated SLNs were developed and loaded with AmB for immunoadjuvant chemotherapy of *Leishmania* infection. The cellular uptake analysis demonstrated that the chitosan-coated AmB-SLN was more efficiently internalized into the J774A.1 cells. The *in vitro* antileishmanial activity revealed their high potency against *Leishmania*-infected cells in which chitosan-coated AmB-SLNs were distinguishably more efficacious over the commercial formulations (AmBisome and Fungizone). An *in vitro* cytokine estimation study revealed that chitosan-coated AmB-SLNs activated the macrophages to impart a specific immune response through enhanced production of TNF-α and IL-12 compared to control group. Furthermore, cytotoxic studies in macrophages and acute toxicity studies in mice evidenced a better safety profile of the developed formulation compared to the marketed formulations. This study indicates that the AmB-SLNs are a safe and efficacious DDS that promises strong competence in antileishmanial chemotherapy and immunotherapy (Jain et al. 2014).

Lopes et al. (2012b) studied oryzalin-containing lipid nanoparticles by an emulsion-solvent evaporation technique. Formulations were revealed to be stable throughout freeze-drying and moist-heat sterilization cycles without significant variations in physicochemical properties and no significant oryzalin losses. Cell viability studies demonstrated that the incorporation of oryzalin in nanoparticles decreases cytotoxicity, thus suggesting this strategy may improve tolerability and therapeutic index of dinitroanilines.

Mazur et al. (2019) prepared diethyldithiocarbamate (DETC) loaded in beeswax nanoparticles containing copaiba oil by the double emulsion/melt technique. In the search of new alternatives for leishmaniasis treatment; DETC has shown an excellent leishmanicidal activity. SLNs provided a protection for DETC, decreasing its cytotoxic effects in macrophages, which led to an improvement in the selectivity against the parasites, which almost doubled from free DETC (11.4) to DETC incorporated in SLNs (18.2). The results demonstrated that SLNs had a direct effect on *L. amazonensis* promastigotes without affecting the viability of the macrophage cells, can be a promising alternative therapy for the cutaneous treatment of *L. amazonensis* (Figure 15.13) (Mazur et al. 2019). Moreover, nanodrug delivery vehicles based on SLNs and their structural details are shown in Table 15.11.

15.7.7 Micro- and Nanoemulsions

The development of nanoemulsions as a delivery system for copaiba and andiroba oils (nanocopa and nanoandi) was studied, in order to test their effect on *L. infantum* and *L. amazonensis* infection. Treatment with nanocopa and nanoandi led to a reduction in *L. infantum* and *L. amazonensis* infection levels according to the results obtained in macrophage culture tests. Treatment with the nanoemulsions showed significant beneficial effects on all the parameters evaluated in lesions

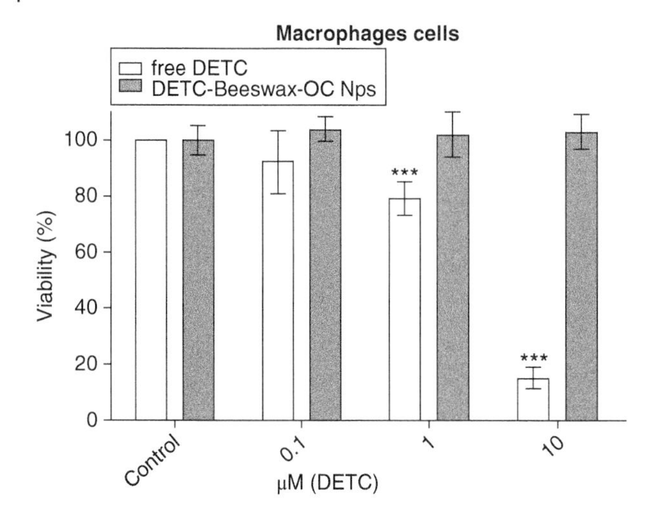

Figure 15.13 Cytotoxicity of free DETC and DETC-Beeswax-OC nanoparticles. *Source:* Adapted from Mazur et al. (2019) with permission from Elsevier.

induced by *L. amazonensis* including the lesion size, parasite burden, and histopathology using BALB/c mice animal models (Figure 15.14). Treatment of *L. infantum*-infected BALB/c mice with the nanoemulsions also showed promising results, reducing parasite burden in spleen and liver and improving histopathological features (Dhorm Pimentel de Moraes et al. 2018).

A new nanoemulsion (NE) formulation loaded with AmB was also prepared by dos Santos et al. which was further evaluated *in vivo* regarding its antileishmanial activity and *in vitro* for any possible hemolytic toxicity. The hemolytic toxicity, evaluated on human red blood cells, for AmB-loaded NE was lower than that observed for the conventional AmB (C-AmB) formulation. In contrast, administration of the AmB-loaded NE, at the same dose, did not result in any sign of acute toxicity, promoting a significant reduction in parasite burden as compared to the C-AmB (dos Santos et al. 2018).

Application of zinc phthalocyanine nanoemulsions for photodynamic therapy (PDT) of leishmania was investigated by de Oliveira de Siqueira et al. (2017). PDT combines light with photosensitizers (PS) for production of reactive oxygen species (ROS) that kills infectious microorganisms such as bacteria, fungi, and protozoa. The toxicity in the dark and the photo-biological activity of the formulations were evaluated *in vitro* for Leishmania and macrophages. The zinc phthalocyanine nanoemulsion is effective in PDT against *Leishmania* spp. Hence their application against skin infections can be a future trend for these topical formulations, avoiding the use of oral or injectable medications, decreasing systemic adverse effects. Table 15.12 lists the micro- and nanoemulsion products reported for treatment of leishmanial infection.

15.7.8 Nanodrugs and Nanosuspensions

Antonia MR Franco et al. (2016) synthesized $Sb_2O_5{\cdot}nH_2O$ nanoparticles by controlled $SbCl_5$ hydrolysis in great excess of water. $Sb_2O_5{\cdot}nH_2O$ hydrosols are proposed as a new form of treatment for CL caused by *L. amazonensis*. The NPs penetrate directly into the affected cells in the lesion, creating a high local concentration of the drug, a precondition to overcoming the parasite resistance to molecular forms of pentavalent antimonials. Similarly, Manandhar et al. (2014) formulated nanoparticles (10–20 nm) from AmB deoxycholate (1–2 lm) applying high-pressure milling homogenization in argon atmosphere and tested its ex vivo efficacy in *Leishmania* infected J774A cell line

Table 15.11 Nanodrug delivery vehicles based on solid lipid nanoparticles and their structural details according to the literature.

Lipid	Technique	Drug	Comment	References
Stearic acid Witepsol1 H12 Witepsol S55 Witepsol W45	Cold homogenization	Saponins		Van de Ven et al. (2009)
Copaiba oil	Double emulsification/ melting	Diethyldithiocarbamate		Mazur et al. (2019)
Triolein Compritol	Emulsification/ ultrasonication Hot melting	Cedrol	*In vivo* findings support >99% reduction in parasite burden using cedrol loaded NLC-C2 formulation in animals treated with two oral doses in 7 d intervals.	Kar et al. (2017)
Witepsol E85 Softisan 154 Gelucire 44/14 Dynasan 118 Compritol 888 Dynasan P60 Sterotex HM Precirol ATO 5 Gelucire 50/13	High-pressure homogenization	Buparvaquone	The NLC showed low *z*-averages (<350 nm), polydispersity (<0.3), and encapsulation efficiency ≈100%.	Monteiro et al. (2017)
Stearic acid	High shear homogenization; Microemulsification	Paromomycin		Kharaji et al. (2016)
Soybean oil Egg lecithin Cholesterol	High-speed stirring	AmB	Fungizone reduced 82 and 69% but lipid nanospheres of AMB reduced 90% and 85%, mannose-anchored lipid nanospheres of AMB reduced 95% and 94% of parasite burden in the liver and the spleen, respectively.	Veerareddy et al. (2009)

(Continued)

Table 15.11 (Continued)

Lipid	Technique	Drug	Comment	References
Stearic acid	High-speed stirring	Paromomycin		Heidari-Kharaji et al. (2016a)
Stearic acid	High-speed stirring	Paromomycin		Heidari-Kharaji et al. (2016b)
Egg phosphatidylcholine	Melt dispersion	AmB	*In vitro* drug release profile of the SLNs showed approx. 40–50% release in 8 h.	Amarji et al. (2007)
Stearic acid	Microemulsification	*Ocotea duckei Vattimo* extract	High drug entrapment efficiency	Marquele-Oliveira et al. (2016)
Stearic acid Cetyl palmitate	Microemulsification; Solvent diffusion	Paromomycin	Mean particle size was reduced to 299.08 nm and the entrapment efficiency was enhanced from immediate release to 24 h.	Ghadiri et al. (2011)
Epikuron™ 200	Nanoprecipitation	AmB		Patel and Patravale (2011)
Epikuron 200 Solutol™ HS 15	Probe sonication-assisted nanoprecipitation	AmB		Chaudhari et al. (2016)
Soy phosphatidylcholine	Solvent emulsification/evaporation	AmB	Cells treated with chitosan-coated AmB-SLNs depicted a 1.68 and 1.84 fold increase in TNF-α and IL-12 production, respectively.	Jain et al. (2014)
Tripalmitin	Solvent emulsification/evaporation	Oryzalin	Results support the superiority of both oryzalin nanoformulations on the reduction of parasitic burden in liver and spleen compared to the control group (84–91%) and similar to Glucantime.	Lopes et al. (2014)
Tripalmitin	Solvent emulsification/evaporation	Oryzalin		Lopes et al. (2012b)
Cottonseed oil Oleic acid Cholesterol Egg phosphatidylcholine	Ultrasonication	Miltefosine		da Gama Bitencourt et al. (2016)

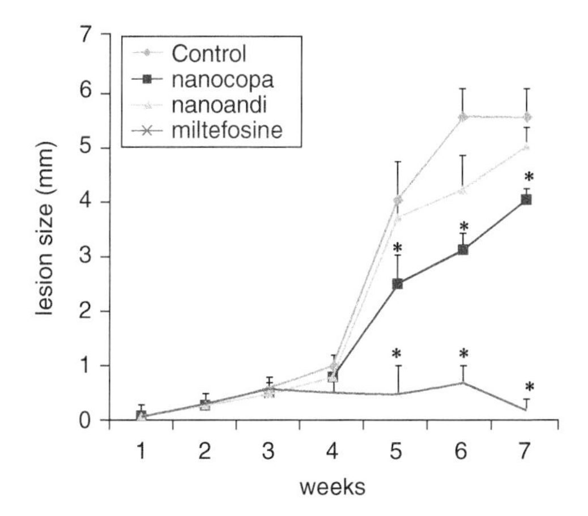

Figure 15.14 Effect of copaiba (nanocopa) and andiroba (nanoandi) releasing nanoemulsions as a delivery system on the lesion size after *L. infantum* and *L. amazonensis* infection in BALB/c mice animal model compared to miltefosine nanoemulsion. The lesion size is expressed as the difference in size between the infected and contralateral non-infected footpads. *Source:* Adapted from Dhorm Pimentel de Moraes et al. (2018) with permission from Elsevier.

and peritoneal macrophage. The efficacy against the *L. donovani* amastigotes in the peritoneal macrophage cell line (infection raised by promastigotes) demonstrated significant inhibition by AmB nanoparticles compared to the conventional AmB formulations. Table 15.13 shows different nanodrug and nanosuspension products proposed for the treatment of leishmanial infection.

15.7.9 Inorganic Nanoparticles

Ahmad et al. (2017) prepared gold nanoparticles (AuNPs), adopting a green and facile method using the aqueous extract of *Rhazya stricta decne* as a source of reducing and stabilizing agents. Interestingly, biogenic AuNPs did not exhibit cytotoxic effect against the THP-1 cells after 24 hours of exposure. The antileishmanial efficacy was studied for 48 hours and the results were expressed as percent inhibition (Figure 15.15a). The findings conclude that phytochemical-stabilized AuNPs could be a safe and effective source of antimicrobial agents.

Moreover, Kalangi et al. (2016) synthesized silver nanoparticles (AgNPs) using *Anethum graveolens* (dill) leaf extract as a reducing agent. The work was aimed in such fashion to study the enhanced antileishmanial activity of miltefosine with silver nanoparticles. AgNPs alone did not show antileishmanial effect on the promastigote stage of the *Leishmania* parasite but in combination with miltefosine, it magnifies the leishmanicidal effect of miltefosine by ~2fold (i.e. AgNPs cut down the IC50 of miltefosine about to half) (Figure 15.15b). These AgNPs can be effective as special category nanodrug carriers with no effect either on the parasites or host cells but thought to be involved in increased availability of drugs via an unknown mechanism.

Prajapati et al. (2011) demonstrated a novel drug formulation of AmB with AmB attached to the functionalized carbon nanotubes (f-CNT–AmB). The *in vivo* toxicity assessment of the compounds in BALB/c mice animal model revealed no hepatic or renal toxicity. Against intracellular amastigotes, the *in vitro* antileishmanial efficacy of f-CNT–AmB was significantly higher than that of AmB. The percentage inhibition of amastigote replication in hamsters treated with f-CNT–AmB was significantly more than that with AmB. Table 15.14 shows various inorganic nanoparticles used as antileishmanial agents as carriers to deliver an antileishmanial drug to target cells and tissues.

Table 15.12 Micro- and nanoemulsion products reported for treatment of leishmanial infection.

Drug	Technique	Comment	References
Amphotericin B	High-pressure homogenization	7.2 times improvement in AUC_{0-48} of Copaiba oil-AmB when used as nano-emulsified carrier in rats. Higher oral bioavailability than free drug.	Gupta et al. (2015a)
Amphotericin B	High-pressure homogenization		dos Santos et al. (2018)
Amphotericin B	Hot homogenization	High encapsulation efficiency (95%) and low particle size	Santos et al. (2012)
Amphotericin B	O/W microemulsification		Silva et al. (2013)
Bassic acid	O/W microemulsification		Lala et al. (2006)
Chalcones	Spontaneous emulsification	Better stability of leishmanicidal activity *in vitro* against *Leishmania amazonensis* amastigote forms. Half maximal inhibitory concentration: $0.32 \pm 0.05\,\mu M$	de Mattos et al. (2015)
Copaiba/andiroba oils	Mechanical stirring	Reduction in *Leishmania infantum* and *Leishmania amazonensis* infection levels	Dhorm Pimentel de Moraes et al. (2018)
Doxorubicin	Emulsion inversion point method	The nanoparticles loaded with DOX were completely internalized into macrophages. Improved efficacy: IC50 <1.9-fold compared to plain drug against intracellular amastigotes.	Kansal et al. (2013)
Fluconazole	Microemulsification	Fluconazole-free samples did not kill normal macrophages according to the cell viability results i.e. cellular viability >92%.	Oliveira et al. (2015)
Zinc phthalocyanine	High-energy method		de Oliveira de Siqueira et al. (2017)

15.8 Conclusion

Tropical diseases like leishmania can be somehow tied up with the neglected diseases in terms of the investments made for drug development purposes. This phenomenon can be inferred following the high expenditures needed for new drug development and the lengthy process of reaching to market from the laboratory starting point. Another reason for this situation is that countries affected by the disease are mostly developing countries with limited resources for drug development by themselves, and low income to make it pleasant for companies from developed countries. In this context, there are few drugs of choice for treating leishmaniasis and there are few new

Table 15.13 Nanodrug and nanosuspension products reported for treatment of leishmanial infection.

Drug	Technique	Comment	References
Amphotericin B	Pressure homogenization		Kayser et al. (2003)
Amphotericin B deoxycholate	High pressure Milling/ Homogenization	The nanoformulation showed high efficacy (ED50: 0.0028–0.0035 $\mu g\,ml^{-1}$) in inhibition of infected macrophage count.	Manandhar et al. (2014)
Aphidicolin	Nanosuspension		Kayser (2000)
Curcumin		Squalenoylcurcumin	Cheikh-Ali et al. (2015)
Curcumin/Miltefosine	Emulsion/solvent evaporation		Tiwari et al. (2017)
Pentavalent antimony	Controlled $SbCl_5$ hydrolysis	A 2.5–3 times higher antiparasitic activity (*in vitro*) of Sb(V) nanohybrid hydrosols, when compared to meglumine antimoniate solution.	Franco et al. (2016)

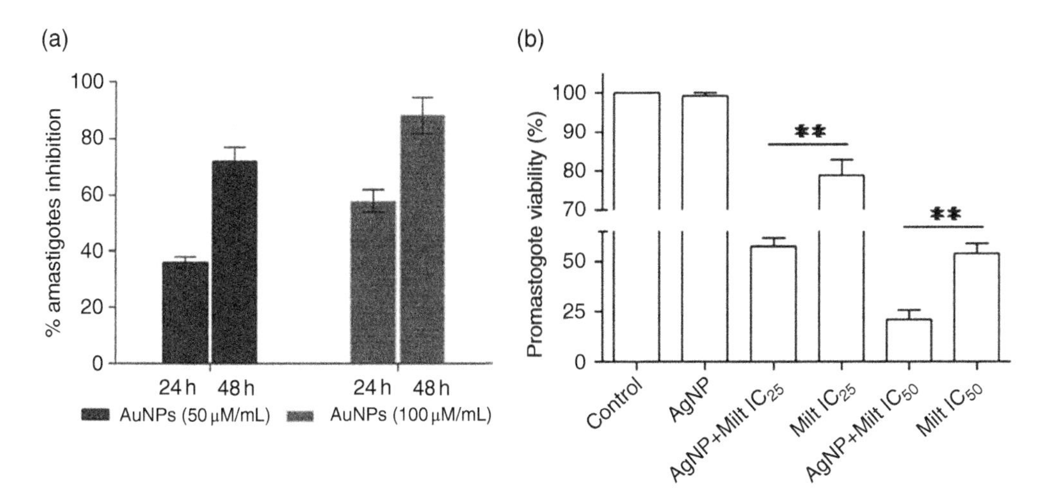

Figure 15.15 (a) Inhibition percentage of intracellular amastigotes in the presence of two different concentrations of gold nanoparticles (AuNPs) synthesized using the aqueous extract of *Rhazya stricta decne* as a source of reducing and stabilizing agents. *Source:* Adapted from Ahmad et al. (2017) with permission from Elsevier. (b) Promastigote viability percentage of silver-nanoparticles (AgNPs) synthesized using *Anethum graveolens* (dill) leaf extract as a reducing agent compared to miltefosine loaded AgNPs. *Source:* Adapted from Kalangi et al. (2016) with permission from Elsevier.

drugs on the horizon, so drug delivery strategies have turned to fruitful choices to use the currently available drugs more efficiently in terms of safety and efficacy.

In contrast to the huge bulk of the literature reporting on the using of nanotechnological approaches and carriers for delivery of antileishmanial agents, few products are currently available

Table 15.14 Inorganic nanoparticles used as antileishmanial agents per se or as carriers to deliver an antileishmanial drug to target cells and tissues.

Metal	Drug	Comment	References
Calcium phosphate	AmB	Higher entrapment of AmB was achieved.	Chaurasia et al. (2016)
Carbon nanotubes	AmB		Prajapati et al. (2012)
Carbon nanotubes	AmB		Prajapati et al. (2011)
Gold	Biogenic gold nanoparticles		Das et al. (2018a)
Gold	Gold nanoparticles	Biogenic Au NPs inhibited 88.88% of intracellular amastigotes at $100\,\mu g\,ml^{-1}$ conc. IC50 = $43\,\mu g\,ml^{-1}$	Ahmad et al. (2017)
Gold	Quercetin	Effective against wild-type (IC50:15 ± 3 mM), sodium stibogluconate resistant strain (IC50 40 ± 8 mM) and the paromomycin resistant (IC50 30 ± 6 mM) strains	Das et al. (2013)
Gold	Terpenoid andrographolide engineered gold nanoparticle	Strong leishmanicidal effect both against wild-type (IC50:19 ± 1.7 mM) and sodium stibogluconate (IC50 55 ± 7.3 mM)/ paromomycin (IC50 41 ± 6 mM) resistant strains.	Das et al. (2018b)
Iron oxide	AmB		Kumar et al. (2017)
Iron oxide	AmB		Kumar et al. (2016b)
Multi-walled carbon nanotubes	AmB	AmB-loaded mannosylated MWCNTs (AmBitubes) were 500 nm in size with tubular structure and good entrapment efficiency (75.46 ± 1.40%).	Pruthi et al. (2012)
Pseudoboehmite	Glucantime		Munhoz Jr. et al. (2016)
Silver	Decanethiol	High leishmanicidal potential of the nanoparticles at concentrations of 0.04 μM.	Isaac-Márquez et al. (2018)
Silver	PfFt-AgNPs		Baiocco et al. (2011)
Sliver	Miltefosine	AgNPs alone (50 μM): No antileishmanial effect on promastigote stage of *Leishmania* AgNPs+ miltefosine (12.5 and 25 μM): ≈ 2-fold increase in the leishmanicidal effect of miltefosine.	Kalangi et al. (2016)
Zn	Zinc sulphide	ZnS NPs IC50 (promastigote): $29.81 \pm 3.15\,\mu g\,ml^{-1}$ Glucantime IC50 (promastigote): $14.75 \pm 4.05\,\mu g\,ml^{-1}$	Sharifi et al. (2018)
Zn	ZnO	93.76% apoptosis in *Leishmania major* after 72 h at $120\,\mu g\,ml^{-1}$ of ZnO nanoparticles concentration	Delavari et al. (2014)

in market, not just due to the limitations enforced by the lengthy regulatory process and so on, but the clarity and eligibility of a part of the literature for further consideration both in terms of their methodology and data presentation and interpretation. However, it can be concluded in summary that a lot of success has been achieved in terms of improved safety adopting nanotechnological approaches also targeting. Surveying the literature and considering the progress and maturity of the reports and the findings, the next generation of the delivery devices for treatment of leishmania can be expected to be available in mid-term using lipid-based nanoparticles as well.

References

Abamor, E.S. (2017). Antileishmanial activities of caffeic acid phenethyl ester loaded PLGA nanoparticles against *Leishmania infantum* promastigotes and amastigotes in vitro. *Asian Pacific Journal of Tropical Medicine* 10: 25–34.

Abamor, E.S. (2018). A new approach to the treatment of leishmaniasis: quercetin-loaded polycaprolactone nanoparticles. *Journal of Turkish Chemical Society, Section A* 5: 1071–1082.

Abamor, E.S., Tosyali, O.A., Bagirova, M., and Allahverdiyev, A. (2018). *Nigella sativa* oil entrapped polycaprolactone nanoparticles for leishmaniasis treatment. *IET Nanobiotechnology* 12: 1018–1026.

Abu Ammar, A., Nasereddin, A., Ereqat, S. et al. (2018). Amphotericin B-loaded nanoparticles for local treatment of cutaneous leishmaniasis. *Drug Delivery and Translational Research* 9: 76–84.

Afzal, I., Sarwar, H.S., Sohail, M.F. et al. (2019). Mannosylated thiolated paromomycin-loaded PLGA nanoparticles for the oral therapy of visceral leishmaniasis. *Nanomedicine* 14: 387–406.

Agrawal, A.K., Agrawal, A., Pal, A. et al. (2002). Superior chemotherapeutic efficacy of amphotericin B in tuftsin-bearing liposomes against *Leishmania donovani* infection in hamsters. *Journal of Drug Targeting* 10: 41–45.

Ahmad, I., Agarwal, A., Pal, A. et al. (1991). Tissue distribution and antileishmanial activity of liposomised amphotericin-B in Balb/c mice. *Journal of Biosciences* 16: 217–221.

Ahmad, A., Wei, Y., Ullah, S. et al. (2017). Synthesis of phytochemical-stabilized gold nanoparticles and their biological activities against bacteria and Leishmania. *Microbial Pathogenesis* 110: 304–312.

Alcântara, L.M., Ferreira, T.C.S., Gadelha, F.R., and Miguel, D.C. (2018). Challenges in drug discovery targeting TriTryp diseases with an emphasis on leishmaniasis. *International Journal for Parasitology: Drugs and Drug Resistance* 8: 430–439.

Alving, C.R. (1983). Delivery of liposome-encapsulated drugs to macrophages. *Pharmaceutical and Therapeutics* 22: 407–424.

Alving, C.R. (1986). Liposomes as drug carriers in leishmaniasis and malaria. *Parasitology Today* 2:101–107.

Amarji, B., Ajazuddin, A., Raghuwanshi, D. et al. (2007). Lipid nano spheres (LNSs) for enhanced oral bioavailability of amphotericin B: development and characterization. *Journal of Biomedical Nanotechnology* 3: 264–269.

Asfaram, S., Teshnizi, S.H., Fakhar, M. et al. (2018). Is urine a reliable clinical sample for the diagnosis of human visceral leishmaniasis? A systematic review and meta-analysis. *Parasitology International* 67: 575–583.

Asthana, S., Jaiswal, A.K., Gupta, P.K. et al. (2013). Immunoadjuvant chemotherapy of visceral leishmaniasis in hamsters using amphotericin B-encapsulated nanoemulsion template-based chitosan nanocapsules. *Antimicrobial Agents and Chemotherapy* 57: 1714–1722.

Asthana, S., Jaiswal, A.K., Gupta, P.K. et al. (2014). Th-1 biased immunomodulation and synergistic antileishmanial activity of stable cationic lipid–polymer hybrid nanoparticle: biodistribution and toxicity assessment of encapsulated amphotericin B. *European Journal of Pharmaceutics and Biopharmaceutics* 89: 62–73.

Asthana, S., Gupta, P.K., Jaiswal, A.K. et al. (2015a). Targeted chemotherapy of visceral leishmaniasis by lactoferrin-appended amphotericin B-loaded nanoreservoir: in vitro and in vivo studies. *Nanomedicine* 10: 1093–1109.

Asthana, S., Gupta, P.K., Jaiswal, A.K. et al. (2015b). Overexpressed macrophage mannose receptor targeted nanocapsules-mediated cargo delivery approach for eradication of resident parasite: in vitro and in vivo studies. *Pharmaceutical Research* 32: 2663–2677.

Baillie, A.J., Coombs, G.H., Dolan, T.F., and Laurie, J. (1986). Non-ionic surfactant vesicles, niosomes, as a delivery system for the anti-leishmanial drug, sodium stibogluconate. *Journal of Pharmacy and Pharmacology* 38: 502–505.

Baiocco, P., Ilari, A., Ceci, P. et al. (2011). Inhibitory effect of silver nanoparticles on trypanothione reductase activity and *Leishmania infantum* proliferation. *ACS Medicinal Chemistry Letters* 2: 230–233.

Barazesh, A., Motazedian, M.H., Sattarahmady, N. et al. (2018). Preparation of meglumine antimonate loaded albumin nanoparticles and evaluation of its anti-leishmanial activity: an in vitro assay. *Journal of Parasitic Diseases* 42: 416–422.

Bavarsad, N., Fazly Bazzaz, B.S., Khamesipour, A., and Jaafari, M.R. (2012). Colloidal, in vitro and in vivo anti-leishmanial properties of transfersomes containing paromomycin sulfate in susceptible BALB/c mice. *Acta Tropica* 124: 33–41.

Bhatia, S. and Goli, D. (2016). *Leishmaniasis: Biology, Control and New Approaches for Its Treatment*. Apple Academic Press.

Biswaro, L.S., Garcia, M.P., da Silva, J.R. et al. (2019). Itraconazole encapsulated PLGA-nanoparticles covered with mannose as potential candidates against leishmaniasis. *Journal of Biomedical Materials Research Part B: Applied Biomaterials* 107B: 680–687.

Borborema, S.E.T., Schwendener, R.A., Osso, J.A. et al. (2011). Uptake and antileishmanial activity of meglumine antimoniate-containing liposomes in *Leishmania (Leishmania) major*-infected macrophages. *International Journal of Antimicrobial Agents* 38: 341–347.

Borborema, S.E.T., Osso Junior, J.A., Tempone, A.G. et al. (2018). Pharmacokinetic of meglumine antimoniate encapsulated in phosphatidylserine-liposomes in mice model: a candidate formulation for visceral leishmaniasis. *Biomedicine and Pharmacotherapy* 103: 1609–1616.

Bose, P.P., Kumar, P., and Dwivedi, M.K. (2016). Hemoglobin guided nanocarrier for specific delivery of amphotericin B to Leishmania infected macrophage. *Acta Tropica* 158: 148–159.

Britta, E.A., da Silva, C.C., Rubira, A.F. et al. (2016). Generating nanoparticles containing a new 4-nitrobenzaldehyde thiosemicarbazone compound with antileishmanial activity. *Materials Science and Engineering: C* 69: 1159–1166.

Brugués, A.P., Naveros, B.C., Calpena Campmany, A.C. et al. (2015). Developing cutaneous applications of paromomycin entrapped in stimuli-sensitive block copolymer nanogel dispersions. *Nanomedicine (London)* 10: 227–240.

Bruni, N., Stella, B., Giraudo, L. et al. (2017). Nanostructured delivery systems with improved leishmanicidal activity: a critical review. *International Journal of Nanomedicine* 12: 5289–5311.

Burza, S., Croft, S.L., and Boelaert, M. (2018). Leishmaniasis. *The Lancet* 392: 951–970.

Calvo, J., Lavandera, J.L., Agüeros, M., and Irache, J.M. (2011). Cyclodextrin/poly(anhydride) nanoparticles as drug carriers for the oral delivery of atovaquone. *Biomedical Microdevices* 13: 1015–1025.

Carneiro, G., Santos, D.C.M., Oliveira, M.C. et al. (2010). Topical delivery and in vivo antileishmanial activity of paromomycin-loaded liposomes for treatment of cutaneous leishmaniasis. *Journal of Liposome Research* 20: 16–23.

Carvalheiro, M., Esteves, M.A., Santos-mateus, D. et al. (2015). Hemisynthetic trifluralin analogues incorporated in liposomes for the treatment of leishmanial infections. *European Journal of Pharmaceutics and Biopharmaceutics* 93: 346–352.

Carvalho, S.G., Siqueira, L.A., Zanini, M.S. et al. (2018). Physicochemical and in vitro biological evaluations of furazolidone-based β-cyclodextrin complexes in *Leishmania amazonensis*. *Research in Veterinary Science* 119: 143–153.

Casa, D.M., Scariot, D.B., Khalil, N.M. et al. (2018). Bovine serum albumin nanoparticles containing amphotericin B were effective in treating murine cutaneous leishmaniasis and reduced the drug toxicity. *Experimental Parasitology* 192: 12–18.

Castillo, J.J., Rindzevicius, T., Wu, K. et al. (2014). Synthesis and characterization of covalent diphenylalanine nanotube-folic acid conjugates. *Journal of Nanoparticle Research* 16: 2525.

Chakraborty, P., Bhaduri, A.N., and Das, P.K. (1990). Sugar receptor mediated drug delivery to macrophages in the therapy of experimental visceral leishmaniasis. *Biochemical and Biophysical Research Communications* 166: 404–410.

Chaubey, P. and Mishra, B. (2014). Mannose-conjugated chitosan nanoparticles loaded with rifampicin for the treatment of visceral leishmaniasis. *Carbohydrate Polymers* 101: 1101–1108.

Chaubey, P., Patel, R.R., and Mishra, B. (2014). Development and optimization of curcumin-loaded mannosylated chitosan nanoparticles using response surface methodology in the treatment of visceral leishmaniasis. *Expert Opinion on Drug Delivery* 11: 1163–1181.

Chaubey, P., Mishra, B., Patel, S.R. et al. (2017). Mannose-conjugated curcumin-chitosan nanoparticles: efficacy and toxicity assessments against *Leishmania donovani*. *International Journal of Biological Macromolecules* 111: 109–120.

Chaudhari, M.B., Desai, P.P., Patel, P.A., and Patravale, V.B. (2016). Solid lipid nanoparticles of amphotericin B (AmbiOnp): in vitro and in vivo assessment towards safe and effective oral treatment module. *Drug Delivery and Translational Research* 6: 354–364.

Chaudhuri, G., Mukhopadhyay, A., and Basu, S.K. (1989). Selective delivery of drugs to macrophages through a highly specific receptor. An efficient chemotherapeutic approach against leishmaniasis. *Biochemical Pharmacology* 38: 2995–3002.

Chaurasia, M., Singh, P.K., Jaiswal, A.K. et al. (2016). Bioinspired calcium phosphate nanoparticles featuring as efficient carrier and prompter for macrophage intervention in experimental leishmaniasis. *Pharmaceutical Research* 33: 2617–2629.

Cheikh-Ali, Z., Caron, J., Cojean, S. et al. (2015). "Squalenoylcurcumin" nanoassemblies as water-dispersible drug candidates with antileishmanial activity. *ChemMedChem* 10: 411–418.

Chowdhary, S.J., Chowdhary, A., and Kashaw, S. (2016). Macrophage targeting: a strategy for leishmaniasis specific delivery. *International Journal of Pharmacy and Pharmaceutical Sciences* 8: 16–26.

Clinical trials 2019, National Institute of Health, US National Library of Medicine, https://clinicaltrials.gov/ct2/results?cond=leishmaniasisandterm=andcntry=andstate=andcity=anddist= (Accessed 12/2/2019)

Costa Lima, S., Rodrigues, V., Garrido, J. et al. (2012). In vitro evaluation of bisnaphthalimidopropyl derivatives loaded into pegylated nanoparticles against *Leishmania infantum* protozoa. *International Journal of Antimicrobial Agents* 39: 424–430.

da Costa-Silva, T.A., Galisteo, A.J., Lindoso, J.A.L. et al. (2017). Nanoliposomal buparvaquone immunomodulates *Leishmania infantum*-infected macrophages and is highly effective in a murine model. *Antimicrobial Agents and Chemotherapy* 61: e02297-16.

Croft, S.L., Davidson, R.N., and Thornton, E.A. (1991). Liposomal amphotericin B in the treatment of visceral leishmaniasis. *Journal of Antimicrobial Chemotherapy* 28 (Suppl B): 111–118.

Daftarian, P.M., Stone, G.W., Kovalski, L. et al. (2013). A targeted and adjuvanted nanocarrier lowers the effective dose of liposomal amphotericin B and enhances adaptive immunity in murine cutaneous leishmaniasis. *The Journal of Infectious Disease* 208: 1914–1922.

Dar, M.J., Din, F.U., and Khan, G.M. (2018). Sodium stibogluconate loaded nano-deformable liposomes for topical treatment of leishmaniasis: macrophage as a target cell. *Drug Delivery* 25: 1595–1606.

Das, S., Roy, P., Mondal, S. et al. (2013). One pot synthesis of gold nanoparticles and application in chemotherapy of wild and resistant type visceral leishmaniasis. *Colloids and Surfaces B: Biointerfaces* 107: 27–34.

Das, S., Ghosh, S., De, A.K., and Bera, T. (2017). Oral delivery of ursolic acid- loaded nanostructured lipid carrier coated with chitosan oligosaccharides: development, characterization, in vitro and in vivo assessment for the therapy of leishmaniasis. *International Journal of Biological Macromolecules* 102: 996–1008.

Das, S., Halder, A., Roy, P., and Mukherjee, A. (2018a). Biogenic gold nanoparticles against wild and resistant type visceral leishmaniasis. *Materials Today: Proceedings* 5: 2912–2920.

Das, S., Halder, A., Mandal, S. et al. (2018b). Andrographolide engineered gold nanoparticle to overcome drug resistant visceral leishmaniasis. *Artificial Cells, Nanomedicine, and Biotechnology* 46 (sup 1): 751–762.

Defaus, S., Gallo, M., Abengózar, M.A. et al. (2017). A synthetic strategy for conjugation of paromomycin to cell-penetrating Tat(48-60) for delivery and visualization into leishmania parasites. *International Journal of Peptides* 2017: 4213037.

Delavari, M., Dalimi, A., Ghaffarifar, F., and Sadraei, J. (2014). In vitro study on cytotoxic effects of ZnO nanoparticles on promastigote and amastigote forms of *Leishmania major* (MRHO/IR/75/ER). *Iranian Journal of Parasitology* 9: 6–13.

Demicheli, C., Ochoa, R., de Silva, J.B.B. et al. (2004). Oral delivery of meglumine antimoniate-β-cyclodextrin complex for treatment of leishmaniasis. *Antimicrobial Agents and Chemotherapy* 48: 100–103. https://doi.org/10.1128/AAC.48.1.100-103.2004.

Dhorm Pimentel de Moraes, A.R., Tavares, G.D., Soares Rocha, F.J. et al. (2018). Effects of nanoemulsions prepared with essential oils of copaiba- and andiroba against *Leishmania infantum* and *Leishmania amazonensis* infections. *Experimental Parasitology* 187: 12–21.

Ehrenfreund-kleinman, T., Golenser, J., and Domb, A.J. (2004). Conjugation of amino-containing drugs to polysaccharides by tosylation: amphotericin B-arabinogalactan conjugates. *Biomaterials* 25: 3049–3057.

Escobar, P., Vera, A.M., Neira, L.F. et al. (2018). Photodynamic therapy using ultradeformable liposomes loaded with chlorine aluminum phthalocyanine against *L. (V.) braziliensis* experimental models. *Experimental Parasitology* 194: 45–52.

Farias Azevedo, L., da Fonseca Silva, P., Porfírio Brandã, o.M. et al. (2018). Polymeric nanoparticle systems loaded with red propolis extract: a comparative study of the encapsulating systems, PCL-Pluronic versus Eudragit® E100-Pluronic. *Journal of Apicultural Research* 57: 255–270.

Ferreira, L.S., Ramaldes, G.A., Nunan, E.A., and Ferreira, L.A. (2004). In vitro skin permeation and retention of paromomycin from liposomes for topical treatment of the cutaneous leishmaniasis. *Drug Development and Industrial Pharmacy* 30: 289–296.

Franco, A.M., Grafova, I., Soares, F.V. et al. (2016). Nanoscaled hydrated antimony (V) oxide as a new approach to first-line antileishmanial drugs. *Journal of Nanomedicine* 11: 6771–6780.

Frézard, F., Martins, P.S., Bahia, A.P. et al. (2008). Enhanced oral delivery of antimony from meglumine antimoniate/β-cyclodextrin nanoassemblies. *International Journal of Pharmaceutics* 347: 102–108.

da Gama Bitencourt, J.J., Pazin, W.M., Ito, A.S. et al. (2016). Miltefosine-loaded lipid nanoparticles: improving miltefosine stability and reducing its hemolytic potential toward erythtocytes and its cytotoxic effect on macrophages. *Biophysical Chemistry* 217: 20–31.

Gaspar, M.M., Calado, S., Pereira, J. et al. (2015). Targeted delivery of paromomycin in murine infectious diseases through association to nano lipid systems. *Nanomedicine: Nanotechnology, Biology and Medicine* 11: 1851–1860.

Ghadiri, M., Vatanara, A., Doroud, D., and Roholamini Najafabadi, A. (2011). Paromomycin loaded solid lipid nanoparticles: characterization of production parameters. *Biotechnology and Bioprocess Engineering* 16: 617–623.

Ghosh, S., Das, S., De, A.K. et al. (2017). Amphotericin B-loaded mannose modified poly(D,L-lactide-co-glycolide) polymeric nanoparticles for the treatment of visceral leishmaniasis: in vitro and in vivo approaches. *RSC Advances* 7: 29575–29590.

Gupta, S. and Vyas, S.P. (2007). Development and characterization of amphotericin B bearing emulsomes for passive and active macrophage targeting. *Journal of Drug Targeting* 15: 206–217.

Gupta, S., Dube, A., and Vyas, S.P. (2007). Antileishmanial efficacy of amphotericin B bearing emulsomes against experimental visceral leishmaniasis. *Journal of Drug Targeting* 15: 437–444.

Gupta, P.K., Jaiswal, A.K., Asthana, S. et al. (2015a). Synergistic enhancement of parasiticidal activity of amphotericin B using copaiba oil in nanoemulsified carrier for oral delivery: an approach for non-toxic chemotherapy. *British Journal of Pharmacology* 172: 3596–3610.

Gupta, P.K., Jaiswal, A.K., Asthana, S. et al. (2015b). Self assembled ionically sodium alginate cross-linked amphotericin B encapsulated glycol chitosan stearate nanoparticles: applicability in better chemotherapy and non-toxic delivery in visceral leishmaniasis. *Pharmaceutical Research* 32: 1727–1740.

Gutiérrez, V., Seabra, A.B., Reguera, R.M. et al. (2016). New approaches from nanomedicine for treating leishmaniasis. *Chemical Society Reviews* 45 (1): 152–168.

Gutierrez-Corbo, C., Dominguez-Asenjo, B., Vossen, L.I. et al. (2017). PEGylated dendritic polyglycerol conjugate delivers doxorubicin to the parasitophorous vacuole in *Leishmania infantum* infections. *Macromolecular Bioscience* 17: 1700098.

Halder, A., Shukla, D., Das, S. et al. (2018). Lactoferrin-modified Betulinic Acid-loaded PLGA nanoparticles are strong anti-leishmanials. *Cytokine* 110: 412–415.

Heath, S., Chance, M.L., and New, R.R. (1984). Quantitative and ultrastructural studies on the uptake of drug loaded liposomes by mononuclear phagocytes infected with *Leishmania donovani*. *Molecular and Biochemical Parasitology* 12: 49–60.

Heidari-Kharaji, M., Taheri, T., Doroud, D. et al. (2016a). Enhanced paromomycin efficacy by solid lipid nanoparticle formulation against Leishmania in mice model. *Parasite Immunology* 38: 599–608.

Heidari-Kharaji, M., Taheri, T., Doroud, D. et al. (2016b). Solid lipid nanoparticle loaded with paromomycin: in vivo efficacy against *Leishmania tropica* infection in BALB/c mice model. *Applied Microbiology and Biotechnology* 100: 7051–7060.

Heurtault, B., Legrand, P., Mosqueira, V. et al. (2001). The antileishmanial properties of surface-modified, primaquine-loaded nanocapsules tested against intramacrophagic *Leishmania donovani* amastigotes in vitro. *Annals of Tropical Medicine and Parasitology* 95: 529–533.

Hunter, C.A., Dolan, T.F., Coombs, G.H., and Baillie, A.J. (1988). Vesicular systems (niosomes and liposomes) for delivery of sodium stibogluconate in experimental murine visceral leishmaniasis. *The Journal of Pharmacy and Pharmacology* 40: 161–165.

Iman, M., Huang, Z., Szoka, F.C., and Jaafari, M.R. (2011). Characterization of the colloidal properties, in vitro antifungal activity, antileishmanial activity and toxicity in mice of a di-stigma-steryl-hemi-succinoyl-glycero-phosphocholine liposome-intercalated amphotericin B. *International Journal of Pharmaceutics* 408: 163–172.

Isaac-Márquez, A.P., Talamás-Rohana, P., Galindo-Sevilla, N. et al. (2018). Decanethiol functionalized silver nanoparticles are new powerful leishmanicidals in vitro. *World Journal of Microbiology and Biotechnology* 34: 38.

Islam, M.Z., Itoh, M., Mirza, R. et al. (2004). Direct agglutination test with urine samples for the diagnosis of visceral leishmaniasis. *American Journal of Tropical Medicine and Hygiene* 70: 78–82.

Islan, G.A., Durán, M., Cacicedo, M.L. et al. (2017). Nanopharmaceuticals as a solution to neglected diseases: is it possible? *Acta Tropica* 170: 16–42.

Italia, J.L., Yahya, M.M., Singh, D., and Ravi Kumar, M.N. (2009). Biodegradable nanoparticles improve oral bioavailability of amphotericin B and show reduced nephrotoxicity compared to intravenous Fungizone. *Pharmaceutical Research* 26: 1324–1331.

Italia, J.L., Kumar, M.N., and Carter, K.C. (2012). Evaluating the potential of polyester nanoparticles for per oral delivery of amphotericin B in treating visceral leishmaniasis. *Journal of Biomedical Nanotechnology* 8: 695–702.

Jain, N.K., Mishra, V., and Mehra, N.K. (2013). Targeted drug delivery to macrophages. *Expert Opinion on Drug Delivery* 10: 353–367.

Jain, V., Gupta, A., Pawar, V.K. et al. (2014). Chitosan-assisted immunotherapy for intervention of experimental leishmaniasis via amphotericin B-loaded solid lipid nanoparticles. *Applied Biochemistry and Biotechnology* 174: 1309–1330.

Jain, K., Verma, A.K., Mishra, P.R., and Jain, N.K. (2015). Characterization and evaluation of amphotericin B loaded MDP conjugated poly(propylene imine) dendrimers. *Nanomedicine: Nanotechnology, Biology and Medicine* 11: 705–713.

Kalangi, S.K., Dayakar, A., Gangappa, D. et al. (2016). Biocompatible silver nanoparticles reduced from *Anethum graveolens* leaf extract augments the antileishmanial efficacy of miltefosine. *Experimental Parasitology* 170: 184–192.

Kalaria, D.R., Sharma, G., Beniwal, V., and Kumar, M.N. (2009). Design of biodegradable nanoparticles for oral delivery of doxorubicin: in vivo pharmacokinetics and toxicity studies in rats. *Pharmaceutical Research* 26: 492–501.

Kansal, S., Tandon, R., Verma, P.R.P. et al. (2013). Development of doxorubicin loaded novel core shell structured nanocapsules for the intervention of visceral leishmaniasis. *Journal of Microencapsulation* 30: 441–450.

Kar, N., Chakraborty, S., De, A.K. et al. (2017). Development and evaluation of a cedrol-loaded nanostructured lipid carrier system for in vitro and in vivo susceptibilities of wild and drug resistant *Leishmania donovani* amastigotes. *European Journal of Pharmaceutical Sciences* 104: 196–211.

Karanikolopoulos, N., Pitsikalis, M., Hadjichristidis, N. et al. (2007). pH-responsive aggregates from double hydrophilic block copolymers carrying zwitterionic groups. Encapsulation of antiparasitic compounds for the treatment of leishmaniasis. *Langmuir* 23: 4214–4224.

Kayser, O. (2000). Nanosuspensions for the formulation of aphidicolin to improve drug targeting effects against leishmania infected macrophages. *International Journal of Pharmaceutics* 196: 253–256.

Kayser, O., Olbrich, C., Yardley, V. et al. (2003). Formulation of amphotericin B as nanosuspension for oral administration. *International Journal of Pharmaceutics* 254: 73–75.

Kharaji, M.H., Doroud, D., Taheri, T., and Rafati, S. (2016). Drug targeting to macrophages with solid lipid nanoparticles harboring paromomycin: an in vitro evaluation against *L. major and L. tropica*. *AAPS PharmSciTech* 17: 1110–1119.

Khazaeli, P., Sharifi, I., Talebian, E. et al. (2014). Antileishmanial effect of itraconazole niosome on in vitro susceptibility of *Leishmania tropica*. *Environmental Toxicology and Pharmacology* 38: 205–211.

Kóczán, G., Ghose, A.C., Mookerjee, A., and Hudecz, F. (2002). Methotrexate conjugate with branched polypeptide influences *Leishmania donovani* infection in vitro and in experimental animals. *Bioconjugate Chemistry* 13: 518–524.

Kole, L., Das, L., and Das, P.K. (1999). Synergistic effect of interferon-gamma and mannosylated liposome-incorporated doxorubicin in the therapy of experimental visceral leishmaniasis. *The Journal of Infectious Diseases* 180: 811–820.

Kothandaraman, G.P., Ravichandran, V., Bories, C. et al. (2017). Anti-fungal and anti-leishmanial activities of pectin-amphotericin B conjugates. *Journal of Drug Delivery Science and Technology* 39: 1–7.

Kumar, A. (2013). *Leishmania and Leishmaniasis*. New York: Springer.

Kumar, R., Sahoo, G.C., Pandey, K. et al. (2014a). Fabrication of iron oxide functionalized with PAMAM dendrimer and glycine for the development of drug delivery carrier against visceral leishmania. *International Journal of Infectious Diseases* 21 (Suppl. 1): 155.

Kumar, R., Sahoo, G.C., Pandey, K. et al. (2014b). Study the effects of PLGA-PEG encapsulated amphotericin B nanoparticle drug delivery system against *Leishmania donovani*. *Drug Delivery* 22: 383–388.

Kumar, R., Sahoo, G.C., Pandey, K. et al. (2016a). Development of PLGA-PEG encapsulated miltefosine based drug delivery system against visceral leishmaniasis. *Materials Science and Engineering: C* 59: 748–753.

Kumar, R., Sahoo, G.C., Pandey, K. et al. (2016b). Development of glycine coated magnetic nanoparticles (GMNPs) advance drug delivery system against visceral leishmaniasis. *International Journal of Infectious Diseases* 45 (Suppl. 1): 365.

Kumar, R., Pandey, K., Sahoo, G.C. et al. (2017). Development of high efficacy peptide coated iron oxide nanoparticles encapsulated amphotericin B drug delivery system against visceral leishmaniasis. *Materials Science and Engineering: C* 75: 1465–1471.

Lage, L.M., Barichello, J.M., Lage, D.P. et al. (2016). An 8-hydroxyquinoline-containing polymeric micelle system is effective for the treatment of murine tegumentary leishmaniasis. *Parasitology Research* 115: 4083–4095.

Lago, J.H.G. and Tempone, A.G. (2018). Natural products as a source of new drugs against leishmania. In: Drug Discovery for Leishmaniasis (ed. D. Thurston) Chapter 9, 179–198. London, UK: The Royal Society of Chemistry.

Lala, S., Nandy, A.K., Mahato, S.B., and Basu, M.K. (2003). Delivery in vivo of 14-deoxy-11-oxoandrographolide, an antileishmanial agent, by different drug carriers. *Indian Journal of Biochemistry and Biophysics* 40: 169–174.

Lala, S., Pramanick, S., Mukhopadhyay, S. et al. (2004). Harmine: evaluation of its antileishmanial properties in various vesicular delivery systems. *Journal of Drug Targeting* 12: 165–175.

Lala, S., Gupta, S., Sahu, N.P. et al. (2006). Critical evaluation of the therapeutic potential of bassic acid incorporated in oil-in-water microemulsions and poly-D,L-lactide nanoparticles against experimental leishmaniasis. *Journal of Drug Targeting* 14: 171–179.

Lopes, R.M., Corvo, M.L., Eleutério, C.V. et al. (2012a). Formulation of oryzalin (ORZ) liposomes: in vitro studies and in vivo fate. *European Journal of Pharmaceutics and Biopharmaceutics* 82: 281–290.

Lopes, R., Eleutério, C.V., Gonçalves, L.M.D. et al. (2012b). Lipid nanoparticles containing oryzalin for the treatment of leishmaniasis. *European Journal of Pharmaceutical Sciences* 45: 442–450.

Lopes, R.M., Gaspar, M.M., Pereira, J. et al. (2014). Liposomes versus lipid nanoparticles: comparative study of lipid-based systems as oryzalin carriers for the treatment of leishmaniasis. *Journal of Biomedical Nanotechnology* 10: 3647–3657.

Manandhar, K.D., Yadav, T.P., Prajapati, V.K. et al. (2014). Nanonization increases the antileishmanial efficacy of amphotericin B: an ex vivo approach. In: *Infectious Diseases and Nanomedicine II. Advances in Experimental Medicine and Biology*, vol. 808 (eds. R. Adhikari and S. Thapa). New Delhi: Springer.

Marquele-Oliveira, F., Torres, E.C., da Silva Barud, H. et al. (2016). Physicochemical characterization by AFM, FT-IR and DSC and biological assays of a promising antileishmania delivery system loaded with a natural Brazilian product. *Journal of Pharmaceutical and Biomedical Analysis* 123: 195–204.

Martinez, A. and Gil, C. (2018). Medicinal chemistry strategies to discover new leishmanicidal drugs. In: *Drug Discovery for Leishmaniasis* (ed. D. Thurston) Chapter 8, 153–178. London, UK: The Royal Society of Chemistry.

de Mattos, C.B., Argenta, D.F., Melchiades Gde, L. et al. (2015). Nanoemulsions containing a synthetic chalcone as an alternative for treating cutaneous leshmaniasis: optimization using a full factorial design. *International Journal of Nanomedicine* 10: 5529–5542.

Mazur, K.L., Feuser, P.E., Valério, A. et al. (2019). Diethyldithiocarbamate loaded in beeswax-copaiba oil nanoparticles obtained by solventless double emulsion technique promote promastigote death in vitro. *Colloids and Surfaces B: Biointerfaces* 176: 507–512.

Mendonça, D.V., Lage, L.M., Lage, D.P. et al. (2016). Poloxamer 407 (Pluronic® F127)-based polymeric micelles for amphotericin B: in vitro biological activity, toxicity and in vivo therapeutic efficacy against murine tegumentary leishmaniasis. *Experimental Parasitology* 169: 34–42.

Mendonça, D.V.C., Martins, V.T., Lage, D.P. et al. (2018). Comparing the therapeutic efficacy of different amphotericin B-carrying delivery systems against visceral leishmaniasis. *Experimental Parasitology* 186: 24–35.

Mendonça, D.V.C., Tavares, G.S.V., Lage, D.P. et al. (2019). In vivo antileishmanial efficacy of a naphthoquinone derivate incorporated into a Pluronic® F127-based polymeric micelle system against *Leishmania amazonensis* infection. *Biomedicine and Pharmacotherapy* 109: 779–787.

Micale, N., Piperno, A., Mahfoudh, N. et al. (2015). A hyaluronic acid–pentamidine bioconjugate as a macrophage mediated drug targeting delivery system for the treatment of leishmaniasis. *RSC Advances* 5: 95545–95550.

Mobasheri, M., Attar, H., Rezayat Sorkhabadi, S.M. et al. (2015). Solubilization behavior of polyene antibiotics in nanomicellar system: insights from molecular dynamics simulation of the amphotericin B and nystatin interactions with polysorbate 80. *Molecules* 21: E6.

Momeni, A., Rasoolian, M., Momeni, A. et al. (2013). Development of liposomes loaded with anti-leishmanial drugs for the treatment of cutaneous leishmaniasis. *Journal of Liposome Research* 23: 134–144.

Monteiro, L.M., Löbenberg, R., Cotrim, P.C. et al. (2017). Buparvaquone nanostructured lipid carrier: 'development of an affordable delivery system for the treatment of Leishmaniasis'. *BioMed Research International* 2017: 9781603.

Moosavian Kalat, S.A., Khamesipour, A., Bavarsad, N. et al. (2014). Use of topical liposomes containing meglumine antimoniate (Glucantime) for the treatment of *L. major* lesion in BALB/c mice. *Experimental Parasitology* 143: 5–10.

Mowbray, C.E. (2018). Anti-leishmanial drug discovery: past, present and future perspectives. In: *Drug Discovery for Leishmaniasis* (ed. D. Thurston) Chapter 2, 24–36. London, UK: The Royal Society of Chemistry.

Mukbel, R.M., Jones, D.E., Ackermann, M., and Waters, R. (2005). *Leishmania amazonensis and Macrophage Interactions : Immune Factors Necessary to Kill the Parasite*. Iowa State University.

Munhoz, A.H. Jr., Martins, J.S., Ribeiro, R.R. et al. (2016). Use of pseudoboehmite nanoparticles for drug delivery system of glucantime®. *Journal of Nano Research* 38: 47–51.

Nahar, M. and Jain, N.K. (2009). Preparation, characterization and evaluation of targeting potential of amphotericin B-loaded engineered PLGA nanoparticles. *Pharmaceutical Research* 26: 2588.

Nahar, M., Mishra, D., Dubey, V., and Jain, N.K. (2009). Development of amphotericin B loaded PLGA nanoparticles for effective treatment of visceral leishmaniasis. In: *13th International Conference on Biomedical Engineering. IFMBE Proceedings*, vol. 23 (eds. C.T. Lim and J.C.H. Goh). Berlin, Heidelberg: Springer.

Nahar, M., Dubey, V., Mishra, D. et al. (2010). In vitro evaluation of surface functionalized gelatin nanoparticles for macrophage targeting in the therapy of visceral leishmaniasis. *Journal of Drug Targeting* 18: 93–105.

Najafian, H.R., Mohebali, M., Rezayat, S.M. et al. (2016). Nanoliposomal miltefosine for the treatment of cutaneous leishmaniasis caused by *Leishmania major* (MRHO/IR/75/ER): the drug preparation and in vitro study. *International Journal of Pharmaceutical Research* 5: 97–107.

Nan, A., Croft, S.L., Yardley, V., and Ghandehari, H. (2004). Targetable water-soluble polymer-drug conjugates for the treatment of visceral leishmaniasis. *Journal of Controlled Release* 94: 115–127.

do Nascimento, T.G., da Silva, P.F., Azevedo, L.F. et al. (2016). Polymeric nanoparticles of brazilian red propolis extract: preparation, characterization, antioxidant and leishmanicidal activity. *Nanoscale Research Letters* 11: 301.

Nazari-Vanani, R., Vais, R.D., Sharifi, F. et al. (2018). Investigation of anti-leishmanial efficacy of miltefosine and ketoconazole loaded on nanoniosomes. *Acta Tropica* 185: 69–76.

Nicoletti, S., Seifert, K., and Gilbert, I.H. (2010). Water-soluble polymer-drug conjugates for combination chemotherapy against visceral leishmaniasis. *Bioorganic and Medicinal Chemistry* 18: 2559–2565.

Nishi, K.K. and Jayakrishnan, A. (2007). Self-gelling primaquine-gum arabic conjugate: an injectable controlled delivery system for primaquine. *Biomacromolecules* 8: 84–90.

de Oliveira de Siqueira, L.B., da Silva Cardoso, V., Rodrigues, I.A. et al. (2017). Development and evaluation of zinc phthalocyanine nanoemulsions for use in photodynamic therapy for Leishmania spp. *Nanotechnology* 28: 065101.

Oliveira, M.B., Calixto, G., Graminha, M. et al. (2015). Development, characterization, and in vitro biological performance of fluconazole-loaded microemulsions for the topical treatment of cutaneous leishmaniasis. *BioMed Research International* 2015: 396894.

Oyama, J., Lera-Nonose, D.S.S.L., Ramos-Milaré, Á.C.F.H. et al. (2019). Potential of Pluronics® P-123 and F-127 as nanocarriers of anti-Leishmania chemotherapy. *Acta Tropica* 192: 11–21.

Papagiannaros, A., Bories, C., Demetzos, C., and Loiseau, P.M. (2005). Antileishmanial and trypanocidal activities of new miltefosine liposomal formulations. *Biomedicine and Pharmacotherapy* 59: 545–550.

Patel, P.A. and Patravale, V.B. (2011). AmbiOnp: solid lipid nanoparticles of amphotericin B for oral administration. *Journal of Biomedical Nanotechnology* 7: 632–639.

Patere, S.N., Pathak, P.O., Kumar Shukla, A. et al. (2016). Surface-modified liposomal formulation of amphotericin B: in vitro evaluation of potential against visceral leishmaniasis. *AAPS PharmSciTech* 18: 710–720.

Peine, K.J., Gupta, G., Brackman, D.J. et al. (2014). Liposomal resiquimod for the treatment of *Leishmania donovani* infection. *Journal of Antimicrobial Chemotherapy* 69: 168–175.

Perez, A.P., Altube, M.J., Schilrreff, P. et al. (2016). Topical amphotericin B in ultradeformable liposomes: formulation, skin penetration study, antifungal and antileishmanial activity in vitro. *Colloids and Surfaces B: Biointerfaces* 139: 190–198.

Prajapati, V.K., Awasthi, K., Gautam, S. et al. (2011). Targeted killing of *Leishmania donovani* in vivo and in vitro with amphotericin B attached to functionalized carbon nanotubes. *Journal of Antimicrobial Chemotherapy* 66: 874–879.

Prajapati, V.K., Awasthi, K., Yadav, T.P. et al. (2012). An oral formulation of amphotericin B attached to functionalized carbon nanotubes is an effective treatment for experimental visceral leishmaniasis. *The Journal of Infectious Diseases* 205: 333–336.

Proulx, M.E., Désormeaux, A., Marquis, J.F. et al. (2001). Treatment of visceral leishmaniasis with sterically stabilized liposomes containing camptothecin. *Antimicrobial Agents and Chemotherapy* 45: 2623–2627.

Pruthi, J., Mehra, N.K., and Jain, N.K. (2012). Macrophages targeting of amphotericin B through mannosylated multiwalled carbon nanotubes. *Journal of Drug Targeting* 20: 593–604.

Qasim, M., Lim, D.J., Park, H., and Na, D. (2014). Nanotechnology for diagnosis and treatment of infectious diseases. *Journal of Nanoscience and Nanotechnology* 14: 7374–7387.

Radwan, M.A., AlQuadeib, B.T., Šiller, L. et al. (2017). Oral administration of amphotericin B nanoparticles: antifungal activity, bioavailability and toxicity in rats. *Drug Delivery* 24: 40–50.

Rajabi, O., Layegh, P., Hashemzadeh, S., and Khoddami, M. (2016). Topical liposomal azithromycin in the treatment of acute cutaneous leishmaniasis. *Dermatologic Therapy* 29: 358–363.

Rathore, A., Jain, A., Gulbake, A. et al. (2011). Mannosylated liposomes bearing amphotericin B for effective management of visceral Leishmaniasis. *Journal of Liposome Research* 21: 333–340.

Reimão, J.Q., Colombo, F.A., Pereira-Chioccola, V.L., and Tempone, A.G. (2012). Effectiveness of liposomal buparvaquone in an experimental hamster model of *Leishmania* (L.) *infantum chagasi*. *Experimental Parasitology* 130: 195–199.

Reis, L.E.S., de Brito, R.C.F., de Oliveira Cardoso, J.M. et al. (2017). Mixed formulation of conventional and pegylated meglumine antimoniate-containing liposomes reduces inflammatory process and parasite burden in *Leishmania infantum*-infected BALB/c mice. *Antimicrobial Agents and Chemotherapy* 11: e00962-17.

Ribeiro, R.R., Ferreira, W.A., Martins, P.S. et al. (2010). Prolonged absorption of antimony(V) by the oral route from non-inclusion meglumine antimoniate-beta-cyclodextrin conjugates. *Biopharmaceutics and Drug Disposition* 31: 109–119.

Ribeiro, T.G., Franca, J.R., Fuscaldi, L.L. et al. (2014). An optimized nanoparticle delivery system based on chitosan and chondroitin sulfate molecules reduces the toxicity of amphotericin B and is effective in treating tegumentary leishmaniasis. *International Journal of Nanomedicine* 9: 5341–5353.

Ribeiro, J.B.P., Miranda-Vilela, A.L., Graziani, D. et al. (2016). Evaluation of the efficacy of systemic miltefosine associated with photodynamic therapy with liposomal chloroaluminium phthalocyanine in the treatment of cutaneous leishmaniasis caused by *Leishmania* (L.) *amazonensis* in C57BL/6 mice. *Photodiagnosis and Photodynamic Therapy* 13: 282–290.

Richter, A.R., Feitosa, J.P.A., Paula, H.C.B. et al. (2018). Pickering emulsion stabilized by cashew gum- poly-l-lactide copolymer nanoparticles: synthesis, characterization and amphotericin B encapsulation. *Colloids and Surfaces B: Biointerfaces* 164: 201–209.

Rodrigues Junior, J.M., Fessi, H., Bories, C. et al. (1995). Primaquine-loaded poly(lactide) nanoparticles: physicochemical study and acute tolerance in mice. *International Journal of Pharmaceutics* 126: 253–260.

Rojas-Aguirre, Y., Castillo, I., Hernández, D.J. et al. (2012). Diversity in the supramolecular interactions of 5,6-dichloro-2-(trifluoromethyl)-1H-benzimidazole with modified cyclodextrins: implications for physicochemical properties and antiparasitic activity. *Carbohydrate Polymers* 87: 471–479.

Roychoudhury, J., Sinha, R., and Ali, N. (2011). Therapy with sodium stibogluconate in stearylamine-bearing liposomes confers cure against SSG-resistant *Leishmania donovani* in BALB/c mice. *PLoS One* 6: e17376.

Santos, C.M., de Oliveira, R.B., Arantes, V.T. et al. (2012). Amphotericin B-loaded nanocarriers for topical treatment of cutaneous leishmaniasis: development, characterization, and in vitro skin permeation studies. *Journal of Biomedical Nanotechnology* 8: 322–329.

dos Santos, D.C.M., de Souza, M.L.S., Teixeira, E.M. et al. (2018). A new nanoemulsion formulation improves antileishmanial activity and reduces toxicity of amphotericin B. *Journal of Drug Targeting* 26: 357–364.

Sarkar, S., Mandal, S., Sinha, J. et al. (2002). Quercetin: critical evaluation as an antileishmanial agent in vivo in hamsters using different vesicular delivery modes. *Journal of Drug Targeting* 10: 573–578.

Schettini, D.A., Ribeiro, R.R., Demicheli, C. et al. (2006). Improved targeting of antimony to the bone marrow of dogs using liposomes of reduced size. *International Journal of Pharmaceutics* 315: 140–147.

Nanda, H.S. and Mishra, N.C. (2011). "Amphotericin B" loaded natural biodegradable nanofibers as a potential drug delivery system against leishmaniasis. *Current Nanoscience* 7: 943–949.

Serrano, D.R., Lalatsa, A., Dea-Ayuela, M.A. et al. (2015). Oral particle uptake and organ targeting drives the activity of amphotericin B nanoparticles. *Nanoparticles Molecular Pharmaceutics* 12: 420–431.

Sett, R., Sarkar, K., and Das, P.K. (1993). Macrophage-directed delivery of doxorubicin conjugated to neoglycoprotein using leishmaniasis as the model disease. *The Journal of Infectious Diseases* 168: 994–999.

Shahnaz, G., Edagwa, B.J., McMillan, J. et al. (2017). Development of mannose-anchored thiolated amphotericin B nanocarriers for treatment of visceral leishmaniasis. *Nanomedicine (London)* 12: 99–115.

Sharifi, F., Sharififar, F., Sharifi, I. et al. (2018). Cytotoxicity, leishmanicidal, and antioxidant activity of biosynthesised zinc sulphide nanoparticles using Phoenix dactylifera. *IET Nanobiotechnology* 12: 264–269.

Shaw, C.D. and Carter, K.C. (2014). Drug delivery: lessons to be learnt from Leishmania studies. *Nanomedicine (London)* 9: 1531–1544.

Shim, Y.H., Kim, Y.C., Lee, H.J. et al. (2011). Amphotericin B aggregation inhibition with novel nanoparticles prepared with poly(epsilon-caprolactone)/poly(n,n-dimethylamino-2-ethyl methacrylate) diblock copolymer. *Journal of Microbiology and Biotechnology* 21: 28–36.

Shukla, A.K., Patra, S., and Dubey, V.K. (2012). Nanospheres encapsulating anti-leishmanial drugs for their specific macrophage targeting, reduced toxicity, and deliberate intracellular release. *Vector Borne and Zoonotic Diseases* 12: 953–960.

Silva, A.E., Barratt, G., Chéron, M., and Egito, E.S. (2013). Development of oil-in-water microemulsions for the oral delivery of amphotericin B. *International Journal of Pharmaceutics* 454: 641–648.

Singh, P.K., Sah, P., Meher, J.G. et al. (2016). Macrophage-targeted chitosan anchored PLGA nanoparticles bearing doxorubicin and amphotericin B against visceral leishmaniasis. *RSC Advances* 6: 71705–71718.

Singh, P.K., Pawar, V.K., Jaiswal, A.K. et al. (2017). Chitosan coated Pluronic F127 micelles for effective delivery of amphotericin B in experimental visceral leishmaniasis. *International Journal of Biological Macromolecules* 105 (Pt 1): 1220–1231.

Singh, P.K., Jaiswal, A.K., Pawar, V.K. et al. (2018). Fabrication of 3-O-sn-phosphatidyl-L-serine anchored PLGA nanoparticle bearing amphotericin B for macrophage targeting. *Pharmaceutical Research* 35: 60.

Singodia, D., Gupta, G.K., Verma, A. et al. (2010). Development and performance evaluation of amphotericin B transfersomes against resistant and sensitive clinical isolates of visceral leishmaniasis. *Journal of Biomedical Nanotechnology* 6: 293–302.

Singodia, D., Verma, A., Khare, P. et al. (2012a). Investigations on feasibility of in situ development of amphotericin B liposomes for industrial applications. *Journal of Liposome Research* 22: 8–17.

Singodia, D., Verma, A., Verma, R.K., and Mishra, P.R. (2012b). Investigations into an alternate approach to target mannose receptors on macrophages using 4-sulfated N-acetyl galactosamine more efficiently in comparison with mannose-decorated liposomes: an application in drug delivery. *Nanomedicine: Nanotechnology, Biology, and Medicine* 8: 468–477.

Sokolsky-papkov, M., Domb, A.J., and Golenser, J. (2006). Impact of aldehyde content on amphotericin B-dextran imine conjugate toxicity. *Biomacromolecules* 7: 1529–1535.

Sousa-Batista, A.J., Cerqueira-Coutinho, C., do Carmo, F.S. et al. (2019). Polycaprolactone antimony nanoparticles as drug delivery system for leishmaniasis. *American Journal of Therapeutics* 26: e12–e17.

Sundar, S. (2003). Diagnosis of kala-azar--an important stride. *The Journal of the Association of Physicians of India* 51: 753–755.

Tavares, G.S.V., Mendonça, D.V.C., Miyazaki, C.K. et al. (2019). A Pluronic® F127-based polymeric micelle system containing an antileishmanial molecule is immunotherapeutic and effective in the treatment against *Leishmania amazonensis* infection. *Parasitology International* 68: 63–72.

Tempone, A.G., Perez, D., Rath, S. et al. (2004). Targeting *Leishmania (L.) chagasi* amastigotes through macrophage scavenger receptors: the use of drugs entrapped in liposomes containing phosphatidylserine. *Journal of Antimicrobial Chemotherapy* 54: 60–68.

Tempone, A.G., Mortara, R.A., de Andrade, H.F. Jr., and Reimão, J.Q. (2010). Therapeutic evaluation of free and liposome-loaded furazolidone in experimental visceral leishmaniasis. *International Journal of Antimicrobial Agents* 36: 159–163.

Tiwari, B., Pahuja, R., Kumar, P. et al. (2017). Nanotized curcumin and miltefosine, a potential combination for treatment of experimental visceral leishmaniasis. *Antimicrobial Agents and Chemotherapy* 61: e01169-16.

Tripathi, P., Dwivedi, P., Khatik, R. et al. (2015). Development of 4-sulfated N-acetyl galactosamine anchored chitosan nanoparticles: a dual strategy for effective management of Leishmaniasis. *Colloids and Surfaces B: Biointerfaces* 136: 150–159.

Tripathi, P., Jaiswal, A.K., Dube, A., and Mishra, P.R. (2017). Hexadecylphosphocholine (Miltefosine) stabilized chitosan modified Ampholipospheres as prototype co-delivery vehicle for enhanced killing of *L. donovani*. *International Journal of Biological Macromolecules* 105 (Pt 1): 625–637.

Van de Ven, H., Vermeersch, M., Shunmugaperumal, T. et al. (2009). Solid lipid nanoparticle (SLN) formulations as a potential tool for the reduction of cytotoxicity of saponins. *Die Pharmazie* 64: 172–176.

Van de Ven, H., Vermeersch, M., Matheeussen, A. et al. (2011). PLGA nanoparticles loaded with the antileishmanial saponin β-aescin: factor influence study and in vitro efficacy evaluation. *International Journal of Pharmaceutics* 420: 122–132.

Veerareddy, P.R., Vobalaboina, V., and Ali, N. (2009). Antileishmanial activity, pharmacokinetics and tissue distribution studies of mannose-grafted amphotericin B lipid nanospheres. *Journal of Drug Targeting* 17: 140–147.

Volpato, H., Scariot, D.B., Soares, E.F.P. et al. (2018). In vitro anti-Leishmania activity of T6 synthetic compound encapsulated in yeast-derived β-(1,3)-d-glucan particles. *International Journal of Biological Macromolecules* 119: 1264–1275.

Want, M.Y., Islamuddin, M., Chouhan, G. et al. (2015). Therapeutic efficacy of artemisinin-loaded nanoparticles in experimental visceral leishmaniasis. *Colloids and Surfaces B: Biointerfaces* 130: 215–221.

Want, M.Y., Yadav, P., and Afrin, F. (2016). Nanomedicines for therapy of visceral leishmaniasis. *Journal of Nanoscience and Nanotechnology* 16: 2143–2151.

Wetzel, D.M. and Phillips, M.A. (2018). Chemotherapy of protozoal infections: amebiasis, giardiasis, trichomoniasis, trypanosomiasis, leishmaniasis, and other protozoal infections. In: *Goodman and Gilman's: The Pharmacological Basis of Therapeutics* (eds. R. Randa Hilal-Dandan, B. Knollmann and L. Brunton) Chapter 54, 987–999. USA: McGraw-Hill Education.

Williams, D.M., Carter, K.C., and Baillie, A.J. (1995). Visceral leishmaniasis in the BALB/c mouse: a comparison of the in vivo activity of five non-ionic surfactant vesicle preparations of sodium stibogluconate. *Journal of Drug Targeting* 3: 1–7.

World Health Organization (WHO) 2019a, Epidemiological situation, https://www.who.int/leishmaniasis/burden/en Accessed 112/2/2019

World Health Organization (WHO) 2019b, Clinical forms of the leishmaniases, https://www.who.int/leishmaniasis/disease/clinical_forms_leishmaniases/en Accessed 112/2/2019

Zadeh-Mehrizi, T., Shafiee Ardestani, M., Haji Molla Hoseini, M. et al. (2018a). Novel nano-sized chitosan amphotericin B formulation with considerable improvement against *Leishmania major. Nanomedicine (London)* 13: 3129–3147.

Zadeh-Mehrizi, T., Shafiee Ardestani, M., Khamesipour, A. et al. (2018b). Reduction toxicity of amphotericin B through loading into a novel nanoformulation of anionic linear globular dendrimer for improve treatment of *Leishmania major. Journal of Materials Science: Materials in Medicine* 29: 125.

Zhou, X. and Chen, Z. (2015). Preparation and performance evaluation of emulsomes as a drug delivery system for silybin. *Archives of Pharmacal Research* 38: 2193–2200.

16

Theranostics and Vaccines: Current Status and Future Expectations

Thais Francine Ribeiro Alves[1], *Fernando Batain*[1], *Cecília Torqueti de Barros*[1], *Kessie Marie Moura Crescencio*[1], *Venâncio Alves do Amaral*[1], *Mariana Silveira de Alcântara Chaud*[2], *Décio Luís Portella*[1,3], *and Marco Vinícius Chaud*[1]

[1] *Laboratory of Biomaterials and Nanotechnology, LaBNUS, University of Sorocaba, Sorocaba, Brazil*
[2] *Department of Cardiology, Municipal Clinics Hospital José Alencar, São Bernardo do Campo, Brazil*
[3] *Department of Surgery and Plastic Surgery, Pontifical Catholic University of São Paulo, Sorocaba, Brazil*

16.1 Introduction

The term theranostic has been used to express more specific and individualized therapies for chronic disease, cancer, neurodegenerative disorders, and various other diseases that have negative impacts on public health (Wang and Cui 2016; Ramanathan et al. 2018). However, a broader and more current concept of nanotheranostics includes diagnostic tools and vaccines (Ciechanover and Kwon 2015) for developing the theranostics nanoparticles, tailored for a specific-target drug release for the treatment and diagnosis of disease at the cellular, molecular, and supramolecular level with fewer side effects.

Various attempts have been made to combine therapeutic and diagnostic properties into an original and useful smart nanomedicine formulation, having the ability of diagnosis, bioactive(s) delivery, and monitoring of therapeutic response (Ahmed et al. 2012). In clinical practices, imaging technologies are routinely implemented for early diagnosis, individualization of treatment, and monitoring of the drug's efficacy. Theranostic drugs or theranostic vaccines not only perform the functions of such technologies but also help to reduce the treatment cost, workload, and moreover, speed up the process of new drug development.

A comprehensive class of disorders such as frontotemporal dementia, amyotrophic lateral sclerosis, Alzheimer's, Parkinson's, and Huntington's diseases, and spinocerebellar ataxia diseases share similarities in the molecular and subcellular level (synoptics abnormalities, neuronal loss, and cerebral deposits) (Scott 2018). Brain cancer and stroke also affect the brain in other subcellular levels which result in short-term and long-term impairments and leads to an emotional, social, and financial burden for the families (Ciechanover and Kwon 2015).

Cancer remains one of the most pressing health concerns and distressing diseases, and if the diagnosis is made late, the treatment takes a long time. Therefore, the most effective treatment and best prognosis depend on tools that decrease the time of diagnosis, identify the location and size of the tumor, choose the most appropriate therapeutic management, and use drugs with minimal

Nanobiotechnology in Diagnosis, Drug Delivery, and Treatment, First Edition. Edited by Mahendra Rai, Mehdi Razzaghi-Abyaneh, and Avinash P. Ingle.

side effects. New-generation individualized diagnosis and therapeutic tools for cancer have received attention in the field of nanomedicine-based theranostics. The emerging trend of theranostic-based strategies with tailored sophisticated nanoparticles, including specific targeting, images at a molecular level, and therapeutic functions, will be beneficial in the selection of the drug delivery system design, planning of treatment, successsful therapy, monitoring of response, and planning of follow-up therapy (Ahmed et al. 2012; Zhang et al. 2019a).

Cardiovascular and neurodegenerative diseases, as well as cancer, are illnesses in progression, so they should also have the same cover, including diagnosis, visualization of cardiac damage, and monitoring of therapeutic effects after drug delivery. It is observed that cardiovascular mortality is primarily due to atherosclerotic plaque destabilization. Therefore, early detection is crucial for preventing myocardial infarction and sudden cardiac death (Dreifuss et al. 2015). The practice of clinical imaging technology has certain limitations which brings problems that need to be solved, including how to accurately distinguish the vulnerability and the rupture of atherosclerotic plaques, as well as monitoring the effect of innovative therapies of heart failure (Srinivas et al. 2015).

One common manifestation of cardiotoxicity is associated with exposure to anticancer therapies and the development of left ventricular systolic dysfunction and overt heart failure. As a result, the need for specialist cardiology input is becoming progressively recognized as an essential resource in the management of both long-term survivors and those undergoing active treatment (Rodeheffer and Redfield 2013). So, the availability of multifunctional nanotheranostics represents a potential solution for these problems. Vaccine development has been a priority in the fight against parasitic, bacterial, and viral infections and, more recently, against chronic disease. Most of the molecular causes underlying neurodegenerative disease, cardiopathies, and cancer are still unclear, although the application of molecular cell biology's techniques has improved the knowledge about chronic illness processes.

Using pathogens as inspiration, biomimetic nanotheranostics interact with an immune system based on a fight with viruses or bacteria. In immune aspects, nanotheranostics have been suggested as a tool to improve vaccine design, safeness, and linear performance, besides, making an immune systems approach as an interlinked network of cells and signaling molecules. Then, it is possible to utilize them as antigen delivery systems to enhance the antibody's processing, or as an immunostimulant to adjuvants in order to activate or strengthen immunity (Liu et al. 2018; Scott 2018).

In the genomic era, the genetic bases have been investigated by linkage analysis, and a restricted subset of causative genes identified by positional cloning (Aalinkeel et al. 2018). To trace the physiological interactions between T cells and tumor cells is always very challenging. A report indicates that labeled T cells with gold nanoparticles as a contrast agent allows examination of the distribution, migration, and kinetics of T-cells activated by antigen-presenting cells (APCs), which are most common aims of immunomodulatory nanotheranostics (Meir et al. 2015). Due in the broadest sense to the theranostics, the clinical practice in immunotherapies is improving, because of better understanding of the essential aspects of the new vaccine concept and its applications in cell-based immune theranostic systems; can be used for multiple purposes, overcoming the barrier mechanisms of poorly understood treatments and their underlying success or failure (Dreifuss et al. 2015).

The most significant fact in the therapeutic vaccine system is the capacity to support the mechanisms of immune surveillance and immune evasion in cancer cells. A novel group of immunomodulatory antibodies has been introduced in the therapeutic clinical setting; these antibodies react against specific immune tolerance and induce regression of tumors (Srinivas et al. 2015). This cancer immunotherapy approach has been showing promising results with more efficacy, less toxicity, capacity for rapid development, and potential for low-cost manufacture and safe administration. As a result of the off-target effect, the approach for delivering vaccines, adjuvants, or antibodies directly to tumor sites is gaining widespread attention (Yang et al. 2009).

Theranostic nanostructures and perfectly engineered nanomaterials increased the protection of drug load from chemical and biological degradation, prolonging the circulation time and promoting the most local concentration of the load in the target tumor cells, as a result of their abnormal vascular architecture, enhanced permeability, and retention effects. (Srinivas et al. 2015; Zhang et al. 2019a). The success rate of the drug delivery system in the treatment of central nervous system disorders has been found to be negligible. Active and passive immunotherapies were reported to have lower efficacy in neurodegenerative illnesses. It has been attributed to the blood-brain barrier and the nonselective distribution of the drugs in the brain, the opsonization of plasma proteins in the proper circulation, and peripheral side effects for the release of antibodies to neutralize abnormal proteins (Ramanathan et al. 2018).

Neuroprotective effects of biodegradable poly(lactic-co-glycolic acid) (PLGA)-ginsenoside Rg3 nanoformulation was evaluated as a potential monotherapy for Alzheimer's disease. Evidence suggested that ginsenoside Rg3 plays an important role in memory and improved cognition, and this has led to speculation that it may be used for the treatment of Alzheimer's disease and other neurological illnesses (Yang et al. 2009; Aalinkeel et al. 2018). The study evaluated the efficacy of PLGA nanoformulation by measuring the quantitative reduction in Aβ plaques, and the down-regulation of β-amyloid A4 precursor protein (AβPP-A4) gene expression levels. The results indicate that such nanotherapeutic approach may be utilized in the treatment of Alzheimer's disease as this biodegradable nanoformulation is likely to have minimal side effects, and thereby have significant clinical translational potential (Aalinkeel et al. 2018). Considering the developments in this technology it is believed that it will lead to the emergence of valuable theranostic probes that may also be applied to another neurodegenerative disease.

The purpose of this chapter is to summarize the biomedical applications of theranostics in the context of various nanoplatforms and discuss some of the promising and emerging theranostics tools.

16.2 The Role of Theranostics and Vaccines in Early Diagnosis and Therapeutic Strategies in Different Diseases

16.2.1 Cancer

Around the world, cancer affects millions of people in all age groups, both sexes, and all social classes. Advances in the development of new drugs, smart drug release strategies, and new tools for more accurate diagnostics that are accessible to the entire population, and above all the public policies to raise awareness among people about prevention and healthy habits have reduced the mortality rate. However, favorable and early prognosis remains far from optimal. Theranostics is a combinatorial tool that associates a specific molecule to optical real-time indicators and precise diagnosis of tumor. Moreover, it dynamically helps the therapeutic approach of oncologic illness. At last, theranostic systems have been developed with the aim of eliminating multistep procedures, reducing delays in starting treatment, and helping in the definition of better-personalized treatment for the cancer patient.

Theranostics aims to explore whether the combination of drugs and vaccination can increase the immune response against melanoma antigens, and immune checkpoint barriers, to mitigate immune-suppression. In 2018, the Nobel Prize in Physiology and Medicine was attributed to the discovery of cancer therapies connected with inhibition of negative immune regulation (https://www.nobelprize.org/prizes/medicine/2018/summary, 2019). Since 2013, cancer immunotherapy has been studied under new perspectives, as a novel and promising approach, and is regarded as advancing the knowledge of the science of cancer treatment (Table 16.1). There has been a burst of interest in the development of nanoparticle-based theranostics for imaging of early-stage cancer and, simultaneously, for the controlled drug delivery system.

Theranostics based on nanoparticles and molecular imaging plays an increasingly important role in noninvasive monitoring of fundamental properties of many types of cancers. High-sensitivity PET (positron emission tomography) or SPECT (single photon emission computed tomography) imaging are valuable techniques that can provide noninvasive longitudinal visualization of pharmacokinetics and tissue deposition. The instrument performance, including MRI (magnetic resonance imaging) and US (optical imaging, ultrasound), and the toolbox for imaging reconstruction software, are indispensable tools for the diagnostic and theranostic-based treatment of cancer.

Etoposide (VP-16), a semi-synthetic derivative of 4′-demethylepipodophyllotoxin, is being used for the treatment of many cancers, including lung cancer and testicular cancer (Aalinkeel et al. 2018). Due to poor water solubility, the etoposide is an alkaloid presented in the form of phosphate salt to form a prodrug easily soluble in water. The etoposide forms a complex with DNA and topoisomerase II, and the interplay among them prevents re-ligation of the DNA strand, and explains the structure-activity relation and molecular basis of drug-resistant mutation, resulting in strand breakage and subsequent death of cancer cells (Montecucco et al. 2015; Srinivas et al. 2015). The etoposide phosphate parenteral administration has been facilitated by a more water-soluble form; however, adverse effects as neutropenia and leukopenia may lead to temporary interruption of treatment, with serious harm to the patient. Children may be more sensitive to the etoposide phosphate, especially allergic effects. However, adults present drowsiness, weakness, loss of appetite,

Table 16.1 Vaccine-theranostics that potentiate active antigens into cells to fight cancer.

Mimetic biomaterial	Function	Action	Results
Synthetic polymers (PLGA, perfluorocarbon)	Immunogenic adjuvant	Type 1 interferon stimulate gen	Turning phagocytes from enemies to allies against cancer
Dendritic cells nanoparticles	Adaptative immune answer	Antigen-negative variant clones	NY-Eso-1 antigen specific stimulate immune response
Tailored dendritic cells nanoparticles	Antigen load cells	Perfluoro carbon targeting ligands	Activation and proliferation $CD8^+$ T cells
Nanobioconjugate engager (MBINE)	Simultaneously targeting HER2	Synergic signaling mediated by calreticulin	PC7A nano-vaccine and anti-PD-1 antibody inhibited TC-1 tumor
Conjugated synthetic (APCs + PLGA + Avidin-Palmitate)	Activate the immune response	Anchor peptide-MHC and costimulatory ligands to the particle surfaces	Stimulated to a much greater extent in T-cell activation
ARTs + BSA	Production of reactive oxygen species	Inhibition of cell cycle in G0/G1 phase	Damaged the mitochondrial integrity and activated mitochondrial-mediated cell apoptosis

Note: NY-Eso-1, New York esophageal squamous cell carcinoma 1; APC, antigen-presenting cells; MHC, major histocompatibility complex; ARTs, artemisinin and its derivatives; BSA, bovine serum albumin; HER2, human epidermal growth factor receptor-type 2.

low white blood cells, and dehydration (Fan et al. 2008). Leukopenia and neutropenia plus adverse effects underscore the need for a nanoparticle delivery system to carry etoposide phosphate to the appropriate cells after systemic administration. Studies confirmed comparable safety and efficacy of oral and intravenous etoposide administration (Rezonja et al. 2013).

For antitumoral drugs such as etoposide or etoposide phosphate, whose water solubility, bioavailability, and safety are the main problems to express specific and individualize therapy, the use of conjugate theranostics can resolve two problems at the same time. Therefore, for the etoposide to be considered a theranostic, it must perform a therapeutic function and simultaneously be a diagnostic tool. Both objectives can be achieved if a conjugation linker is responsive at extracellular or intracellular biological or chemical stimuli (Srinivas et al. 2015). Conjugated drugs can be triggered to release in a particular environment when the conjugation linker is responsive to extracellular or intracellular biological or chemical stimuli. Conjugation of hydrophobic low-molecular-weight drugs to water-soluble polymers can improve the solubility and efficacy through passive or active targeting with reduced *in vivo* toxicity.

Srinivas et al. (2015) used indium both as a carrier to deliver etoposide phosphate to tumors cells and as a SPECT imaging agent through the incorporation of ^{111}In. Etoposide phosphate was successfully encapsulated together with indium in nanoparticles and exhibited dose-dependent cytotoxicity and induction of apoptosis in cultured H460 cancer cells via G2/M cell cycle arrest (Srinivas et al. 2015). In a mouse xenograft lung cancer model, etoposide phosphate/indium nanoparticles induce tumor cell apoptosis, leading to significant enhancement of tumor growth inhibition compared to the free drug (Srinivas et al. 2015).

Gemcitabine is a depleting chemotherapeutic agent of myeloid-derived suppressor cells. Zhang et al. (2019b) encapsulated gemcitabine, including myeloid-derived suppressor cells into nanoscaffold lipid-coated calcium phosphate, followed by grafting a high density of polyethylene glycol (PEG) chain. In this case, the calcium phosphate can act as a carrier scaffold for monophosphate metabolite of gemcitabine through the formation of nano-coprecipitate dispersion. The reduction of tumor-induced immunosuppression by lipid-coated calcium phosphate-gemcitabine monophosphate can boost the endogenous adaptive antitumor immunity and improve antitumor efficacy (Zhang et al. 2019b).

Polymer-drug conjugates for delivery of a single therapeutic agent have been used for a long time, but only in recent years have specific polymers extended their use to the release of multi-agent therapy and multi-agent function. The ideal polymer carrier should be non-reactive in blood, nontoxic, nonimmunogenic, and should have suitable loading capacity. Some representative polymers used in clinical applications include PEG, N-(2-hydroxypropyl) methacrylamide (HPMA) copolymers, and polyglutamic acid (PGA). Therapeutic protein PEGylation has led to the development of numerous drugs, many of them have been approved and entered in the market such as PEG-granulocyte colony-stimulating factor (Neulasta™) (Duncan 2006). Methotrexate-conjugated PEGylated dendrimers showed differential patterns of deposition and activity after subcutaneous and intravenous administration in lymph nodes for specific targeting. These dendrimers demonstrated improved activity in lymph nodes compared to free methotrexate (MTX) (Kaminskas et al. 2015).

HPMA offers several advantages over other polymers, such as capability to functionalize via its side-chain hydroxyl group to incorporate antibodies or drugs, imaging agents, and specific target-ligands. The article focuses on recent advances in HPMA-based nanotheranostics. Particular attention is placed on polymer-drug conjugates, self-assembled nanoparticles, and other examples of HPMA-based nanotherapeutics. HPMA-drug conjugates demonstrated promising anticancer activity against prostate cancer stem cells. Many cancer cells overexpress certain surface receptors that can be selectively targeted on HPMA drugs; through the attachment of dexamethasone to the HPMA copolymer, a theranostic was synthesized. Ehling et al. utilized fluorescence molecular tomography with microcomputed tomography to make a hybrid imaging system to monitor the biodistribution of HPMA drugs (Greco and Vicent 2009; Ehling et al. 2014; Tucker and Sumerlin 2014).

In another study, MTX was used as the model anticancer drug to generate MTX-PGA-based nanoparticles as a theranostic platform. The results obtained indicated that the theranostic nanoparticles (MTX-PGA) are not only effective release vehicles for tumor-specific delivery of anticancer therapeutics such as MTX aimed at RFR (riboflavin receptor) (+ve) tumors but also have an added feature of simultaneous diagnosis (Dumoga et al. 2017).

16.2.2 Cardiovascular Disease

Cardiovascular diseases are invariably related to atherosclerosis and its sequelae, including myocardial infarction and cerebrovascular accidents. Cardiovascular disease and ischemic stroke are the main cause of disability-adjusted life year (DALY) and death in the worldwide (Benjamin et al. 2018). Conventional vascular therapies for cardiovascular disease commonly include oral, parenteral, intravenous, or intra-arterial injection, and administration of small molecules, which affect both diseased and healthy cells. However, the nonconventional and alternative treatments usually include intracardiac injections, grafts, scaffolds for tissue regeneration, and cardiac allograft vasculopathy (Riaz et al. 2016; Gupta et al. 2017; Alves et al. 2018). The drug instability and rapid drug clearance from circulation are limiting factors to obtaining good results in cardiovascular treatments (Gupta et al. 2017; Kim et al. 2017). The use of small molecule drugs in association with stents, balloon devices, and vascular grafts have been used to improve therapeutic efficacy, but these approaches still lack imaging exams (angiography, computed tomography angiography, magnetic resonance angiography, and duplex ultrasound) (Kim et al. 2017).

In general, cardiovascular disease is also an aging-associated chronic disease, and despite the signs of progress in understanding their etiology, unsolved questions still exist, and they will not be resolved or prevented by the unique advancement of personalized medicine. However, due to the molecular mechanisms of vascular disease, theranostic agents have been used in the experimental cardiac clinic. The first goals of theranostics are for individualized diagnosis, personalized treatments, and providing strategies to design specific nanomaterials. Nanotheranostics has selective target potential to achieve damaged cells and tissues and to monitor injury sites, delivery kinetics, drug distribution, and release, as well as therapeutic efficacy monitoring. The use of theranostic agents can perform localized cytotoxicity with little collateral damage, but the potential benefits of theranostics for patients include enhanced therapeutic effects with minimum adverse effects for conventional systemic administrations.

Atherosclerotic lesions offer a plethora of potential targets, including specific inflammatory cell types. Moreover, diseases such as myocardial infarction and acute coronary syndrome overproduce reactive oxygen species (ROS), thus generated ROS have become a critical target for diagnostics, therapeutics, and theranostics (Wang et al. 2014). The potential applications of cardiac theranostic nanomedicine formulations range from the noninvasive assessment of the biodistribution to the target site accumulation of low molecular-weight drugs. Moreover, it also includes visualization of drug distribution, drug release at the heart target site, the optimization of strategies relying on triggered drug release, and the prediction and real-time monitoring of therapeutic responses.

The cellular biomarkers specific to endothelial cells (Integrin $\alpha V\beta 3$, CAMS [chorioallantoic membrane]), platelets (integrin's GPIIb-IIIa, GPIa-IIa, GPVI, GPIb-IX-V, p-selectin), activated macrophages (scavenger receptors, oxidized LDL [low-density lipoprotein] receptors), and smooth muscle cells can be used as cellular targets for identification of these diseases at early stages. Non-cellular biomarkers include ECM (extracellular matrix) proteins, lipoproteins, annexins, fibrin, MMPs (matrix metalloproteinases), and tissue factor. Theranostic targeted nanoparticles with specific cellular and non-cellular biomarkers have been mostly designed and developed. The nanoparticles can be functionalized with specific ligands/antibodies/peptides to target the specific vascular regions affected (Tang et al. 2012; Gupta et al. 2017; Kim et al. 2017).

YKL-40 (40 kDa heparin- and chitin-binding glycoprotein) is an acute phase protein expressed by several cell types of the immune system by chemical vapor deposition (CVD) patients, including macrophages. Some studies have shown that serum YKL-40 is increased in patients with acute myocardial infarction, and it is associated with both early and late phases in the development of atherosclerosis. Moreover, YKL-40 stimulates the maturation of monocytes to macrophages and then is hidden by macrophages during the late stages of differentiation (Zhang et al. 2018). Nanotheranostic agents can increase the circulation time of therapeutic or diagnostic agents, which ultimately helps to reduce the dose and off-target toxicity of the drugs. The nanoparticle's origins can be labeled as inorganics such as gold, silicon, silica, silver, and metal oxides for alternative nanocarriers, such as perfluorocarbon nanoemulsions. Lipid-based nanoparticles, micelles, polymeric nanoparticles, gel-like nanoparticles, and dendrimers have been used as a vehicle to carry drugs in cardiovascular treatments and diagnosis (Tang et al. 2012).

Macrophages have been the most studied and used target because of their pathogenic roles, phagocytic property, and abundance in the atherosclerosis lesion (Zhang et al. 2019a). The macrophages can phagocytose the theranostic nanoparticles into the atherosclerosis lesion site when the nanomaterials are modified with molecules such as dextran or mannose. There are two strategies for the treatment of atherosclerosis targeting macrophages which mainly include macrophage ablation and macrophage repair. Macrophage ablation therapy can affect homeostasis and lead to infections, while non-ablation methods reduce the pathogenic activity of macrophages. For example, phosphatidylserine (PS) presenting liposomes were developed for the modulation of cardiac macrophages in post-myocardial infarction (Patel and Janjic 2015; Zhang et al. 2019a).

The nanotheranostic agents (nanocomposites, nanoconjugates, and nanocomplexes) with optical and magnetic properties associated with contrast agents have been used as vehicles for the delivery of therapeutic molecules, or directly in treatment. For example, the combination of optical and plasmonic properties of gold nanoparticles, with interactions between drugs and polymers, can be considered in cardiovascular diagnosis and treatment (Bejarano et al. 2018). Doxorubicin-loaded hyaluronic acid–polypyrrole nanoparticles (DOX-HA–PPyNPs) were designed with pH-responsive properties as a smart theranostic platform for noninvasive fluorescence imaging and therapy of proliferating macrophages in atherosclerotic lesions. A cytotoxicity test and confocal microscopy study performed revealed that the therapeutic effect of DOX-HA–PPyNPs on the proliferating macrophage cells was enhanced in part due to CD44-mediated endocytosis of the nanoparticles (Park et al. 2014).

Targeted dendrimers incorporating LyP-1, a cyclic 9-amino acid peptide, were developed with the aim of binding it to p32 proteins on macrophages. After incorporating ^{64}Cu into the dendrimers for PET imaging, authors observed that the conjugated LyP-1 was accumulated in the aorta and atherosclerotic plaque-containing regions. The hypothesis is that the subcutaneous injection of targeted dendrimers could produce a sustained delivery of conjugated therapeutics to atherosclerotic plaque (Seo et al. 2014). Kaminskas et al. (2015) reported that subcutaneous administration of drug-conjugated dendrimers provides an opportunity to improve MTX deposition in downstream tumor-burdened lymph nodes.

Solid lipid nanoparticles (SLNs) with nucleoside-lipids loaded with iron oxide particles and therapeutic agents (theranostics) were developed. The insertion of nucleoside-lipids allowed the formation of stable SLNs loaded with prostacyclinable to inhibit platelet aggregation. Compared to the clinically used contrast agents (Feridex), the SLNs have higher magnetization properties with 2.6-fold higher transverse relaxivity at 4.7 T. The relaxivity of the contrast agent is expressed in how the relaxation rate of a solution that is triggered by solute concentration. It is noteworthy that the insertion of the fragile prostacyclin into the SLNs maintains its bioactivity as shown by complete inhibition of platelet aggregation (Bejarano et al. 2018).

As a theranostic, polymer-lipid hybrid nanoparticles conjugated with C11 (polypeptide with high affinity and specificity for collagen IV) co-delivering ultra-small superparamagnetic iron oxide (USPIO) and paclitaxel (UP-NP-C11) were developed by Dong et al. (2016). After binding C11-collagen IV, paclitaxel was released to treat the atherosclerotic plaque, and simultaneously, the macrophages engulfed the USPIO in atherosclerotic plaque. Table 16.2 shows the details of various nanotherapy used for imaging in cardiovascular illness as nanomedical theranostics.

Duivenvoorden et al. (2014) developed a statin-loaded reconstituted high-density lipoprotein (rHDL)-based nanoparticle for delivery of statins at atherosclerotic lesions. Atorvastatin calcium (AT)-loaded dextran sulfate (DXS)-coated core-shell rHDL was fabricated by encapsulating spherical PLGA nanoparticles into liposomes, followed by incubation with apoA-I, then decoration with DXS on the surface. These nanoparticles were developed for targeted drug AT to macrophages and suppression of inflammation via the high affinity of DXS with SR-AI (scavenger

Table 16.2 Evaluation of nanotherapy by imaging in cardiovascular illness as nanomedical theranostic.

Illness	Agent	Action	Evaluation	Result
Atherosclerosis USPIO induced	p38 kinase inhibitor	Immune response	MRI: decreased accumulation of USPIO in the aortic root and arch area	USPIO was associated with macrophages
Vascular injury/ Stenosis inhibition	rapamycin-loaded $\alpha v\beta 3$ – integrin – targeted PNp	PNp uptake in injured arteries	Improved therapeutic effect and adjust local drug concentrations by controlling the external magnetic field.	Patent and less stenotic arteries
Peripheral arterial disease	Gold nanoparticle	Improving perfusion	Laser Doppler perfusion imaging-noinvasive	VEGF-conjugated gold NP increased perfusion of ischemic musculature
Limiting reperfusion injury	Adenosine-liposome tailored	Reperfusion therapy	Fluorescence and TEM images	Accumulation of liposomes in infarcted areas, enhanced circulation time, reduced side effects
Thrombosis	PPACK – PFC NPc	Anti-thrombotic	Proton MRI and 19F-MRI	NP specifically accumulated at thrombotic sites

Note: USPIO, ultrasmall particles of iron oxide; MRI, magnetic resonance contrast; PNp, paramagnetic nanoparticle; VEGF, vascular endothelial growth factor; Np, Nanoparticle; TEM, transmission electronic microscopy; PPACK, Phe[D]-Pro-Arg-Chloromethylketone; PFC, perfluorocarbon; NPc, nanoparticle conjugated.

receptor type AI) as well as depletion of intracellular cholesterol by apolipoprotein A-I (apoA-I), mediated cholesterol efflux (Zhao et al. 2017).

High-density lipoprotein-like magnetic nanostructures (HDL-MNSs) offer prospects for diagnosis via noninvasive MRI for anatomic detection and serve as effective cholesterol efflux agents to address atherosclerotic vascular lesions. Nandwana et al. (2017) synthesized HDL-MNS by coating phospholipids and apoA1 onto MNSs to mimic the surface composition of natural HDL. The MNSs are composed of Fe_3O_4 nanoparticles, biocompatible T_2 contrast agents for MRI. The internalization of HDL-MNS particles by macrophage cells was confirmed by transmission electronic microscopy (TEM)/EDS (energy dispersive spectrometry) and ICP-MS (inductively coupled plasma-mass spectrometry), demonstrating the ability of HDL-MNS to label the relevant cell type in atherosclerotic plaques. The macrophages that internalized HDL-MNS were analyzed by MRI and data showed a higher T_2 contrast enhancement than commercial T_2 contrast agents (Nandwana et al. 2017).

Wu et al. (2018) synthesized a theranostic nanoparticle comprising an iron oxide(core) and cerium oxide (shell) as a potential theranostic nanomaterial for ROS-related inflammatory diseases. The iron oxide-cerium oxide core-shell nanoparticles (IO-CO) showed potential agent for diagnosis and treatment of ROS-related inflammatory diseases such as atherosclerotic.

Zhao et al. (2018) synthesized targeting theranostic nanoprobes by encapsulation of the PCM (primary cardiomyocyte) conjugated, 17β-estradiol(E2)-loaded, and PFP (perfluoropentane) (PCM-E2/PFPs). *In vitro* and *in vivo* studies confirmed that PCM-E2/PFPs can be used as an amplifiable imaging contrast agent when exposed to low-intensity focused ultrasound (LIFU). Besides that, it accelerated the release of E2, enhancing the therapeutic efficacy of the E2 and preventing systemic side effects. The combined PCM-E2/PFPs and LIFU treatment also significantly increased cardiac targeting and circulation time and suppressed cardiac hypertrophy (Zhao et al. 2018).

The ability to detect and treat CVD is the main concern of clinical medicine. With the advantage of nanotechnology and the generation of multifunctional agents (theranostics), it becomes possible to diagnose and to treat the disease. There are many advantages to this approach, such as the ability to determine agent localization, release, or efficacy with low invasive treatment and great results.

16.2.3 Neurodegenerative Diseases

The screening of neurodegenerative disorders and their early detection is an essential step in slowing the progression of the disease and having time to choose more effective treatment. The size of theranostic nanoparticles for clinical treatment in neurodegenerative diseases has been limited to 220 nm when they are administered by parenteral route. However, for the other administration routes like naso-brain delivery, the size of nanoparticles should be in the range of 10–100 nm. An ideal theranostic nanoparticle required for the treatment, diagnosis, and to measure the neurological disorders evolution should be designed with the utmost precision. The theranostic nanoparticles for neurodegenerative disease need associated properties and particular purpose to overcome biological barriers crossing the blood-brain barrier, from the bloodstream to the brain, or naso-brain by olfactory bulb way.

Among hundreds of other drugs, the metallic nanoparticles have been pointed out as promising theranostic agents for neurodegenerative disease and cerebral tumor for earlier diagnosis as well as good responsive therapeutics (Kang et al. 2015; Aioub et al. 2018; Li et al. 2018; Ramanathan et al. 2018). Oxidative and nitrative stress accompanies with many diseases, including neurodegenerative disorders and sometimes age-related disorders. Cerium oxide nanotheranostics are being investigated for their efficacy in several neurodegenerative disorders.

The fine ceria oxide powder was prepared by the precipitation method to build the oxygen sensors in the thick film formed with a particle size of the 120 nm. Thus, the prepared thick film showed excellent adhesion to alumina substrate and also reported to have promising levels of neuroprotection. The alumina-ceria-oxide nanoparticles have potential theranostic ability for neuronal disorders. This study also confirmed the significant role of particle size in shortening response time (Izu et al. 2004). Redox nanoparticles have many potential applications in nanotheranostics and this subject has been studied a lot. The nano-size and surface hydrophobicity of cerium oxide redox-active nanoparticles make them capable of penetrating altered cells at a molecular level, crossing the blood-brain barrier.

The preparation and theranostic application of redox-active nanoparticles, and concentrates on nitroxide radicals for Alzheimer's disease and Parkinson's disease, was revisited by Sadowska-Bartosz and Bartosz (2018), in the excellent revision about redox theranostic nanoparticles showed low toxicity and ability to penetrate the blood-brain barrier will be more efficient in the treatment of neurodegenerative disease (Sadowska-Bartosz and Bartosz 2018).

The antioxidant properties can help remove ROS and thus decrease cell damage caused by free radicals. Due to biomimetic redox-active nanoparticles having long-term blood circulation and low toxicity, they also could protect against ionizing radiation (Kang et al. 2015). Once the nanotheranostic reaches the tissue of interest, an additional control bonding method between the drug and nanotheranostic system can be manipulated to a precise tune for drug release stimuli-responsiveness (pH, temperature, magnetic field, shear stress, and enzymes). When nanotheranostics have peroxalate bonds a cleavage will release an antioxidant molecule, such as vanillyl alcohol (Kim et al. 2017).

In neurodegenerative disease, nanotheranostics have a wide area of studies for the application of multidisciplinary knowledge. Using nanoparticles with nanocarriers carefully engineered and bioinspired for delivery antibodies in site-targeted brain illnesses such as Alzheimer's, Parkinson's, Huntington's, and multiple sclerosis could be cured or have the progression delayed.

16.3 Vaccine

Theranostic vaccine is an important conjugation concept that refers to the integration of diagnostics, bioagent release, and monitorization of therapeutic benefits. There are many advantages to combining therapeutic functionalities with molecular images, which mainly include planning of treatment and monitoring of disease regression. Both of these are based on the molecular characteristics of the illness. A practical alternative in theranostic vaccines that is gaining attention is using nano-engineered particles, functioning as drug-delivery systems or as antigen-delivery systems themselves, affecting off-target and the approach for releasing adjuvants and antibodies' tumor specific-sites (Liu et al. 2018).

The immune-therapeutic theranostic strategies developed based on DNA vaccine has created new perspectives in the treatment of cancer, autoimmune diseases (Zhang and Nandakumar 2018), allergies (Hobernik and Bros 2018), and infectious diseases (Maslow 2017). The DNA vaccine might have some prerequisites which are responsible for its functionalities as a nanocarrier, then to the direction the DNA vector to APCs. A nuclear localization signal should contain a promoter that facilitates the transcriptional targeting of APCs and avoid the undesired expression of antigen in myeloid-derived suppressor cells (Poecheim et al. 2015; Hobernik and Bros 2018).

Despite the aggregate therapeutic values related to the specificity, safety, stability, and alternatives of the routes for the administration of the theranostic vaccine, the cost-benefit relation must be considered from different economic points of view, especially in endemic or pandemic situations. The theranostic vaccine has, in general, a biomimetic carrier in the vesicular form based on lipids or polymers. Biomimetic vesicles are spherical structures with lipid bilayers of nanometric scale, which can be used to carry bioactive materials through membranes and the bloodstream (Schwendener 2014; Kour et al. 2018).

Biological membranes consist of a complex mixture of different lipid types and various integral or associated proteins. The phospholipids are the main lipids of membranes, and they are amphiphilic molecules and have excellent biocompatibility characteristics. Due to physicochemical properties, when phospholipids are dispersed in aqueous solutions they tend to form membranes. The vesicles formed by phospholipids are a suitable group for biomaterial engineering and vaccine carriers. A variety of nanovesicles have been used for application as theranostic vaccines, including exosomes, mimetic nanovesicles of exosomes, liposomes, virosomes, and archeosomes.

Exosomes are small (30–150 nm) endosomal-derived vesicles, that are secreted by most cells into the extracellular space, can enter the bloodstream, and travel to distant organs and tissues (Kim et al. 2017; Alves et al. 2019). Being rich in protein, lipids, and nucleic acids, the exosomes are involved in several physiological processes, such as cell–cell communication, signal transduction transport of genetic material, and modulation of the immune response (Wang et al. 2014). Delivery of exosomes by immune and tumor cells and their multifaceted roles reinforce the importance of these vesicles as ideal candidates for therapeutic, diagnostic, and prognostic targets (Jang et al. 2013). Admyre et al. (2006), for the first time, showed that exosomes can directly stimulate human peripheral CD8+ T cells in an antigen-specific manner and ELISPOT (enzyme-linked immunospot assay) is a suitable method for detecting exosome-induced peripheral T-cell responses and could also induce tumor necrosis factor (TNF)-α production (Admyre et al. 2006).

Devhare and Ray (2017) comment about the use of exosomes for the transport of antigen; biocompatible physiological exosomes released by antigen-presenting dendritic cells are known to express major molecule histocompatibility complexes of class I and II that play significant roles in the activation of lymphocyte and immune cell functions. Though the immunogenic properties of pathogen-derived exosomal antigens have been tested in several preclinical models and diseases of parasitic and viral origin, the circulation of viral antigens through the exosomes in the serum of the host with no pathogen load detected in peripheral circulation (NV [non-viremic]) suggested the importance of this study for a novel vaccine approach (Montaner et al. 2014; Devhare and Ray 2017). This exosomal protein-mediated antigenic activity was very similar to the antigenic activity contained in commercially available vaccines (Porcilis PRRSV [porcine reproductive and respiratory syndrome virus] vaccine, Intervet, Boxmeer, The Netherlands) (Devhare and Ray 2017).

Liposomes are biomimetic nanovesicles produced synthetically as bioactive carriers for the cells. The similarities of the small unilamellar liposome with exosomes may allow the developing of liposomes as antigen-carrier systems. The liposome is surface-functionalized using the methodologies developed in the field of technology for vaccines, drug delivery, imaging agents, and genes. In addition, the use of cationic liposomes allows the adsorption of polyanions, such as DNA and RNA (Antimisiaris et al. 2018; Zhi et al. 2018). Cationic liposomes are generally very suitable for transfection, since they preferentially interact with the anionic cell membrane, thereby facilitating the incorporation of the transfection complex into the cell by endocytosis (Hobernik and Bros 2018). Since the ability of liposomes to induce immune responses to incorporated or associated antigens has been reported, liposomes and liposome-derived nanovesicles, such as aracheosomes and virosomes, have become essential carrier systems.

Virosomes are liposomes usually prepared by combination of phospholipids, viral spicule glycoproteins, and other viral proteins. The virosomes incorporate the antigen on the surface of the vesicle and thus imitate a virus, and then the virosomes can be, biomimetically, considered as a type of viral particle (Tregoning et al. 2018). By mimicking a virus, virosomes can aid in antigen uptake in APCs, promote cellular activation and trafficking within the lymphatic system (Blom et al. 2017).

Virosomes with antigen exposed to the surface can also increase antibody responses by improving the 3D structure of the antigen, increasing antigen density, which leads to increased cross-linking of B cell receptors (Tregoning et al. 2018). Progress was made with influenza virosomes, which are small unilamellar lipid vesicles (SUVs) with peak projections of the surface glycoproteins of influenza hemagglutinin and neuraminidase (Schwendener 2014), when only the fusogenic properties of hemagglutinin are its main characteristics (Hamilton et al. 2012). These virosomes allow the presentation of antigen in the context of major histocompatibility complex (MHC)-I and MHC-II and induce B- and T-cell responses (Blom et al. 2017).

Aracheosomes are liposomes prepared with glycolipids (glycerol-ether polar lipids of archaea) from archaebacteria (archaea) and cholesterol. Archaea lipids have unique characteristics obtained from eukaryotes and bacteria; these are amphiphilic of isoprenoid chains, which offer numerous pharmaceutical advantages and excellent physicochemical stability (Kour et al. 2018). Aracheosomes increase the presentation of antigen to APC by means of the MHC-I and MHC-II pathways, without producing an inflammatory response which can release biotics to specific cell types, including macrophages, phagocytes, and antigen-processing cells (Patel and Sprott 1999; Patel and Chen 2005; Kour et al. 2018). Systemic immunization of antigen-containing aracheosomes yields strong immune responses with the specific systemic immunity, characterized by the high population of CD8 + responses of cytotoxic T lymphocytes (Patel and Chen 2005; Schwendener 2014).

The enormous versatility of liposomes and the corresponding archaeosomes and virosomes confer them as highly valuable transport systems for theranostics. However, enhanced antigen stability and presentation to immunocompetent cells is dependent on their specific properties such as composition, size, and surface properties. These nanocarriers can also overcome biological barriers such as skin and mucosa and provide a controlled release and slowing down of antigens. Together with the ability to induce potent immune responses provided by nanocarriers, mimetic nanovesicle-based theranostics provide properties that are critical to the development of innovative treatments for diseases that so far have no cure. These release systems are expected to be increasingly successfully applied shortly, leading to significant improvements in the development of theranostics vaccines.

16.4 Conclusions

The development of theranostic nanoparticles highlights recent approaches in the development of therapeutic tools to enable real-time diagnosis and treatment of chronic diseases in more different areas. Nanotheranostics as nanocarriers administered via the noninvasive mucosal route as a buccal and nasal way have been shown to play an important role as well as preventive and curative action, due to their tunable and unique biological and physicochemical properties. Apart from engineering, the design of new biomaterials or composites for novel therapies in immune-mediated diseases restoring drugs or antigen-specific or antigen-nonspecific tolerance remains an attractive objective for the treatment of chronic and progressive diseases such as cancer, cardiovascular, infectious, and neurodegenerative diseases.

The vaccine for chronic inflammatory autoimmune diseases is the major challenge of antigen-specific immunotherapy. Plasmid-based reverse genetic methods have provided scientists with a unique opportunity to study different aspects of the biology and pathogenesis of these viruses. Even though the vast majority of studies have been conducted in small animal models that have significantly different anatomies compared to humans. However, further extensive, prudent toxicological and pharmacological studies are necessary before considering the clinical translation of theranostics. Moreover, the clinical trials should be carefully designed to verify the feasibility of such potential theranostic candidates. The translation may be accelerated through tethering of biologically active substances such as peptide and nucleic acids to the application of the ex-vivo human model. So collaborative efforts between basic and applied scientists and clinicians are necessary to implement and translating this theranostic technology.

References

Aalinkeel, R., Kutscher, H.L., Singh, A. et al. (2018). Neuroprotective effects of a biodegradable poly(lactic-co-glycolic acid)-ginsenoside Rg3 nanoformulation: a potential nanotherapy for Alzheimer's disease. *Journal of Drug Targeting* 26 (2): 182–193.

Admyre, C., Johansson, S.M., Paulie, S., and Gabrielsson, S. (2006). Direct exosome stimulation of peripheral human T cells detected by ELISPOT. *European Journal of Immunology* 36 (7): 1772–1781.

Ahmed, N., Fessi, H., and Elaissari, A. (2012). Theranostic applications of nanoparticles in cancer. *Drug Discovery Today* 17 (17–18): 928–934.

Aioub, M., Austin, L.A., and El-Sayed, M.A. (2018). Gold nanoparticles for cancer diagnostics, spectroscopic imaging, drug delivery, and plasmonic photothermal therapy. In: *Inorganic Frameworks as Smart Nanomedicines*, vol. 1 (ed. A. Grumezescu), 41–91. Elsevier.

Alves, T.F.R., de Souza, J.F., Severino, P. et al. (2018). Dense lamellar scaffold as biomimetic materials for reverse engineering of myocardial tissue: preparation, characterization and physiomechanical properties. *Journal of Material Science and Engineering* 07 (05): 494–502.

Alves, T.F., Souza, J.F., Amaral, V. et al. (2019). Biomimetic dense lamellar scaffold based on a colloidal complex of the polyaniline (PANi) and biopolymers for electroactive and physiomechanical stimulation of the myocardial. *Colloids and Surface A* 579: 123650.

Antimisiaris, S.G., Mourtas, S., and Marazioti, A. (2018). Exosomes and exosome-inspired vesicles for targeted drug delivery. *Pharmaceutics* 10 (4): E218.

Bejarano, J., Navarro-Marquez, M., Morales-Zavala, F. et al. (2018). Nanoparticles for diagnosis and therapy of atherosclerosis and myocardial infarction: evolution toward prospective theranostic approaches. *Theranostics* 8 (17): 4710–4732.

Benjamin, E.J., Virani, S.S., Callaway, C.W. et al. (2018). Heart disease and stroke statistics - update: a report from the American Heart Association. *Circulation* 137: 67–492.

Blom, R.A.M., Amacker, M., van Dijk, R.M. et al. (2017). Pulmonary delivery of virosome-bound antigen enhances antigen-specific CD4+ T cell proliferation compared to liposome-bound or soluble antigen. *Frontiers in Immunology* 8 (4): 1–17.

Ciechanover, A. and Kwon, Y.T.A. (2015). Degradation of misfolded proteins in neurodegenerative diseases: therapeutic targets and strategies. *Experimental and Molecular Medicine* 47: e147.

Devhare, P.B. and Ray, R.B. (2017). A novel role of exosomes in the vaccination approach. *Annals of Translational Medicine* 5 (1): 23–25.

Dong, Y., Chen, H., Chen, C. et al. (2016). Polymer-lipid hybrid theranostic nanoparticles co-delivering ultrasmall superparamagnetic iron oxide and paclitaxel for targeted magnetic resonance imaging and therapy in atherosclerotic plaque. *Journal of Biomedical Nanotechnology* 12 (6): 1245–1257.

Dreifuss, T., Betzer, O., Shilo, M. et al. (2015). A challenge for theranostics: is the optimal particle for therapy also optimal for diagnostics. *Nanoscale* 7: 15175–15184.

Duivenvoorden, R., Tang, J., Cormode, D.P. et al. (2014). A statin-loaded reconstituted high-density lipoprotein nanoparticle inhibits atherosclerotic plaque inflammation. *Nature Communications* 5: 1–12.

Dumoga, S., Rai, Y., Bhatt, A.N. et al. (2017). Block copolymer based nanoparticles for theranostic intervention of cervical cancer: synthesis, pharmacokinetics, and in vitro/in vivo evaluation in HeLa xenograft models. *ACS Applied Materials and Interfaces* 9 (27): 22195–22211.

Duncan, R. (2006). Polymer conjugates as anticancer nanomedicines. *Nature Reviews Cancer* 6: 688–701.
Ehling, J., Storm, G., Deckers, R. et al. (2014). Characterizing EPR-mediated passive drug targeting using contrast-enhanced functional ultrasound imaging. *Journal of Controlled Release* 182: 83–89.
Fan, J.R., Peng, A.L., Chen, H.C. et al. (2008). Cellular processing pathways contribute to the activation of etoposide-induced DNA damage responses. *DNA Repair* 7 (3): 452–463.
Greco, F. and Vicent, M.J. (2009). Combination therapy: opportunities and challenges for polymer-drug conjugates as anticancer nanomedicines. *Advanced Drug Delivery Reviews* 61 (13): 1203–1213.
Gupta, M.K., Lee, Y., Boire, T.C. et al. (2017). Recent strategies to design vascular theranostic nanoparticles. *Nanotheranostics* 1 (2): 166–177.
Hamilton, B.S., Whittaker, G.R., and Daniel, S. (2012). Influenza virus-mediated membrane fusion: determinants of hemagglutinin fusogenic activity and experimental approaches for assessing virus fusion. *Viruses* 4 (7): 1144–1168.
Hobernik, D. and Bros, M. (2018). DNA vaccines-how far from clinical use? *International Journal of Molecular Sciences* 19 (11): 3605.
Izu, N., Shin, W., Matsubara, I., and Murayama, N. (2004). Development of resistive oxygen sensors based on cerium oxide thick film. *Journal of Electroceramics* 13: 703–706.
Jang, S.C., Kim, O.Y., Yoon, C.M. et al. (2013). Bioinspired exosome-mimetic nanovesicles for targeted delivery of chemotherapeutics to malignant tumors. *ACS Nano* 7 (9): 7698–7710.
Kaminskas, L.M., McLeod, V.M., Ascher, D.B. et al. (2015). Methotrexate-conjugated PEGylated dendrimers show differential patterns of deposition and activity in tumor-burdened lymph nodes after intravenous and subcutaneous administration in rats. *Molecular Pharmaceutics* 12 (2): 432–443.
Kang, H., Mintri, S., Menon, A.V. et al. (2015). Pharmacokinetics, pharmacodynamics and toxicology of theranostic nanoparticles. *Nanoscale* 7 (45): 18848–18862.
Kim, K.S., Song, C.G., and Kang, P.M. (2017). Targeting oxidative stress using nanoparticles as a theranostic strategy for cardiovascular diseases. *Antioxidants and Redox Signaling* 30 (5): 733–746.
Kour, P., Rath, G., Sharma, G., and Goyal, A.K. (2018). Recent advancement in nanocarriers for oral vaccination. *Artificial Cells, Nanomedicine and Biotechnology* 46 (Sup3): S1102–S1114.
Li, Y., Wei, Q., Ma, F. et al. (2018). Surface-enhanced Raman nanoparticles for tumor theranostics applications. *Acta Pharmaceutica Sinica B* 8 (3): 349–359.
Liu, Y., Wang, X., Hussain, M. et al. (2018). Theranostics applications of nanoparticles in cancer immunotherapy. *Medical Sciences* 6 (4): 100–106.
Maslow, J.N. (2017). Vaccines for emerging infectious diseases: lessons from MERS coronavirus and Zika virus. *Human Vaccines and Immunotherapeutics* 13 (12): 2918–2930.
Meir, R., Shamalov, K., Betzer, O. et al. (2015). Nanomedicine for cancer immunotherapy: tracking cancer-specific t-cells in vivo with gold nanoparticles and CT imaging. *ACS Nano* 9 (6): 6363–6372.
Montaner, S., Galiano, A., Trelis, M. et al. (2014). The role of extracellular vesicles in modulating the host immune response during parasitic infections. *Frontiers in Immunology* 5: 433–444.
Montecucco, A., Zanetta, F., and Biamonti, G. (2015). Molecular mechanisms of etoposide. *EXCLI Journal* 14 (19): 95–108.
Nandwana, V., Ryoo, S.R., Kanthala, S. et al. (2017). High-density lipoprotein-like magnetic nanostructures (HDL-MNS): theranostic agents for cardiovascular disease. *Chemistry of Materials* 29 (5): 2276–2282.
Park, D., Cho, Y., Goh, S.H., and Choi, Y. (2014). Hyaluronic acid-polypyrrole nanoparticles as pH-responsive theranostics. *Chemical Communications* 50 (95): 15014–15017.
Patel, G. and Chen, W. (2005). Archaeosome immunostimulatory vaccine delivery system. *Current Drug Delivery* 2 (4): 407–421.

Patel, S.K. and Janjic, J.M. (2015). Macrophage targeted theranostics as personalized nanomedicine strategies for inflammatory diseases. *Theranostics* 5 (2): 150–172.

Patel, G.B. and Sprott, G.D. (1999). Archaeobacterial ether lipid liposomes (Archaeosomes) as novel vaccine and drug delivery systems. *Critical Reviews in Biotechnology* 194: 317–357.

Poecheim, J., Heuking, S., Brunner, L. et al. (2015). Nanocarriers for DNA vaccines: co-delivery of TLR-9 and NLR-2 ligands leads to synergistic enhancement of proinflammatory cytokine release. *Nanomaterials* 5 (4): 2317–2334.

Ramanathan, S., Archunan, G., Sivakumar, M. et al. (2018). Theranostic applications of nanoparticles in neurodegenerative disorders. *International Journal of Nanomedicine* 13: 5561–5576.

Rezonja, R., Knez, L., Cufer, T., and Mrhar, A. (2013). Oral treatment with etoposide in small cell lung cancer-dilemmas and solutions. *Radiology and Oncology* 47 (1): 1–13.

Riaz, N., Wolden, S.L., Gelblum, D.Y., and Eric, J. (2016). Percutaneous endoscopic gastrostomy in oropharyngeal cancer patients treated with intensity-modulated radiotherapy with concurrent chemotherapy. *HHS Public Access* 118 (24): 6072–6078.

Rodeheffer, R.J. and Redfield, M.M. (2013). Pharmacologic therapy of systolic ventricular dysfunction and heart failure. In: *Mayo Clinic Cardiology Concise Textbook*, 4e (eds. M.A.L. Joseph and G. Murphy), 838. Oxford: Mayo Clinic Scientific Press.

Sadowska-Bartosz, I. and Bartosz, G. (2018). Redox nanoparticles: synthesis, properties and perspectives of use for treatment of neurodegenerative diseases. *Journal of Nanobiotechnology* 16 (1): 1–16.

Schwendener, R.A. (2014). Liposomes as vaccine delivery systems: a review of the recent advances. *Therapeutic Advances in Vaccines* 2 (6): 159–182.

Scott, E. (2018). Immune theranostics. In: *National Academy of Engineering. Frontiers of Engineering: Reports on Leading-Edge Engineering*, 1e, vol. 1. Washington, DC: National Academies Press.

Seo, J.W., Baek, H., Mahakian, L.M. et al. (2014). Cu-labeled LyP-1-dendrimer for PET-CT imaging of atherosclerotic plaque. *Bioconjugate Chemistry* 25 (2): 231–239.

Srinivas, R., Satterle, A., Wang, Y. et al. (2015). Theranostic etoposide phosphate/indium nanoparticles for cancer therapy and imaging. *Nanoscale* 7 (44): 8542–18551.

Tang, J., Lobatto, M.E., Read, J.C. et al. (2012). Nanomedical theranostics in cardiovascular disease. *Current Cardiovascular Imaging Reports* 5 (1): 19–25.

Tregoning, J.S., Russell, R.F., and Kinnear, E. (2018). Adjuvanted influenza vaccines. *Humman Vaccines and Immunotherapeutics* 14 (3): 550–564.

Tucker, B.S. and Sumerlin, B.S. (2014). Poly(N-(2-hydroxypropyl) methacrylamide)-based nanotherapeutics. *Polymer Chemistry* 5 (5): 1566–1572.

Wang, J. and Cui, H. (2016). Nanostructure-based theranostic systems. *Theranostics* 6 (9): 1274–1276.

Wang, D., Lin, B., and Ai, H. (2014). Theranostic nanoparticles for cancer and cardiovascular applications. *Pharmaceutical Research* 31 (6): 1390–1406.

Wu, Y., Yang, Y., Zhao, W. et al. (2018). Novel iron oxide-cerium oxide core-shell nanoparticles as a potential theranostic material for ROS related inflammatory diseases. *Journal of Materials Chemistry B* 6 (30): 4937–4951.

Yang, L., Hao, J., Zhang, J. et al. (2009). Ginsenoside Rg3 promotes beta-amyloid peptide degradation by enhancing gene expression of neprilysin. *Journal of Pharmacy and Pharmacology* 61 (3): 375–380.

Zhang, N. and Nandakumar, K.S. (2018). Recent advances in the development of vaccines for chronic inflammatory autoimmune diseases. *Vaccine* 36 (23): 3208–3220.

Zhang, H., Zhou, W., Cao, C. et al. (2018). Amelioration of atherosclerosis in apolipoprotein e-deficient mice by combined RNA interference of lipoprotein-associated phospholipase A 2 and YKL-40. *PLoS One* 13 (8): 1–12.

Zhang, Y., Koradia, A., Kamato, D. et al. (2019a). Treatment of atherosclerotic plaque: perspectives on theranostics. *Journal of Pharmacy and Pharmacology* 71: 1029–1043.

Zhang, Y., Bush, X., Yan, B., and Chen, J.A. (2019b). Gemcitabine nanoparticles promote antitumor immunity against melanoma. *Biomaterials* 189: 48–59.

Zhao, Y., Jiang, C., He, J. et al. (2017). Multifunctional dextran sulfate-coated reconstituted high density lipoproteins target macrophages and promote beneficial antiatherosclerotic mechanisms. *Bioconjugate Chemistry* 28 (2): 438–448.

Zhao, X., Luo, W., Hu, J. et al. (2018). Cardiomyocyte-targeted and 17β-estradiol-loaded acoustic nanoprobes as a theranostic platform for cardiac hypertrophy. *Journal of Nanobiotechnology* 16 (1): 1–14.

Zhi, D., Bai, Y., Yang, J. et al. (2018). A review on cationic lipids with different linkers for gene delivery. *Advances in Colloid and Interface Science* 253: 117–140.

17

Toxicological Concerns of Nanomaterials Used in Biomedical Applications

Avinash P. Ingle[1], *Indarchand Gupta*[2,3], *and Mahendra Rai*[3]

[1] *Department of Biotechnology, Engineering School of Lorena, University of Sao Paulo, Lorena, SP, Brazil*
[2] *Department of Biotechnology, Government Institute of Science, Nipat-Niranjan Nagar, Aurangabad, Maharashtra, India*
[3] *Nanobiotechnology Laboratory, Department of Biotechnology, SGB Amravati University, Amravati, Maharashtra, India*

17.1 Introduction

Nanotechnology is considered one of the most promising technologies of the twentieth century and it has tremendous potential to revolutionize various sectors, including the biomedical field. Nanotechnology deals with the manipulation of matter on atomic and molecular levels (Thiruvengadam et al. 2018; Seqqat et al. 2019). Various nanomaterials including nanoparticles are the building blocks of nanotechnology and according to the International Organization for Standardization, nanoparticles are the structures whose sizes in one, two, or three dimensions are within the range from 1 to 100 nm (Sukhanova et al. 2018). Nanomaterials particularly exhibit unique, fascinating, and useful characteristics which are completely different from their bulk counterparts. As the size of nanomaterials (nanoparticles) decreases, its different properties like surface to volume ratio, chemical reactivity, etc. increases. On one hand, these properties make them suitable for various applications including biomedical applications, however on the other hand, their extremely small size and high reactivity facilitates an increase in biological responses such as cellular uptake and delivery efficiency, which ultimately gives rise to toxicological concern in biological systems (Khan et al. 2019; Yang et al. 2019).

Over the last few years, nanomaterials have attracted considerable interest for their use in different biomedical applications due to these aforementioned novel properties. The important biomedical applications include diagnosis (bioimaging, biosensors, etc.), targeted drug delivery, and treatment (hyperthermia, photoablation therapy, etc.) for many diseases. To date, various attempts have been made for the development of novel nanomaterials which could possess excellent properties such as chemical stability, nontoxicity, biocompatibility, etc. (McNamara and Tofail 2017). In this context, a variety of nanomaterials, e.g. metal and metal oxide nanoparticles, polymeric nanoparticles, carbon-based nanoparticles, liposomes, solid lipid nanoparticles, dendrimers, micelles, nanocapsules, and so on have been developed and successfully used in different biomedical applications (Gim et al. 2019; Jeong et al. 2019; Mitragotri and Stayton 2019). Among these nanomaterials, metal nanoparticles were reported to have certain toxic effects on normal cells and tissues.

Nanobiotechnology in Diagnosis, Drug Delivery, and Treatment, First Edition. Edited by Mahendra Rai, Mehdi Razzaghi-Abyaneh, and Avinash P. Ingle.

It is well investigated that the nanomaterials used in biomedical applications usually trigger numerous toxic effects in various body parts. Various systems like respiratory, central nervous, and cardiac systems were found to be unique targets for the potential toxicity of nanomaterials. The interaction of nanomaterials with a variety of biomolecules, cells, tissues, and organs generates adverse biological outcomes through disruption of the cell membrane, protein unfolding, fibrillation, thiol crosslinking, and inhibition of the activity of several important enzymes required for normal metabolic pathway processes, apoptosis, generation of reactive oxygen species (ROS), and DNA damage which collectively leads to cell death (Xia et al. 2008; Hong et al. 2016; Teleanu et al. 2019). Moreover, it was proposed that the toxic effects of nanomaterials depend on their various structural and physicochemical properties such as size, shape, charge, chemical composition, coating, shell, concentration, aggregation state, time spam of their interaction with living matter, their stability in biological fluids, and their ability of accumulation in tissues and organs (Mala and Celsia 2018; Nam and An 2019; Adewale et al. 2019).

Therefore, it is extremely important to have a complete understanding of the interactions between all factors and mechanisms involved in the toxicity of nanomaterials to develop efficient, novel, biocompatible, and safe nanomaterials for the diagnosis and treatment of various human diseases. Considering all the facts, the present chapter aims to discuss various aspects related to the toxicological concerns of nanomaterials. In this chapter, we have focused on various factors influencing the toxicity of nanomaterials, their actual *in vivo* and *in vitro* toxic effects, and the role of nanomaterial surface modification to avoid their toxicity.

17.2 Factors Influencing the Toxicity of Nanoparticles

As discussed earlier, nanomaterials are the materials whose sizes in one, two, or three dimensions are within the range from 1 to 100 nm. Moreover, all the different types of nanomaterials have enormous promising applications in the biomedical field. It was proposed that the novel and unique physicochemical properties such as size, shape, surface charge, the composition of the core or shell, etc., possessed by these nanomaterials make them suitable candidates for various medical applications (Sukhanova et al. 2018). However, on the other hand, it was reported that these nanomaterials sometimes cause toxicity to the human being and their same physicochemical properties were found to be responsible for that.

The *in vitro* toxicological studies of various nanomaterials, including various nanoparticles, performed using different models help to study their efficacy on individual cell components and individual tissues. However, *in vivo* studies performed to date help to estimate the toxicity for nanomaterials on individual organs or the body as a whole. Moreover, from all such studies, it was reported that the toxicity of nanomaterials usually depends on their structural and physicochemical properties (e.g. size, shape, composition, surface chemistry, aggregation, etc.), concentration, time spam of their interaction with living matter, their stability in biological fluids, and their ability of accumulation in tissues and organs (Mala and Celsia 2018; Nam and An 2019). Therefore, it is suggested that the development of effective, safe, and biocompatible nanomaterials for diagnosis and treatment of human diseases is only possible having a complete understanding of the interactions between all these factors and the possible mechanisms of nanomaterial toxicity. The important factors which influence the toxicity of nanomaterials are discussed in the following sections. Figure 17.1 shows a schematic illustration of different factors affecting the toxicity of nanomaterials.

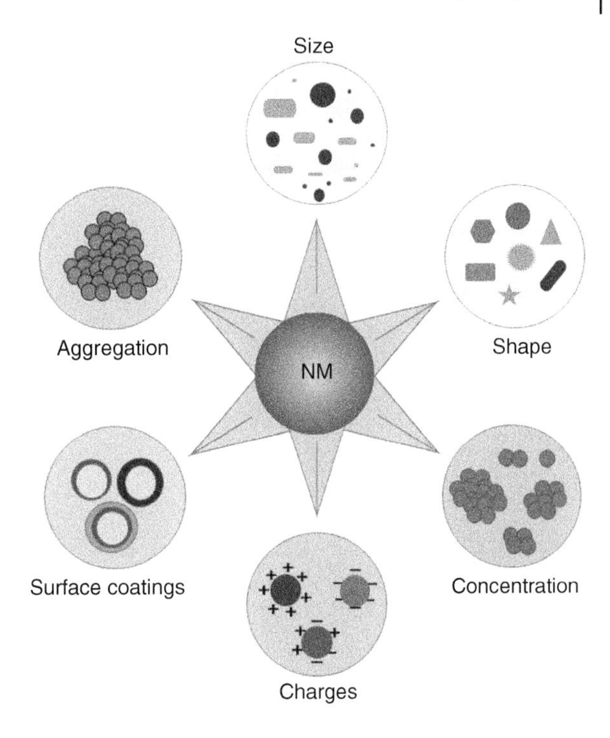

Figure 17.1 The schematic illustration of different factors affecting the toxicity of nanomaterials.

17.2.1 Size

It is well-known fact that reduction in the size of any materials also changes its properties; moreover, it is proposed that smaller particles offer a greater surface to interact with other molecules present in the surrounding environment. Similarly, particles with smaller size and higher surface area can easily be metabolized and distributed to the entire system. Nanomaterials are usually transported in the body through two channels: one is passive diffusion and another is active transport via pumps. Such nanomaterials are uptaken by receptor-mediated or non-receptor-mediated endocytosis and their internalization is governed by particle size (Aillon et al. 2009). However, it was found that nanomaterials having a size of about 40–50 nm can only be taken up by receptor-mediated endocytosis. Therefore, it is reported that the size of nanomaterials plays a crucial role in toxicity. Nanomaterials with larger size cannot enter the biological systems; nanomaterials having size less than 50 nm can be easily taken and distributed throughout the body, whereas nanomaterials in the size range of 100–200 nm are filtered by the reticuloendothelial system, which ultimately blocks their further distribution and they are eliminated by the spleen and liver. Therefore, toxicity in the liver and spleen is more prevalent (Mala and Celsia 2018). In addition, it was confirmed that the variation in the size of nanomaterials is an important reason for the differential distribution of nanomaterials in different systems of the body. According to Asgharian and Price (2007) in the respiratory tract, nanomaterials having a size less than 10 nm are accumulated in the trachea, whereas those having a size range between 10 and 20 nm are deposited in alveoli.

Many studies have been performed on the size-dependent toxicity of various nanomaterials in different systems. Recently, Perde-Schrepler et al. (2019) evaluated the size-dependent cytotoxicity and genotoxicity of silver nanoparticles (AgNPs) of four sizes (5, 25, 50, and 110 nm) on HEI-OC1 cochlear cells and HaCaT keratinocytes. The results obtained revealed that after four hours of exposure, only the AgNPs of 50 nm were observed in both the above-mentioned cell lines, and the 5 nm AgNPs were observed in the HaCaT cells. However, after 24 hours of exposure, AgNPs of all sizes were observed in both cell lines. In another study, Chen et al. (2006) demonstrated the size-dependent cytotoxicity of silver AgNPs and gold nanoparticles (AuNPs) in Zebrafish. Apart from these, several other reports are also available which suggest the size-dependent toxicity of various nanomaterials (Gliga et al. 2014; Zhou et al. 2016).

17.2.2 Shape

Apart from size, shape is another important factor that influences the toxicity of nanomaterials. As mentioned, nanomaterials are usually internalized by the process of endocytosis. Hence, it was reported that nanomaterials having spherical shape can more easily be taken inside by endocytosis (Champion and Mitragotri 2006). Moreover, one of the available reports demonstrated that nanomaterials having spherical shapes are comparatively less toxic than other shapes. In this context, Mala and Celsia (2018) reviewed that carbon nanotubes are more toxic because they have fourfold more ability to block the potassium channel than spherical particles. Similarly, Hsiao and Huang (2011) reported that ZnO nanoparticles in rod shape are more toxic than spherical nanoparticles. Moreover, recently Dong et al. (2019a) demonstrated the shape-dependent toxicity of nanosized alumina which are widely used in numerous commercial applications, hence there are huge chances of the release of these nanoparticles into the environment, posing a threat to the health of wildlife and humans. In their study, the efficacy of flake and rod shape alumina nanoparticles was studied on metabolic profiles of astrocytes in rat cerebral cortex. The finding thus obtained revealed that both shapes of nanoparticles showed dose-dependent significant cytotoxicity and apoptosis after 72 hours of exposure. However, among these two, a significantly stronger response was reported in the case of rod-shaped alumina nanoparticles compared to flake-shaped. These effects were mostly due to the higher ROS accumulation and inflammation induction which was subsequently confirmed from increased concentrations of IL-1β, IL-2, and IL-6.

17.2.3 Concentration or Dose

Like size and shape, concentration and contact time are the other factors which influence the toxicity of nanomaterials; it also varies depending on the organism. Various reports are available which explained the dose-dependent cytotoxicity and genotoxicity of nanomaterials. Kovrižnych et al. (2013) studied the toxicity of 31 different types of nanoparticles which mainly includes metal and metal oxide nanoparticles. The study observed the dose-dependent nature of nanoparticles, where the LC50 of the copper nanoparticle to fish was about 4.2 ppm at 48 hours, whereas it was only 3.8 ppm at 96 hours. Moreover, it was reported that among all the tested nanomaterials, AgNPs were found to more toxic, followed by copper, than other metal and metal oxide nanomaterials. The LC50 of AgNPs and copper nanomaterials were 2.9 and 4.2 ppm, respectively, at 48 hours, whereas for all other nanomaterials the LC50 was recorded above 100 ppm. In another study, Duan et al. (2013a) evaluated the toxicity of silica nanomaterials at various concentrations ranging from 25 to 200 ppm in zebrafish embryo. Different kind of behavioral patterns were reported at different concentrations. It was observed that silica nanomaterials in the concentration range of 25–50 ppm showed hyperactivity in larvae, whereas at higher concentration, i.e. above 100 ppm, it showed hypoactivity. Naqvi et al. (2010) studied the toxicity of iron oxide nanoparticles at various concentrations in murine macrophage (J774) cells. The MTT assay results showed more than 95% of cells were viable at lower concentrations (25–200 $\mu g\,ml^{-1}$) after three hours of exposure. On the other hand, the viability of these cells was reported to reduce to 55%–65% at higher concentrations (i.e. 300–500 $\mu g\,ml^{-1}$) after six hours of exposure. Similarly, Vecchio et al. (2012) and Paknejadi et al. (2018) demonstrated the concentration-dependent toxicity of citrate-capped AuNPs in *Drosophila melanogaster* and AgNPs on normal human skin fibroblast cell lines, respectively.

17.2.4 Surface Charge

It is well studied that the charge present on the surface of nanomaterials plays an important role in their interaction with biomolecules like proteins, DNA, etc. It was revealed through many reports that positively charged nanomaterials rapidly interact with proteins and they are more easily taken up inside cells than negative or neutral charge. When the permeation and interaction of positively charged nanomaterials is high, their toxicity is also comparatively higher. Bhattacharjee et al. (2010) presented that silica nanomaterials having positive charges were found to be significantly more toxic than negatively charged and uncharged silica nanomaterials. Similarly, Hühn et al. (2013) comparatively studied the toxicity of both positively and negatively charged AuNPs. The results obtained showed that positively charged AuNPs were absorbed by cells in higher amounts and more rapidly than negatively charged AuNPs, and were also found to be more toxic. According to Alexis et al. (2008) positively charged nanoparticles have an enhanced ability for opsonization, i.e. adsorption of proteins facilitating phagocytosis, including antibodies and complement components, from blood and biological fluids. However, Sukhanova et al. (2012) reported that binding of proteins to nanoparticles alters their structure, which ultimately leads to the loss of their enzymatic activity which creates a disturbance in different biological processes, and precipitation of ordered polymeric structures, e.g. amyloid fibrils. In another study, Jeon et al. (2018) synthesized fluorophore-conjugated polystyrene nanoparticles and their surface was modified with seven different types of functional groups, and evaluated the role of surface charge in the cellular uptake of nanoparticles. Further, phagocytic differentiated THP-1 cells and non-phagocytic A549 cells were incubated with these nanoparticles for four hours, and their cellular uptake was analyzed using fluorescence intensity and confocal microscopy. The results obtained suggested a good positive correlation with the surface charge (zeta potential) of nanoparticles in both cell lines. Hence, the authors concluded that surface charge is one of the major parameters influencing the cellular uptake efficiency of nanoparticles. Apart from these, there are some other reports by El-Badawy et al. (2011) and Kasemets et al. (2019), among others. They also demonstrated that surface charge plays an important role in explaining cellular toxicity. In addition, there were many studies performed which proved that positively charged nanomaterials are more toxic than other charges (Mala and Celsia 2018).

17.2.5 Chemical Composition and Surface Coating

Like other parameters discussed above, the chemical composition of nanomaterials and their crystal structure also reportedly influence the toxicity of nanomaterials. Yang et al. (2009) comparatively studied the toxicity of silicon dioxide (SiO_2) and zinc oxide (ZnO) NPs on mouse fibroblasts and the findings obtained revealed that both nanoparticles showed different mechanisms of toxicity. ZnONPs reported to cause oxidative stress, whereas SiO_2NPs alter the DNA structure. This study proved that the toxicity of nanomaterials certainly depends on their chemical composition.

Moreover, it is proposed that the degradation of nanoparticles is also determined by their chemical composition and the environmental conditions such as pH or ionic strength. Moreover, the composition of the core of nanomaterials was also found to influence their toxicity. Generally, nanomaterials like nanoparticles exert their toxic effect on cells through the interaction of metal ions released from the core of nanoparticles. Some metal ions, such as silver and cadmium, are most toxic and hence cause damage to the cells. However, there are some metal ions (iron and zinc) which are biologically useful at specific concentrations but beyond those concentrations, they exert toxic effects and damage cellular pathways. In this context, it is proposed that modification of the surface by coating different polymeric molecules makes it possible to manage the toxicity of nanoparticles. Lu et al. (2010) evaluated the cyto-, geno-, and phototoxicity of AgNPs with different surface coating on human skin HaCaT keratinocytes. The results obtained revealed that the citrate-coated colloidal AgNPs at the concentration of $100\,\mu g\,ml^{-1}$ do not show any geno-, cyto-, and phytotoxic effect. However, AgNPs coated with citrate powder were found to be toxic. This confirmed that the coating of the AgNPs with a biodegradable polymer prevents their toxicity. In another study, Nguyen et al. (2013) comparatively studied the toxicity of uncoated and standard citrate and polyvinylpyrrolidone (PVP)-coated AgNPs in J774A.1 macrophage and HT29 epithelial cells. In the case of uncoated AgNPs, it was observed that the cell viability was decreased by 20–40% at $1\,\mu g\,ml^{-1}$ after 24 hours of exposure. However, in the case of coated AgNPs, a decrease in cell viability was reported at $25\,\mu g\,ml^{-1}$ or higher concentrations, but PVP-coated AgNPs induced higher cytotoxicity than citrate-coated AgNPs. Further analysis revealed that uncoated AgNPs at $1\,\mu g\,ml^{-1}$ showed decreased expression of selected cytokines like TNF-α, IL-1β, and IL-12 (p70) in J774A.1 and IL-8 in HT29 cells. Whereas, both citrate and PVP-coated AgNPs showed increased expression of these cytokines at higher concentrations ($25\,\mu g\,ml^{-1}$). Moreover, PVP-coated AgNPs mostly reported increasing the cytokine levels.

Likewise, the presence or absence of a shell onto the surface of nanoparticles also changes their optical, magnetic, and electrical properties. Usually, the shell is applied to nanoparticles to improve their biocompatibility and solubility in water and biological fluids. It also helps to avoid aggregation of nanoparticles and increase their stability. Hence, the application of shell can decrease the toxicity of nanoparticles by increasing their ability to selectively interact with different types of cells and biological molecules. Moreover, it was reported that the shell significantly influences the pharmacokinetics, patterns of distribution, and accumulation of nanoparticles in the body (Arami et al. 2015). As discussed earlier, mosts toxic effects of nanoparticles are due to the formation of free radicals. In this context, application of shell on nanoparticles not only solves this negative effect, but also provides stability to nanoparticles, increases their resistance to environmental factors, decreases the release of toxic substances from them, or makes them tissue-specific (Peng et al. 2013). Various attempts have been made to study the role of the application of shells on nanoparticles in their toxicity and it was proved that the presence of shell on nanoparticles greatly influences their toxicity (Sukhanova et al. 2018).

Similarly, the crystal structure of nanoparticles also influences the toxicity of nanoparticles. Gurr et al. (2005) studied the relationship between crystal structure and toxicity of titanium oxide nanoparticles (TiO_2NPs) using a human bronchial epithelium cell line with different types of crystal lattice. It was observed that nanoparticles having a rutile-like crystal structure (prism-shaped TiO2 crystals) cause oxidative damage of DNA, lipid peroxidation, and formation of micronuclei, which specifies abnormal chromosome segregation during mitosis. On the other hand, nanoparticles having an anatase-like crystal structure (octahedral TiO_2 crystals) were found to have no toxicity. According to Zhang et al. (2003), the crystal structure of nanoparticles can differ depending on the environment, e.g. after interaction with water, biological fluids, or other dispersion media. In their study, they observed that the crystal lattice of zinc sulfide nanoparticles is rearranged into a more ordered structure after their contact with water.

17.2.6 Aggregation

Aggregation is another one of the most important factors which influences cellular toxicity of nanomaterials, because aggregation of nanoparticles affects various other parameters associated with nanomaterials such as charge, size, and surface chemistry (Adewale et al. 2019). All these factors (charge, size, and surface chemistry), discussed above, greatly influence the toxicity caused by different nanomaterials (Moore et al. 2015). Yang et al. (2007) reported that cationic and oligo-cationic species lead to aggregation of nanoparticles, however negatively charged nanoparticles like AuNPs repel each other and inhibit aggregation (Basu et al. 2007). In another study, Cui et al. (2012) also studied the effect of AuNP aggregation in human cervical carcinoma (HeLa) cell activity. The results obtained confirmed that aggregation of AuNPs affects the various associated factors, including the size of AuNPs which ultimately showed a differential toxicity pattern. Similarly, Albanese and Chan (2011) developed a novel technique to produce transferrin-coated AuNP aggregates of different sizes and further studied their uptake and toxicity in three different cell lines viz. HeLa, A549, and MDA-MB 435. They observed that the aggregation did not stimulate any unique toxic response of AuNPs and the uptake patterns for single and aggregated AuNPs were different. The analysis of results showed that there was about a 25% reduction in uptake of aggregated AuNPs in the case of HeLa and A549 cells compared to single and monodisperse AuNPs. On the contrary, about a twofold increased uptake of aggregated AuNPs was observed in MDA-MB 435 cells. These findings confirmed that cell type and the mechanism of interactions also play a significant role in cellular uptake and toxicity. Moreover, it was observed that the presence of various capping agents such as lipids, polymers, bovine serum albumin (BSA), proteins, etc. in cellular medium helps to avoid the aggregation of nanoparticles (Adewale et al. 2019).

17.3 Why Is Toxicity Evaluation of Nanomaterials Necessary?

It is well-known fact that due to recent advances and developments in the field of nanotechnology, different nanomaterials have attracted a great deal of attention, and hence are being widely used in many medicinal and pharmacological applications. Despite the many proposed advantages of nanomaterials, increasing concerns have been expressed on their possible toxicological effects on human health (Zhao and Castranova 2011). As mentioned earlier, unique properties of nanomaterials such as small size, shape, morphology, composition, distribution, dispersion, surface area, surface chemistry, and reactivity are reported to play a crucial role in their toxicity and hence the evaluation of their toxicity is complex (Mala and Celsia 2018; Nam and An 2019). The toxicity tests which are currently available for testing different classical chemical compounds may not be useful for the safety evaluation of nanomaterials. To date, there is no promising toxicity test or validated procedure available for the safety evaluation of nanomaterials. Indeed, the National Center for Toxicological Research and National Institute of Standards and Technology are currently working on the development of positive standards for toxicity testing of nanomaterials (Slikker et al. 2012; NIST 2013; Howard et al. 2014; FDA 2016); but still no such widely accepted nanomaterial standards are available which can be used as positive controls. In this context, it is believed that the establishment of standard reference materials and standard dispersion protocols would be of great importance for the evaluation of nanotoxicity because it can provide common standards to follow for nanotoxicity testing. Moreover, the utilization of good laboratory practices in addition to standard protocols would be more useful as far as nanotoxicity is concerned. It is observed that no single vehicle is enough for all kinds of nanomaterials and those obtained from different sources. Therefore, the dispersion characteristics of any nanomaterial can be optimized only by experimenting with different vehicles and with proper experimental observations. Hence, only the development of careful toxicology testing strategies for nanomaterials will help researchers to increase the general understanding of the potential health effects of nanomaterials (Sahu and Hayes 2017). Still, extensive studies are required to identify relevant biomarkers and tests to better understand the toxicological effects of nanomaterials on biological systems, particularly on human beings.

17.4 Toxicity of Nanomaterials Used in Biomedical Applications

As discussed earlier, evaluation of toxicity of nanomaterials which are intended for use in various human healthcare applications is very important. As far as preclinical studies on toxicity are concerned, its testing is usually performed using *in vitro* cell culture systems, and such studies at the cell culture level are used as basic standards for *in vivo* tests. To date, different *in vivo* models such as zebrafish, mice, rabbits, etc., have been studied for the assessment of nanomaterial toxicity, however among all these models currently, zebrafish is most preferably used due to certain advantages over other models (Mala and Celsia 2018).

17.4.1 *In vitro* Toxicity

The *in vitro* assays for the assessment of nanomaterial toxicity are considered sensitive, rapid, cost-effective, and less time consuming compared to *in vivo* animal tests. In this context, utilization of human cell culture systems showed the potential toward the elimination of the necessity of animal models, and these studies were found to have enhanced efficiencies in testing toxicity of nanomaterials (Drasler et al. 2017). However, the use of *in vitro* cell culture systems has certain limitations which are mainly associated with their limited metabolic capacity and the biotransformation of a chemical as compared to *in vivo* systems (Sahu and Hayes 2017; Franco and Lavado 2019). It is proven that unique physical and chemical properties exhibited by nanomaterials created the need to develop and validate the *in vitro* assays for the evaluation of their potential toxicity. Moreover, it is believed that the observations obtained from *in vitro* studies associated with exposure-effect relationships may be predictive for the toxicological and pharmacological activities of *in vivo* models, provided they are appropriately validated (Huntjens et al. 2006; Checkley et al. 2015). However, some of the reports revealed that the same nanomaterial fabricated using different sources can show contradictory results (Buford et al. 2007).

To date, various studies have been performed on the assessment of toxicity of a variety of nanomaterials which are commonly used in health care using different human cell lines. One of the reports demonstrated that iron oxide nanomaterials cause toxicological effects to human aortic endothelial cells through inflammation (Zhu et al. 2011). In another study by Lovrić et al. (2005), dose-dependent toxicity of cadmium telluride (CdTe) quantum dots was reported in human umbilical vein endothelial cells, MCF-7 cells, and bovine corneal stromal cells. However, as mentioned earlier, Perde-Schrepler et al. (2019) demonstrated the size-dependent toxicity of AgNPs on HEI-OC1 cochlear cells and HaCaT keratinocytes. In this study, the authors evaluated the effects of AgNPs of four sizes (5, 25, 50, and 110 nm) on the aforementioned cell lines. The cells were treated with different concentrations of AgNPs and various effects such as uptake of silver ions, ROS production genotoxicity, etc. were determined using different assays. Initially, it was observed that the intracellular concentration of silver increased proportionally with the incubation time and the size of the AgNPs. The uptake of silver ions was higher in HEI-OC1 cells compared to HaCaT. However, the presence of only 50 nm AgNPs was recorded in both cell lines after four hours of exposure and of 5 nm AgNPs only in the HaCaT cells after 24 hours of exposure. All the AgNPs with varied sizes thus were reported to reduce the viability proportionally with the concentration, with HEI-OC1 cells being more affected. Moreover, the toxicity of AgNPs was found to decrease with the size of AgNPs, and ROS production was found to be dose and size-dependent, mainly in the cochlear cells. Further, the assessment of genotoxicity revealed a higher level of DNA lesions in HEI-OC1 cells after treatment with small-sized AgNPs.

Cha and Myung (2007) evaluated the toxicity of zinc, iron, and silica nanomaterials on different cell lines such as the human brain, liver, stomach, and lung cell lines. The results obtained revealed that these nanomaterials altered the mitochondrial membrane permeability and hence the activity was reduced from 100% to 75%. Moreover, Pandurangan and Kim (2015) studied the toxicity of zinc oxide nanomaterials on bronchial epithelial cells and A549 alveolar adenocarcinoma cells. Authors reported the high concentration of zinc ions inside the cells which are released from zinc oxide nanomaterials by unknown mechanisms. The presence of zinc ions in cells reported to induce toxicity via generating oxidative stress and destroying the membrane potential of mitochondria. The disruption of oxidative phosphorylation affects energy transduction which leads to cell death. Magrez et al. (2006) studied the toxicity of multiwalled carbon nanotubes in different human lung cancer cell lines, viz. H596, H446, and Calu-1. The authors tried to analyze the difference between the *in vitro* and *in vivo* results concerning the route of entry of nanomaterials and it was found that the host response to the circulating nanomaterials is usually produced by Kupffer cells of the liver. Apart from these, Geiser et al. (2005) and Radomski et al. (2005) evaluated the liver toxicity of nanomaterials in their respective studies using secondary cell lines like immortal Hep2G and BRL3A, respectively.

In another recent study, Bengalli et al. (2019) reviewed different *in vitro* toxicological concerns of TiO_2:SiO_2 nanocomposites in various human cell lines. Simeone and Costa (2019) studied the cytotoxicity of metal oxides in different types of *in vitro* systems. Likewise, various other reports are available which demonstrated the *in vitro* toxicological effects of a variety of nanomaterials in various human cell lines.

17.4.2 *In vivo* Toxicity

The nanomaterial used in biomedical applications can be entered in the human body through different routes depends on the administration route selected for particular disease, e.g. ingestion, dermal exposure, intravenous injections, etc. Once the nanomaterials enter the bloodstream, they are transported to different parts of the body through blood circulation due to their extremely small size. Some of the reports available on *in vivo* studies demonstrated that the lung, liver, and kidney are the important distribution sites and targeted body organs for nanomaterial exposure. Wang et al. (2013) have reviewed the *in vivo* metabolism of nanomaterials with a special focus on blood circulation and their clearance profiles in various body organs like lung, liver, and kidney. As discussed earlier, zebrafish, mice, and rabbits are the most preferably used *in vivo* models for the toxicity of nanomaterials. The toxicity of nanomaterials has been briefly discussed below depending on the different body organs.

17.4.2.1 Toxicity to the Liver

The liver is the most important primary organ responsible for the metabolism and detoxification of various chemical substances like xenobiotics. Usually, different toxicants present in the blood are filtered by the liver before being distributed to other body parts. The continuous flow of blood containing a high level of toxicants leads to the delivery of high concentrations of the toxicant to the liver. Therefore, constant exposure of toxicants in high levels and high metabolic activity makes the liver a major target organ of different toxicants, including nanomaterials. Once the nanomaterial enters the bloodstream through any of the above-discussed exposure routes, it may be translocated to the liver (Sahu and Hayes 2017). Some of the studies available suggest that nanomaterials like nanoparticles can be entrapped by the reticuloendothelial system, indicating the liver and spleen as the main target organs (Dong et al. 2019b). From these facts it is clear that nanomaterials can be potential hepatotoxicants (liver toxicants), and hence, evaluation of hepatotoxicity is an important testing strategy for the safety assessment of nanomaterials. Although the toxicity of nanomaterials in the liver (hepatotoxicity) is an important concern, very few reports are available on this aspect.

In this context, Kermanizadeh et al. (2014) reviewed the toxicological effects of engineered nanomaterials on the liver. In this article, the authors mentioned that the liver acts as a secondary exposure site where nanomaterials were reported to accumulate preferentially, i.e. >90% of translocated nanomaterials accumulated compared with other organs. Balasubramanian et al. (2010) studied the biodistribution of AuNPs; in this study adult Wistar rats were exposed to 20 nm AuNPs via an intravenous route, and their localization was examined after a specific time interval (1 day, 1 week, and 2 months). The data recorded showed that administered AuNPs very rapidly accumulated in the liver (after 1 day = $49.4 \pm 50.4\,\text{ng}\,\text{g}^{-1}$; after 1 week = $64.8 \pm 39.7\,\text{ng}\,\text{g}^{-1}$; after 2 months = $72.2 \pm 40.5\,\text{ng}\,\text{g}^{-1}$). The authors also observed that the accumulation of AuNPs in the liver caused severe changes in the expression of genes in the organ involved in the detoxification, lipid metabolism, and the cell cycle. Similarly, in another study demonstrating the intravenous administration of varied size of silica and AuNPs, i.e. amorphous silica nanomaterials and AuNPs in male Wistar rats revealed that amorphous silica nanomaterials showed DNA damage in the liver; however, no genotoxicity was observed in Wistar rats exposed with AuNPs (Downs et al. 2012).

In another report, it was observed that MW-CNTs exert toxic effects in the spleen and liver of mice. As far as the spleen is concerned, it found to increase micronuclei frequency in addition to chromosomal aberrations and the promotion of allergic response. Apart from these, activation of cyclooxygenase enzymes via suppression of systemic immune function was also observed in the spleen. However, the case of liver-altered gene expression was noted in the liver (Ji et al. 2009; Patlolla et al. 2010). Similarly, *in vivo* exposure of AuNPs was reported to cause apoptosis and acute inflammation in the liver, in addition to bioaccumulation in organs; effect on penetration ability in sperm head and tail regions was also recorded (Wiwanitkit et al. 2009; Lasagna-Reeves et al. 2010). Cho et al. (2009) presented that 13 nm-sized polyethylene glycol (PEG)-coated AuNPs cause toxicity in the liver by accumulating in the liver and spleen up to seven days after injection. The toxicological effects were governed by the induction of apoptosis and inflammation in the liver.

Moreover, Kermanizadeh et al. (2013) evaluated the toxicological efficacy of different nanomaterials which mainly included two ZnO materials, i.e. coated (100 nm) and uncoated (130 nm), two multiwalled carbon nanotubes (MW-CNTs), AgNPs (<20 nm), and one positively charged rutile TiO_2NPs (10 nm) on the liver of C57/BL6 mice. The results obtained showed that most of these nanomaterials induce a neutrophil influx into the liver after only six hours of intravenous injection. However, it was observed that the neutrophils were only affected in the initial phases of the immune response against the nanomaterials and after 48 hours they returned to the control level. Further, no significant effect was recorded in the level of intracellular glutathione in case of exposure of AgNPs and the TiO_2NPs. But, upregulation of interleukin 10 (IL-10), Chemokine (C-X-C motif) ligand 2 (CXCL2), and intercellular adhesion molecule-1 (ICAM-1) mRNA, as well as a sharp decline in the level of complement proteins factor (C3) and interleukin 6 (IL-6), was reported in the livers of mice treated with these nanomaterials.

In addition to the studies discussed here, some of the recent studies have also proposed that a variety of nanomaterials are reported as having toxicological effects on the liver. Dong et al. (2019b) studied the *in vivo* toxic effects of several AuNPs with core sizes ranging from 4 to 152 nm on mice through intravenous administration. The authors reported the size dependent toxicity; the small size (4 and 15 nm) PEGylated AuNPs found to circulate for a longer time in the blood compared to higher size AuNPs (50–152 nm). Valentini et al. (2019) studied dose-dependent toxic effects of TiO_2NPs in the liver and kidney of rats. For this, rats were exposed to different doses of TiO_2NPs and sacrificed, respectively, for four days, one month, and two months after the treatment. It was observed that thus administered TiO_2NPs accumulated and induces different morphological and physiological alterations in the liver and kidney. As far as the liver is concerned, these alterations mainly affect the hepatocytes located around the centrilobular veins which are the site of oxidative stress. In addition, it was also reported that these nanoparticles also disrupt the function of various enzymatic markers of liver and kidney functions such as AST (aspartate aminotransferase) and uric acid at higher doses.

Most of the researchers reported various toxicological effects of different nanomaterials to the liver, however, it was proposed that there is the possibility to develop specific nanomaterials that do not have any acute or chronic toxicological effect to liver. In this context, Bailly et al. (2019) demonstrated the efficacy of laser-ablated dextran-coated AuNP (AuNPd) in a small animal model via intravenous administration. The data recorded in the form of observations showed that thus tested AuNPs were rapidly eliminated from the blood circulation system and accumulated preferentially in liver and spleen, without inducing any kind of toxicity in liver or kidney. But despite accumulation in the liver and spleen, the authors did not detect any sign of acute and chronic toxicities in liver, spleen, and kidneys, which was confirmed from the presence of IL-6 level. Therefore, the authors proposed that thus-developed modified AuNPs can be safely used in biomedical applications. All the studies mentioned confirmed that in general, nanomaterials exert a wide range of toxicological effects on the liver, but it is also true that there is a possibility of careful modification of nanomaterials to avoid their toxicity.

17.4.2.2 Toxicity to Kidney

Like the liver, nanomaterials are also reported to cause toxicity to the kidney (nephrotoxicity). It is considered an important primary organ for clearance of nanomaterials. To date, limited studies have been performed on the evaluation of nephrotoxicity which is confirmed from scant literature (Kumar et al. 2017). However, some of the studies available that demonstrate nephrotoxicity were performed using different types of nanomaterials in different *in vivo* models. Therefore, the conclusions drawn from these studies may not be comparable (Sahu and Hayes 2017). One of the studies performed demonstrated kidney toxicity (commonly known as nephrotoxicity) of ZnONPs in rats. The results obtained showed that these tested ZnONPs cause mitochondrial and cell membrane damage in rat kidney (Yan et al. 2012). In another study, Liao and Liu (2012) reported induction of renal proximal tubule necrosis in kidneys of rats after exposure with nano copper (copper nanoparticles) which was analyzed by renal gene expression profiles.

It is well known, and also discussed earlier, that mesoporous silica nanoparticles were reported to have promising biomedical applications and hence are preferably used in the management of different diseases nowadays. Therefore, it becomes necessary to study the toxicological effects of such nanoparticles to different body organs. In previous reports, it was hypothesized that mesoporous silica nanoparticles usually altered TLR4/MyD88/NF-κB, JAK2/STAT3, and Nrf2/ARE/HO-1 signaling pathways which lead to the induction of oxidative stress, inflammation, and fibrosis. In this context, Mahmoud et al. (2019) recently investigated the hepatic and nephrotoxicity of these nanoparticles in rats. For this, rats were administered with different concentrations of these nanoparticles for 30 days. The results showed various kinds of functional and histologic alterations in the liver and kidney of rats which mainly include increased levels of ROS, lipid peroxidation, suppressed antioxidants, and Nrf2/HO-1 signaling. In addition, upregulation in the expression of TLR4, MyD88, NF-κB p65, and caspase-3, and increased serum pro-inflammatory cytokines was also recorded. Furthermore, these nanoparticles were also found to activate the JAK2/STAT3 signaling pathway, downregulated peroxisome proliferator-activated receptor gamma (PPARγ), and promoted fibrosis evidenced by the increased collagen expression and deposition.

17.4.2.3 Toxicity to Brain

Toxicity to brain is referred to as neurotoxicity and involves any reversible or irreversible adverse effect caused to structure, function, or chemistry of the nervous system, during development or at maturity, produced by any physical or chemical means. The adverse effect also represents any kind of change caused due to any treatment involving the administration of some materials which affects the normal function of the brain (Teleanu et al. 2019). Costa et al. (2017) proposed that normal neurological adverse effects related to morphological changes mainly include neuronopathy, axonopathy, myelinopathy, and gliopathy. However, according to Teleanu et al. (2018), the common mechanisms to date used to investigate for neurotoxicity mainly include the excessive production of ROS leading to oxidative stress, the release of cytokines causing neuroinflammation, and dysregulations of apoptosis leading to neuronal death. Although all this information is available, there is not enough information available about the neurotoxicity of nanomaterials. Considering these facts, some researchers attempted to investigate the neurotoxicity of different nanoparticles.

As discussed earlier, polymeric nanoparticles have been the most commonly used organic nanomaterials in biomedical applications including neurological disorders due to their various unique properties, which have already been discussed in earlier sections. Apart from having several advantages, polymeric nanoparticles also have some disadvantages like aggregation, and their toxicity associated with the degradation processes and their residual materials (Banik et al. 2016; Singh et al. 2017). Yuan et al. (2015) investigated the *in vivo* neurotoxicity of polysorbate 80-modified chitosan nanoparticles in rats after their exposure via intravenous injection. The dose-dependent accumulation of the tested nanoparticles was observed in the frontal cortex and cerebellum leading to neuronal apoptosis, slight inflammatory response, increased oxidative stress, and body weight loss. Similarly, in another study, the authors studied the neurotoxic effects of cisplatin-containing liposomal formulations in rats. In a comparative study it was reported that exposure of rats to commercially available drug-free liposomal formulation or a formulation containing lower doses of cisplatin induced mild to severe hemorrhage, necrosis, edema, and macrophage infiltration. Hence, the authors concluded that thus-reported neurotoxicity was mainly due to the intrinsic toxicity of the liposomes combined with the neurotoxic effects of cisplatin (Huo et al. 2012). Moreover, another kind of organic nanomaterials, dendrimers, reportedly exerted toxic effects related to the innate immune response in zebrafish embryos and larvae. It was proposed that the neurotoxic effects may be due to a decrease in the locomotor function of the larvae after the administration of certain dendrimers (Calienni et al. 2017).

Like organic nanomaterials, several inorganic nanomaterials also have promising applications in biomedicine. Among the inorganic nanomaterials, there is a possibility of translocation of several inorganic nanoparticles like gold, silver, iron oxide, titanium oxide, silica, etc. into the brain through the blood–brain barrier after entering the body. Once such nanoparticles entered in the brain they may accumulate in the brain due to limited provision of excretion and may cause damage to neuronal cells and function impairments (Song et al. 2016a). Teleanu et al. (2019) reviewed the neurotoxic effects of a variety of inorganic nanoparticles and proposed that different factors such as size, shape, surface coatings of nanoparticles, release rates of metal ions, and interactions with specific cells and proteins greatly affect their neurotoxicity. However, some of the common neurotoxic effects associated with inorganic nanoparticles include alterations in synaptic transmissions and nerve conduction, leading to neuroinflammation, apoptosis, immune cell infiltration, oxidative stress, neuroinflammation, genotoxicity, dysregulated neurotransmitters, synaptic plasticity, and disrupted signaling pathways (Song et al. 2016b, Valdiglesias et al. 2016). But these effects may vary depending on each kind of inorganic nanoparticle and their properties. Moreover, carbon nanotubes were also reported to exert certain neurotoxic effects such as inhibition of cell proliferation, apoptosis, alterations in mitochondrial membrane potential, promoting ROS formation, lipid peroxidization, astrocyte function reduction, anxiety, and depression (Gholamine et al. 2017; Shi et al. 2017). Similarly, quantum dots have also gained considerable interest worldwide for their use in biomedicine for various applications like drug delivery, cell labeling and tracking, targeted cancer therapy, and bioimaging due to their outstanding properties (Aswathi et al. 2018). Some of the studies performed in the recent past indicated that quantum dots can be neurotoxic and may exert different toxicological effects, however their toxicity depends on their morphological characteristics, surface chemistry, and purity (Zhao et al. 2015; Wu et al. 2016). Some of the common neurotoxic effects by quantum dots include increased oxidative stress and cell function damages. Apart from these behavioral changes, learning and memory impairments are some specific neurotoxic effects exerted by quantum dots (Wu et al. 2016).

17.4.2.4 Toxicity to Skin

Skin is another important organ of the body that can act as a primary route of environmental and/or occupational human exposure for chemicals. The different sections present in the epidermis, such as stratum corneum, sebum membrane, lipids, and tight intercellular junctions between keratinocytes usually serve as a physical barrier to protect the host against toxins and pathogens. But, it was observed that nanomaterials like metallic nanoparticles, due to their extremely small size, can pass through the stratum corneum of healthy skin, hair follicles, and sebaceous glands (Wang et al. 2018). Nowadays a variety of nanomaterials such as AgNPs, AuNPs, nanoemulsions, nanocapsules, nanocrystals, dendrimers, fullerenes, liposomes, hydrogels, etc. have been commonly used in the cosmetic industry due to their excellent capability to increase solubility, transparency, and color of cosmetic products (Aziz et al. 2019). In addition, some nanomaterials are also being used in the prevention and treatment of various skin diseases including skin cancer due to their unique properties (Prow et al. 2011; Dianzani et al. 2014). There are two different kinds of opinions about the penetration of nanomaterials through skin: some of the reports available suggested that nanomaterials can be easily penetrated through skin (Borm et al. 2004); on the contrary, few studies reported that nanoparticle cannot penetrate through the intact skin of human, pig, or mouse (Gamer et al. 2006; Baroli et al. 2007).

As already discussed, various factors associated with nanomaterials such as size, shape, concentration, etc. play a crucial role in deciding the toxicity of nanomaterials. In this context, Tak et al. (2015) reported that rod-shaped AgNPs can more easily penetrate deep layers of the stratum corneum compared to triangular or spherical AgNPs in mice. Moreover, a higher concentration of silver was observed in the blood in the case of rod-shaped AgNPs applied through topical application, however comparatively less concentration of silver was reported in the case of triangular and spherical-shaped AgNPs. Adachi et al. (2013) proposed that not all metallic nanoparticles are systemically absorbed by skin exposure. Their study revealed that there were no significant differences in the concentration of titanium in primary organs of mice after TiO_2NPs were topically applied for four weeks.

Mortensen et al. (2008) evaluated the *in vivo* penetration of quantum dot nanoparticles through the skin in the murine model by irradiation of ultraviolet radiation (UVR). The results revealed poor penetration of tested nanomaterials. However, in another study, it was also observed that the condition of the skin (flexed or non-flexed) does not affect the penetration, but in the case of tape-stripped skin, quantum dots were found only on the surface of the viable epidermis. Therefore, the authors claimed that these types of nanomaterials can only penetrate the dermal layer by abrasion of the skin (Zhang et al. 2008). In this context, Jatanaa and DeLouisea (2014) reviewed various reports focusing on the understanding of nanomaterial skin interactions with a specific emphasis on the effects of UVR skin exposure.

In some of the reports, it was observed that co-exposure of skin to UVR and nanomaterials like ZnONPs or TiO_2NPs exerts a synergistic effect on phototoxicity by generating higher levels of ROS and simultaneously decreases the antioxidant-enzyme activity. It was also suggested that such nanoparticles can surpass the antioxidant system by generating hydroxyl radicals to oxidize proteins and nitrogenize tyrosine residues of proteins in skin cells (Tu et al. 2012; Tyagi et al. 2016). Apart from these, DNA damage was observed in human epidermal keratinocytes after six-hour exposure of ZnONPs at 8 and 14 $g\,ml^{-1}$ of concentrations. The SEM micrographs recorded to study the internalization of ZnONPs showed the internalization at only 14 $g\,ml^{-1}$ indicating that even if the nanoparticles could not enter the nucleus, they still caused DNA damage during mitosis (Sharma et al. 2011).

Nevertheless, there are several other reports which proposed the possible mechanisms involved in the toxicity of metallic nanoparticles with skin cells and those mainly include membrane damage, oxidative stress, DNA damage, epigenetic modulation, autophagy, apoptosis, etc. (Wang et al. 2018).

17.4.2.5 Toxicity to Heart

Like other biomedical applications, nanomaterials obtained from various sources have been promisingly used nowadays in tackling different cardiac diseases (Prajnamitra et al. 2019). Therefore, like all other body organs, nanomaterials showed the ability to cause toxicity to the heart, commonly referred to as cardiotoxicity, but very few studies are available on this aspect. Du et al. (2013) evaluated the various tests associated with cardiovascular toxicity such as hematologic parameters, inflammatory reactions, oxidative stress, endothelial dysfunction, and presence or absence of myocardial enzymes in serum after exposure of rats with silica nanoparticles by intratracheal administration. The results showed that these tested nanoparticles can easily pass through the alveolar-capillary barrier into the systemic circulation. However, the authors also proposed that the cardiovascular toxicity was dependent on particle size and dosage of silica nanoparticles. Oxidative stress played a key role in inflammatory reaction and endothelial dysfunction. Similarly, in another study, Duan et al. (2013b) investigated the cardiovascular effect of silica nanoparticles both *in vitro* (in endothelial cells) and *in vivo* (in zebrafish). Among the proposed various possible mechanisms of cardiovascular toxicity, authors reported that oxidative stress and apoptosis were major factors for endothelial cell dysfunction. Moreover, it was also proposed that silica nanoparticle exposure can be a potential risk factor for cardiovascular system failure. Yang et al. (2016) reported chronic cardiac toxicity in mice exposed to different sizes of AuNPs. It was observed that

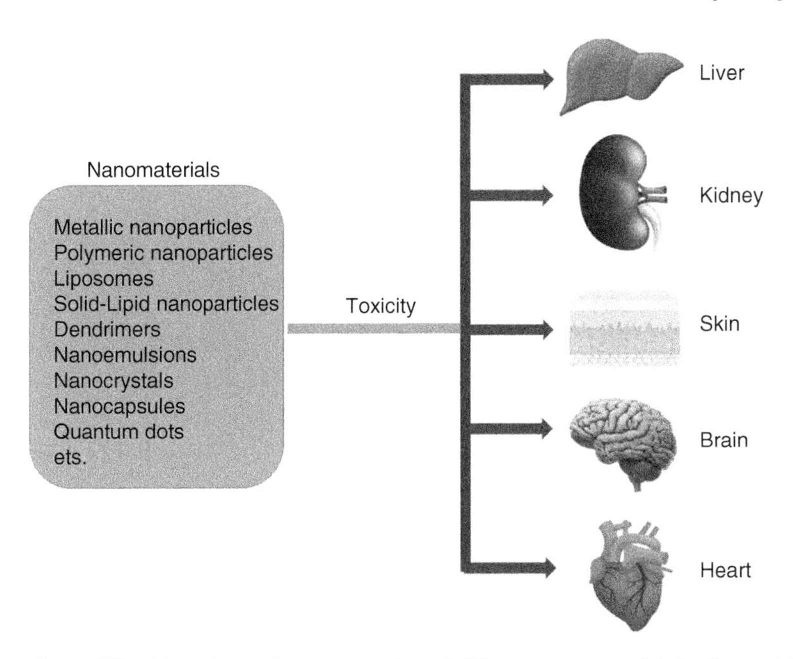

Figure 17.2 The schematic representation of different nanomaterials having toxicity in various body organs.

AuNPs did not affect systolic function, but a significant increase in left ventricular end-diastolic inner dimension, left ventricular mass, and heart weight/body weight was recorded in mice exposed to 10 nm AuNPs after two weeks.

Moreover, Kan et al. (2018) reviewed the cardiovascular effects of engineered nanomaterials and proposed a few possible mechanisms as (i) such nanomaterials may translocate from lung to circulatory system and directly induce pathological changes in cardiovascular tissue; (ii) engineered nanomaterials may trigger a lung-mediated systemic oxidative stress/inflammatory response, which affects normal cardiovascular function; or (iii) these nanoparticles may affect the normal cardiovascular function through a neurogenic pathway. Apart from these, considering the complex nature of engineered nanoparticles authors proposed that such nanomaterials may not just utilize only one of the proposed mechanisms but instead act via some combination of all the three mechanisms to bring potential changes in the cardiovascular system.

From all the aforementioned studies it can be concluded that certain nanomaterials can have potential cardiotoxicity in different *in vivo* models. But various factors such as a source of nanomaterials, different methods of their characterization, experimental conditions, etc., play important roles in deciding their cardiotoxicity, and hence more studies are extensively required in context to draw a concrete conclusion about nanomaterial-derived cardiotoxicity. Figure 17.2 represents a schematic illustration of different nanomaterials having toxicity in various body organs.

17.5 Surface Engineering to Avoid Nanotoxicity

It is well-known and proven from several studies that nanomaterials used in a wide range of biomedical applications exert many acute to chronic types of harmful effects in different body organs. The morphological and physicochemical properties, including surface chemistry of such nanomaterials, were found to play a crucial role in deciding their toxicity. Even though nanomaterials showed different toxicological effects at certain concentrations and conditions, it is possible to reduce their toxic nature through their surface engineering by coating specific types of compounds (Arias et al. 2018; Auclair et al. 2019). For instance, Rivet et al. (2012) observed neuronal cell death after exposure to polydimethylamine-functionalized iron oxide nanoparticles; on the other hand, iron oxide nanoparticles surface-coated with amino silane and dextran were reported to have minimal toxicity. Chia and Leong (2016) demonstrated that uncoated (naked) ZnONPs showed significant toxicity to colorectal cell lines. However, their surface coating with silica significantly helps in reducing their toxicity through the prevention of the dissociation of ZnONPs to zinc ions in both neutral and acidic conditions. But authors also proposed that the application of silica-coated ZnONPs at higher concentrations can still induce undesirable cytotoxicity to mammalian gut cells. As described above, Calienni et al. (2017) demonstrated that dendrimers have various neurotoxic effects such as alterations in mitochondrial activity, apoptosis, neuronal differentiation, and gene expression related to oxidative stress and DNA damage in both *in vitro* and *in vivo* systems. Zeng et al. (2016) demonstrated that the different particles and the surface group density are the important characteristics that influence the cytotoxicity of dendrimers. Therefore, Vidal et al. (2017) proved from their study that surface functionalization of dendrimers with PEG or folate can help to reduce their neurotoxicity.

Recently, Auclair et al. (2019) evaluated the role of different surface coatings such as citrate, silicate, PVP, and branched polyethyleneimine in the modification of AgNPs, and also studied their behavior in the environment, bioavailability, and toxicity in freshwater mussels, *Elliptio complanata*. The results obtained after exposure of test models with all these surface-modified AgNPs at different concentrations showed that the surface coatings on AgNPs could influence its availability and the chronic toxicity. Arias et al. (2018) reviewed surface coatings of iron oxide nanoparticles by several compounds such as synthetic and natural polymers, organic surfactants, inorganic compounds, and bioactive molecules that play an important role in the nanomaterial performance and their toxicity. It was proposed that the coating of nanomaterials like iron oxide nanoparticles with the aforementioned compounds greatly reduce their toxicity, thereby increasing their bioavailability. Apart from these, several other reports are available which all indicated and confirmed that surface engineering of nanomaterials with specific types of the compound can help in the reduction of their toxicological effects in different *in vivo* models.

17.6 Conclusion and Future Perspectives

Nanoparticles have many applications in the field of medical science. Especially they have shown promising applications as nanomedicine for diagnosis and treatment of various diseases. However, with the increased use of nanoparticles in nanomedicine, the concerns regarding their toxicity are also rising. Although nanomaterials are being used due to their novel properties such as smaller size and increased reactivity, those same properties can contribute to their toxicity. Hence, there is an increasing demand to test the toxicity of nanoparticles before their use as nanomedicine. The smaller size of nanoparticles shows increased toxicity due to an increase in their surface area. The nanoparticle dissolution, dosage, aggregation, surface coating, and composition play a role in exerting their toxic effect. After entry into the body, nanomedicine may get accumulated in various tissues and organs. Our body has few mechanisms to excrete the nanomaterials, hence they remain accumulated in many cases. In such incidences, they induce harmful effects on the tissues of concern and organs like skin, liver, kidney, lung, heart, brain, etc. disturbing their normal functions.

Taking into consideration all such issues, the toxicity study of nanomedicine should be given prime importance before releasing them as a product in the market for consumer use. Prior and proper evaluation of nanomaterials for their toxic effects will thus help to create a safer world through sustainable use of nanotechnology.

In the present scenario, nanomedicine seems promising. But they may pose a serious complication in the form of harmful effects to humans and the environment. Still, the available data is insufficient to draw a firm conclusion regarding the factors responsible for their toxic effects. Additionally, the contribution of each factor in deciding the particular toxic effects is still not known. Hence, in the near future, there is a requirement for studies focusing on the in-depth analysis of each of the parameters of nanoparticle toxicity. The dissolution of nanomaterials like nanoparticles in the surrounding medium is the major issue of nanotoxicity. Hence, there is a great need for reduction of nanoparticle dissolution. There are few reports which suggest that surface engineering of certain nanomaterials helps to solve problems like dissolution and aggregation of nanoparticles which greatly helps in the reduction of toxicity concerning nanoparticles. This is because the decreased dissolution will keep the nanoparticle intact and thus it can be available to play its designated role. The fate and transport of the nanomaterials in the biological systems need special attention as they can elucidate the toxic effects of nanomaterials.

Acknowledgments

MR is thankful to University Grants Commission, New Delhi, for the award of Basic Science Research faculty fellowship (F 18-1/2011 (BSR); 30/12/2016).

References

Adachi, K., Yamada, N., Yoshida, Y., and Yamamoto, O. (2013). Subchronic exposure of titanium dioxide nanoparticles to hairless rat skin. *Experimental Dermatology* 22 (4): 278–283.

Adewale, O.B., Davids, H., Cairncross, L., and Roux, S. (2019). Toxicological behavior of gold nanoparticles on various models: influence of physicochemical properties and other factors. *International Journal of Toxicology* 38 (5): 357–384.

Aillon, K.L., Xie, Y., El-Gendy, N. et al. (2009). Effects of nanomaterial physicochemical properties on *in vivo* toxicity. *Advanced Drug Delivery Reviews* 61 (6): 457–466.

Albanese, A. and Chan, W.C.W. (2011). Effect of gold nanoparticle aggregation on cell uptake and toxicity. *ACS Nano* 5 (7): 5478–5489.

Alexis, F., Pridgen, E., Molnar, L.K., and Farokhzad, O.C. (2008). Factors affecting the clearance and biodistribution of polymeric nanoparticles. *Molecular Pharmaceutics* 5 (4): 505–515.

Arami, H., Khandhar, A., Liggitt, D., and Krishnan, K.M. (2015). *In vivo* delivery, pharmacokinetics, biodistribution and toxicity of iron oxide nanoparticles. *Chemical Society Reviews* 44 (23): 8576–8607.

Arias, L.S., Pessan, J.P., Vieira, A.P.M. et al. (2018). Iron oxide nanoparticles for biomedical applications: a perspective on synthesis, drugs, antimicrobial activity, and toxicity. *Antibiotics (Basel)* 7 (2): 46. https://doi.org/10.3390/antibiotics7020046.

Asgharian, B. and Price, O.T. (2007). Deposition of ultrafine (nano) particles in the human lung. *Inhalation Toxicology* 19 (13): 1045–1054.

Aswathi, M., Ajitha, A., Akhina, H. et al. (2018). Quantum dots: a promising tool for biomedical application. *JSM Nanotechnology & Nanomedicine* 6 (2): 1066.

Auclair, J., Turcotte, P., Gagnon, C. et al. (2019). The influence of surface coatings of silver nanoparticles on the bioavailability and toxicity to *Elliptio complanata* mussels. *Journal of Nanomaterials* 2019 (7843025) https://doi.org/10.1155/2019/7843025.

Aziz, Z.A.A., Mohd-Nasir, H., Ahmad, A. et al. (2019). Role of nanotechnology for design and development of cosmeceutical: application in makeup and skin care. *Frontiers in Chemistry* 7 (739) https://doi.org/10.3389/fchem.2019.00739.

Bailly, A.L., Correard, F., Popov, A. et al. (2019). *In vivo* evaluation of safety, biodistribution and pharmacokinetics of laser synthesized gold nanoparticles. *Scientific Reports* 9 (12890) https://doi.org/10.1038/s41598-019-48748-3.

Balasubramanian, S.K., Jittiwat, J., Manikandan, J. et al. (2010). Biodistribution of gold nanoparticles and gene expression changes in the liver and spleen after intravenous administration in rats. *Biomaterials* 13: 2034–2042.

Banik, B.L., Fattahi, P., and Brown, J.L. (2016). Polymeric nanoparticles: the future of nanomedicine. *Wiley Interdisciplinary Reviews. Nanomedicine and Nanobiotechnology* 8: 271–299.

Baroli, B., Ennas, M.G., Loffredo, F. et al. (2007). Penetration of metallic nanoparticles in human full-thickness skin. *Journal of Investigative Dermatology* 127: 1701–1712.

Basu, S., Ghosh, S.K., Kundu, S. et al. (2007). Biomolecule induced nanoparticle aggregation: effect of particle size on interparticle coupling. *Journal of Colloid and Interface Science* 313 (2): 724–734.

Bengalli, R., Ortelli, S., Blosi, M. et al. (2019). *In vitro* toxicity of TiO2:SiO2 nanocomposites with different photocatalytic properties. *Nanomaterials* 9 (7): 1041. https://doi.org/10.3390/nano9071041.

Bhattacharjee, S., de Haan, L.H.J., Evers, N.M. et al. (2010). Role of surface charge and oxidative stress in cytotoxicity of organic monolayer-coated silicon nanoparticles towards macrophage NR8383 cells. *Particle and Fibre Toxicology* 7 (25) http://www.particleandfibretoxicology.com/content/7/1/25.

Borm, P.J., Schins, R.P., and Albrecht, C. (2004). Inhaled particles and lung cancer. Part B: paradigms and risk assessment. *International Journal of Cancer* 110: 3–14.

Buford, M.C., Hamilton, R.F., and Andrij-Holian, A. (2007). A comparison of dispersing media for various engineered carbon nanoparticles. *Particle and Fibre Toxicology* 4 (6) https://doi.org/10.1186/1743-8977-4-6.

Calienni, M.N., Feas, D.A., Igartua, D.E. et al. (2017). Nanotoxicological and teratogenic effects: a linkage between dendrimer surface charge and zebrafish developmental stages. *Toxicology and Applied Pharmacology* 337: 1–11.

Cha, K.E. and Myung, H. (2007). Cytotoxic effects of nanoparticles assessed *in vitro* and *in vivo*. *Journal of Microbiology and Biotechnology* 17 (9): 1573–1578.

Champion, J.A. and Mitragotri, S. (2006). Role of target geometry in phagocytosis. *Proceedings of the National Academy of Sciences of the United States of America* 103 (13): 4930–4934.

Checkley, S., MacCallum, L., Yates, J. et al. (2015). Bridging the gap between *in vitro* and *in vivo*: dose and schedule predictions for the ATR inhibitor AZD6738. *Scientific Reports* 5: 13545–13552.

Chen, Z., Meng, H., Xing, G. et al. (2006). Acute toxicological effects of copper nanoparticles *in vivo*. *Toxicology Letters* 163 (2): 109–120.

Chia, S.L. and Leong, D.T. (2016). Reducing ZnO nanoparticles toxicity through silica coating. *Heliyon* 2 (10): e00177. https://doi.org/10.1016/j.heliyon.2016.e00177.

Cho, W.S., Cho, M., Jeong, J. et al. (2009). Acute toxicity and pharmacokinetics of 13 nm-sized PEG-coated gold nanoparticles. *Toxicology and Applied Pharmacology* 236 (1): 16–24.

Costa, L.G., Pellacani, C., and Guizzetti, M. (2017). *In vitro* and alternative approaches to developmental neurotoxicity. In: *Reproductive and Developmental Toxicology*, 2e (ed. R.C. Gupta), 241–253. Cambridge, MA, USA: Academic Press.

Cui, W., Li, J., Zhang, Y. et al. (2012). Effects of aggregation and the surface properties of gold nanoparticles on cytotoxicity and cell growth. *Nanomedicine: Nanotechnology, Biology and Medicine* 8 (1): 46–53.

Dianzani, C., Zara, G.P., Maina, G. et al. (2014). Drug delivery nanoparticles in skin cancers. *BioMed Research International, vol.* 895986 https://doi.org/10.1155/2014/895986.

Dong, L., Tang, S., Deng, F. et al. (2019a). Shape-dependent toxicity of alumina nanoparticles in rat astrocytes. *Science of the Total Environment* 690: 158–166.

Dong, Y.C., Hajfathalian, M., Maidment, P.S.N. et al. (2019b). Effect of gold nanoparticle size on their properties as contrast agents for computed tomography. *Scientific Reports* 9 (14912) https://doi.org/10.1038/s41598-019-50332-8.

Downs, T.R., Crosby, M.E., Hu, T. et al. (2012). Silica nanoparticles administrated at the maximum tolerated dose induce genotoxic effects through an inflammatory reaction while gold nanoparticles do not. *Mutation Research* 745: 38–50.

Drasler, B., Sayre, P., Steinhäuser, K.G. et al. (2017). *In vitro* approaches to assess the hazard of nanomaterials. *NanoImpact* 8: 99–116.

Du, Z., Zhao, D., Jing, L. et al. (2013). Cardiovascular toxicity of different sizes amorphous silica nanoparticles in rats after intratracheal instillation. *Cardiovascular Toxicology* 13 (3): 194–207.

Duan, J., Yu, Y., Shi, H. et al. (2013a). Toxic effects of silica nanoparticles on zebrafish embryos and larvae. *PLoS One* 8 (9): e74606. https://doi.org/10.1371/journal.pone.0074606.

Duan, J., Yu, Y., Li, Y. et al. (2013b). Cardiovascular toxicity evaluation of silica nanoparticles in endothelial cells and zebrafish model. *Biomaterials* 34 (23): 5853–5862.

El-Badawy, A.M., Silva, R.G., Morris, B. et al. (2011). Surface charge-dependent toxicity of silver nanoparticles. *Environmental Science & Technology* 45: 283–287.

FDA (Food and Drug Administration), 2016, 'Global Summit on Regulatory Science (GSRS16) Nanotechnology Standards and Applications', Available at: https://www.fda.gov/media/99865/download (accessed 24 December, 2019).

Franco, M.E. and Lavado, R. (2019). Applicability of *in vitro* methods in evaluating the biotransformation of polycyclic aromatic hydrocarbons (PAHs) in fish: advances and challenges. *Science of the Total Environment* 671: 685–695.

Gamer, A.O., Leibold, E., and van Ravenzwaay, B. (2006). The *in vitro* absorption of microfine zinc oxide and titanium dioxide through porcine skin. *Toxicology In Vitro* 20: 301–307.

Geiser, M., Rothen-Rutishauser, B., Kapp, N. et al. (2005). Ultrafine particles cross cellular membranes by nonphagocytic mechanisms in lungs and in cultured cells. *Environmental Health Perspectives* 113 (11): 1555–1560.

Gholamine, B., Karimi, I., Salimi, A. et al. (2017). Neurobehavioral toxicity of carbon nanotubes in mice: focus on brain-derived neurotrophic factor messenger RNA and protein. *Toxicology and Industrial Health* 33: 340–350.

Gim, S., Zhu, Y., Seeberger, P.H., and Delbianco, M. (2019). Carbohydrate-based nanomaterials for biomedical applications. *Wiley Interdisciplinary Reviews: Nanomedicine and Nanobiotechnology* 11 (5): e1558. https://doi.org/10.1002/wnan.1558.

Gliga, A.R., Skoglund, S., Wallinder, I.O. et al. (2014). Size-dependent cytotoxicity of silver nanoparticles in human lung cells: the role of cellular uptake, agglomeration and Ag release. *Particle and Fibre Toxicology* 11 (11) https://doi.org/10.1186/1743-8977-11-11.

Gurr, J.R., Wang, A.S.S., Chen, C.H., and Jan, K.Y. (2005). Ultrafine titanium dioxide particles in the absence of photoactivation can induce oxidative damage to human bronchial epithelial cells. *Toxicology* 213 (1–2): 66–73.

Hong, W., Li, L., Liang, J. et al. (2016). Investigating the environmental factors affecting the toxicity of silver nanoparticles in *Escherichia coli* with dual fluorescence analysis. *Chemosphere* 155: 329–335.

Howard, P.C., Tong, W., Weichold, F. et al. (2014). Global summit on regulatory science 2013. *Regulatory Toxicology and Pharmacology* 70: 728–732.

Hsiao, I.L. and Huang, Y.J. (2011). Effects of various physicochemical characteristics on the toxicities of ZnO and TiO2 nanoparticles toward human lung epithelial cells. *Science of the Total Environment* 409 (7): 1219–1228.

Hühn, D., Kantner, K., Geidel, C. et al. (2013). Polymer-coated nanoparticles interacting with proteins and cells: focusing on the sign of the net charge. *ACS Nano* 7 (4): 3253–3263.

Huntjens, D.R.H., Spalding, D.J.M., Danhof, M., and Della Pasqua, O.E. (2006). Correlation between *in vitro* and *in vivo* concentration–effect relationships of naproxen in rats and healthy volunteers. *British Journal of Pharmacology* 148 (4): 396–404.

Huo, T., Barth, R.F., Yang, W. et al. (2012). Preparation, biodistribution and neurotoxicity of liposomal cisplatin following convection enhanced delivery in normal and f98 glioma bearing rats. *PLoS One* 7 (e48752) https://doi.org/10.1371/journal.pone.0048752.

Jatanaa, S. and DeLouisea, L.A. (2014). Understanding engineered nanomaterial skin interactions and the modulatory effects of UVR skin exposure. *Wiley Interdisciplinary Reviews: Nanomedicine and Nanobiotechnology* 6 (1): 61–79.

Jeon, S., Clavadetscher, J., Lee, D.K. et al. (2018). Surface charge-dependent cellular uptake of polystyrene nanoparticles. *Nanomaterials* 8 (1028) https://doi.org/10.3390/nano8121028.

Jeong, H.H., Choi, E., Ellis, E., and Lee, T.C. (2019). Recent advances in gold nanoparticles for biomedical applications: from hybrid structures to multi-functionality. *Journal of Materials Chemistry B* 7: 3480–3496.

Ji, Z., Zhang, D., Li, L. et al. (2009). The hepatotoxicity of multi-walled carbon nanotubes in mice. *Nanotechnology* 20: 445101. https://doi.org/10.1088/0957-4484/20/44/445101.

Kan, H., Pan, D., and Castranova, V. (2018). Engineered nanoparticle exposure and cardiovascular effects: the role of a neuronal-regulated pathway. *Inhalation Toxicology* 30 (9–10): 335–342.

Kasemets, K., Käosaar, S., Vija, H. et al. (2019). Toxicity of differently sized and charged silver nanoparticles to yeast *Saccharomyces cerevisiae* BY4741: a nano-biointeraction perspective. *Nanotoxicology* 13 (8): 1041–1059.

Kermanizadeh, A., Brown, D.M., Hutchison, G., and Stone, V. (2013). Engineered nanomaterial impact in the liver following exposure via an intravenous route- the role of polymorphonuclear leukocytes and gene expression in the organ. *Journal of Nanomedicine & Nanotechnology* 4 (1): 1000157. https://doi.org/10.4172/2157-7439.1000157.

Kermanizadeh, A., Gaiser, B.K., Johnston, H. et al. (2014). Toxicological effect of engineered nanomaterials on the liver. *British Journal of Pharmacology* 171: 3980–3987.

Khan, I., Saeed, K., and Kha, I. (2019). Nanoparticles: properties, applications and toxicities. *Arabian Journal of Chemistry* 12 (7): 908–931.

Kovrižnych, J.A., Sotníková, R., Zeljenková, D. et al. (2013). Acute toxicity of 31 different nanoparticles to zebrafish (*Danio rerio*) tested in adulthood and in early life stages-comparative study. *Interdisciplinary Toxicology* 6 (2): 67–73.

Kumar, V., Sharma, N., and Maitra, S.S. (2017). *In vitro* and *in vivo* toxicity assessment of nanoparticles. *International Nano Letters* 7: 243–256.

Lasagna-Reeves, C., Gonzalez-Romero, D., Barria, M.A. et al. (2010). Bioaccumulation and toxicity of gold nanoparticles after repeated administration in mice. *Biochemical and Biophysical Research Communications* 393 (4): 649–655.

Liao, M. and Liu, H. (2012). Gene expression profiling of nephrotoxicity from copper nanoparticles in rats after repeated oral administration. *Environmental Toxicology and Pharmacology* 34 (1): 67–80.

Lovrić, J., Cho, S., Winnik, F., and Maysinger, D. (2005). Unmodified cadmium telluride quantum dots induce reactive oxygen species formation leading to multiple organelle damage and cell death. *Chemical Biology* 12 (11): 1227–1234.

Lu, W., Senapati, D., Wang, S. et al. (2010). Effect of surface coating on the toxicity of silver nanomaterials on human skin keratinocytes. *Chemical Physics Letters* 487 (1–3) https://doi.org/10.1016/j.cplett.2010.01.027.

Magrez, A., Kasas, S., Salicio, V. et al. (2006). Cellular toxicity of carbon-based nanomaterials. *Nano Letter* 6: 1121–1125.

Mahmoud, A.M., Desouky, E.M., Hozayen, W.G. et al. (2019). Mesoporous silica nanoparticles trigger liver and kidney injury and fibrosis via altering TLR4/NF-κB, JAK2/STAT3 and Nrf2/HO-1 signalling in rats. *Biomolecules* 9 (528) https://doi.org/10.3390/biom9100528.

Mala, R. and Celsia, R.A.S. (2018). Toxicity of nanomaterials to biomedical applications: A review. In: *Fundamental Biomaterials: Ceramics* (eds. S. Thomas, P. Balakrishnan and M.M. Sreekala), 439–473. UK: Woodhead Publishing.

McNamara, K. and Tofail, S.A.M. (2017). Nanoparticles in biomedical applications. *Advances in Physics: X* 2 (1): 54–88.

Mitragotri, S. and Stayton, P. (2019). Organic nanoparticles for drug delivery and imaging. *MRS Bulletin* 39: 219–223.

Moore, T.L., Rodriguez-Lorenzo, L., Hirsch, V. et al. (2015). Nanoparticle colloidal stability in cell culture media and impact on cellular interactions. *Chemical Society Reviews* 44 (17): 6287–6305.

Mortensen, L.J., Oberdorster, G., Pentland, A.P., and Delouise, L.A. (2008). *In vivo* skin penetration of quantum dot nanoparticles in the murine model: the effect of UVR. *Nano Letters* 8: 2779–2787.

Nam, S.H. and An, Y.J. (2019). Size- and shape-dependent toxicity of silver nanomaterials in green alga *Chlorococcum infusionum*. *Ecotoxicology and Environmental Safety* 168: 388–393.

Naqvi, S., Samim, M., Abdin, M.Z. et al. (2010). Concentration-dependent toxicity of iron oxide nanoparticles mediated by increased oxidative stress. *International Journal of Nanomedicine* 5: 983–989.

Nguyen, K.C., Seligy, V.C., Massarsky, A. et al. (2013). Comparison of toxicity of uncoated and coated silver nanoparticles. *Journal of Physics: Conference Series* 429 (012025) https://doi.org/10.1088/1742-6596/429/1/012025.

NIST (National Institute of Standards and Technology) (2013), American National Standards Institute, website providing links to existing lists of nanomaterial standards, available at: http://nanostandards.ansi.org/tiki-index.php (accessed 20 June 2017).

Paknejadi, M., Bayat, M., Salimi, M., and Razavilar, V. (2018). Concentration- and time-dependent cytotoxicity of silver nanoparticles on normal human skin fibroblast cell line. *Iranian Red Crescent Medical Journal* 20 (10): e79183. https://doi.org/10.5812/ircmj.79183.

Pandurangan, M. and Kim, D. (2015). *In vitro* toxicity of zinc oxide nanomaterials: a review. *Journal of Nanoparticle Research* 17 (3): 158.

Patlolla, A.K., Hussain, S.M., Schlager, J.J. et al. (2010). Comparative study of the clastogenicity of functionalized and non-functionalized multiwalled carbon nanotubes in bone marrow cells of Swiss-Webster mice. *Environmental Toxicology* 25 (6): 608–621.

Peng, L., He, M., Chen, B. et al. (2013). Cellular uptake, elimination and toxicity of CdSe/ZnS quantum dots in HepG2 cells. *Biomaterials* 34 (37): 9545–9558.

Perde-Schrepler, M., Florea, A., Brie, I. et al. (2019). Size-dependent cytotoxicity and genotoxicity of silver nanoparticles in cochlear cells *in vitro*. *Journal of Nanomaterials* 2019 (6090259) https://doi.org/10.1155/2019/6090259.

Prajnamitra, R.P., Chen, H.C., Lin, C.J. et al. (2019). Nanotechnology approaches in tackling cardiovascular diseases. *Molecules* 24 https://doi.org/10.3390/molecules24102017.

Prow, T.W., Grice, J.E., Lin, L.L. et al. (2011). Nanoparticles and microparticles for skin drug delivery. *Advanced Drug Delivery Reviews* 63 (6): 470–491.

Radomski, A., Jurasz, P., Alonso-Escolano, D. et al. (2005). Nanoparticle-induced platelet aggregation and vascular thrombosis. *British Journal of Pharmacology* 146: 882–893.

Rivet, C.J., Yuan, Y., Borca-Tasciuc, D.A., and Gilbert, R.J. (2012). Altering iron oxide nanoparticle surface properties induce cortical neuron cytotoxicity. *Chemical Research in Toxicology* 25 (1): 153–161.

Sahu, S.C. and Hayes, A.W. (2017). Toxicity of nanomaterials found in human environment: a literature review. *Toxicology Research and Application* 1: 1–13.

Seqqat, R., Blaney, L., Quesada, D. et al. (2019). Nanoparticles for environment, engineering, and nanomedicine. *Journal of Nanotechnology* 2019 (2850723) https://doi.org/10.1155/2019/2850723.

Sharma, V., Singh, S.K., Anderson, D. et al. (2011). Zinc oxide nanoparticle induced genotoxicity in primary human epidermal keratinocytes. *Journal of Nanoscience and Nanotechnology* 11 (5): 3782–3788.

Shi, D., Mi, G., and Webster, T.J. (2017). The synthesis, application, and related neurotoxicity of carbon nanotubes. In: *Neurotoxicity of Nanomaterials and Nanomedicine* (eds. X. Jiang and H. Gao), 259–284. Cambridge, MA, USA: Academic Press.

Simeone, F.C. and Costa, A.L. (2019). Assessment of cytotoxicity of metal oxide nanoparticles on the basis of fundamental physical-chemical parameters: a robust approach to grouping. *Environmental Science: Nano* 6: 3102–3112.

Singh, N., Joshi, A., Toor, A.P., and Verma, G. (2017). Drug delivery: advancements and challenges. In: *Nanostructures for Drug Delivery* (eds. E. Andronescu and A.M. Grumezescu), 865–886. Amsterdam, The Netherlands: Elsevier.

Slikker, W. Jr., Miller, M.A., Valdez, M.L., and Hamburg, M.A. (2012). Advancing global health through regulatory science research: summary of the global summit on regulatory science research and innovation. *Regulatory Toxicology and Pharmacology* 62: 471–473.

Song, B., Zhang, Y., Liu, J. et al. (2016a). Is neurotoxicity of metallic nanoparticles the cascades of oxidative stress? *Nanoscale Research Letters* 11: 291. https://doi.org/10.1186/s11671-016-1508-4.

Song, B., Zhang, Y., Liu, J. et al. (2016b). Unraveling the neurotoxicity of titanium dioxide nanoparticles: focusing on molecular mechanisms. *Beilstein Journal of Nanotechnology* 7: 645–654.

Sukhanova, A., Poly, S., Shemetov, A., and Nabiev, I.R. (2012). Quantum dots induce charge-specific amyloid-like fibrillation of insulin at physiological conditions. In: *International Society for Optics and Photonics*, vol. 8548 (eds. S.H. Choi, J.H. Choy, U. Lee and V.K. Varadan), 85485F.

Sukhanova, A., Bozrova, S., Sokolov, P. et al. (2018). Dependence of nanoparticle toxicity on their physical and chemical properties. *Nanoscale Research Letters* 3 (44) https://doi.org/10.1186/s11671-018-2457-x.

Tak, Y.K., Pal, S., Naoghare, P.K. et al. (2015). Shape-dependent skin penetration of silver nanoparticles: does it really matter? *Scientific Reports* 5: 16908. https://doi.org/10.1038/srep16908.

Teleanu, D., Chircov, C., Grumezescu, A. et al. (2018). Impact of nanoparticles on brain health: an up to date overview. *Journal of Clinical Medicine* 7: 490.

Teleanu, D.M., Chircov, C., Grumezescu, A.M., and Teleanu, R.I. (2019). Neurotoxicity of nanomaterials: an up-to-date overview. *Nanomaterials* 9 (96) https://doi.org/10.3390/nano9010096.

Thiruvengadam, M., Rajakumar, G., and Chung, M. (2018). Nanotechnology: current uses and future applications in the food industry. *3 Biotech* 8 (1): 74. https://doi.org/10.1007/s13205-018-1104-7.

Tu, M., Huang, Y., Li, H.L., and Gao, Z.H. (2012). The stress caused by nitrite with titanium dioxide nanoparticles under UVA irradiation in human keratinocyte cell. *Toxicology* 299 (1): 60–68.

Tyagi, N., Srivastava, S.K., Arora, S. et al. (2016). Comparative analysis of the relative potential of silver, zinc-oxide and titanium-dioxide nanoparticles against UVB-induced DNA damage for the prevention of skin carcinogenesis. *Cancer Letters* 383 (1): 53–61.

Valdiglesias, V., Fernández-Bertólez, N., Kiliç, G. et al. (2016). Are iron oxide nanoparticles safe? Current knowledge and future perspectives. *Journal of Trace Elements in Medicine and Biology* 38: 53–63.

Valentini, X., Rugira, P., Frau, A. et al. (2019). Hepatic and renal toxicity induced by TiO_2 nanoparticles in rats: a morphological and metabonomic study. *Journal of Toxicology* 2019 (5767012) https://doi.org/10.1155/2019/5767012.

Vecchio, G., Galeone, A., Brunetti, V. et al. (2012). Concentration-dependent, size-independent toxicity of citrate capped AuNPs in *Drosophila melanogaster*. *PLoS One* 7 (1): e29980. https://doi.org/10.1371/journal.pone.0029980.

Vidal, F., Vasquez, P., Cayuman, F.R. et al. (2017). Prevention of synaptic alterations and neurotoxic effects of PAMAM dendrimers by surface functionalization. *Nanomaterials* 8 https://doi.org/10.3390/nano8010007.

Wang, B., He, X., Zhang, Z. et al. (2013). Metabolism of nanomaterials *in vivo*: blood circulation and organ clearance. *Accounts of Chemical Research* 46 (3): 761–769.

Wang, M., Lai, X., Shao, L., and Li, L. (2018). Evaluation of immunoresponses and cytotoxicity from skin exposure to metallic nanoparticles. *International Journal of Nanomedicine* 13: 4445–4459.

Wiwanitkit, V., Sereemaspun, A., and Rojanathanes, R. (2009). Effect of gold nanoparticles on spermatozoa: the first world report. *Fertility and Sterility* 91 (1): e7–e8.

Wu, T., Zhang, T., Chen, Y., and Tang, M. (2016). Research advances on potential neurotoxicity of quantum dots. *Journal of Applied Toxicology* 36: 345–351.

Xia, T., Kovochich, M., Liong, M. et al. (2008). Comparison of the mechanism of toxicity of zinc oxide and cerium oxide nanoparticles based on dissolution and oxidative stress properties. *ACS Nano* 2: 2121–2134.

Yan, G., Huang, Y., Bu, Q. et al. (2012). Zinc oxide nanoparticles cause nephrotoxicity and kidney metabolism alterations in rats. *Journal of Environmental Science and Health, Part A. Toxic/Hazardous Substances and Environmental Engineering* 47 (4): 577–588.

Yang, Y., Matsubara, S., Nogami, M., and Shi, J. (2007). Controlling the aggregation behaviour of gold nanoparticles. *Materials Science and Engineering B* 140 (3): 172–176.

Yang, H., Liu, C., Yang, D. et al. (2009). Comparative study of cytotoxicity, oxidative stress and genotoxicity induced by four typical nanomaterials: the role of particle size, shape and composition. *Journal of Applied Toxicology* 29 (1): 69–78.

Yang, C., Tian, A., and Lia, Z. (2016). Reversible cardiac hypertrophy induced by PEG-coated gold nanoparticles in mice. *Scientific Reports* 6 (20203): 1–12.

Yang, L., Zhou, Z., Song, J., and Chen, X. (2019). Anisotropic nanomaterials for shape-dependent physicochemical and biomedical applications. *Chemical Society Reviews* 48: 5140–5176.

Yuan, Z.Y., Hu, Y.L., and Gao, J.Q. (2015). Brain localization and neurotoxicity evaluation of polysorbate 80-modified chitosan nanoparticles in rats. *PLoS One* 10 (8): e0134722. https://doi.org/10.1371/journal.pone.0134722.

Zeng, Y., Kurokawa, Y., Win-Shwe, T.T. et al. (2016). Effects of PAMAM dendrimers with various surface functional groups and multiple generations on cytotoxicity and neuronal differentiation using human neural progenitor cells. *Journal of Toxicological Sciences* 41: 351–370.

Zhang, H., Gilbert, B., Huang, F., and Banfield, J.F. (2003). Water-driven structure transformation in nanoparticles at room temperature. *Nature* 424 (6952): 1025–1029.

Zhang, L.W., Yu, W.W., Colvin, V.L., and Monteiro-Riviere, N.A. (2008). Biological interactions of quantum dot nanoparticles in skin and in human epidermal keratinocytes. *Toxicology and Applied Pharmacology* 228 (2): 200–211.

Zhao, J. and Castranova, V. (2011). Toxicology of nanomaterials used in nanomedicine. *Journal of Toxicology and Environmental Health Part B* 14 (8): 593–632.

Zhao, Y., Wang, X., Wu, Q. et al. (2015). Translocation and neurotoxicity of CdTe quantum dots in RMEs motor neurons in nematode *Caenorhabditis elegans*. *Journal of Hazardous Materials* 283: 480–489.

Zhou, G., Li, Y., Ma, Y. et al. (2016). Size-dependent cytotoxicity of yttrium oxide nanoparticles on primary osteoblasts *in vitro*. *Journal of Nanoparticle Research* 18 (135) https://doi.org/10.1007/s11051-016-3447-5.

Zhu, M.T., Wang, B., Wang, Y. et al. (2011). Endothelial dysfunction and inflammation induced by iron oxide nanoparticle exposure: risk factors for early atherosclerosis. *Toxicology Letters* 203 (2): 62–171.

Index

Nanobiotechnology in Diagnosis, Drug Delivery, and Treatment, First Edition. Edited by Mahendra Rai, Mehdi Razzaghi-Abyaneh, and Avinash P. Ingle.

e

f

g

h

i

k

l